MW01377776

Plant Stress Biology

Plant Stress Biology

Edited by Margot Scott

Larsen & Keller
www.larsen-keller.com

Plant Stress Biology
Edited by Margot Scott
ISBN: 979-8-88836-077-4 (Hardback)

© 2023 Larsen & Keller

Larsen & Keller
Published by Larsen and Keller Education,
5 Penn Plaza,
19th Floor,
New York, NY 10001, USA

Cataloging-in-Publication Data

Plant stress biology / edited by Margot Scott.
p. cm.
Includes bibliographical references and index.
ISBN 979-8-88836-077-4
1. Plants--Effect of stress on. 2. Plant physiology. 3. Crops--Effect of stress on.
4. Crops--Physiology. I. Scott, Margot.
QK754 .A25 2023
581.7--dc23

This book contains information obtained from authentic and highly regarded sources. Copyright for all individual chapters remain with the respective authors as indicated. All chapters are published with permission under the Creative Commons Attribution License or equivalent. A wide variety of references are listed. Permission and sources are indicated; for detailed attributions, please refer to the permissions page and list of contributors. Reasonable efforts have been made to publish reliable data and information, but the authors, editors and publisher cannot assume any responsibility for the validity of all materials or the consequences of their use.

Trademark Notice: Registered trademark of products or corporate names are used only for explanation and identification without intent to infringe.

For more information regarding Larsen and Keller Education and its products, please visit the publisher's website www.larsen-keller.com

Table of Contents

Preface

Every book is initially just a concept; it takes months of research and hard work to give it the final shape in which the readers receive it. In its early stages, this book also went through rigorous reviewing. The notable contributions made by experts from across the globe were first molded into patterned chapters and then arranged in a sensibly sequential manner to bring out the best results.

Plants are exposed to a wide range of environmental stresses that reduce and limit crop productivity. Biotic stress and abiotic stress are the two types of environmental stresses that affect plants. Some examples of abiotic stress are radiation, salinity, floods, drought, extreme temperatures, and heavy metals. These factors contribute towards the depletion of important crops across the world. Biotic stress, on the other hand, refers to the damage caused by various pathogens including weeds, fungi, insects, herbivores, nematodes and bacteria. Plants develop various strategies to combat the threats of biotic and abiotic stressors. On detecting environmental stress, plants produce the necessary cellular responses. The role of signaling pathways is crucial as it helps in detecting the external stress environment. This leads to plants producing an appropriate response, which may be biochemical or physiochemical in nature. The genetic code of the plants stores information with respect to the defense mechanisms that are activated when faced with stresses. This book includes some of the vital pieces of works being conducted across the world, on various topics related to plant stress biology. It will provide comprehensive knowledge to the readers.

It has been my immense pleasure to be a part of this project and to contribute my years of learning in such a meaningful form. I would like to take this opportunity to thank all the people who have been associated with the completion of this book at any step.

Editor

1

Insight into the Role of Epigenetic Processes in Abiotic and Biotic Stress Response in Wheat and Barley

Lingyao Kong [1], Yanna Liu [1,2], Xiaoyu Wang [1] and Cheng Chang [1,*]

[1] College of Life Sciences, Qingdao University, Qingdao 266071, China; konglingyao@126.com (L.K.); Lyn19801259593@163.com (Y.L.); 2017020884@qdu.edu.cn (X.W.)

[2] National Key Facility for Crop Gene Resources and Genetic Improvement, Institute of Crop Science, Chinese Academy of Agricultural Sciences, Beijing 100081, China

* Correspondence: cc@qdu.edu.cn.

Abstract: Environmental stresses such as salinity, drought, heat, freezing, heavy metal and even pathogen infections seriously threaten the growth and yield of important cereal crops including wheat and barley. There is growing evidence indicating that plants employ sophisticated epigenetic mechanisms to fine-tune their responses to environmental stresses. Here, we provide an overview of recent developments in understanding the epigenetic processes and elements—such as DNA methylation, histone modification, chromatin remodeling, and non-coding RNAs—involved in plant responses to abiotic and biotic stresses in wheat and barley. Potentials of exploiting epigenetic variation for the improvement of wheat and barley are discussed.

Keywords: epigenetic; abiotic and biotic stress; wheat; barley; DNA methylation; histone modification; chromatin remodeling; non-coding RNAs

---

1. Introduction

As two founding crops of the agricultural revolution that took place 10,000 years ago in the Fertile Crescent, bread wheat (*Triticum aestivum* L. ssp. *aestivum*) and cultivated barley (*Hordeum vulgare* L. ssp. *vulgare*) are widely cultivated in the world and provide more than 20% of the caloric intake for one-half of the world's population [1–4]. There is a crucial need to improve the production of wheat and barley for a growing population. However, environmental stresses such as salinity, drought, heat, freezing, heavy metal and even pathogen infections seriously threaten the growth and yield of wheat and barley under field conditions [5–12]. For instance, the majority of crops are highly sensitive to salinity, and the average yield of all important glycophytic crops decreased by 50%–80% at medium salinity conditions [5]. The accumulation of salt in the soil solution reduces the absorption of water and nutrients, leading to osmotic stress, ion toxicity, nutrient imbalance, and even water deficit [6,7]. More than 5% of Na^+ can cause the clay to expand excessively when wet, severely restricting the movement of air and water, then resulting in poor drainage [7]. Currently, of the 230 million hectares (ha) of irrigated land in the world, 45 million ha (19.5%) have been threatened by salinity [7]. As global climate conditions continue to deteriorate, drought and heat always go hand in hand, which leads to higher crop losses. For barley, Xie et al. reported that the yield could be reduced by between 3% and 17% in those harsh conditions [8]. For wheat, the optimum temperature is about 21 °C at reproductive growth stage. Temperatures in excess of 33 °C in this stage result in a decrease of leaf photosynthesis, an accumulation of peroxides, and serious yield loss [9]. For heavy metal, cadmium (Cd) at low concentrations (0.3–0.8 mg kg^{-1}) in soils could inhibit regular cell division, decrease photosynthesis and impair antioxidant enzyme activity [10,11]. In all major wheat-growing areas, lead

(Pb) accumulation is generally accompanied by Cd pollution, seriously threatening crop yield and safety [12]. In acid soils, aluminum (Al) ions severely inhibit root growth and reduce the absorption of water and nutrients, resulting in crop yields [13]. In addition to these abiotic stresses, biotic stress also seriously damage grain yield and quality. It has been conservatively estimated that fungal pathogens alone are responsible for 15% to 20% yield losses per annum [14,15]. Among them, rust, the blotches and head blight/scab are the most devastating diseases leading to great yield loss in bread wheat [15]. Therefore, how to improve plant resistance against abiotic and biotic stresses in wheat and barley is the focus of attention for breeders.

Through evolution, plants have acquired highly sophisticated systems to cope with various environmental stresses. The past decade has seen unprecedented progress in understanding the signaling pathways controlling the plant responses to stresses, which has been summarized in prior reviews [16–18]. Activation of these signaling pathways usually results in the dramatic transcriptional reprogramming to initiate a set of stress responses [19–21]. There is increasing evidence indicating that this transcriptional reprogramming and regulation of stress-responsive genes involves diverse epigenetic processes and elements, such as DNA methylation, histone modification, chromatin remodeling and non-coding RNAs [22–24]. Here, we summarized the most recent progress on studies of epigenetic regulation of plant responses to abiotic and biotic stresses in wheat and barley, and discussed the potentials of exploiting natural and induced epigenetic variation for the improvement of wheat and barley.

2. DNA Methylation

As a type of DNA chemical modification, DNA methylation regulates the chromatin structure, DNA stability, and even gene expression without changing the DNA sequence. Under the action of DNA methyltransferase, the cytosine C5 position is covalently bonded with a methyl group, which is one of the most common modifications of DNA in eukaryotic cells [25–27]. In plants, cytosine methylation is detected in the context of CG, CHG, and CHH (where H is any nucleotide except G) [28,29], in which CG is the most abundant and widespread methylation site [30]. It has been revealed that the DNA de novo C methylation in *Arabidopsis* is catalyzed by methyltransferase DOMAINS REARRANGED METHYLTRANFERASE 1 (DRM1) and DRM2 [29], while the maintenance of DNA methylation in mitosis and meiosis relies on the METHYLTRANSFERASE 1 (MET1) [31], CHROMOMETHYLASE 2/3 (CMT2/3) [27,31]. High-resolution DNA methylation profiling in *Arabidopsis* and rice revealed that DNA methylation could take place in many chromatin regions, including intergenic transposable elements (TE), gene promoters and even gene-body [32–35]. Many studies revealed that DNA methylation in TEs is required for maintaining genome integrity [32–35]. Furthermore, DNA methylation at promoters generally represses gene expression, whereas methylation in gene-body DNA appears to be associated with active gene expression in *Arabidopsis* [32–35].

As an important epigenetic process, DNA methylation generally regulates plant responses to environmental stresses such as salinity and heavy metal stress in wheat and barley [36,37]. For instance, a recent report showed that DNA methylation could regulate the expression of a salinity-responsive gene in bread wheat [25]. A reduction in global DNA methylation level was observed in two bread wheat cultivars (salinity tolerant wheat cultivar SR3 and salinity susceptible wheat cultivar JN177) upon exposure to salinity stress [25]. Notably, the methylation level at the promoter of a stress-responsive gene TaFLS1 (encoding a flavonol synthase) was lower in the salinity tolerant wheat cultivar SR3 than in salinity susceptible wheat cultivar JN177, which is opposite to the higher gene expression in SR3 than in JN177, suggesting that DNA methylation might get involved in regulation of wheat salinity tolerance [25]. Besides, the modulation of metal-stress response by DNA methylation is reported in wheat and barley [37,38]. In heavy metal detoxifications, the multidrug resistance-associated protein

(MRP) type ATP-binding cassette (ABC) transporters play important roles, which also involve in other plant biological processes such as pathogen response and development [39–42]. In addition, AtABCC5 also get involved in drought stress response by altering guard cell movement in *Arabidopsis* [43,44]. Eighteen MRP-type ABC transporter genes were identified from the wheat genome [41]. Shafiq et al. found that the expression of TaABCCs and HEAVY METAL ATPASE 2 (TaHMA2) was higher in the heavy metal-resistant wheat varieties than in heavy metal-sensitive varieties upon exposure to the heavy metal stresses [37]. Furthermore, DNA methylation levels at the promoter of TaABCCs and TaHMA2 were lower in heavy metal-resistant varieties than in heavy metal-sensitive varieties under heavy metal stresses, suggesting that DNA methylation is associated with metal stress tolerance in wheat [37]. In barley, Al-activated citrate transporter1 (HvAACT1) is a major gene in charge of citrate efflux from roots for external Al detoxification in the rhizosphere [45,46]. Although the expression of HvAACT1 was not altered by Al treatment, its expression was significantly higher in Al-tolerant accessions than in Al-sensitive accession, indicating that HvAACT1 is a principal gene regulated Al tolerance in barley [47]. It is intriguing that in several European barley accessions, DNA methylation level at a multiretrotransposon-like (MRL) sequence, localized at the upstream genomic sequence of HvAACT1, is associated with the expression of HvAACT1 [38]. DNA demethylation in MRL resulted in the enhanced expression of HvAACT1, especially in the zone of root apical [39]. Meanwhile, low-level expression of HvAACT1 was found associated with a higher degree of DNA methylation in MRL, suggesting that the DNA methylation regulates the HvAACT1 expression, which was responsible for Al tolerance in barley [38]. Compared with extensive studies on the role of DNA methylation in regulation of abiotic stress tolerance, understanding of DNA methylation regulating defense response to pathogens is limited. A recent study reported that DNA methylation, particularly CHH methylation, gets involved in the regulation of defense responses to *Bgt* (*Blumeria graminis* f. sp. *tritici*), the causal agent of wheat powdery mildew, in wheat diploid progenitor *Aegilops tauschii* [48]. Upon Bgt infection, abundant differentially methylated regions (DMRs) were associated with CHH hypomethylation [48]. WGBS (whole-genome bisulfite sequencing) further revealed that TAGs (genes near transposable elements) with CHH-hypomethylated DMRs were enriched in genes with annotation for 'response to stress' functions, such as receptor kinase, peroxidase, and pathogenesis-related genes, suggesting that DNA methylation is involved in the regulation of plant defense responses in crops [48]. In addition, several instances indicated that the sensitivity of transcription factors (TFs) to DNA methylation can affect the binding of TF to chromatin [49]. In *Arabidopsis*, O'Malley et al. found that a regulatory relationship may exist between specific DNA methyltransferase and TF family [49]. Although similar results have not been found in wheat and barley, it is important for us to understand how DNA methylation plays a vital role in plant responses to stress.

To balance the genomic methylation level and fine-tune gene expression, DNA demethylases were employed to remove 5-methylcytosine and replace it with unmethylated cytosine [50]. In plants, DNA demethylation occurs in two ways: passive demethylation and active demethylation. During DNA replication, methylated cytosines are replaced with unmodified cytosines in passive demethylation [51,52]. Active DNA demethylation is mediated by specific DNA glycosylases, which hydrolyze the N-glycosidic bond between ribose and base [51,52]. In the past decade, several DNA glycosylases, including DMEMER (DME) and REPRESSOR OF SILENCING 1 (ROS1), involved in the active DNA demethylation were well studied in the model plant *Arabidopsis* [53,54]. Recently, exploration on the DNA glycosylase in wheat and barley has also emerged. For instance, *HvDME* encodes a DME-family DNA glycosylase in barley (Table 1) [55]. The expression of *HvDME* is markedly induced in drought-treated barley seedlings, especially in the drought-tolerant cultivars, suggesting a potential role of DNA demethylation in the regulation of barley responses to drought stress (Table 1) [55].

Table 1. The Epigenetics elements involved in stress response in wheat and barley.

Epigenetic Category	Element Name	Element Category	Species	Biological Function and Evidence	Reference
DNA methylation	HvDME	DNA glysosylase	Barley	*HvDME* expression is induced by drought stress, which is correlated with the differential DNA methylation patterns within the gene.	[55]
Histone modification	TaGCN5	Histone acetyltransferase	Wheat	The expression of the wheat *TaGCN5* gene in Arabidopsis *gcn5* mutant plants complemented the heat and salt tolerance.	[56,57]
	TaHDA6	Histone deacetylase	Wheat	TaHDA6 represses histone acetylation at promoters of defense-related genes and thus negatively regulates their expressions as well as plant defense responses to *Bgt*.	[58]
Chromatin remodeling	TaCHR729	Chromatin remodeling factor	Wheat	TaCHR729 promotes H3K4me3 at the *TaKCS6* promoters and positively regulates the *TaKCS6* expression and cuticular wax biosynthesis, thereby affecting twheat-*Bgt* interaction.	[59]
Non-coding RNA	TalncRNA18, TalncRNA73, TalncRNA106, TalncRNA108	LncRNA	Wheat	They exhibit differential expression and target wheat defense-related genes in response to *Pst* infection	[60]

3. Histone Modification

As one of the most common types of epigenetic regulation, diverse histone modification manners have been characterized, including acetylation, methylation, butyrylation, propionylation, crotonylation, malonylation, succinylation, 2-hydroxyisobutyrylation and β-hydroxybutyrylation [61–63]. Among these histone modifications, histone methylation/demethylation and acetylation/deacetylation have been widely studied, which regulates many biological processes in plants, including development and responses to biotic and abiotic stresses [62,63]. The majority of histone methylation takes place on the lysine residue of histone H3, such as H3K4me3, H3K36me3, H3K79me3, H3K9me2 and H3K27me3, in which H3K4me3 and H3K27me3 are highly conserved epigenetic marks for gene activation and repression, respectively [64–66]. Histone methylation is dynamically regulated by the histone methyltransferases (HMTs) and histone demethylases (HDMs) [64–66]. For instance, the repressive H3K27me3 modification is mediated by HMT complexes PRC1 and PRC2 recruited by various DNA-binding proteins. The first HMT recruiters in *Arabidopsis* are the members of the BBR/BPC family, that were shown to be responsible for establishing the silencing mark [67,68]. As another common type of histone modification—histone acetylation—is reversible and dynamically maintained by the antagonistic action of histone acetyltransferases (HATs) and histone deacetylases (HDACs) [56,69]. It is generally realized that the acetylation neutralizes the positive charge of lysine side chains on histones and reduces its interaction with the negatively charged DNA backbone, and thus relax the chromatin structure and promote gene transcription [70,71]. Indeed, H3K4ac and H3K9ac are often associated with gene activation, thereby modulating numerous biological processes such as stress responses in higher plants such as model plant *Arabidopsis* [72–74]. Increasing evidence from studies in *Arabidopsis* revealed that histone acetylations such as H3K4ac and H3K9ac are usually connected with histone methylation including H3K4me3, simultaneously regulating gene expression [71]. Therefore, histone post-transcriptional modifications are cross-talked, which fine-tunes the gene expression and response in various important biological processes in eukaryotes [75].

Previous studies revealed that various HATs and HDACs modulate plant gene expression in response to environmental stresses in wheat and barley. For instance, TaGCN5, a wheat ortholog of *Arabidopsis* histone acetyltransferases AtGCN5, plays an important role in regulating wheat defense response to heat and salt stresses (Table 1) [56,57]. The expression of *TaGCN5* was induced by treatment with heat and salt in bread wheat (Table 1) [56,57]. In *Arabidopsis*, GCN5 protein is recruited to the promoters of *HSFA3* and *UVH6* (UV-HYPERSENSITIVE 6) in response to heat stress, and enriched at the promoters of *CTL1*, *PGX3* and *MYB54* (involved in tolerance of salt stress) under salt stress as well [56,57]. At the same time, GCN5 was revealed to facilitate the acetylation of H3K9 and H3K14, which is associated with activation of *HSFA3, UVH6, CTL1, PGX3*, and *MYB54* in *Arabidopsis* [56,57]. Interestingly, the expression of *HSFA3, UVH6, CTL1, PGX3*, and *MYB54* were constitutively increased in *35S:TaGCN5/gcn5* transgenic *Arabidopsis*, compared with wild-type and *gcn5* plants, suggesting that GCN5-mediated histone acetylation responding to abiotic stress tolerance might be conserved in *Arabidopsis* and bread wheat (Table 1) [56,57]. In addition to abiotic stresses, biotic stresses such as the fungal infection also initiate the plant responses partly controlled by histone modifications. Recently, Sharma et al. characterized the wheat histone acetylation at the promoters of defense-related genes upon infection of *Puccinia triticina*, the causal agent of wheat leaf rust. In this study, two near-isogenic wheat lines (NILs), leaf rust-susceptible NIL and resistant NIL, were employed and the expression levels of six defense-related genes were analyzed as well [73]. Among the six genes, *N-acetyltransferase* is activated by enrichment of H3K4ac and H3K9ac at its promoter in leaf rust-susceptible NIL, and repressed by the histone deacetylation in leaf rust-resistant NIL [71]. In contrast, enrichment of H3K4ac and H3K9ac are largely correlated with higher expression of *Peroxidase 12* in both NILs. The expression of other remaining four genes (*WRKY 70, ASR1, Peroxidase 12* and *Sarcosine oxidase*) was not correlated with histone acetylation [67,73]. These results suggested that histone acetylation indeed get involved in the regulation of wheat response to *P. triticina* infection, whose underlying mechanism remains further study. Recently, the wheat histone deacetylase TaHDA6 was found to interact with the wheat

WD40-repeat protein TaHOS15 and was recruited to the promoters of defense-related genes, where TaHDA6 mediated histone deacetylation (Table 1) [58]. The decrease of the transcription level of TaHDA6 results in the enhanced transcription of defense-related genes, thus strengthening resistance to *Bgt* infection, suggesting that TaHDA6 fine-tunes the acetylation levels of these wheat defense-related genes (Table 1) [58]. In barley, the senescence-associated gene *HvS40* exhibited enhanced H3K9ac at its promoter and coding regions during the early response to drought stress [76]. Interestingly, histone modifications such as histone methylation and acetylation were found usually accompanied by DNA methylation in response to environmental stresses in the model plant *Arabidopsis* [77,78]. However, it remains to be studied about the interplay of histone modifications with other epigenetic regulation such as DNA methylation in governing resistance against environmental stresses in wheat and barley.

4. Chromatin Remodeling

Besides DNA methylation and histone modifications, chromatin structure and gene expression may also be affected by chromatin remodeling, a process that disrupts histone-DNA interactions resulting in the altered accessibility of specific DNA regions to transcription machinery [79,80]. Chromatin remodeling factor (CHR), including the SWI/SNF ATPases, the imitation switch (ISWI) ATPases, and the chromodomain and helicase-like domain (CHD) ATPases subfamilies, could mediate either the ATP-dependent chromatin remodeling or the posttranslational histone modifications [57,58,73–84]. The ATP-dependent chromatin remodeling complexes could alter nucleosome composition, and positioning and thus regulate DNA accessibility and gene expression. In contrast, the posttranslational histone modifications could alter the interaction between nucleosomes, and thus affect the chromatin compactness and structure in model plant *Arabidopsis* [85–87]. Chromatin remodeling has been well documented to regulate plant growth, development and response to environmental stresses in *Arabidopsis* and rice [88,89]. Besides, the investigation on the role of chromatin remodeling in regulating plant responses to stresses in wheat and barley is emerging [88,89]. For instance, the wheat CHD-type chromatin remodeling factor TaCHR729 was reported recently to interact with the *TaKCS6* promoter-associated bHLH type transcription factor 1 (TaKPAB1) and thereby bind to the promoter regions of wheat *3-KETOACYL-CoA SYNTHASE* (*TaKCS6*), which encodes a key enzyme in the wheat cuticular wax biosynthesis (Table 1) [59]. Interestingly, TaCHR729 was found to promote H3K4me3 at the promoter region of *TaKCS6* and positively regulate the *TaKCS6* transcription (Table 1) [59]. Consistently, silencing of *TaCHR729* attenuated the biosynthesis of wheat cuticular wax and germination of *Bgt* conidia, suggesting that the wheat chromatin remodeling factor TaCHR729 regulate the wheat-powdery mildew interaction through mediating histone methylation and fine-tuning the cuticular wax biosynthesis (Table 1) [59].

Although the study of chromosome remodeling in response to stress in wheat and barley is very limited, the research in *Arabidopsis* is abundant, in-depth and worthy reference. For instance, ABA signal transduction which responds to abiotic stresses, such as drought, salinity and freezing, is regulated by chromatin remodeling in *Arabidopsis*. The clade A PP2C phosphatase HAB1 (HYPERSENSITIVE TO ABA 1, a core component in ABA signal pathway) interacts with SWI3B physically, a core subunit of the putative SWI/SNF complex in *Arabidopsis* [90]. As a phosphatase, HAB1 may directly dephosphorylate SWI/SNF complexes including SWI3B in an ABA-dependent manner [90]. Another study has shown that several chromatin regulators, such as BRM SWI/SNF ATPase, could be phosphorylated by of SnRK2 type kinases (another core component in ABA signal pathway) as the substrates [91,92]. These results suggest that the phosphorylation and dephosphorylation states of SWI/SNF complexes may modulate the response to environmental stresses by ABA signal pathway and further molecular mechanisms need to be studied in wheat and barley.

5. Non-coding RNAs

As important epigenetic elements, non-coding RNAs (ncRNAs) widely regulate plant multiple processes, including growth, development and even responses to environmental stresses. Based on the size, ncRNA can be divided into short-chain non-coding RNAs and long-chain non-coding RNA [93–95]. In the past few decades, enormous studies in animals and plants have revealed that short-chain non-coding RNAs, such as microRNAs (miRNA) and small interfering RNAs (siRNAs), participate in both transcriptional and post-transcriptional regulation of gene expression [93–95]. MicroRNAs and siRNAs usually contain 18-24 nucleotides (nt) and were classified as small RNA (sRNA). In contrast, non-coding RNA with more than 200 nucleotides (nt) length has been generally defined as long non-coding RNAs (lncRNAs) [93,95,96]. In eukaryotic cells, miRNAs coding genes are transcribed to generate primary miRNAs (pri-miRNAs), which are then cleaved and processed into mature miRNAs under the action of DICER-LIKE proteins (DCLs). In *Arabidopsis*, ARGONAUTE (AGO) family proteins such as AGO1 then bind the nascent mature miRNAs and guide the target-specific post-transcriptional gene silencing (PTGS) [97]. Unlike microRNA, short/small interfering RNAs (siRNAs) are generated from long linear double-stranded RNAs with 20-24 nt length and are transcribed by RNA polymerase IV (RNAPIV) from transposons and repetitive regions. So far, several subclasses of siRNAs have been identified, some of which function in PTGS and others function in transcriptional gene silencing (TGS) [98,99]. LncRNAs are widespread in all species and take part in gene expression regulation at transcription and post-transcription, epigenetic level [100,101]. LncRNAs share similarities with mRNAs in the structure and biogenesis process, and they are transcribed by RNA polymerase II (RNAPII) and poly-adenylated [102]. Like mRNAs, lncRNAs own multiple exons and are subjected to alternative splicing. However, lncRNAs are short of discernable coding potential [102]. More recently, lncRNAs are revealed to have adjusting functions in the major biological processes, including development, vernalization, and environmental stress adaptation by direct and indirect manners in *Arabidopsis* [103,104].

Increasing evidence from *Arabidopsis* studies revealed that ncRNAs such as siRNAs and lncRNAs regulate plant stress-responsive gene expression through multiple epigenetic mechanisms, including DNA methylation, histone modification and genome topology changes [97]. For instance, siRNAs and lncRNAs both participate into the DNA de novo cytosine methylation via the RNA-directed DNA methylation (RdDM) pathway in *Arabidopsis* [105]. *Arabidopsis* RNAPIV-generated siRNAs could load to Argonaute 4 (AGO4) and interact with lncRNAs generated by RNAPII to constitute a siRNA–AGO4–lncRNA silencing complex, which subsequently recruits the DMT domains rearranged methyltransferase 2 (DRM2) to mediate DNA de novo cytosine methylation [106]. *Arabidopsis* mutant deficient in NRPD2, an essential subunit of RNAPIV were hypersensitive to heat stress, suggesting that RdDM pathway is essential to the regulation of plant stress responses [107]. Besides, some lncRNAs were revealed to regulate histone modifications in *Arabidopsis* [108]. For instance, the cold-induced lncRNA *COOLAIR*, a group of long antisense RNAs expressed from the *FLOWERING LOCUS C* (*FLC*) locus, promote the replacement of H3K36me3 with H3K27me3, as well as the H3K4me2 demethylation, at *FLC* locus during cold exposure [109,110]. Similarly, another cold-induced lncRNA *COLDAIR* could interact with polycomb repressive complex 2 (PRC2) to facilitate H3K27me3 enrichment at *FLC* [111–113]. In addition to DNA methylation and histone modification, genome topology is also regulated by ncRNAs in *Arabidopsis*. For instance, *Arabidopsis* lncRNA APOLO promotes the chromatin loop formation at the *PINOID* (*PID*) locus, encoding a key regulator of polar auxin transport, which is further regulated by RdDM and ultimately determines the *PID* expression patterns [114]. With the development of high-throughput sequencing technology and computational methods, the research of ncRNA has been carried out gradually in wheat and barley [60,115–118]. For instance, Zhang et al. found that four lncRNAs (*TalncRNA18*, *TalncRNA73*, *TalncRNA106*, and *TalncRNA108*) exhibited differential expression upon infection of *Puccinia striiformis* f. sp. *tritici* (*Pst*), suggesting that these lncRNAs may get involved in regulation of wheat defense responses to *Pst*. However, detailed epigenetic mechanism in the regulation of wheat and barley stress responses remain to be explored in the future research.

6. Concluding Remarks

In this review, we discuss the recent advance in the understanding of epigenetic regulation of plant responses to abiotic and biotic stresses in wheat and barley (summarized in Figure 1). Under non-stress condition, the expression of stress responsive genes is repressed. Upon sensing the environmental stresses, plants such as wheat and barley initiate the stress-responsive signaling, which resulted in the epigenetic remodeling involving DNA methylation, histone modification and chromatin remodeling. DNA methylation is regulated by DMT and DDM, while histone modifications include histone acetylation/deacetyaltion and methylation/demethylation mediated by HAT/HDAC and HMT/HDM enzymes. In addition, CHR-mediated chromatin remodeling and ncRNA-regulated epigenetic processes, including RdDM, histone modification as well as genome topology changes, also regulate gene expression in response to the environmental stresses. These epigenetic processes orchestrate the plant stresses responses and fine-tune the balance of plant growth and defense in wheat and barley.

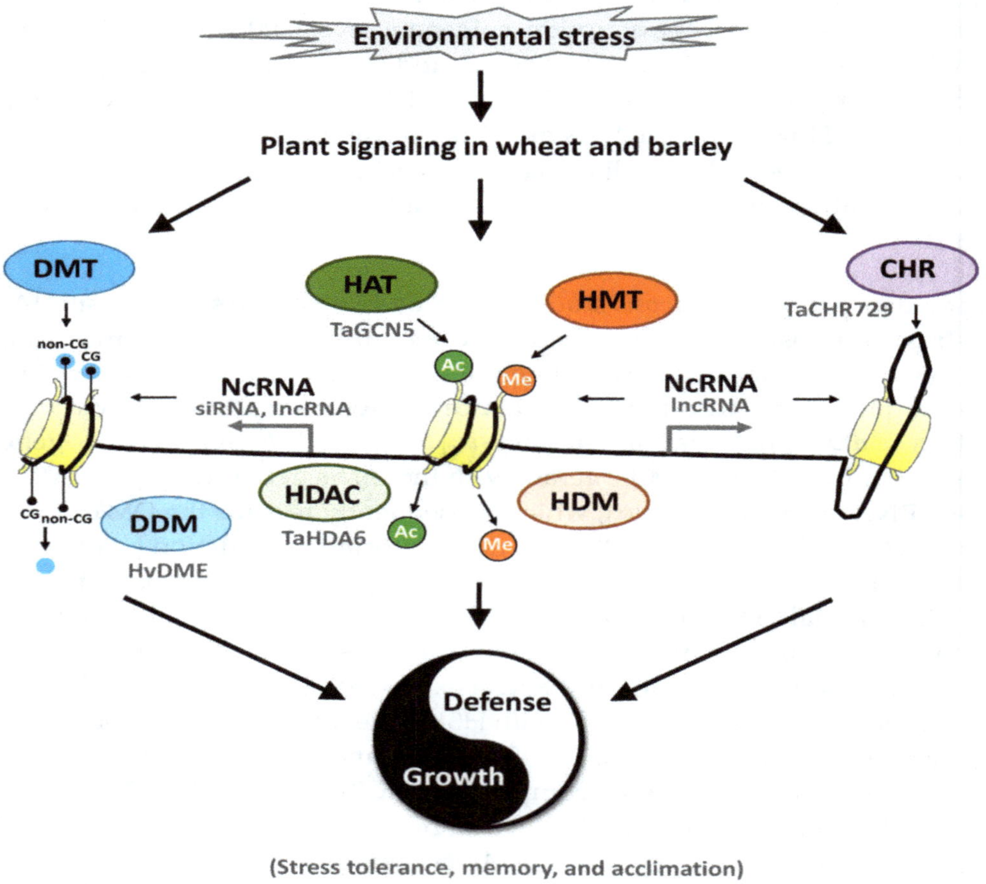

Figure 1. A general model for the role of epigenetic elements and processes in stress responses in wheat and barley. DNA methylation is regulated by DMT and DDM, while histone modifications include histone acetylation/deacetyaltion and methylation/demethylation mediated by HAT/HDAC and HMT/HDM enzymes. In addition, CHR-mediated chromatin remodeling and ncRNA-regulated epigenetic processes, including RdDM, histone modification as well as genome topology changes, also regulate gene expression in response to the environmental stresses. Ac, histone acetylation marker; Me, histone methylation marker; CHR, Chromatin remodeling complex/factor; DDM, DNA demethylase; DMT, DNA methyltransferase; HAT, histone acetyltransferase; HDAC, histone deacetylase; HDM, histone demethylase; HMT, histone methyltransferase; lncRNA, long non-coding RNA; NcRNA, non-coding RNA; RdDM, RNA-directed DNA methylation; siRNA, small interfering RNA.

Although past decades have seen progress in understanding the epigenetic mechanisms controlling wheat and barley stress responses, we still have a long way to go towards fully understanding the epigenetic mechanisms regulating plant responses to environmental stresses in wheat and barley. In *Arabidopsis*, the MAP kinase MPK3, a key component in defense signaling, directly phosphorylates the histone deacetylase HD2B, thereby regulating the intra-nuclear compartmentalization of HD2B, as well as the reprogramming of defense gene expression and innate immunity [119]. However, detailed steps from signaling to epigenetic modification in response to environmental stresses remain to be uncovered in wheat and barley. In addition, *Arabidopsis* studies revealed that multiple epigenetic processes such as DNA methylation, histone modifications and chromatin remodeling regulate transcriptional memory to environmental stresses, including heat, freezing, drought and even pathogen infection [120–122]. Such stress memory greatly improves plant stress adaptation, and also prepares their offspring for future environmental challenges [120–122]. However, stress memory and its epigenetic mechanisms in wheat and barley remain to be explored in the future research.

With the advance of molecular technologies, our knowledge of the mechanisms of epigenetic responses to environmental stresses is rapidly growing, which could certainly lead to the more efficient improvement of cereal crops [123]. In *Arabidopsis*, epigenetic recombinant inbred line (epiRIL) populations were constructed and exhibited discernible phenotypic variation, including altered resistance against pathogen infection [123–126]. Creating similar epiRILs in wheat and barley would provide substantial resources not only for identifying ideal epigenetic variation in crops, but also for fully using the potential of epigenetics in crop improvement [123–126]. Besides, genome-editing enzymes such as transcription activator-like effector nucleases (TALENs) and CRISPR-Cas9 system have been used to engineer epigenomes in a sequence-specific manner in mammalian systems [127–130]. In *Arabidopsis*, Johnson et al. directed DNA methylation to target DNA sequences and caused expected phenotype changes through fusing ZFNs with the SRA domain-containing protein SUVH9—a protein integral to RNA-directed DNA methylation (RdDM) [131]. The development of methodologies to create epiRIL, generate epimutagenesis, and engineer epigenomes in a site-specific manner, would provide new avenues for generating epigenetic diversity and harnessing epigenetic variation for the improvement of agricultural traits in wheat and barley.

Author Contributions: C.C. and L.K. wrote this manuscript with help from X.W. and Y.L. All authors have read and agreed to the published version of the manuscript.

Acknowledgments: This work was supported by the National Natural Science Foundation of China (31701986, 31701412), the Natural Science Foundation of Shandong Province (ZR2017BC109) and the Qingdao Science and Technology Bureau Fund (18-2-2-51-jch, 17-1-1-50-jch).

References

1. International Wheat Genome Sequencing Consortium. Shifting the limits in wheat research and breeding using a fully annotated reference genome. *Science* **2018**, *361*, eaar7191. [CrossRef] [PubMed]
2. Avni, R.; Nave, M.; Barad, O.; Baruch, K.; Twardziok, S.O.; Gundlach, H.; Hale, I.; Mascher, M.; Spannagl, M.; Wiebe, K.; et al. Wild emmer genome architecture and diversity elucidate wheat evolution and domestication. *Science* **2017**, *97*, 93–97. [CrossRef] [PubMed]
3. Pourkheirandish, M.; Hensel, G.; Kilian, B.; Senthil, N.; Chen, G.; Sameri, M.; Azhaguvel, P.; Sakuma, S.; Dhanagond, S.; Sharma, R.; et al. Evolution of the grain dispersal system in barley. *Cell* **2015**, *162*, 527–539. [CrossRef] [PubMed]
4. International Barley Sequencing Consortium (IBSC). A physical, genetic and functional sequence assembly of the barley genome. *Nature* **2012**, *491*, 711–716. [CrossRef] [PubMed]
5. Panta, S.; Flowers, T.J.; Lane, P.; Doyle, R.; Haros, G.; Shabala, S. Halophyte agriculture: Success stories. *Environ. Exp. Bot.* **2014**, *107*, 71–83. [CrossRef]
6. Zörb, C.; Geilfus, C.M.; Dietz, K.J. Salinity and crop yield. *Plant Biol.* **2019**, *21* (Suppl. 1), 31–38.

7. Hanin, M.; Ebel, C.; Ngom, M.; Laplaze, L.; Masmoudi, K. New insights on plant salt tolerance mechanisms and their potential use for breeding. *Front. Plant Sci.* **2016**, *7*, 1787. [CrossRef]
8. Xie, W.; Xiong, W.; Pan, J.; Ali, T.; Cui, Q.; Guan, D.; Meng, J.; Mueller, N.D.; Lin, E.; Davis, S.J. Decreases in global beer supply due to extreme drought and heat. *Nat. Plants* **2018**, *4*, 964–973. [CrossRef]
9. Wang, X.; Xin, C.; Cai, J.; Zhou, Q.; Dai, T.; Cao, W.; Jiang, D. Heat priming induces tran-generational tolerance to high temperature stress in wheat. *Front. Plant Sci.* **2016**, *7*, 501. [CrossRef]
10. Chen, D.; Chen, D.; Xue, R.; Long, J.; Lin, X.; Lin, Y.; Jia, L.; Zeng, R.; Song, Y. Effects of boron, silicon and their interactions on cadmium accumulation and toxicity in rice plants. *J. Hazard. Mater.* **2019**, *367*, 447–455. [CrossRef]
11. Rizwan, M.; Ali, S.; Rehman, M.Z.; Rinklebe, J.; Tsang, D.C.W.; Bashir, A.; Maqbool, A.; Tack, F.M.G.; Ok, Y.S. Cadmium phytoremediation potential of Brassica crop species: A review. *Sci. Total Environ.* **2018**, *631*, 1175–1191. [CrossRef] [PubMed]
12. Aprile, A.; Sabella, E.; Francia, E.; Milc, J.; Ronga, D.; Pecchioni, N.; Ferrari, E.; Luvisi, A.; Vergine, M.; De Bellis, L. Combined Effect of Cadmium and Lead on Durum Wheat. *Int. J. Mol. Sci.* **2019**, *20*, 5891. [CrossRef] [PubMed]
13. Wu, L.; Yu, J.; Shen, Q.; Huang, L.; Wu, D.; Zhang, G. Identification of microRNAs in response to aluminum stress in the roots of Tibetan wild barley and cultivated barley. *BMC Genomics* **2018**, *19*, 560. [CrossRef] [PubMed]
14. Twamley, T.; Gaffney, M.; Feechan, A. A microbial fermentation mixture primes for resistance against powdery mildew in wheat. *Front. Plant Sci.* **2019**, *10*, 1241. [CrossRef]
15. Figueroa, M.; Hammond-Kosack, K.E.; Solomon, P.S. A review of wheat diseases-a field perspective. *Mol. Plant Pathol.* **2018**, *19*, 1523–1536. [CrossRef]
16. Lee, S.C.; Luan, S. ABA signal transduction at the crossroad of biotic and abiotic stress responses. *Plant Cell Environ.* **2012**, *35*, 53–60. [CrossRef]
17. Ku, Y.S.; Sintaha, M.; Cheung, M.Y.; Lam, H.M. Plant hormone signaling crosstalks between biotic and abiotic stress responses. *Int. J. Mol. Sci.* **2018**, *19*, 3206. [CrossRef]
18. Kumar, M.; Kesawat, M.S.; Ali, A.; Lee, S.C.; Gill, S.S.; Kim, H.U. Integration of abscisic acid signaling with other signaling pathways in plant stress responses and development. *Plants* **2019**, *8*, 592. [CrossRef]
19. Shahid, S. To Be or Not to Be Pathogenic: Transcriptional reprogramming dictates a fungal pathogen's response to different hosts. *Plant Cell* **2019**, *32*, 289. [CrossRef]
20. Kim, H.; Shim, D.; Moon, S.; Lee, J.; Bae, W.; Choi, H.; Kim, K.; Ryu, H. Transcriptional network regulation of the brassinosteroid signaling pathway by the BES1-TPL-HDA19 co-repressor complex. *Planta* **2019**, *250*, 1371–1377. [CrossRef]
21. Wu, M.S.; Ding, X.; Fu, X.; Lozano-Duran, R. Transcriptional reprogramming caused by the geminivirus *Tomato yellow leaf curl virus* in local or systemic infections in *nicotiana benthamiana*. *BMC Genomics* **2019**, *20*, 542. [CrossRef] [PubMed]
22. Chen, R.; Li, M.; Zhang, H.Y.; Duan, L.J.; Sun, X.J.; Jiang, Q.Y.; Zhang, H.; Hu, Z. Continuous salt stress-induced long non-coding RNAs and DNA methylation patterns in soybean roots. *BMC Genomics* **2019**, *20*, 730. [CrossRef] [PubMed]
23. Deleris, A.; Halter, T.; Navarro, L. DNA methylation and demethylation in plant immunity. *Annu. Rev. Phytopathol.* **2016**, *54*, 579–603. [CrossRef] [PubMed]
24. Kim, J.M.; To, T.K.; Matsui, A.; Tanoi, K.; Kobayashi, N.I.; Matsuda, F.; Habu, Y.; Ogawa, D.; Sakamoto, T.; Matsunaga, S.; et al. Acetate-mediated novel survival strategy against drought in plants. *Nat. Plants* **2017**, *3*, 17097. [CrossRef] [PubMed]
25. Wang, M.; Qin, L.; Xie, C.; Li, W.; Yuan, J.; Kong, L.; Yu, W.; Xia, G.; Liu, S. Induced and constitutive DNA methylation in a salinity-tolerant wheat introgression line. *Plant Cell Physiol.* **2014**, *55*, 1354–1365. [CrossRef] [PubMed]
26. Niederhuth, C.E.; Bewick, A.J.; Ji, L.; Alabady, M.S.; Kim, K.D.; Li, Q.; Rohr, N.A.; Rambani, A.; Burke, J.M.; Udall, J.A.; et al. Widespread natural variation of DNA methylation within angiosperms. *Genome Biol.* **2016**, *17*, 194. [CrossRef]
27. Yaari, R.; Katz, A.; Domb, K.; Harris, K.D.; Zemach, A.; Ohad, N. RdDM-independent de novo and heterochromatin DNA methylation by plant CMT and DNMT3 orthologs. *Nat. Commun.* **2019**, *10*, 1613. [CrossRef]

28. Law, J.A.; Jacobsen, S.E. Establishing, maintaining and modifying DNA methylation patterns in plants and animals. *Nat. Rev. Genet.* **2010**, *11*, 204–220. [CrossRef]
29. Henderson, I.R.; Jacobsen, S.E. Epigenetic inheritance in plants. *Nature* **2007**, *447*, 418–424. [CrossRef]
30. Park, K.; Kim, M.Y.; Vickers, M.; Park, J.S.; Hyun, Y.; Okamoto, T.; Zilberman, D.; Fischer, R.L.; Feng, X.; Choi, Y.; et al. DNA demethylation is initiated in the central cells of *Arabidopsis* and rice. *Proc. Natl. Acad. Sci. USA* **2016**, *113*, 15138–15143. [CrossRef]
31. Liu, H.; Zhang, H.; Dong, Y.; Hao, Y.; Zhang, X. DNA METHYLTRANSFERASE1-mediated shoot regeneration is regulated by cytokinin-induced cell cycle in *Arabidopsis*. *New Phytol.* **2018**, *217*, 219–232. [CrossRef] [PubMed]
32. Li, X.; Zhu, J.; Hu, F.; Ge, S.; Ye, M.; Xiang, H.; Zhang, G.; Zheng, X.; Zhang, H.; Zhang, S.; et al. Single-base resolution maps of cultivated and wild rice methylomes and regulatory roles of DNA methylation in plant gene expression. *BMC Genomics* **2012**, *13*, 300. [CrossRef] [PubMed]
33. Zhang, X.; Yazaki, J.; Sundaresan, A.; Cokus, S.; Chan, S.W.; Chen, H.; Henderson, I.R.; Shinn, P.; Pellegrini, M.; Jacobsen, S.E.; et al. Genome-wide high-resolution mapping and functional analysis of DNA methylation in *Arabidopsis*. *Cell* **2006**, *126*, 1189–1201. [CrossRef] [PubMed]
34. Zilberman, D.; Gehring, M.; Tran, R.K.; Ballinger, T.; Henikoff, S. Genome-wide analysis of *Arabidopsis thaliana* DNA methylation uncovers an interdependence between methylation and transcription. *Nat. Genet.* **2007**, *39*, 61–69. [CrossRef] [PubMed]
35. Simmen, M.W.; Leitgeb, S.; Charlton, J.; Jones, S.J.; Harris, B.R.; Clark, V.H.; Bird, A. Nonmethylated transposable elements and methylated genes in a chordate genome. *Science* **1999**, *283*, 1164–1167. [CrossRef] [PubMed]
36. Zhang, M.; Kimatu, J.N.; Xu, K.; Liu, B. DNA cytosine methylation in plant development. *J. Genet. Genom.* **2010**, *37*, 1–12. [CrossRef]
37. Shafiq, S.; Zeb, Q.; Ali, A.; Sajjad, Y.; Nazir, R.; Widemann, E.; Liu, L. Lead, cadmium and zinc phytotoxicity alter DNA methylation levels to confer heavy metal tolerance in wheat. *Int. J. Mol. Sci.* **2019**, *20*, 4676. [CrossRef]
38. Kashino-Fujii, M.; Yokosho, K.; Yamaji, N.; Yamane, M.; Saisho, D.; Sato, K.; Ma, J.F. Retrotransposon insertion and DNA methylation regulate aluminum tolerance in european barley accessions. *Plant Physiol.* **2018**, *178*, 716–727. [CrossRef]
39. Nagy, R.; Grob, H.; Weder, B.; Green, P.; Klein, M.; Frelet-Barrand, A.; Schjoerring, J.K.; Brearley, C.; Martinoia, E. The *Arabidopsis* ATP-binding cassette protein AtMRP5/AtABCC5 is a high affinity inositol hexakisphosphate transporter involved in guard cell signaling and phytate storage. *J. Biol. Chem.* **2009**, *284*, 33614–33622. [CrossRef]
40. Park, J.; Song, W.Y.; Ko, D.; Eom, Y.; Hansen, T.H.; Schiller, M.; Lee, T.G.; Martinoia, E.; Lee, Y. The phytochelatin transporters AtABCC1 and AtABCC2 mediate tolerance to cadmium and mercury. *Plant J.* **2012**, *69*, 278–288. [CrossRef]
41. Bhati, K.K.; Sharma, S.; Aggarwal, S.; Kaur, M.; Shukla, V.; Kaur, J.; Mantri, S.; Pandey, A.K. Genome-wide identification and expression characterization of ABCC-MRP transporters in hexaploid wheat. *Front. Plant Sci.* **2015**, *6*, 488. [CrossRef] [PubMed]
42. Ji, H.; Peng, Y.; Meckes, N.; Allen, S.; Stewart, C.N., Jr.; Traw, M.B. ATP-dependent binding cassette transporter G family member 16 increases plant tolerance to abscisic acid and assists in basal resistance against *Pseudomonas syringae* DC3000. *Plant Physiol.* **2014**, *166*, 879–888. [CrossRef] [PubMed]
43. Gaedeke, N.; Klein, M.; Ansorge, M.; Kolukisaoglu, H.U.; Forestier, C.; Becker, D.; Schulz, B.; Mueller-Roeber, B.; Martinoia, E. The *Arabidopsis thaliana* ABC-transporter AtMRP5 controls root growth and stomata movement. *EMBO J.* **2001**, *20*, 1875–1887. [CrossRef] [PubMed]
44. Wanke, D.; Kolukisaoglu, H.U. An update on the ABCC transporter family in plants: Many genes, many proteins, but how many functions? *Plant Biol.* **2010**, *12*, 15–25. [CrossRef]
45. Ma, Y.; Li, C.; Ryan, P.R.; Shabala, S.; You, J.; Liu, J.; Liu, C.; Zhou, M. A new allele for aluminium tolerance gene in barley (*Hordeum vulgare* L.). *BMC Genomics* **2016**, *17*, 186. [CrossRef]
46. Zhou, G.; Delhaize, E.; Zhou, M.; Ryan, P.R. The barley *MATE* gene, *HvAACT1*, increases citrate efflux and Al^{3+} tolerance when expressed in wheat and barley. *Ann. Bot.* **2013**, *112*, 603–612. [CrossRef]
47. Furukawa, J.; Yamaji, N.; Wang, H.; Mitani, N.; Murata, Y.; Sato, K.; Katsuhara, M.; Takeda, K.; Ma, J.F. An aluminum-activated citrate transporter in barley. *Plant Cell Physiol.* **2007**, *48*, 1081–1091. [CrossRef]

48. Geng, S.; Kong, X.; Song, G.; Jia, M.; Guan, J.; Wang, F.; Qin, Z.; Wu, L.; Lan, X.; Li, A.; et al. DNA methylation dynamics during the interaction of wheat progenitor *Aegilops tauschii* with the obligate biotrophic fungus *Blumeria graminis* f. sp. *tritici*. *New Phytol.* **2019**, *221*, 1023–1035. [CrossRef]
49. O'Malley, R.C.; Huang, S.C.; Song, L.; Lewsey, M.G.; Bartlett, A.; Nery, J.R.; Galli, M.; Andrea Gallavotti, A.; Ecker, J.R. Cistrome and epicistrome features shape the regulatory DNA landscape. *Cell* **2016**, *165*, 1280–1292. [CrossRef]
50. Tirnaz, S.; Batley, J. DNA Methylation: Toward crop disease resistance improvement. *Trends Plant Sci.* **2019**, *24*, 1137–1150. [CrossRef]
51. Zhu, J.K. Active DNA demethylation mediated by DNA glycosylases. *Annu. Rev. Genet.* **2009**, *43*, 143–166. [CrossRef] [PubMed]
52. Gehring, M.; Reik, W.; Henikoff, S. DNA demethylation by DNA repair. *Trends Genet.* **2009**, *25*, 82–90. [CrossRef] [PubMed]
53. Nie, W.F.; Lei, M.; Zhang, M.; Tang, K.; Huang, H.; Zhang, C.; Miki, D.; Liu, P.; Yang, Y.; Wang, X.; et al. Histone acetylation recruits the SWR1 complex to regulate active DNA demethylation in *Arabidopsis*. *Proc. Natl. Acad. Sci. USA* **2019**, *116*, 16641–16650. [CrossRef] [PubMed]
54. Tang, K.; Lang, Z.; Zhang, H.; Zhu, J.K. The DNA demethylase ROS1 targets genomic regions with distinct chromatin modifications. *Nat. Plants* **2016**, *2*, 16169. [CrossRef]
55. Kapazoglou, A.; Drosou, V.; Argiriou, A.; Tsaftaris, A.S. The study of a barley epigenetic regulator, HvDME, in seed development and under drought. *BMC Plant Biol.* **2013**, *13*, 172. [CrossRef]
56. Zheng, M.; Liu, X.; Lin, J.; Liu, X.; Wang, Z.; Xin, M.; Yao, Y.; Peng, H.; Zhou, D.X.; Ni, Z.; et al. Histone acetyltransferase GCN5 contributes to cell wall integrity and salt stress tolerance by altering the expression of cellulose synthesis genes. *Plant J.* **2019**, *97*, 587–602.
57. Hu, Z.; Song, N.; Zheng, M.; Liu, X.; Liu, Z.; Xing, J.; Ma, J.; Guo, W.; Yao, Y.; Peng, H.; et al. Histone acetyltransferase GCN5 is essential for heat stress-responsive gene activation and thermotolerance in *Arabidopsis*. *Plant J.* **2015**, *84*, 1178–1191. [CrossRef]
58. Liu, J.; Zhi, P.; Wang, X.; Fan, Q.; Chang, C. Wheat WD40-repeat protein TaHOS15 functions in a histone deacetylase complex to fine-tune defense responses to *Blumeria graminis* f.sp. *tritici*. *J. Exp. Bot.* **2019**, *70*, 255–268. [CrossRef]
59. Wang, X.; Zhi, P.; Fan, Q.; Zhang, M.; Chang, C. Wheat CHD3 protein TaCHR729 regulates the cuticular wax biosynthesis required for stimulating germination of *Blumeria graminis* f.sp. *tritici*. *J. Exp. Bot.* **2019**, *70*, 701–713. [CrossRef]
60. Zhang, H.; Chen, X.; Wang, C.; Xu, Z.; Wang, Y.; Liu, X.; Ji, W. Long non-coding genes implicated in response to stripe rust pathogen stress in wheat (*Triticum aestivum* L.). *Mol. Biol. Rep.* **2013**, *40*, 6245–6253. [CrossRef]
61. Chang, Y.N.; Zhu, C.; Jiang, J.; Zhang, H.; Zhu, J.K.; Duan, C.G. Epigenetic regulation in plant abiotic stress responses. *J. Integr. Plant Biol.* **2019**. [CrossRef] [PubMed]
62. Liu, G.; Khan, N.; Ma, X.; Hou, X. Identification, Evolution, and Expression Profiling of Histone Lysine Methylation Moderators in *Brassica rapa*. *Plants* **2019**, *8*, 526. [CrossRef] [PubMed]
63. Liu, C.; Lu, F.; Cui, X.; Cao, X. Histone methylation in higher plants. *Annu. Rev. Plant Biol.* **2010**, *61*, 395–420. [CrossRef] [PubMed]
64. Wang, L.; Chen, H.; Li, J.; Shu, H.; Zhang, X.; Wang, Y.; Tyler, B.M.; Dong, S. Effector gene silencing mediated by histone methylation underpins host adaptation in an oomycete plant pathogen. *Nucleic Acids Res.* **2019**. [CrossRef] [PubMed]
65. Thorstensen, T.; Grini, P.E.; Aalen, R.B. SET domain proteins in plant development. *Biochim. Biophys. Acta* **2011**, *1809*, 407–420. [CrossRef] [PubMed]
66. Black, J.C.; Van Rechem, C.; Whetstine, J.R. Histone lysine methylation dynamics: Establishment, regulation, and biological impact. *Mol. Cell* **2012**, *48*, 491–507. [CrossRef] [PubMed]
67. Xiao, J.; Jin, R.; Yu, X.; Shen, M.; Wagner, J.D.; Pai, A.; Song, C.; Zhuang, M.; Klasfeld, S.; He, C.; et al. Cis and trans determinants of epigenetic silencing by Polycomb repressive complex 2 in Arabidopsis. *Nat. Genet.* **2017**, *49*, 1546–1552. [CrossRef]
68. Hecker, A.; Brand, L.H.; Peter, S.; Simoncello, N.; Kilian, J.; Harter, K.; Gaudin, V.; Wanke, D. The Arabidopsis GAGA-Binding Factor BASIC PENTACYSTEINE6 Recruits the POLYCOMB- REPRESSIVE COMPLEX1 component LIKE HETEROCHROMATIN PROTEIN1 to GAGA DNA motifs. *Plant Physiol.* **2015**, *168*, 1013–1024. [CrossRef]

69. Jie, Y.; Yuan, L.; Yen, M.R.; Zheng, F.; Ji, R.; Peng, T.; Gu, D.; Yang, S.; Cui, Y.; Chen, P.Y.; et al. SWI3B and HDA6 interact and are required for transposon silencing in *Arabidopsis*. *Plant J.* **2019**. [CrossRef]
70. Imhof, A.; Wolffe, A.P. Transcription: Gene control by targeted histone acetylation. *Curr. Biol.* **1998**, *8*, R422–R424. [CrossRef]
71. Sharma, C.; Kumar, S.; Saripalli, G.; Jain, N.; Raghuvanshi, S.; Sharma, J.B.; Prabhu, K.V.; Sharma, P.K.; Balyan, H.S.; Gupta, P.K. H3K4/K9 acetylation and *Lr28*-mediated expression of six leaf rust responsive genes in wheat (*Triticum aestivum*). *Mol. Genet. Genomics* **2019**, *294*, 227–241. [CrossRef] [PubMed]
72. Zhou, J.; Wang, X.; He, K.; Charron, J.B.; Elling, A.A.; Deng, X.W. Genome-wide profiling of histone H3 lysine 9 acetylation and dimethylation in *Arabidopsis* reveals correlation between multiple histone marks and gene expression. *Plant Mol. Biol.* **2010**, *72*, 585–595. [CrossRef] [PubMed]
73. Bilichak, A.; Ilnystkyy, Y.; Hollunder, J.; Kovalchuk, I. The progeny of *Arabidopsis thaliana* plants exposed to salt exhibit changes in DNA methylation, histone modifications and gene expression. *PLoS ONE* **2012**, *7*, e30515. [CrossRef] [PubMed]
74. Rymen, B.; Kawamura, A.; Lambolez, A.; Inagaki, S.; Takebayashi, A.; Iwase, A.; Sakamoto, Y.; Sako, K.; Favero, D.S.; Ikeuchi, M.; et al. Histone acetylation orchestrates wound-induced transcriptional activation and cellular reprogramming in Arabidopsis. *Commun. Biol.* **2019**, *2*, 404. [CrossRef]
75. Liu, S.; Liu, G.; Cheng, P.; Xue, C.; Zhou, Y.; Chen, X.; Ye, L.; Qiao, Z.; Zhang, T.; Gong, Z. Genome-wide profiling of histone lysine butyrylation reveals its role in the positive regulation of gene transcription in rice. *Rice (NY)* **2019**, *12*, 86. [CrossRef]
76. Janack, B.; Sosoi, P.; Krupinska, K.; Humbeck, K. Knockdown of WHIRLY1 affects drought stress-induced leaf senescence and histone modifications of the senescence-associated gene *HvS40*. *Plants* **2016**, *5*, 37. [CrossRef]
77. Kim, J.M.; To, T.K.; Ishida, J.; Morosawa, T.; Kawashima, M.; Matsui, A.; Toyoda, T.; Kimura, H.; Shinozaki, K.; Seki, M. Alterations of lysine modifications on the histone H3 N-tail under drought stress conditions in *Arabidopsis thaliana*. *Plant Cell Physiol.* **2008**, *49*, 1580–1588. [CrossRef]
78. Luo, M.; Liu, X.; Singh, P.; Cui, Y.; Zimmerli, L.; Wu, K. Chromatin modifications and remodeling in plant abiotic stress responses. *Biochim. Biophys. Acta* **2012**, *1819*, 129–136. [CrossRef]
79. Narlikar, G.J.; Sundaramoorthy, R.; Owen-Hughes, T. Mechanisms and functions of ATP-dependent chromatin-remodeling enzymes. *Cell* **2013**, *154*, 490–503. [CrossRef]
80. Chen, W.; Zhu, Q.; Liu, Y.; Zhang, Q. Chromatin Remodeling and Plant Immunity. *Adv. Protein Chem. Struct. Biol.* **2017**, *106*, 243–260.
81. Archacki, R.; Yatuseich, R.; Buszewicz, D.; Krzyczmonik, K.; Patryn, J.; Iwanicka-Nowicka, R.; Biecek, P.; Wilczynski, B.; Koblowska, M.; Jerzmanowski, A. *Arabidopsis* SWI/SNF chromatin remodeling complex binds both promoters and terminators to regulate gene expression. *Nucleic Acids Res.* **2017**, *45*, 3116–3129. [CrossRef] [PubMed]
82. Burgio, G.; La Rocca, G.; Sala, A.; Arancio, W.; Di Gesù, D.; Collesano, M.; Sperling, A.S.; Armstrong, J.A.; van Heeringen, S.J.; Logie, C. Genetic identification of a network of factors that functionally interact with the nucleosome remodeling ATPase ISWI. *PLoS Genet.* **2008**, *4*, e1000089. [CrossRef] [PubMed]
83. Zou, B.; Sun, Q.; Zhang, W.; Ding, Y.; Yang, D.L.; Shi, Z.; Hua, J. The *Arabidopsis* chromatin-remodeling factor CHR5 regulates plant immune responses and nucleosome occupancy. *Plant Cell Physiol.* **2017**, *58*, 2202–2216. [CrossRef] [PubMed]
84. Yang, R.; Hong, Y.; Ren, Z.; Tang, K.; Zhang, H.; Zhu, J.K.; Zhao, C. A role for PICKLE in the regulation of cold and salt stress tolerance in *Arabidopsis*. *Front. Plant Sci.* **2019**, *10*, 900. [CrossRef]
85. Fenley, A.T.; Anandakrishnan, R.; Kidane, Y.H.; Onufriev, A.V. Modulation of nucleosomal DNA accessibility via charge-altering post-translational modifications in histone core histone core. *Epigenet. Chromatin* **2018**, *11*, 11. [CrossRef]
86. Tessarz, P.; Kouzarides, T. Histone core modifications regulating nucleosome structure and dynamics. *Nat. Rev. Mol. Cell Biol.* **2014**, *15*, 703–708. [CrossRef]
87. Buszewicz, D.; Archacki, R.; Palusiński, A.; Kotliński, M.; Fogtman, A.; Iwanicka-Nowicka, R.; Sosnowska, K.; Kuciński, J.; Pupel, P.; Olędzki, J.; et al. HD2C histone deacetylase and a SWI/SNF chromatin remodelling complex interact and both are involved in mediating the heat stress response in *Arabidopsis*. *Plant Cell Environ.* **2016**, *39*, 2108–2122. [CrossRef]
88. Han, S.K.; Wu, M.F.; Cui, S.; Wagner, D. Roles and activities of chromatin remodeling ATPases in plants. *Plant J.* **2015**, *83*, 62–77. [CrossRef]

89. Han, S.K.; Wagner, D. Role of chromatin in water stress responses in plants. *J. Exp. Bot.* **2014**, *65*, 2785–2799. [CrossRef]
90. Saez, A.; Rodrigues, A.; Santiago, J.; Rubio, S.; Rodriguez, P.L. HAB1-SWI3B interaction reveals a link between abscisic acid signaling and putative SWI/SNF chromatin-remodeling complexes in *Arabidopsis*. *Plant Cell* **2008**, *20*, 2972–2988. [CrossRef]
91. Umezawa, T.; Sugiyama, N.; Takahashi, F.; Anderson, J.C.; Ishihama, Y.; Peck, S.C.; Shinozaki, K. Genetics and phosphoproteomics reveal a protein phosphorylation network in the abscisic acid signaling pathway in *Arabidopsis thaliana*. *Sci. Signal* **2013**, *6*, rs8. [CrossRef]
92. Wang, P.; Xue, L.; Batelli, G.; Lee, S.; Hou, Y.J.; Van Oosten, M.J.; Zhang, H.; Tao, W.A.; Zhu, J.K. Quantitative phosphoproteomics identifies SnRK2 protein kinase substrates and reveals the effectors of abscisic acid action. *Proc. Natl. Acad. Sci. USA* **2013**, *110*, 11205–11210. [CrossRef]
93. Hou, J.; Lu, D.; Mason, A.S.; Li, B.; Xiao, M.; An, S.; Fu, D. Non-coding RNAs and transposable elements in plant genomes: Emergence, regulatory mechanisms and roles in plant development and stress responses. *Planta* **2019**, *250*, 23–40. [CrossRef]
94. Brant, E.J.; Budak, H. Plant small non-coding RNAs and their roles in biotic stresses. *Front. Plant Sci.* **2018**, *9*, 1038. [CrossRef]
95. Wang, J.; Meng, X.; Dobrovolskaya, O.B.; Orlov, Y.L.; Chen, M. Non-coding RNAs and their roles in stress response in plants. *Genomics Proteomics Bioinform.* **2017**, *15*, 301–312. [CrossRef]
96. Mach, J. The long-noncoding RNA ELENA1 functions in plant immunity. *Plant Cell* **2017**, *29*, 916. [CrossRef]
97. Ravichandran, S.; Ragupathy, R.; Edwards, T.; Domaratzki, M.; Cloutier, S. MicroRNA-guided regulation of heat stress response in wheat. *BMC Genomics* **2019**, *20*, 488. [CrossRef]
98. Sun, F.; Guo, W.; Du, J.; Ni, Z.; Sun, Q.; Yao, Y. Widespread, abundant, and diverse TE-associated siRNAs in developing wheat grain. *Gene* **2013**, *522*, 1–7. [CrossRef]
99. Yu, Z.; Wang, X.; Mu, X.; Zhang, L. RNAi mediated silencing of dehydrin gene WZY2 confers osmotic stress intolerance in transgenic wheat. *Funct. Plant Biol.* **2019**, *46*, 877–884. [CrossRef]
100. Sun, X.; Zhang, H.; Sui, N. Regulation mechanism of long non-coding RNA in plant response to stress. *Biochem. Biophys. Res. Commun.* **2018**, *503*, 402–407. [CrossRef]
101. Song, Y.; Zhang, D. The Role of Long Noncoding RNAs in Plant Stress Tolerance. *Methods Mol. Biol.* **2017**, *1631*, 41–68.
102. Quinn, J.J.; Chang, H.Y. Unique features of long non-coding RNA biogenesis and function. *Nat. Rev. Genet.* **2016**, *17*, 47–62. [CrossRef]
103. Khemka, N.; Singh, V.K.; Garg, R.; Jain, M. Genome-wide analysis of long intergenic non-coding RNAs in chickpea and their potential role in flower development. *Sci. Rep.* **2016**, *6*, 33297. [CrossRef]
104. Kim, E.; Sung, S. Long noncoding RNA: Unveiling hidden layer of gene regulatory networks. *Trends Plant Sci.* **2012**, *17*, 16–21. [CrossRef]
105. Matzke, M.A.; Mosher, R.A. RNA-directed DNA methylation: An epigenetic pathway of increasing complexity. *Nat. Rev. Genet.* **2014**, *15*, 394–408. [CrossRef]
106. Wierzbicki, A.T. The role of long non-coding RNA in transcriptional gene silencing. *Curr. Opin. Plant Biol.* **2012**, *15*, 517–522. [CrossRef]
107. Popova, O.V.; Dinh, H.Q.; Aufsatz, W.; Jonak, C. The RdDM pathway is required for basal heat tolerance in *Arabidopsis*. *Mol. Plant* **2013**, *6*, 396–410. [CrossRef]
108. Wang, J.; Meng, X.; Yuan, C.; Harrison, A.P.; Chen, M. The roles of cross-talk epigenetic patterns in *Arabidopsis thaliana*. *Brief. Funct. Genomics* **2016**, *15*, 278–287. [CrossRef]
109. Csorba, T.; Questa, J.I.; Sun, Q.; Dean, C. Antisense COOLAIR mediates the coordinated switching of chromatin states at FLC during vernalization. *Proc. Natl. Acad. Sci. USA* **2014**, *111*, 16160–16165. [CrossRef]
110. Michaels, S.D.; Amasino, R.M. FLOWERING LOCUS C encodes a novel MADS domain protein that acts as a repressor of flowering. *Plant Cell* **1999**, *11*, 949–956. [CrossRef]
111. Marquardt, S.; Raitskin, O.; Wu, Z.; Liu, F.; Sun, Q.; Dean, C. Functional consequences of splicing of the antisense transcript COOLAIR on FLC transcription. *Mol. Cell* **2014**, *54*, 156–165. [CrossRef] [PubMed]
112. Heo, J.B.; Sung, S. Vernalization-mediated epigenetic silencing by a long intronic noncoding RNA. *Science* **2011**, *331*, 76–79. [CrossRef] [PubMed]
113. Movahedi, A.; Sun, W.; Zhang, J.; Wu, X.; Mousavi, M.; Mohammadi, K. RNA-directed DNA methylation in plants. *Plant Cell Rep.* **2015**, *34*, 1857–1862. [CrossRef]

114. Ariel, F.; Jegu, T.; Latrasse, D.; Romero-Barrios, N.; Christ, A.; Benhamed, M.; Crespi, M. Noncoding transcription by alternative RNA polymerases dynamically regulates an auxin-driven chromatin loop. *Mol. Cell* **2014**, *7*, 383–396. [CrossRef]
115. Shumayla; Shailesh, S.; Mehak, T.; Shivi, T.; Kashmir, S.; Santosh, K.U. Survey of high throughput RNA-Seq data reveals potential roles for lncRNAs during development and stress response in bread wheat. *Front. Plant Sci.* **2017**, *8*, 1019. [CrossRef]
116. Cagirici, H.B.; Alptekin, B.; Budak, H. RNA sequencing and co-expressed long non-coding RNA in modern and wild wheats. *Sci. Rep.* **2017**, *7*, 10670. [CrossRef]
117. Huang, Y.; Li, L.; Smith, K.P.; Muehlbauer, G.J. Differential transcriptomic responses to Fusarium graminearum infection in two barley quantitative trait loci associated with Fusarium head blight resistance. *BMC Genomics* **2016**, *17*, 387. [CrossRef]
118. Unver, T.; Tombuloglu, H. Barley long non-coding RNAs (lncRNA) responsive to excess boron. *Genomics* **2019**, *112*, 1945–1955. [CrossRef]
119. Latrasse, D.; Jégu, T.; Li, H.; de Zelicourt, A.; Raynaud, C.; Legras, S.; Gust, A.; Samajova, O.; Veluchamy, A.; Rayapuram, N.; et al. MAPK-triggered chromatin reprogramming by histone deacetylase in plant innate immunity. *Genome Biol.* **2017**, *18*, 131. [CrossRef]
120. Friedrich, T.; Faivre, L.; Bäurle, I.; Schubert, D. Chromatin-based mechanisms of temperature memory in plants. *Plant Cell Environ.* **2019**, *42*, 762–770. [CrossRef]
121. Jaskiewicz, M.; Conrath, U.; Peterhänsel, C. Chromatin modification acts as a memory for systemic acquired resistance in the plant stress response. *EMBO Rep.* **2011**, *12*, 50–55. [CrossRef]
122. Lämke, J.; Bäurle, I. Epigenetic and chromatin-based mechanisms in environmental stress adaptation and stress memory in plants. *Genome Biol.* **2017**, *18*, 124. [CrossRef]
123. Springer, N.M.; Schmitz, R.J. Exploiting induced and natural epigenetic variation for crop improvement. *Nat. Rev. Genet.* **2017**, *18*, 563–575. [CrossRef]
124. Zhang, Y.Y.; Latzel, V.; Fischer, M.; Bossdorf, O. Understanding the evolutionary potential of epigenetic variation: A comparison of heritable phenotypic variation in epiRILs, RILs, and natural ecotypes of *Arabidopsis thaliana*. *Heredity* **2018**, *121*, 257–265. [CrossRef]
125. Reinders, J.; Wulff, B.B.; Mirouze, M.; Marí-Ordóñez, A.; Dapp, M.; Rozhon, W.; Bucher, E.; Theiler, G.; Paszkowski, J. Compromised stability of DNA methylation and transposon immobilization in mosaic *Arabidopsis* epigenomes. *Genes Dev.* **2009**, *23*, 939–950. [CrossRef]
126. Johannes, F.; Porcher, E.; Teixeira, F.K.; Saliba-Colombani, V.; Simon, M.; Agier, N.; Bulski, A.; Albuisson, J.; Heredia, F.; Audigier, P. Assessing the impact of transgenerational epigenetic variation on complex traits. *PLoS Genet.* **2009**, *5*, e1000530. [CrossRef]
127. Miller, J.C.; Patil, D.P.; Xia, D.F.; Paine, C.B.; Fauser, F.; Richards, H.W.; Shivak, D.A.; Bendaña, Y.R.; Hinkley, S.J.; Scarlott, N.A.; et al. Enhancing gene editing specificity by attenuating DNA cleavage kinetics. *Nat. Biotechnol.* **2019**, *37*, 945–952. [CrossRef]
128. Gupta, D.; Bhattacharjee, O.; Mandal, D.; Sen, M.K.; Dey, D.; Dasgupta, A.; Kazi, T.A.; Gupta, R.; Sinharoy, S.; Acharya, K.; et al. CRISPR-Cas9 system: A new-fangled dawn in gene editing. *Life Sci.* **2019**, *232*, 116636. [CrossRef]
129. Paschon, D.E.; Lussier, S.; Wangzor, T.; Xia, D.F.; Li, P.W.; Hinkley, S.J.; Scarlott, N.A.; Lam, S.C.; Waite, A.J.; Truong, L.N.; et al. Diversifying the structure of zinc finger nucleases for high-precision genome editing. *Nat. Commun.* **2019**, *10*, 1133. [CrossRef]
130. Nakamura, S.; Watanabe, S.; Ando, N.; Ishihara, M.; Sato, M. Transplacental gene delivery (TPGD) as a noninvasive tool for fetal gene manipulation in mice. *Int. J. Mol. Sci.* **2019**, *20*, 5926. [CrossRef]
131. Johnson, L.M.; Du, J.; Hale, C.J.; Bischof, S.; Feng, S.; Chodavarapu, R.K.; Zhong, X.; Marson, G.; Pellegrini, M.; Segal, D.J.; et al. SRA-and SET-domain-containing proteins link RNA polymerase V occupancy to DNA methylation. *Nature* **2014**, *507*, 124–128. [CrossRef]

2

Advances in Omics Approaches for Abiotic Stress Tolerance in Tomato

Juhi Chaudhary [1], Praveen Khatri [2], Pankaj Singla [2], Surbhi Kumawat [2], Anu Kumari [2], Vinaykumar R [3], Amit Vikram [3], Salesh Kumar Jindal [4], Hemant Kardile [5], Rahul Kumar [6], Humira Sonah [2,*] and Rupesh Deshmukh [2,*]

1 Department of Biology, Oberlin College, Oberlin, OH 44074, USA; juhi.chaudhary@gmail.com
2 National Agri-Food Biotechnology Institute (NABI), Mohali, Punjab 140306, India; p.khatri2712@gmail.com (P.K.); pankajsingla2614@gmail.com (P.S.); surbhikumawat002@gmail.com (S.K.); anuk991@gmail.com (A.K.)
3 Department of Vegetable Science, Dr. Yashwant Singh Parmar University of Horticulture and Forestry, Solan, Himachal Pradesh 173230, India; agrivinay123@gmail.com (V.R.); amitsolan@gmail.com (A.V.)
4 Department of Vegetable Science, Punjab Agricultural University, Ludhiana, Punjab 141004, India; saleshjindal@pau.edu
5 Division of Crop Improvement, ICAR-Central Potato Research Institute (CPRI), Shimla, Himachal Pradesh 171001, India; kufrihemant@gmail.com
6 Department of Plant Science, University of Hyderabad, Hyderabad 500046, India; Rksl@uohyd.ac.in
* Correspondence: biohuma@gmail.com (H.S.); rupesh0deshmukh@gmail.com (R.D.)

Abstract: Tomato, one of the most important crops worldwide, has a high demand in the fresh fruit market and processed food industries. Despite having considerably high productivity, continuous supply as per the market demand is hard to achieve, mostly because of periodic losses occurring due to biotic as well as abiotic stresses. Although tomato is a temperate crop, it is grown in almost all the climatic zones because of widespread demand, which makes it challenge to adapt in diverse conditions. Development of tomato cultivars with enhanced abiotic stress tolerance is one of the most sustainable approaches for its successful production. In this regard, efforts are being made to understand the stress tolerance mechanism, gene discovery, and interaction of genetic and environmental factors. Several omics approaches, tools, and resources have already been developed for tomato growing. Modern sequencing technologies have greatly accelerated genomics and transcriptomics studies in tomato. These advancements facilitate Quantitative trait loci (QTL) mapping, genome-wide association studies (GWAS), and genomic selection (GS). However, limited efforts have been made in other omics branches like proteomics, metabolomics, and ionomics. Extensive cataloging of omics resources made here has highlighted the need for integration of omics approaches for efficient utilization of resources and a better understanding of the molecular mechanism. The information provided here will be helpful to understand the plant responses and the genetic regulatory networks involved in abiotic stress tolerance and efficient utilization of omics resources for tomato crop improvement.

Keywords: proteomics; metabolomics; ionomics; genotyping by sequencing; genome-wide association study; quantitative trait loci

1. Introduction

Tomato, one of the most valuable fruit and vegetable crops worldwide, is integral to the human diet. Due to the diverse range of its utility in raw, cooked, and processed food, the global demand for tomato has increased tremendously in recent years [1]. Although tomato is a temperate crop, it is being cultivated in diverse climatic zones, which makes the cultivation of tomato more challenging. Often, crop productivity and yield are severely affected by changing environmental conditions and abiotic stresses such as drought, salinity, and heat. Therefore, most of the non-conventional tomato cropping areas have adopted greenhouse-based cultivation for maintaining an uninterrupted supply throughout the year. Not only is the cost higher in greenhouse cultivation but there is also rapid accumulation of nitrate, phosphate, and salinity observed in the soil, which ultimately leads to soil degradation and groundwater or surface water pollution. Therefore, improvement of stress tolerance in tomato cultivars is sustainable and economically more desirable. Abiotic stress conditions imposed by extreme water and temperature regimes, nutritional imbalance in the soil substrate, elemental toxicity, and high salinity are the major factors limiting tomato production. The abiotic stresses become more complex under field conditions where more than one stressor typically coincide. Therefore, the development of sustainable, high-yielding varieties with improved tolerance to various abiotic stresses is a prerequisite to meet the global food demand [2]. Numerous efforts have been undertaken to address single stress traits under controlled conditions, but this approach is not always practical because the plant response is different in the field where multiple factors and stresses are imposed simultaneously [3]. In the past decade, conventional breeding has led to significant advances for a broad set of traits including biotic and abiotic stress tolerance, yield components and quality-related traits [4]. However, traditional varieties are susceptible to multiple stresses at different locations. Therefore, considering the genetic complexity and environmental interactions, application of more comprehensive and multidisciplinary approaches offers a better strategy to improve stress tolerance in modern crops [5–7].

Recent advances in the field of genomics have accelerated the successful development of new varieties. Molecular markers are based on the polymorphism identified in any given DNA sample, and they have dramatically increased the ability to characterize genetic diversity in the germplasm pool for essentially any crop species. Molecular markers have several advantages over the morphological or biochemical markers. These advantages include easy assay, reproducibility, convenience of use, high availability, stability regardless of environmental or external factors, and representation throughout entire genomes [8]. DNA markers have been widely used for versatile applications in genetics, molecular biology, genomics, and breeding in plants, including tomato. The most widely used applications of molecular markers in plant breeding include mapping of genes and quantitative trait loci, germplasm evaluation, population characterization, diversity studies, genomic fingerprinting, and marker-assisted breeding [9]. The efficiency of marker-assisted selection (MAS) for any trait during breeding requires precise information of map position and the molecular markers [10,11]. High-resolution mapping of QTLs, validation of linked markers, and marker conversion are the steps involved in the development of markers for MAS [12,13]. MAS has been extensively applied to the breeding of disease-resistant varieties in tomato [14]. MAS has been accelerated with the relatively recent adaptation of genotyping methods based on single nucleotide polymorphisms (SNPs) [15,16]. The availability of high-throughput marker genotyping systems and plentiful, well distributed markers make it possible to perform MAS more efficiently. In tomato, whole genome resequencing of 84 tomato accessions has identified over millions SNPs distributed throughout the entire genome [17]. Such a resource will be helpful for mapping studies as well as molecular biology research focusing on understanding the genetic regulation of different traits in tomato.

Recent technological advances have created several omics branches dealing specifically with the molecular components of cellular biology. To date, the major omics approaches include genomics, transcriptomics, proteomics, metabolomics, phenomics, and ionomics [5,18,19] (Figure 1). Omics approaches provide a holistic view of molecules at the cellular, tissues, or organism level. The integration of different omics providing many-dimensional biological information is being approached through a relatively new branch of life science known as system biology [20,21]. A recent development in DNA sequencing technology has accelerated genomics and transcriptomic research in plants and all other domains of life, including animals, fungi, and insects. Other omics branches like proteomics, metabolomics, and ionomics are not yet explored sufficiently as compared to genomics and transcriptomics. Since tomato has high economic importance and commercial value, it requires the integration of multi-disciplinary knowledge to design climate-smart varieties for high and stable yield in adverse climatic conditions. In this context, the present review provides in-depth information on recent advances in different omics branches, and methods for efficient exploration of available resources in tomato are discussed. The integration of different omics tools, techniques, and approaches to advance tomato research has also been addressed.

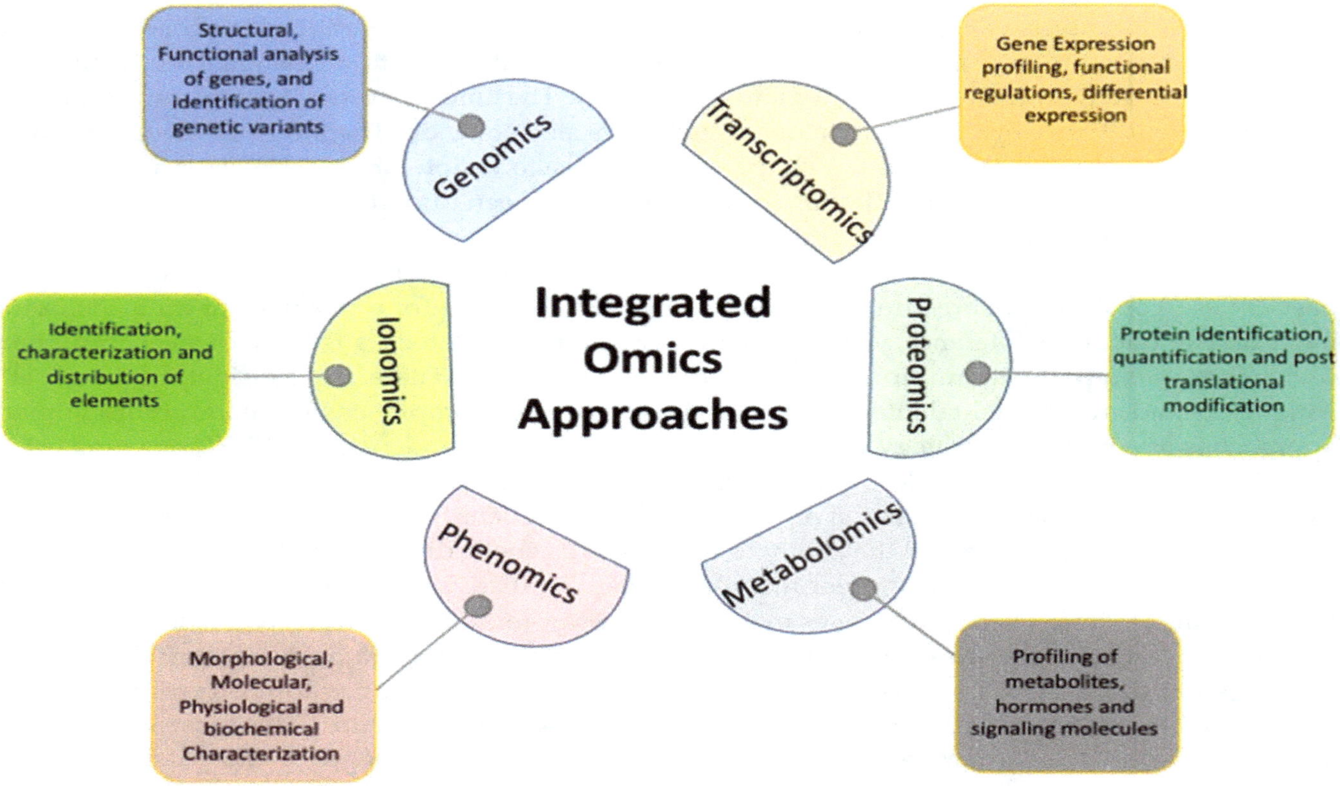

Figure 1. Different omics branches being used individually or in an integrated manner in plant science.

2. Tomato Genomics for Abiotic Stress Tolerance

Whole Genome Sequencing and Resequencing

Sequencing of the entire genome is most efficient in accelerating molecular research. In plants, Arabidopsis was the first genome to be sequenced by an international consortium [22]. Its genome sequence helped in understanding genome organization, regulation, and evolution. Furthermore, with the invention of next-generation sequencing (NGS) technologies allowing parallel sequencing of millions of molecules simultaneously, whole genome sequencing became significantly cheaper and faster than the conventional methods [23]. Subsequently, many crop genomes, including tomato, have been sequenced using both Sanger's and NGS methods [24–26]. The high quality, well-annotated tomato reference genome sequence is routinely used for genomics and transcriptomic studies. The more evident benefits acquired from genome sequencing include the catalog of annotated gene models, genome organization, syntenic information, repeats, and most importantly the basis to identify genetic variations (The Tomato Genome Consortium, 2012) [17]. The reference tomato genome also serves as the basis for the annotation of other Solanaceae species genomes.

With the availability of the high-quality reference genome, sequencing of the entire genome (resequencing) for many genotypes of the species is much easier and more cost-effective. Earlier efforts of resequencing eight tomato genotypes have identified more than 4 million SNPs, over a hundred thousand InDels, and seven thousand copy-number variations [27]. *Solanum pennellii* is known for its unique morphology as well as its extreme stress tolerance; therefore, it has been crossed with *S. lycopersicum* for the improvement of several agronomic traits. Bolger et al., 2014 performed high-quality genome sequencing of introgressed lines of *S. pennellii* x *S. lycopersicum* in order to identify the candidate genes associated with stress tolerance [28].

Furthermore, resequencing of two tomato landraces, COR and LUC, selected based on traits related to drought tolerance and fruit quality, identified hundreds of thousands of SNPs and hundreds of structural changes [29]. The sequence variation is expected to explain the high drought tolerance and adaptability to low water regimes in these genotypes. Further investigation of candidate genes identified about 122 genes with high effect SNPs (Non-Synonymous). Since both of the genotypes are drought tolerant, the genes with common variation have been selected as most promising candidates. The list of promising candidates includes heat shock proteins like Solyc05g055200, Solyc08g078720, Solyc09g011710, and Cation/H+ antiporters like Solyc03g032240, and Solyc09g010530 [29]. In addition, resequencing of plant pathogens also helps to understand co-evolution, particularly the rapidly evolving gene-for-gene system of virulence and resistance factors in the pathogen and host plant, respectively [30–32]. In recent years, pan-genome sequencing has become increasingly significant because it adds depth and completeness to the reference genome. Recently, genome sequences of 725 accessions were utilized for tomato pan-genome sequencing which revealed that 4873 genes were not identified in the reference genome. The study further performed presence/absence variation analyses in order to comprehend substantial gene loss and intense negative selection of genes and promoters during tomato domestication and improvement [33].

3. Molecular Markers Resources in Tomato

The whole genome sequence of tomato has been explored extensively for the development of molecular markers. Microsatellites or simple sequence repeats (SSRs) are one of the promising marker systems suitable for the labs where SNP genotyping is not feasible. Genome-wide identification of microsatellites and subsequent marker development creates a valuable resource for breeding programs [8].

With the advent of cost-efficient and high-throughput genotyping methods, SNP genotyping methods are gaining wide popularity. Among the several SNP-based genotyping methods, the genotyping by sequencing (GBS) approach is a highly multiplexed system for constructing RRL (reduced representation libraries), molecular marker discovery, and genotyping for crop improvement [34,35]. Due to low cost and advancing technologies, GBS has been applied to several crop species [36,37]. For example, a tomato GBS study led to the discovery of 8,784 SNPs based on an NGS approach. Of these SNPs, 88% are frequently observed in tomato germplasm [38].

Even though GBS is a simplified and cost-effective approach, its use is restricted because of the computational and data analysis expertise required. It may be widely used in the future with the development of computational packages and pipelines [16].

4. Identification of Loci Governing Abiotic Stress through QTL Mapping and GWAS

Genetic fingerprinting, linkage maps, and QTL mapping are marker-based approaches that require extensive genotype data. Linkage mapping and association mapping have led to the detection of QTL by identifying marker-trait associations [39]. A lot of focus has been given to mapping QTLs for several abiotic stresses such as salinity, drought, and low temperatures in tomato; however, other stresses like high temperatures, limited nutritional regimes, and environmental pollutants (heavy metals, ozone) still need to be explored. QTL mapping was performed using a linkage map of 1345 markers spaced at an average interval of 1.68 cM, representing 524 unique map positions. The genetic map covers more than 84% of the 900 Mb tomato genome and measures 2,156 cM [40]. The study identified QTLs regulating seed germination under different stresses. Several QTL mapping studies have been performed in tomato particularly to identify loci governing stress tolerance (Table 1). Similarly, genomics advances facilitated more complex approaches involving multi-parental populations like nested association mapping (NAM) and Multi-parent advanced generation inter-cross (MAGIC) (Figure 2). Up to now, very few studies exploring NAM and MAGIC populations in tomato have been published [40].

Table 1. Significant quantitative trait loci (QTL) mapping studies performed to identify loci governing abiotic stress tolerance in Tomato.

Sr.No.	Stress	Trait	QTL	Chromosome	Position (cM)	LOD Score	R (%)	References
1	Cold Tolerance	RGR (Relative Germination Ratio)	qRGI-1-1	1	40.4	5.45	19.55	[41]
			qRGI-1-2	1	47.2	2.53	8.52	
			qRGI-4-1	4	10.4	2.06	6.02	
			qRGI-9-1	9	7.8	2.12	5.95	
			qRGI-12-1	12	8	4.26	11.33	
		CI (Chilling index)	qCI-1-1	1	9.8	3.25	0.95	
			qCI-2-1	2	18	2.96	10.34	
			qCI-3-1	3	0	3.01	10.31	
			qCI-9-1	9	26.8	2.35	7.31	
2	cold stress	Seed Germination	cld1.1	1	7	7.41	30.95	[42]
			cld1.2	1	8.8	4.27	17.24	
			cld1.3	1	15.4	2.27	9.78	
			cld4.1	4	13.8	2.06	9.13	
			cld4.1	4	20.7	2.98	13.15	
			cld8.1	8	2.5	1.26	5.76	
3	Salt Stress	Seed Germination	slt1.1	1	7	2.66	10.86	
			slt1.2	1	8.8	3.53	13.58	
			slt2.1	2	18.7	1.2	6.4	
			slt5.1	5	16	1.52	8.32	
			slt7.1	7	4.5	1.52	7.16	
			slt9.1	9	21.4	2.01	6.3	
			slt12.1	12	7.1	2.4	12.41	
4	Salt Tolerance	Seedling Stage	Stlq4	4			63.6	[43]
			Stlq6	6			64.8	
			Stlq9a	9			61	
			Stlq9b	9			63.6	
			Stlq12a	12			63.3	
			Stlq12a	12			61	
			Stlq12b	12			64.4	
5	Heat tolerance	Pollen viability	qPV11	11	19.4		36.3	[44]
		Pollen Number	qPN7	7	134.7		18.6	
		Style protrusion	qSP1	1	16		19.5	
			qSP3	3	80.4		28	
		Anther length	qAL1	1	70		15.5	
			qAL2	2	80.8		11.6	
			qAL7	7	134.7		25.2	
		Style length	qSL1	1	16		22.7	
			qSL2	2	80.8		10.5	
			qSL3	3	75.8		15.8	
		Flowers per inflorescence	qFPI1	1	40		38.7	
		Inflorescence number	qIN1	1	39		21.9	
			qIN8	8	95.3		13.4	

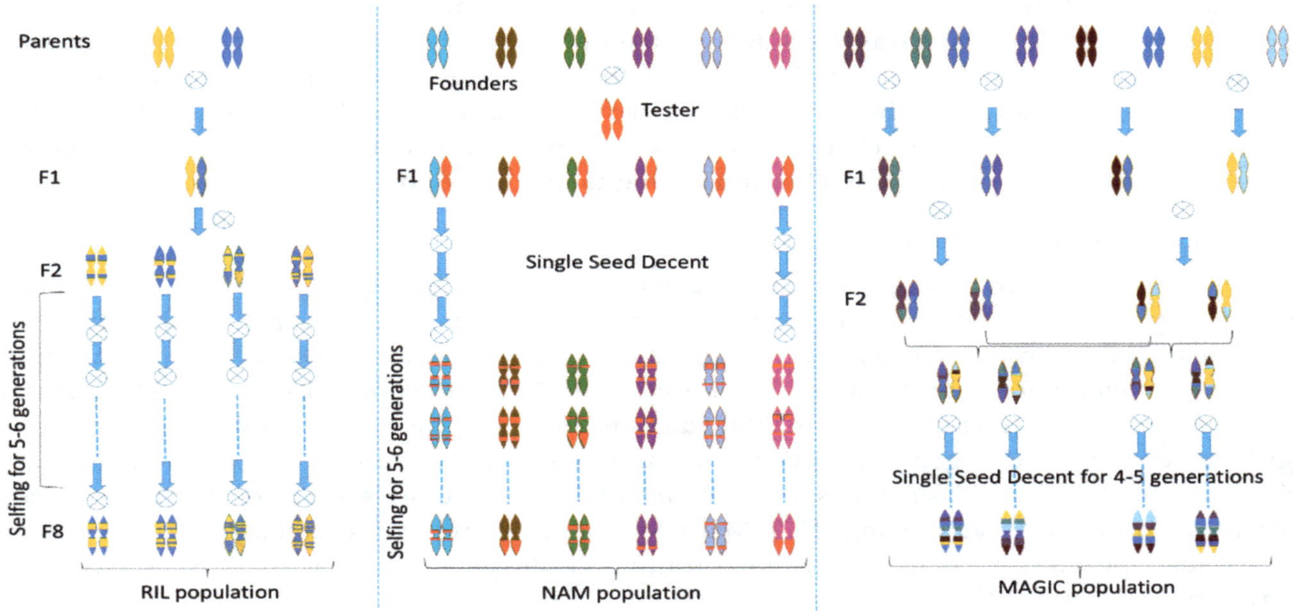

Figure 2. Examples of strategies for the development of conventionally used mapping population such as Recombinant inbred lines (RILs) and more current strategies such as nested association mapping (NAM) and multi-parent advanced generation inter-cross (MAGIC).

In contrast, a GWAS (Genome-wide association studies) approach has an advantage over linkage mapping as it explores the genetic diversity and recombination events present in germplasm collections and provides higher mapping resolution [18]. Therefore, GWAS has been routinely used to detect SNPs for agronomic traits in a world-wide tomato germplasm collection [45]. For instance, in tomato GWAS was performed using 182 SSR markers to identify the chromosome regions associated with 28 different volatile molecules defining tomato flavor [46]. Furthermore, GWAS studies have been done for fruit metabolic traits and other traits but there is no study of GWAS for abiotic stress in tomato yet.

5. Genomic Selection (GS) for Abiotic Stress in Tomato

The decreasing cost of SNP assays has made it possible to genotype large number of experimental lines allowing the implementation of the GS approach in crop breeding programs. The GS approach is efficient in simultaneously tracking all the loci contributing to trait development, irrespective of the magnitude of their individual effect. The GS approach overcomes the limitation of QTL mapping-based breeding where tracking/identification of small effect QTLs is difficult. Importantly, the small effect QTLs may collectively produce larger effects on economically important abiotic traits. [47]. Most economically important traits are complex and affected by unexpected trait expression because of epistatic interactions [48]. Therefore, GS is the best approach to predict genetic values for selection by utilizing all available molecular markers in combination with the phenotypic data of a training population. In this approach, a model that is used to establish and evaluate genotypic and phenotypic data to assess the phenotypic variation based on their whole genome genotypes (genetic composition) [19]. To determine breeding values, different GS algorithms like non-linear regressions (RKHS and RF), Bayesian approaches (Bayes A and B), and penalized regressions (RR, LASSO, and EN) have been developed. Among the available approaches, non-Linear regression is considered the best approach for prediction accuracies [49].

In tomato, GEBV (genomic estimated breeding value) is used majorly for yield and flavor improvement; fruit weight and SSC (soluble solid content) was calculated and gives the highest predictability in tomato [50]. Furthermore, tomato fruit quality was studied to analyze the accuracy of genomic selection for several metabolic and quality traits. In this study, GS has been performed for

45 phenotypic traits using a set of 163 tomato accessions as a training population (TNP). The overall conclusion was that several parameters such as the number and density of markers and the size of TNP affect the accuracy of prediction [51]. Overall, the use of high throughput phenotyping together with genomic information can help to enhance prediction accuracy and accelerate genetic gains by shortening the breeding cycle. Therefore, GS has a clear-cut advantage over MAS and association mapping for complex traits and notably contributing to the development and release of new cultivars.

6. Advances in Transcriptomics

A wide range of environmental factors challenge plants, including tomato, for optimum growth and development. In response, the plant often activates defense mechanisms to mitigate adverse conditions. Understanding the gene regulatory cascades for such responses is very important for the effective management of abiotic stress. Therefore, collection and comparison of the transcriptome of different tissue types, and developmental stages is the best strategy to investigate plant response regulation and to identify genes involved in stress tolerance mechanisms. Thus, understanding the transcriptome of different tissue types or developmental stages would lead to a deeper understanding of corresponding phenotypic change [52]. Many tools and techniques are available for the evaluation of the transcriptomic data to get expression profiling for the gene-by-gene as well as collectively for many genes at a time [53].

Microarray has been used to identify the differentially expressed transcript of genes in response to abiotic stresses, including salinity and ABA, however very few studies have been undertaken for cold, drought, and oxidative stress [54,55]. Numerous studies have been conducted in tomato using transcriptomic approaches to identify genes having significant role in stress tolerance mechanisms (Table 2) as well as for the understanding of diverse molecular mechanisms (Figure 3). The Tomato Expression Database (TED) was developed (http://ted.bti.cornell.edu) which includes raw gene expression data derived from the public tomato cDNA microarray as well as experimental design and array information in compliance with the MIAME guidelines and provides web interfaces for researchers to retrieve data for their own analysis and use. In addition, the Tomato Digital Expression Database contains raw and normalized digital expression (EST abundance) data derived from analysis of the complete public tomato EST collection containing >150 000 ESTs derived from 27 different non-normalized EST libraries [56].

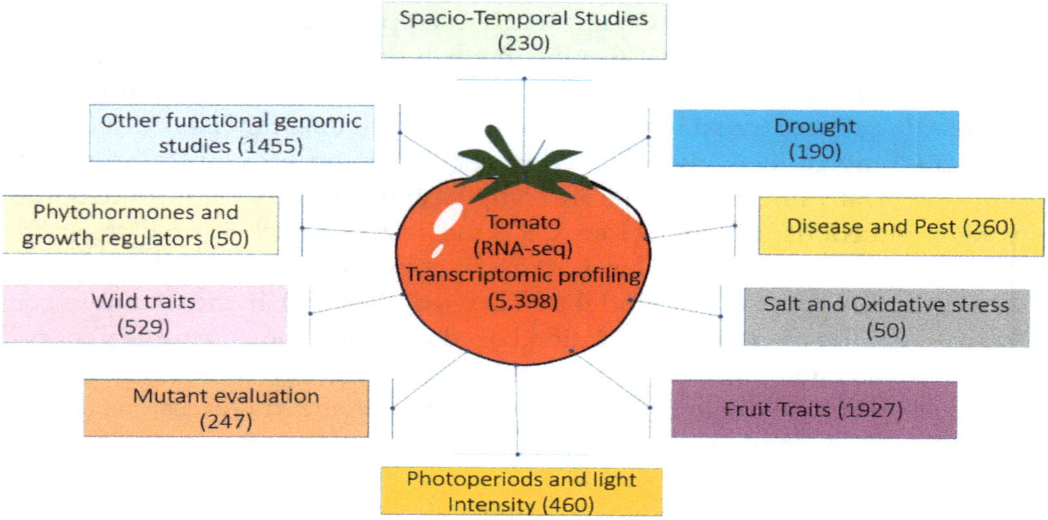

Figure 3. Transcriptomic resources generated through RNA-seq approaches in tomato being used for different studies. The values provided in the parenthesis indicate approximate number of RNA-seq sequenced libraries publicly available at SRA database (www.ncbi.nlm.nih.gov/sra).

With the rapid development in next-generation sequencing, RNA sequencing (RNA-seq) has become the most cost-effective, efficient, and high throughput transcriptomic technology. Unlike microarray, the RNA-seq approach is not only confined to compare the transcripts levels, but it is also useful in novel gene discovery and spliced forms, especially in non-model plants. The impact of drought stress on gene expression has been analyzed with high-throughput transcriptomics in various plants such as rice [57], maize [58], and poplar [59]. In one of the studies, 400 drought responsive genes have been identified using microarray profiling of two drought tolerant tomato lines and a drought-sensitive cultivar (M82) [60]. The information generated with the microarray profiling will be helpful to identify regulatory genes and the molecular pathways involved in the drought tolerance mechanism in tomato.

Table 2. Major Transcriptomic analysis for abiotic stress tolerance in tomato.

Trait	Platform	DEG	Key Point	References
Microarray gene expression data of tomato to study meta-analysis of stress response	Affymetrix tomato Genome Array	835	Expression profile of different genes under different conditions, Meta-analysis to characterize the candidate genes for abiotic stress.	[61]
Temporal stage of fruit development To study the transcriptome profiling of ERF family genes.	Tomato Gene Chip arrays	57	Over expression of ERF family genes in tomato has been shown to confer increased resistance to abiotic stresses.	[62]
Tomato leaf responses to exogenous ABA	Illumina RNA-sequencing	2787	Exogenous ABA has potential to up-regulate many genes related to stress tolerance.	[63]
Solanum lycopersicum cultivars (WT) and MT (Micro-Tom)	RNA-sequencing	619	BR-deficient (Brassinosteroids) Micro-Tom showed lower drought and osmotic stress tolerance. BR signaling is tightly connected with gene networks related to abiotic stress and development	[64]
Micro-Tom seedling	RNA sequencing IlluminaGAIIx Platform	6643	Salt and oxidative stresses regulate tomato cytokinin level and transcriptomic responses.	[65]
Different stages of cultivated and wild tomato (Root, stem, leaf, flower, fruit and seedling)	RNA sequencing Illumina high-throughput sequencing	Upregulated- 126 Downregulated-87	These DEG associated with salt resistance, drought resistance and fruit nutrition.	[66]

Analysis of differential co-expression may also assist in investigating key regulatory steps in the metabolic pathway. Gene co-expression networks can be conveniently constructed using data generated by high-throughput gene expression profiling generated through RNA sequencing or microarray

(Supplementary Figure S1). Co-expression analysis derived from the microarray and RNA-Seq data reveals that co-expression estimates are stable even when network constructed from mixed data of both. Furthermore, the tomato expression atlas (TEA; http://tea.solgenomics.net/) offers simultaneous visualization of groups of genes at a cell/tissue level of resolution within an organ in order to facilitate candidate gene identification [67]. Differential co-expression may assist to evaluate the regulatory steps in metabolic pathways. The gene co-expression, for instance, occurred in two metabolic pathways involved in lycopene and flavonoid biosynthesis [68]. In a recent study, graft-healing-related gene networks were identified. The study concluded that a large proportion of modules represented asymmetric expression networks from different pathways that were related to position. Additionally, auxin and sugar transport and signaling-related genes expression increased above the cut while stress response-related genes were found to be upregulated below the cut. The study concluded that some modules were related to graft union formation, among which oxidative detoxification genes were co-expressed along with both wounding response and cell wall organization genes [69]. Development in transcriptomics has driven the discovery of novel regulators in response to several abiotic stresses, for example, GLYCOALKALOID METABOLISM (GAME) 9, also called as JRE4, an AP2/ERF transcription factor, was identified in tomato as the main regulator of the SGA pathway [70–72]. Klee and Jiovannani (2011) elaborated the involvement of ethylene and the transcription factors with the ripening process and how the network controls the tomato fruit ripening and quality [73].

7. Tomato Proteomics: Applicability and Challenges

The proteomics approach commonly utilizes two-dimensional (2-D) gel electrophoresis, mass spectrometry (MS), matrix-assisted laser desorption ionization–time of flight (MALDI TOF), western blots, and ELISA in combination with bioinformatics tools to identify proteins and map their interactions in a cellular context. The precise quantification of proteins and peptides is challenging in fruit and root tissue, creating a significant bottleneck in proteomic studies. Nevertheless, MS-based methods in combination with computational tools are capable of processing hundreds of peptide transitions simultaneously as well as enabling good reproducibility, predictability, and accuracy. Furthermore, label-free quantitation procedures are convenient as well as provides reliable data for global protein expression studies; this method has been used extensively for assessment of tomato lines [74,75].

In tomato, a comparative proteome analysis was performed to study the impact of aluminum on cotyledon development of tomato [76]. Forty-nine proteins were differentially accumulated, affected by Al stress in tomato [76]. In another study, proteomic analysis of tomato seedlings subjected to Silicone and salt stress, a total of 40 differentially expressed proteins were identified using 2D-gel electrophoresis. Furthermore, twenty-four proteins found to be associated with the stress were up-regulated by Si supplements and down-regulated in salt-stressed root libraries [77]. A total of 52 differentially expressed proteins and many novel proteins were identified in tomato leaves in response to waterlogging stress. These proteins are involved with various processes such as photosynthesis, disease resistance, stress and defense mechanisms, energy and metabolism and protein biosynthesis [78]. Chilling tolerance with hot water treatment was studied in tomato fruits, the study concluded by using a hot water treatment to induce chilling injury tolerance. The chilling injury tolerance is thought to be due to the prevention of protein denaturation and activation of the antioxidant compounds [79]. Moreover, proteomic analysis for several stresses have been performed in tomato. Proteomics used for the identification of proteins responsive to NaCl and NaHCO3 stress [80], temperature stress [81], and drought tolerance [82]. Proteins related to temperature-related stress has been extensively in tomato, for example, 67 differentially expressed proteins were identified in tomato seedlings in response to high temperature using 2D-gel electrophoresis and MALDI-TOF/TOF MS ([83]. In a recent study, proteomic analysis was used to dissect the changes in contrasting tomato varieties under low-temperature stress [84]. Although a number of proteomics analyses have been performed to study abiotic stress response in tomato, data analysis and interpretation is still a bottleneck on the road of in-depth proteomics studies in crops. However, public databases with improved protein annotations are becoming available. The new proteomics advancements will help identify more regulatory candidate proteins and will ultimately direct to the development of stress-tolerant crops with higher yield and quality.

8. Tomato Metabolomics

Metabolomics is one of the promising approaches which provide a biochemical snapshot of an organism's phenotype. Metabolomics allows the systematic identification and quantification of low-molecular-weight molecules that are closely associated with important toxicological and nutritional characteristics. Knowledge of genes, proteins, and transcriptomes are not enough to identify a cell completely, it is necessary to study the wide range of primary and secondary metabolites present in a cell. Numerous studies have been performed to understand the role of metabolites under high salinity and drought stress conditions in plants. Techniques such as GC-MS (Gas Chromatography-Mass Spectrometry) [85], CE-MS (capillary electrophoresis–Mass spectrometry [86] and NMR (nuclear magnetic resonance) [87] have been used to study metabolites for stress response in plants. In tomato, few studies have been performed for both biotic and abiotic stress responses, among them very few for abiotic stresses.

Different drought-tolerant tomato varieties were studied to test the effects of water stress on flavonoids and caffeoyl derivatives. In five cultivars of cherry tomato varieties, water stress resulted in decreased shikimate and phenolic compounds [88]. Low oxygen stress can be induced by storage of fruit and vegetables under a closed atmosphere. In another study, the metabolic response of plant organs to low oxygen levels was examined and cultured tomato cells were used for the metabolic study to low oxygen. It was revealed that low oxygen stress altered the metabolic profile of tomato cells by accumulating the glycolysis intermediates in addition to increased lactate and sugar alcohols [89].

The integration of metabolomics, linkage mapping studies, and metabolome-based genome-wide association studies (mGWAS) provide comprehensive insight into the extent of natural variation in metabolism and its genetic and biochemical control in tomato [90]. Recently, Nunes-Nesi et al., 2019 conducted a study to identify leaf mQTL in tomato that are potentially important with respect to stress responses and plant physiology. The study identified 42 positive and 76 negative mQTL which are involved in the regulation of leaf primary carbon and nitrogen metabolism [91]. Indeed, metabolomic studies in tomato have increased our understanding of several metabolite networks and pathways related to many economic traits. The application of metabolomics to study abiotic stress will help us to elucidate underlying molecular mechanisms associated with stress and would surely lead to developing tolerant tomato plants with enhanced yield.

9. Tomato Ionomics

Ionomics is the study of the accumulation of essential and non-essential elements (metals, metalloids, and nonmetals). It can be applied to various types of plant species. There are various factors such as plant species, variety, organ, and environment which affect the ionome profile. Ionomics have been used to understand the role of mineral elements in plants by using high-throughput technologies like Inductively Coupled Plasma-Mass Spectrometry (ICP-MS) and ICP-Atomic Emission Spectrometry (ICP-AES) [92].

A wide range of studies have been done in the field of ionomics mainly on silicon (Si). Most of the dicots and particularly the Solanaceae family, take up small quantities of silicon and accumulate less than 0.5% in their tissue. Silicon has been found to improve drought tolerance and delay in wilting and benefit certain plants when they are under stress. In addition, many reports showed that the plants growing under heavy metal stress in the presence of silicon had reduced ROS (reactive oxygen species) contents which is indicative of enhanced stress response [93–95]. A study was conducted to investigate effect and mechanism of exogenous Si on salt tolerance in tomato, silicon was found to be responsible for the decreased concentration of Na and Cl in roots, stem, and leaves without any disturbance in translocation from root to shoot [96]. Furthermore, hydroponics tomato analysis revealed Si alleviated the effect of salinity stress on plant photosynthesis, chlorophyll concentration, and water content of leaf [97]. The fruit analysis of several tomato cultivars revealed that Si addition has a significant effect on arsenic uptake [98]. There are several elements in their different biochemical forms yet to be studied extensively enough to understand their precise role in plants.

10. Phenomic Advances for Abiotic Stress Tolerance in Tomato

Phenomics is the study of high-throughput analysis of phenotypic variation which is a complex web of interactions between genotype, phenotype, and environment. The genome and phenome (a set of all phenotypes) studies performed with individuals or with large populations are complementary to each other [99]. The plants with tolerant phenotypes are good genomic resources and also become a target to identify the alleles using high throughput sequencing. Advances in sequencing technologies have improved genotyping efficiencies while phenotypic characterization has progressed more slowly in the past decade which limits the characterization of quantitative traits especially those related to stress tolerance [100]. However, there are developments in phenotyping methods which allows the identification of specific characteristic. The use of advanced imaging systems, sensors, automations, and computational resources for the phenotyping in plants make phenomics a high-throughput approach capable of handling thousands of genotypes for the evaluation of hundreds of phenotypic parameters simultaneously [100–102]. There are a number of phenomics platforms available such as scan analyzer 3D, used to investigate physiological parameters in tomato plants under drought conditions [103].

Due to the complications in phenomics data collection, it requires the collaboration between scientists from diverse area of expertise. In addition, phenomic data collection is expensive and time-consuming. Integrated technical advances would therefore aid to lower the associated costs and enhance phenomic throughput.

11. Integration of Omics Technologies

The recent progress in omics approaches (genomics, transcriptomes, proteomics, metabolomics, ionomics, and phenomics) has generated a huge amount of data which can be used to identify novel genetic and chemical elements controlling various physiological processes [104]. However, using only one approach is not sufficient to understand the complexity of stress response in plants which requires the integration of various approaches to comprehend the complex stress response (Figure 1). In addition, the analysis of high throughput data from various omics approaches is one of the biggest challenges to interpreting the response mechanism(s). Although there is a range of software tools for basic data analysis to meet these omics data challenges, there is still the requirement of collaborative approaches to understand the complex physiological and biochemical responses under stress conditions. To develop novel climate-smart crop varieties, efficient integration of different tools, techniques, and approaches looks like a promising strategy [105,106].

Understanding the precise molecular function of genes is a challenging task for the plant molecular biologist. In this regard, the QTL mapping and GWAS approaches provide information of regulatory loci governing a particular trait which can be used further to identify candidate genes. The candidate gene identification reduces the molecular biology efforts by reducing the number of genes that need to be functionally evaluated. The choice of technique to identify the genomic loci is a crucial step. GWAS can be applied to any set of germplasm and detects regulatory loci for several traits simultaneously that show variation [107]. Both approaches, QTL mapping and GWAS identify a chromosomal region that is associated with a particular trait. QTL regions are quite large, and harboring too many genes makes it difficult to identify the candidate gene, but the combination of QTL and GWAS approaches has been successfully used to identify candidate genes in soybean [107]. Furthermore, the integration of RNA sequencing profiles with gene expression profiles can provide vital clues for the identification of functions of unknown genes [108,109]. Therefore, combining QTL and GWAS with transcriptome profiling will be helpful to identify differentially expressed candidate genes [110,111]. A study performed in tomato by integrating information of QTLs, eQTLs, and differentially expressed genes identified candidate genes regulating water stress tolerance [112].

Efficient adaptation of computational techniques by the plant breeder largely depends on features such as user-friendly interface, easy accessibility, online tutorials and manuals, and interactive options. In this regard, several user-friendly databases useful to integrate omics scale data from different approaches have been developed for tomato, as described in Table 3.

Table 3. Online databases developed for tomato for integrated omics.

Sr.No	Database	URL	Description/Applications	References
1.	KaFtom	www.pgb.kazusa.or.jp	Database for Micro Tom full length cDNA clones, Full length cDNA libraries for EST sequencing.	[113]
2.	MiBASE	http: //omictools.com	Database for Micro Tom ESTs and tomato Unigenes EST Sequencing, ESTAnnotations, SNPs, SSRs, Gene ontology, Metabolic pathways of Gene expressions And Sequence similarities.	[114]
3.	Tomatoma (Micro Tom Database)	http: //tomatoma.nbrp.jp	Micro Tom mutant Resources, Metabolite information, Phenotype information, TILLING.	[115]
4.	Tomatomics	http: //omictool.com/ tomatomics-tool	Full length mRNA sequences, Gene structures, Expression Profiles and functional annotations of genes.	[116]
5.	TGRD (Tomato Genomic Resources Database)	http://omictool.com/trgd-tool	Interactive browsing of tomato genes, micro RNAs, simple sequence repeats (SSRs), important quantitative trait loci.	[117]
6.	TFGD (Tomato Functional Genomic Database)	http://ted.bti.cornell.edu	Microarray Expression Database, Metabolite profile Data analysis, RNA Seq. Data	[118]
7.	KaTomics DB (Kazusa Tomato Genomic Database)	www.kazusa.or.jp	Database for DNA markers, SNP annotations, and genome sequences	[119]
8.	MoToDB Metabolome Database	http:// appliedbioinformatics.wur.nl/moto/	LC-MS (Liquid Chromatography Mass Spectrometry)	[120]
9.	CoxPathDB	http://cox-path-db.kazusa.or.jp/ tomato	To predict function of tomato genes from result of functional enrichment analyses of co-expressed genes.	[121]
10.	Sol Genomics Network	http: //solgenomics.net	Browse the tomato genome, Find the sequence similarity, and Download annotations.	[122]

12. Conclusions

Abiotic stress is one of the major limiting factors in plant growth and yield. Various omics tools and techniques have been developed to understand the molecular mechanisms of plants responses in abiotic stress conditions. Under stress conditions, plants modulate themselves to adopt the existing stresses by controlling gene regulation, proteins, and metabolites. It is essential to elucidate the functions of newly identified stress-responsive genes to understand the abiotic stress responses of plants. To identify changes by various tools and techniques like genomics, transcriptomics, metabolomics, ionmics, and phenomics have been devised to allow the understanding of genetic makeup in depth, their signaling cascade, and their adaptability under stress conditions. In tomato, genomics, ionomics, and transcriptomics have been developed for abiotic stress, but the other major branches like metabolomics, proteomics, and phenomics are as of yet lingering behind. Diverse study of omics tools and integrated approaches discussed in this review tell us about the current situations and future points for successful management of abiotic stress in tomato.

Author Contributions: J.C., H.S., and R.D. conceived the idea, J.C., A.K., P.K., and S.K. prepared the first draft of the manuscript. P.S., and V.K. provided valuable input A.V., S.K.J., H.K., R.K., H.S., and R.D. reviewed the manuscript with valuable inputs. All authors contributed to finalizing the manuscript. All authors read and approved the final manuscript.

Acknowledgments: Authors are thankful to Department of Biotechnology, Government of India for the Ramalingaswami Fellowship Award to H.S. and R.D., and to University Grants Commission, India for Ph.D. fellowship to S.K. Authors would like to thank Dennis Yungbluth and Liz Prenger from University of Missouri, USA for language editing.

References

1. Chaudhary, J.; Alisha, A.; Bhatt, V.; Chandanshive, S.; Kumar, N.; Mir, Z.; Kumar, A.; Yadav, S.K.; Shivaraj, S.M.; Sonah, H.; et al. Mutation Breeding in Tomato: Advances, Applicability and Challenges. *Plants* **2019**, *8*, 128. [CrossRef]
2. Patil, G.; Do, T.; Vuong, T.D.; Valliyodan, B.; Lee, J.D.; Chaudhary, J.; Shannon, J.G.; Nguyen, H.T. Genomic-assisted haplotype analysis and the development of high-throughput SNP markers for salinity tolerance in soybean. *Sci. Rep.* **2016**, *6*, 19199. [CrossRef]
3. Rizhsky, L.; Liang, H.; Shuman, J.; Shulaev, V.; Davletova, S.; Mittler, R. When Defense Pathways Collide. The Response of Arabidopsis to a Combination of Drought and Heat Stress. *Plant Physiol.* **2004**, *134*, 1683–1696. [CrossRef]
4. Lin, T.; Zhu, G.; Zhang, J.; Xu, X.; Yu, Q.; Zheng, Z.; Zhang, Z.; Lun, Y.; Li, S.; Wang, X.; et al. Genomic analyses provide insights into the history of tomato breeding. *Nat. Genet.* **2014**, *46*, 1220–1226. [CrossRef] [PubMed]
5. Chaudhary, J.; Patil, G.B.; Sonah, H.; Deshmukh, R.K.; Vuong, T.D.; Valliyodan, B.; Nguyen, H.T. Expanding Omics Resources for Improvement of Soybean Seed Composition Traits. *Front. Plant Sci.* **2015**, *6*, 504. [CrossRef] [PubMed]
6. Sharma, A.; Kailasrao Deshmukh, R.; Jain, N.; Kumar Singh, N. Combining qtl mapping and transcriptome profiling for an insight into genes for grain number in rice (*Oryza sativa* L.). *Indian J. Genet. Plant Breed.* **2011**, *71*, 115.
7. Chopperla, R.; Singh, S.; Tomar, R.; Mohanty, S.; Khan, S.; Reddy, N.; Padaria, J.C.; Solanke, A.U. Isolation and allelic characterization of finger millet (*Eleusine coracana* L.) small heat shock protein echsp17. 8 for stress tolerance. *Indian J. Genet. Plant Breed.* **2018**, *78*, 95–103. [CrossRef]
8. Sonah, H.; Deshmukh, R.K.; Sharma, A.; Singh, V.P.; Gupta, D.K.; Gacche, R.N.; Rana, J.C.; Singh, N.K.;

Sharma, T.R. Genome-wide distribution and organization of microsatellites in plants: An insight into marker development in brachypodium. *PLoS ONE* **2011**, *6*, e21298. [CrossRef] [PubMed]

9. Zargar, S.M.; Raatz, B.; Sonah, H.; Nazir, M.; Bhat, J.A.; Dar, Z.A.; Agrawal, G.K.; Rakwal, R. Recent advances in molecular marker techniques: Insight into QTL mapping, GWAS and genomic selection in plants. *J. Crop. Sci. Biotechnol.* **2015**, *18*, 293–308. [CrossRef]
10. Francia, E.; Tacconi, G.; Crosatti, C.; Barabaschi, D.; Bulgarelli, D.; Dall'Aglio, E.; Valè, G. Marker assisted selection in crop plants. *Plant Cell Tissue Organ Cult.* **2005**, *82*, 317–342. [CrossRef]
11. Kumari, S.; Mir, R.R.; Tyagi, S.; Balyan, H.S.; Gupta, P.K. Validation of QTL for grain weight using MAS-derived pairs of NILs in bread wheat (*Triticum aestivum* L.). *J. Plant Biochem. Biotechnol.* **2019**, *28*, 336–344. [CrossRef]
12. Collard, B.C.Y.; Jahufer, M.Z.Z.; Brouwer, J.B.; Pang, E.C.K. An introduction to markers, quantitative trait loci (QTL) mapping and marker-assisted selection for crop improvement: The basic concepts. *Euphytica* **2005**, *142*, 169–196. [CrossRef]
13. Agarwal, G.; Clevenger, J.; Pandey, M.K.; Wang, H.; Shasidhar, Y.; Chu, Y.; Fountain, J.C.; Choudhary, D.; Culbreath, A.K.; Liu, X.; et al. High-density genetic map using whole-genome resequencing for fine mapping and candidate gene discovery for disease resistance in peanut. *Plant Biotechnol. J.* **2018**, *16*, 1954–1967. [CrossRef] [PubMed]
14. Foolad, M.R. *Current Status of Breeding Tomatoes for Salt and Drought Tolerance*; Springer Science and Business Media LLC: Berlin, Germany, 2007; pp. 669–700.
15. Giancola, S.; McKhann, H.I.; Bérard, A.; Camilleri, C.; Durand, S.; Libeau, P.; Roux, F.; Reboud, X.; Gut, I.G.; Brunel, D. Utilization of the three high-throughput SNP genotyping methods, the GOOD assay, Amplifluor and TaqMan, in diploid and polyploid plants. *Theor. Appl. Genet.* **2006**, *112*, 1115–1124. [CrossRef]
16. Sonah, H.; Bastien, M.; Iquira, E.; Tardivel, A.; Légaré, G.; Boyle, B.; Normandeau, E.; Laroche, J.; LaRose, S.; Jean, M.; et al. An Improved Genotyping by Sequencing (GBS) Approach Offering Increased Versatility and Efficiency of SNP Discovery and Genotyping. *PLoS ONE* **2013**, *8*, e54603. [CrossRef]
17. 100 Tomato Genome Sequencing Consortium; Aflitos, S.; Schijlen, E.; de Jong, H.; de Ridder, D.; Smit, S.; Finkers, R.; Wang, J.; Zhang, G.; Li, N.; et al. Exploring genetic variation in the tomato (Solanum section Lycopersicon) clade by whole-genome sequencing. *Plant J.* **2014**, *80*, 136–148.
18. Fukushima, A.; Kusano, M.; Redestig, H.; Arita, M.; Saito, K. Integrated omics approaches in plant systems biology. *Curr. Opin. Chem. Biol.* **2009**, *13*, 532–538. [CrossRef]
19. Shah, T.; Xu, J.; Zou, X.; Cheng, Y.; Nasir, M.; Zhang, X. Omics Approaches for Engineering Wheat Production under Abiotic Stresses. *Int. J. Mol. Sci.* **2018**, *19*, 2390. [CrossRef]
20. Chaudhary, J.; Deshmukh, R.; Mir, Z.A.; Bhat, J.A. Metabolomics: An emerging technology for soybean improvement. In *Biotechnology Products in Everyday Life*; Springer: Berlin, Germany, 2019; pp. 175–186.
21. Hong, J.; Yang, L.; Zhang, D.; Shi, J. Plant Metabolomics: An Indispensable System Biology Tool for Plant Science. *Int. J. Mol. Sci.* **2016**, *17*, 767. [CrossRef]
22. Berardini, T.Z.; Reiser, L.; Li, D.; Mezheritsky, Y.; Muller, R.; Strait, E.; Huala, E. The Arabidopsis information resource: Making and mining the "gold standard" annotated reference plant genome. *Genesis* **2015**, *53*, 474–485. [CrossRef]
23. Goodwin, S.; McPherson, J.D.; McCombie, W.R. Coming of age: Ten years of next-generation sequencing technologies. *Nat. Rev. Genet.* **2016**, *17*, 333–351. [CrossRef]
24. Varshney, R.K.; Nayak, S.N.; May, G.D.; Jackson, S.A. Next-generation sequencing technologies and their implications for crop genetics and breeding. *Trends Biotechnol.* **2009**, *27*, 522–530. [CrossRef]
25. Consortium, T.G. The tomato genome sequence provides insights into fleshy fruit evolution. *Nature* **2012**, *485*, 635. [CrossRef] [PubMed]
26. Kumar, R.; Khurana, A. Functional genomics of tomato: Opportunities and challenges in post-genome NGS era. *J. Biosci.* **2014**, *39*, 917–929. [CrossRef] [PubMed]
27. Causse, M.; Desplat, N.; Pascual, L.; Le Paslier, M.C.; Sauvage, C.; Bauchet, G.; Bérard, A.; Bounon, R.; Tchoumakov, M.; Brunel, D.; et al. Whole genome resequencing in tomato reveals variation associated with introgression and breeding events. *BMC Genom.* **2013**, *14*, 791. [CrossRef] [PubMed]

28. Bolger, A.; Scossa, F.; Bolger, M.E.; Lanz, C.; Maumus, F.; Tohge, T.; Quesneville, H.; Alseekh, S.; Sørensen, I.; Lichtenstein, G.; et al. The genome of the stress-tolerant wild tomato species *Solanum pennellii*. *Nat. Genet.* **2014**, *46*, 1034–1038. [CrossRef] [PubMed]
29. Tranchida-Lombardo, V.; Cigliano, R.A.; Anzar, I.; Landi, S.; Palombieri, S.; Colantuono, C.; Bostan, H.; Termolino, P.; Aversano, R.; Batelli, G.; et al. Whole-genome re-sequencing of two Italian tomato landraces reveals sequence variations in genes associated with stress tolerance, fruit quality and long shelf-life traits. *DNA Res.* **2017**, *25*, 149–160. [CrossRef]
30. Patil, V.U.; Girimalla, V.; Sagar, V.; Bhardwaj, V.; Chakrabarti, S. Draft genome sequencing of rhizoctonia solani anastomosis group 3 (ag3-pt) causing stem canker and black scurf of potato. *Am. J. Potato Res.* **2018**, *95*, 87–91. [CrossRef]
31. Yang, H.; Zhao, T.; Jiang, J.; Wang, S.; Wang, A.; Li, J.; Xu, X. Mapping and screening of the tomato *Stemphylium lycopersici* resistance gene, Sm, based on bulked segregant analysis in combination with genome resequencing. *BMC Plant Boil.* **2017**, *17*, 266. [CrossRef]
32. Arsenault-Labrecque, G.; Sonah, H.; Lebreton, A.; Labbé, C.; Marchand, G.; Xue, A.; Belzile, F.; Knaus, B.J.; Grünwald, N.J.; Bélanger, R.R. Stable predictive markers for *Phytophthora sojae* avirulence genes that impair infection of soybean uncovered by whole genome sequencing of 31 isolates. *BMC Boil.* **2018**, *16*, 80. [CrossRef]
33. Gao, L.; Gonda, I.; Sun, H.; Ma, Q.; Bao, K.; Tieman, D.M.; Burzynski-Chang, E.A.; Fish, T.L.; Stromberg, K.A.; Sacks, G.L.; et al. The tomato pan-genome uncovers new genes and a rare allele regulating fruit flavor. *Nat. Genet.* **2019**, *51*, 1044–1051. [CrossRef] [PubMed]
34. Elbasyoni, I.S.; Lorenz, A.; Guttieri, M.; Frels, K.; Baenziger, P.; Poland, J.; Akhunov, E. A comparison between genotyping-by-sequencing and array-based scoring of SNPs for genomic prediction accuracy in winter wheat. *Plant Sci.* **2018**, *270*, 123–130. [CrossRef] [PubMed]
35. Eltaher, S.; Sallam, A.; Belamkar, V.; Emara, H.A.; Nower, A.A.; Salem, K.F.M.; Poland, J.; Baenziger, P.S. Genetic Diversity and Population Structure of F3:6 Nebraska Winter Wheat Genotypes Using Genotyping-By-Sequencing. *Front. Genet.* **2018**, *9*, 76. [CrossRef] [PubMed]
36. Poland, J.A.; Rife, T.W. Genotyping-by-Sequencing for Plant Breeding and Genetics. *Plant Genome* **2012**, *5*, 92. [CrossRef]
37. Kim, C.; Guo, H.; Kong, W.; Chandnani, R.; Shuang, L.S.; Paterson, A.H. Application of genotyping by sequencing technology to a variety of crop breeding programs. *Plant Sci.* **2016**, *242*, 14–22. [CrossRef]
38. Sim, S.C.; Durstewitz, G.; Plieske, J.; Wieseke, R.; Ganal, M.W.; Van Deynze, A.; Hamilton, J.P.; Buell, C.R.; Causse, M.; Wijeratne, S.; et al. Development of a Large SNP Genotyping Array and Generation of High-Density Genetic Maps in Tomato. *PLoS ONE* **2012**, *7*, e40563. [CrossRef]
39. Cockram, J.; Mackay, I. *Genetic Mapping Populations for Conducting High-Resolution Trait Mapping in Plants In Plant Genetics and Molecular Biology*; Springer: Cham, Germany, 2018; pp. 109–138.
40. Pascual, L.; Desplat, N.; Huang, B.E.; Desgroux, A.; Bruguier, L.; Bouchet, J.P.; Le, Q.H.; Chauchard, B.; Verschave, P.; Causse, M. Potential of a tomato magic population to decipher the genetic control of quantitative traits and detect causal variants in the resequencing era. *Plant Biotechnol. J.* **2015**, *13*, 565–577. [CrossRef]
41. Liu, Y.; Zhou, T.; Ge, H.; Pang, W.; Gao, L.; Ren, L.; Chen, H. SSR Mapping of QTLs Conferring Cold Tolerance in an Interspecific Cross of Tomato. *Int. J. Genom.* **2016**, *2016*, 1–6. [CrossRef]
42. Foolad, M.R.; Subbiah, P.; Zhang, L. Common qtl affect the rate of tomato seed germination under different stress and nonstress conditions. *Int. J. Plant Genom.* **2007**, *2007*, 97–106. [CrossRef]
43. Li, J.; Liu, L.; Bai, Y.; Zhang, P.; Finkers, R.; Du, Y.; Visser, R.G.; van Heusden, A.W.J.E. Seedling salt tolerance in tomato. *Euphytica* **2011**, *178*, 403–414. [CrossRef]
44. Xu, J.; Driedonks, N.; Rutten, M.J.; Vriezen, W.H.; de Boer, G.J.; Rieu, I.J.M.B. Mapping quantitative trait loci for heat tolerance of reproductive traits in tomato (*Solanum lycopersicum*). *Mol. Breed.* **2017**, *37*, 58. [CrossRef]
45. Pasam, R.K.; Sharma, R.; Malosetti, M.; Van Eeuwijk, F.A.; Haseneyer, G.; Kilian, B.; Graner, A. Genome-wide association studies for agronomical traits in a world wide spring barley collection. *BMC Plant Boil.* **2012**, *12*, 16. [CrossRef] [PubMed]
46. Zhang, J.; Zhao, J.; Xu, Y.; Liang, J.; Chang, P.; Yan, F.; Li, M.; Liang, Y.; Zou, Z. Genome-Wide Association Mapping for Tomato Volatiles Positively Contributing to Tomato Flavor. *Front. Plant Sci.* **2015**, *6*, 617. [CrossRef] [PubMed]

47. Crossa, J.; Pérez-Rodríguez, P.; Cuevas, J.; Montesinos-López, O.; Jarquín, D.; Campos, G.D.L.; Burgueño, J.; González-Camacho, J.M.; Pérez-Elizalde, S.; Beyene, Y.; et al. Genomic Selection in Plant Breeding: Methods, Models, and Perspectives. *Trends Plant Sci.* **2017**, *22*, 961–975. [CrossRef]
48. Deshmukh, R.; Sonah, H.; Patil, G.; Chen, W.; Prince, S.; Mutava, R.; Vuong, T.; Valliyodan, B.; Nguyen, H.T. Integrating omic approaches for abiotic stress tolerance in soybean. *Front. Plant Sci.* **2014**, *5*, 244. [CrossRef]
49. Shikha, M.; Kanika, A.; Rao, A.R.; Mallikarjuna, M.G.; Gupta, H.S.; Nepolean, T. Genomic Selection for Drought Tolerance Using Genome-Wide SNPs in Maize. *Front. Plant Sci.* **2017**, *8*, 63. [CrossRef]
50. Yamamoto, E.; Matsunaga, H.; Onogi, A.; Kajiya-Kanegae, H.; Minamikawa, M.; Suzuki, A.; Shirasawa, K.; Hirakawa, H.; Nunome, T.; Yamaguchi, H.; et al. A simulation-based breeding design that uses whole-genome prediction in tomato. *Sci. Rep.* **2016**, *6*, 19454. [CrossRef]
51. Duangjit, J.; Causse, M.; Sauvage, C. Efficiency of genomic selection for tomato fruit quality. *Mol. Breed.* **2016**, *36*, 29. [CrossRef]
52. Shinde, S.; Behpouri, A.; McElwain, J.C.; Ng, C.K.Y. Genome-wide transcriptomic analysis of the effects of sub-ambient atmospheric oxygen and elevated atmospheric carbon dioxide levels on gametophytes of the moss, *Physcomitrella patens*. *J. Exp. Bot.* **2015**, *66*, 4001–4012. [CrossRef]
53. Wirta, V. *Mining the Transcriptome-Methods and Applications*; KTH: Stockholm, Sweden, 2006.
54. Albert, E.; Duboscq, R.; Latreille, M.; Santoni, S.; Beukers, M.; Bouchet, J.P.; Bitton, F.; Gricourt, J.; Poncet, C.; Gautier, V.; et al. Allele-specific expression and genetic determinants of transcriptomic variations in response to mild water deficit in tomato. *Plant J.* **2018**, *96*, 635–650. [CrossRef]
55. Iovieno, P.; Punzo, P.; Guida, G.; Mistretta, C.; Van Oosten, M.J.; Nurcato, R.; Bostan, H.; Colantuono, C.; Costa, A.; Bagnaresi, P.; et al. Transcriptomic Changes Drive Physiological Responses to Progressive Drought Stress and Rehydration in Tomato. *Front. Plant Sci.* **2016**, *7*, 371. [CrossRef] [PubMed]
56. Fei, Z. Tomato Expression Database (TED): A suite of data presentation and analysis tools. *Nucleic Acids Res.* **2006**, *34*, D766–D770. [CrossRef] [PubMed]
57. Oono, Y.; Yazawa, T.; Kanamori, H.; Sasaki, H.; Mori, S.; Handa, H.; Matsumoto, T. Genome-Wide Transcriptome Analysis of Cadmium Stress in Rice. *BioMed Res. Int.* **2016**, *2016*, 1–9. [CrossRef] [PubMed]
58. Sa, K.J.; Choi, I.Y.; Park, D.H.; Lee, J.K. Comparative Gene Expression Analysis of Seed Development in Waxy and Dent Corn (*Zea mays* L.). *Plant Breed. Biotechnol.* **2018**, *6*, 337–353. [CrossRef]
59. Garcia, B.J.; Labbé, J.L.; Jones, P.; Abraham, P.E.; Hodge, I.; Climer, S.; Jawdy, S.; Gunter, L.; Tuskan, G.A.; Yang, X.; et al. Phytobiome and Transcriptional Adaptation of *Populus deltoides* to Acute Progressive Drought and Cyclic Drought. *Phytobiomes J.* **2018**, *2*, 249–260. [CrossRef]
60. Gong, P.; Zhang, J.; Li, H.; Yang, C.; Zhang, C.; Zhang, X.; Khurram, Z.; Zhang, Y.; Wang, T.; Fei, Z.; et al. Transcriptional profiles of drought-responsive genes in modulating transcription signal transduction, and biochemical pathways in tomato. *J. Exp. Bot.* **2010**, *61*, 3563–3575. [CrossRef]
61. Ashrafi-Dehkordi, E.; Alemzadeh, A.; Tanaka, N.; Razi, H. Meta-analysis of transcriptomic responses to biotic and abiotic stress in tomato. *PeerJ* **2018**, *6*, e4631. [CrossRef]
62. Sharma, M.K.; Kumar, R.; Solanke, A.U.; Sharma, R.; Tyagi, A.K.; Sharma, A.K. Identification, phylogeny, and transcript profiling of ERF family genes during development and abiotic stress treatments in tomato. *Mol. Genet. Genom.* **2010**, *284*, 455–475. [CrossRef]
63. Wang, Y.; Tao, X.; Tang, X.M.; Xiao, L.; Sun, J.L.; Yan, X.F.; Li, D.; Deng, H.Y.; Ma, X.R. Comparative transcriptome analysis of tomato (*Solanum lycopersicum*) in response to exogenous abscisic acid. *BMC Genom.* **2013**, *14*, 841. [CrossRef]
64. Lee, J.; Shim, D.; Moon, S.; Kim, H.; Bae, W.; Kim, K.; Kim, Y.H.; Rhee, S.K.; Hong, C.P.; Hong, S.Y.; et al. Genome-wide transcriptomic analysis of BR-deficient Micro-Tom reveals correlations between drought stress tolerance and brassinosteroid signaling in tomato. *Plant Physiol. Biochem.* **2018**, *127*, 553–560. [CrossRef]
65. Keshishian, E.A.; Hallmark, H.T.; Ramaraj, T.; Plačková, L.; Sundararajan, A.; Schilkey, F.; Novák, O.; Rashotte, A.M. Salt and oxidative stresses uniquely regulate tomato cytokinin levels and transcriptomic response. *Plant Direct* **2018**, *2*, e00071. [CrossRef] [PubMed]
66. Dai, Q.; Geng, L.; Lu, M.; Jin, W.; Nan, X.; He, P.A.; Yao, Y. Comparative transcriptome analysis of the different tissues between the cultivated and wild tomato. *PLoS ONE* **2017**, *12*, 0172411. [CrossRef] [PubMed]
67. Fernandez-Pozo, N.; Zheng, Y.; Snyder, S.I.; Nicolas, P.; Shinozaki, Y.; Fei, Z.; Catala, C.; Giovannoni, J.J.; Rose, J.K.; Mueller, L.A. The tomato expression atlas. *Bioinformatics* **2017**, *33*, 2397–2398. [CrossRef] [PubMed]

68. Fukushima, A.; Nishizawa, T.; Hayakumo, M.; Hikosaka, S.; Saito, K.; Goto, E.; Kusano, M. Exploring tomato gene functions based on coexpression modules using graph clustering and differential coexpression approaches. *Plant Physiol.* **2012**, *158*, 1487–1502. [CrossRef] [PubMed]
69. Xie, L.; Dong, C.; Shang, Q. Gene co-expression network analysis reveals pathways associated with graft healing by asymmetric profiling in tomato. *BMC Plant Boil.* **2019**, *19*, 373. [CrossRef] [PubMed]
70. Thagun, C.; Imanishi, S.; Kudo, T.; Nakabayashi, R.; Ohyama, K.; Mori, T.; Kawamoto, K.; Nakamura, Y.; Katayama, M.; Nonaka, S.; et al. Jasmonate-responsive ERF transcription factors regulate steroidal glycoalkaloid biosynthesis in tomato. *Plant Cell Physiol.* **2016**, *57*, 961–975. [CrossRef]
71. Cárdenas, P.D.; Sonawane, P.D.; Pollier, J.; Bossche, R.V.; Dewangan, V.; Weithorn, E.; Tal, L.; Meir, S.; Rogachev, I.; Malitsky, S.; et al. GAME9 regulates the biosynthesis of steroidal alkaloids and upstream isoprenoids in the plant mevalonate pathway. *Nat. Commun.* **2016**, *7*, 10654. [CrossRef]
72. Itkin, M.; Rogachev, I.; Alkan, N.; Rosenberg, T.; Malitsky, S.; Masini, L.; Meir, S.; Iijima, Y.; Aoki, K.; De Vos, R.; et al. GLYCOALKALOID METABOLISM1 Is Required for Steroidal Alkaloid Glycosylation and Prevention of Phytotoxicity in Tomato. *Plant Cell* **2011**, *23*, 4507–4525. [CrossRef]
73. Klee, H.J.; Giovannoni, J.J. Genetics and Control of Tomato Fruit Ripening and Quality Attributes. *Annu. Rev. Genet.* **2011**, *45*, 41–59. [CrossRef]
74. Neilson, K.A.; Ali, N.A.; Muralidharan, S.; Mirzaei, M.; Mariani, M.; Assadourian, G.; Lee, A.; Van Sluyter, S.C.; Haynes, P.A. Less label, more free: Approaches in label-free quantitative mass spectrometry. *Proteomics* **2011**, *11*, 535–553. [CrossRef]
75. Mora, L.; Bramley, P.M.; Fraser, P.D.; Soler, L.M. Development and optimisation of a label-free quantitative proteomic procedure and its application in the assessment of genetically modified tomato fruit. *Proteomics* **2013**, *13*, 2016–2030. [CrossRef] [PubMed]
76. Zhou, S.; Sauvé, R.; Thannhauser, T.W. Proteome changes induced by aluminium stress in tomato roots. *Plant Signal. Behav.* **2009**, *60*, 1849–1857. [CrossRef] [PubMed]
77. Muneer, S.; Jeong, B.R. Proteomic analysis of salt-stress responsive proteins in roots of tomato (*Lycopersicon esculentum* L.) plants towards silicon efficiency. *Plant Growth Regul.* **2015**, *77*, 133–146. [CrossRef]
78. Ahsan, N.; Lee, D.G.; Lee, S.H.; Kang, K.Y.; Bahk, J.D.; Choi, M.S.; Lee, I.J.; Renaut, J.; Lee, B.H. A comparative proteomic analysis of tomato leaves in response to waterlogging stress. *Physiol. Plant.* **2007**, *131*, 555–570. [CrossRef]
79. Salazar-Salas, N.Y.; Valenzuela-Ponce, L.; Vega-Garcia, M.O.; Pineda-Hidalgo, K.V.; Vega-Alvarez, M.; Chavez-Ontiveros, J.; Delgado-Vargas, F.; Lopez-Valenzuela, J.A. Protein changes associated with chilling tolerance in tomato fruit with hot water pre-treatment. *Postharvest Boil. Technol.* **2017**, *134*, 22–30. [CrossRef]
80. Gong, B.; Zhang, C.; Li, X.; Wen, D.; Wang, S.; Shi, Q.; Wang, X. Identification of nacl and nahco3 stress responsive proteins in tomato roots using itraq-based analysis. *Biochem. Biophys. Res. Commun.* **2014**, *446*, 417–422. [CrossRef]
81. Muneer, S.; Ko, C.H.; Wei, H.; Chen, Y.; Jeong, B.R. Physiological and Proteomic Investigations to Study the Response of Tomato Graft Unions under Temperature Stress. *PLoS ONE* **2016**, *11*, e0157439. [CrossRef]
82. Tamburino, R.; Vitale, M.; Ruggiero, A.; Sassi, M.; Sannino, L.; Arena, S.; Costa, A.; Batelli, G.; Zambrano, N.; Scaloni, A.; et al. Chloroplast proteome response to drought stress and recovery in tomato (*Solanum lycopersicum* L.). *BMC Plant Boil.* **2017**, *17*, 40. [CrossRef]
83. Sang, Q.; Shan, X.; An, Y.; Shu, S.; Sun, J.; Guo, S. Proteomic Analysis Reveals the Positive Effect of Exogenous Spermidine in Tomato Seedlings' Response to High-Temperature Stress. *Front. Plant Sci.* **2017**, *8*, 555. [CrossRef]
84. Alam Khan, T.; Yusuf, M.; Ahmad, A.; Bashir, Z.; Saeed, T.; Fariduddin, Q.; Hayat, S.; Mock, H.P.; Wu, T.; Khan, T.A. Proteomic and physiological assessment of stress sensitive and tolerant variety of tomato treated with brassinosteroids and hydrogen peroxide under low-temperature stress. *Food Chem.* **2019**, *289*, 500–511. [CrossRef]
85. Kaspar, S.; Peukert, M.; Mock, H.P.; Svatos, A.; Matros, A.; Mock, H. MALDI-imaging mass spectrometry—An emerging technique in plant biology. *Proteomics* **2011**, *11*, 1840–1850. [CrossRef] [PubMed]
86. Lee, Y.J.; Perdian, D.C.; Song, Z.; Yeung, E.S.; Nikolau, B.J. Use of mass spectrometry for imaging metabolites in plants. *Plant J.* **2012**, *70*, 81–95. [CrossRef] [PubMed]

87. Schripsema, J. Application of nmr in plant metabolomics: Techniques, problems and prospects. *Hytochem. Anal. Int. J. Plant Chem. Biochem. Tech.* **2010**, *21*, 14–21. [CrossRef] [PubMed]
88. Sánchez-Rodríguez, E.; Moreno, D.A.; Ferreres, F.; Rubio-Wilhelmi, M.D.M.; Ruiz, J.M. Differential responses of five cherry tomato varieties to water stress: Changes on phenolic metabolites and related enzymes. *Phytochemistry* **2011**, *72*, 723–729. [CrossRef] [PubMed]
89. Ampofo-Asiama, J.; Baiye, V.; Hertog, M.; Waelkens, E.; Geeraerd, A.; Nicolai, B.J.P.B. The metabolic response of cultured tomato cells to low oxygen stress. *Plant Biol.* **2014**, *16*, 594–606. [CrossRef] [PubMed]
90. Zhu, G.; Wang, S.; Huang, Z.; Zhang, S.; Liao, Q.; Zhang, C.; Lin, T.; Qin, M.; Peng, M.; Yang, C.; et al. Rewiring of the Fruit Metabolome in Tomato Breeding. *Cell* **2018**, *172*, 249–261. [CrossRef]
91. Nunes-Nesi, A.; Alseekh, S.; Silva, F.M.D.O.; Omranian, N.; Lichtenstein, G.; Mirnezhad, M.; González, R.R.R.; Garcia, J.S.Y.; Conte, M.; Leiss, K.A.; et al. Identification and characterization of metabolite quantitative trait loci in tomato leaves and comparison with those reported for fruits and seeds. *Metabolomics* **2019**, *15*, 46. [CrossRef]
92. Salt, D.E.; Baxter, I.; Lahner, B. Ionomics and the Study of the Plant Ionome. *Annu. Rev. Plant Boil.* **2008**, *59*, 709–733. [CrossRef]
93. Bhat, J.A.; Shivaraj, S.M.; Singh, P.; Navadagi, D.B.; Tripathi, D.K.; Dash, P.K.; Solanke, A.U.; Sonah, H.; Deshmukh, R. Role of Silicon in Mitigation of Heavy Metal Stresses in Crop Plants. *Plants* **2019**, *8*, 71. [CrossRef]
94. Ahmad, P.; Tripathi, D.K.; Deshmukh, R.; Singh, V.P.; Corpas, F.J. Revisiting the role of ROS and RNS in plants under changing environment. *Environ. Exp. Bot.* **2019**, *161*, 1–3. [CrossRef]
95. Kim, Y.H.; Khan, A.L.; Waqas, M.; Lee, I.J. Silicon Regulates Antioxidant Activities of Crop Plants under Abiotic-Induced Oxidative Stress: A Review. *Front. Plant Sci.* **2017**, *8*, 282. [CrossRef] [PubMed]
96. Li, H.; Zhu, Y.; Hu, Y.; Han, W.; Gong, H. Beneficial effects of silicon in alleviating salinity stress of tomato seedlings grown under sand culture. *Acta Physiol. Plant.* **2015**, *37*, 71. [CrossRef]
97. Haghighi, M.; Pessarakli, M. Influence of silicon and nano-silicon on salinity tolerance of cherry tomatoes (*Solanum lycopersicum* L.) at early growth stage. *Sci. Hortic.* **2013**, *161*, 111–117. [CrossRef]
98. Marmiroli, M.; Pigoni, V.; Savo-Sardaro, M.; Marmiroli, N. The effect of silicon on the uptake and translocation of arsenic in tomato (*Solanum lycopersicum* L.). *Environ. Exp. Bot.* **2014**, *99*, 9–17. [CrossRef]
99. Ichihashi, Y.; Sinha, N.R. From genome to phenome and back in tomato. *Curr. Opin. Plant Boil.* **2014**, *18*, 9–15. [CrossRef] [PubMed]
100. White, J.W.; Andrade-Sanchez, P.; Gore, M.A.; Bronson, K.F.; Coffelt, T.A.; Conley, M.M.; Feldmann, K.A.; French, A.N.; Heun, J.T.; Hunsaker, D.J.; et al. Field-based phenomics for plant genetics research. *Field Crops Res.* **2012**, *133*, 101–112. [CrossRef]
101. Ubbens, J.R.; Stavness, I. Deep Plant Phenomics: A Deep Learning Platform for Complex Plant Phenotyping Tasks. *Front. Plant Sci.* **2017**, *8*, 8. [CrossRef]
102. Tardieu, F.; Cabrera-Bosquet, L.; Pridmore, T.; Bennett, M. Plant Phenomics, From Sensors to Knowledge. *Curr. Boil.* **2017**, *27*, R770–R783. [CrossRef]
103. Laxman, R.H.; Hemamalini, P.; Bhatt, R.M.; Sadashiva, A.T. Non-invasive quantification of tomato (*Solanum lycopersicum* L.) plant biomass through digital imaging using phenomics platform. *Indian J. Plant Physiol.* **2018**, *23*, 369–375. [CrossRef]
104. Cohen, H.; Aharoni, A.; Szymanski, J.; Dominguez, E. Assimilation of 'omics' strategies to study the cuticle layer and suberin lamellae in plants. *J. Exp. Bot.* **2017**, *68*, 5389–5400. [CrossRef]
105. Chaudhary, J.; Shivaraj, S.; Khatri, P.; Ye, H.; Zhou, L.; Klepadlo, M.; Dhakate, P.; Kumawat, G.; Patil, G.; Sonah, H.; et al. Approaches, Applicability, and Challenges for Development of Climate-Smart Soybean. In *Genomic Designing of Climate-Smart Oilseed Crops*; Springer Science and Business Media LLC: Berlin, Germany, 2019; pp. 1–74.
106. Shivaraj, S.M.; Dhakate, P.; Sonah, H.; Vuong, T.; Nguyen, H.T.; Deshmukh, R. Progress Toward Development of Climate-Smart Flax: A Perspective on Omics-Assisted Breeding. In *Genomic Designing of Climate-Smart Oilseed Crops*; Springer Science and Business Media LLC: Berlin, Germany, 2019; pp. 239–274.
107. Sonah, H.; O'Donoughue, L.; Cober, E.; Rajcan, I.; Belzile, F. Identification of loci governing eight agronomic traits using a gbs-gwas approach and validation by qtl mapping in soya bean. *Plant Biotechnol. J.* **2015**, *13*, 211–221. [CrossRef] [PubMed]

108. Chen, W.; Yao, Q.; Patil, G.B.; Agarwal, G.; Deshmukh, R.K.; Lin, L.; Wang, B.; Wang, Y.; Prince, S.J.; Song, L.; et al. Identification and Comparative Analysis of Differential Gene Expression in Soybean Leaf Tissue under Drought and Flooding Stress Revealed by RNA-Seq. *Front. Plant Sci.* **2016**, *7*, 827. [CrossRef] [PubMed]
109. Sonah, H.; Zhang, X.; Deshmukh, R.K.; Borhan, M.H.; Fernando, W.G.D.; Bélanger, R.R. Comparative Transcriptomic Analysis of Virulence Factors in *Leptosphaeria maculans* during Compatible and Incompatible Interactions with Canola. *Front. Plant Sci.* **2016**, *7*, 86. [CrossRef] [PubMed]
110. Deshmukh, R.; Singh, A.; Jain, N.; Anand, S.; Gacche, R.; Singh, A.; Gaikwad, K.; Sharma, T.; Mohapatra, T.; Singh, N. Identification of candidate genes for grain number in rice (*Oryza sativa* L.). *Funct. Integr. Genom.* **2010**, *10*, 339–347. [CrossRef]
111. Guo, T.; Yang, J.; Li, D.; Sun, K.; Luo, L.; Xiao, W.; Wang, J.; Liu, Y.; Wang, S.; Wang, H.; et al. Integrating GWAS, QTL, mapping and RNA-seq to identify candidate genes for seed vigor in rice (*Oryza sativa* L.). *Mol. Breed.* **2019**, *39*, 87. [CrossRef]
112. Albert, E.; Sauvage, C.; Bouchet, J.P.; Bitton, F.; Beukers, M.; Carretero, Y.; Causse, M. Integration of qtl, eqtl and allele specific expression to unravel genotype by watering regime interaction in cultivated tomato. In Proceedings of the Plant and Animal Genome Conference (PAG), San Diego, CA, USA, 14–18 January 2017.
113. Aoki, K.; Yano, K.; Suzuki, A.; Kawamura, S.; Sakurai, N.; Suda, K.; Kurabayashi, A.; Suzuki, T.; Tsugane, T.; Watanabe, M.; et al. Large-scale analysis of full-length cDNAs from the tomato (*Solanum lycopersicum*) cultivar Micro-Tom, a reference system for the Solanaceae genomics. *BMC Genom.* **2010**, *11*, 210. [CrossRef]
114. Yano, K.; Watanabe, M.; Yamamoto, N.; Tsugane, T.; Aoki, K.; Sakurai, N.; Shibata, D. MiBASE: A database of a miniature tomato cultivar Micro-Tom. *Plant Biotechnol.* **2006**, *23*, 195–198. [CrossRef]
115. Shikata, M.; Hoshikawa, K.; Ariizumi, T.; Fukuda, N.; Yamazaki, Y.; Ezura, H. TOMATOMA Update: Phenotypic and Metabolite Information in the Micro-Tom Mutant Resource. *Plant Cell Physiol.* **2015**, *57*, e11. [CrossRef]
116. Kudo, T.; Kobayashi, M.; Terashima, S.; Katayama, M.; Ozaki, S.; Kanno, M.; Saito, M.; Yokoyama, K.; Ohyanagi, H.; Aoki, K.; et al. TOMATOMICS: A Web Database for Integrated Omics Information in Tomato. *Plant Cell Physiol.* **2017**, *58*, e8. [CrossRef]
117. Suresh, B.V.; Roy, R.; Sahu, K.; Misra, G.; Chattopadhyay, D. Tomato Genomic Resources Database: An Integrated Repository of Useful Tomato Genomic Information for Basic and Applied Research. *PLoS ONE* **2014**, *9*, e86387. [CrossRef]
118. Fei, Z.; Joung, J.G.; Tang, X.; Zheng, Y.; Huang, M.; Lee, J.M.; McQuinn, R.; Tieman, D.M.; Alba, R.; Klee, H.J.; et al. Tomato Functional Genomics Database: A comprehensive resource and analysis package for tomato functional genomics. *Nucleic Acids Res.* **2010**, *39*, D1156–D1163. [CrossRef] [PubMed]
119. Shirasawa, K.; Hirakawa, H. DNA marker applications to molecular genetics and genomics in tomato. *Breed. Sci.* **2013**, *63*, 21–30. [CrossRef] [PubMed]
120. Moco, S.; Bino, R.J.; Vorst, O.; Verhoeven, H.A.; De Groot, J.; Van Beek, T.A.; Vervoort, J.; De Vos, C.R. A Liquid Chromatography-Mass Spectrometry-Based Metabolome Database for Tomato1. *Plant Physiol.* **2006**, *141*, 1205–1218. [CrossRef] [PubMed]
121. Narise, T.; Sakurai, N.; Obayashi, T.; Ohta, H.; Shibata, D. Co-expressed Pathways DataBase for Tomato: A database to predict pathways relevant to a query gene. *BMC Genom.* **2017**, *18*, 437. [CrossRef]
122. Shinozaki, Y.; Nicolas, P.; Fernandez-Pozo, N.; Ma, Q.; Evanich, D.J.; Shi, Y.; Xu, Y.; Zheng, Y.; Snyder, S.I.; Martin, L.B.B.; et al. High-resolution spatiotemporal transcriptome mapping of tomato fruit development and ripening. *Nat. Commun.* **2018**, *9*, 364. [CrossRef]

3

Genome-Wide Analysis of the β-Amylase Gene Family in *Brassica* L. Crops and Expression Profiles of *BnaBAM* Genes in Response to Abiotic Stresses

Dan Luo, Ziqi Jia, Yong Cheng, Xiling Zou and Yan Lv *

Key Laboratory of Biology and Genetic Improvement of Oil Crops, Ministry of Agriculture, Oil Crops Research Institute of the Chinese Academy of Agricultural Sciences, Wuhan 430062, China; ld17191091496@126.com (D.L.); jiaziqi0507@163.com (Z.J.); chengyong@caas.cn (Y.C.); zouxiling@caas.cn (X.Z.)
* Correspondence: lvyan01@caas.cn

Abstract: The β-amylase (BAM) gene family, known for their property of catalytic ability to hydrolyze starch to maltose units, has been recognized to play critical roles in metabolism and gene regulation. To date, *BAM* genes have not been characterized in oil crops. In this study, the genome-wide survey revealed the identification of 30 *BnaBAM* genes in *Brassica napus* L. (*B. napus* L.), 11 *BraBAM* genes in *Brassica rapa* L. (*B. rapa* L.), and 20 *BoBAM* genes in *Brassica oleracea* L. (*B. oleracea* L.), which were divided into four subfamilies according to the sequence similarity and phylogenetic relationships. All the *BAM* genes identified in the allotetraploid genome of *B. napus*, as well as two parental-related species (*B. rapa* and *B. oleracea*), were analyzed for the gene structures, chromosomal distribution and collinearity. The sequence alignment of the core glucosyl-hydrolase domains was further applied, demonstrating six candidate β-amylase (BnaBAM1, BnaBAM3.1-3.4 and BnaBAM5) and 25 β-amylase-like proteins. The current results also showed that 30 *BnaBAMs*, 11 *BraBAMs* and 17 *BoBAMs* exhibited uneven distribution on chromosomes of *Brassica* L. crops. The similar structural compositions of *BAM* genes in the same subfamily suggested that they were relatively conserved. Abiotic stresses pose one of the significant constraints to plant growth and productivity worldwide. Thus, the responsiveness of *BnaBAM* genes under abiotic stresses was analyzed in *B. napus*. The expression patterns revealed a stress-responsive behaviour of all members, of which *BnaBAM3s* were more prominent. These differential expression patterns suggested an intricate regulation of *BnaBAMs* elicited by environmental stimuli. Altogether, the present study provides first insights into the *BAM* gene family of *Brassica* crops, which lays the foundation for investigating the roles of stress-responsive *BnaBAM* candidates in *B. napus*.

Keywords: *Brassica* L. crops; *BAM* gene family; starch metabolism; abiotic stresses

---

1. Background

Abiotic stresses reduce plant growth, productivity and quality, which leads to great economic loss for the farming community. To deal with abiotic stresses, plants have evolved adaptive mechanisms in their growth and development processes [1]. Starch is the major carbohydrate storage in plants, which is mainly produced in chloroplast during the day and mobilized during the following night to guarantee a steady supply of carbon and energy [2,3]. Previous studies have shown that starch metabolism is a key determinant in the stress response. For instance, the starch content of leaf is found to be decreased under drought [4–7], osmotic [8,9], extreme temperature [10,11], salt [12,13], and oxidative stress conditions [14]. These changes are accompanied by released sugars and other derived metabolites, that function as compatible solutes to protect against the damage caused by abiotic stress [11,15]. However, osmotic and cold stresses also trigger an increase in starch content, as well as the related soluble sugars [2,16–18], suggesting that starch may play different roles in plant species under stress conditions [3].

The general degradation pathway of starch involves α-amylase (AMY), β-amylase, limit dextrinase (PUL), β-glucosidase, and α-glucan phosphorylase (PHO) [19]. β-amylases (BAMs) have been recognized to be responsible for starch degradation and gene regulation [20,21]. The catalytic activity of BAM is found to elevate in a series of starch-deficient mutants of *Arabidopsis* (*Arabidopsis thaliana* (L.) Heynh.), these enzymes are subsequently cloned, including AtBAM1, AtBAM2, AtBAM3, AtBAM5 [22–24]. Only these four members (AtBAM1, AtBAM2, AtBAM3, AtBAM5) were catalytically active, which show highly conserved amino acid motifs known to be involved in catalysis [7]. AtBAM7 and AtBAM8 show to be catalytically inactive in vitro [25,26], which are localized in the nucleus and contain a BRASSINAZOLE RESISTANT1 (BZR1)-type DNA binding domain attached to the N-terminus of the BAM domain [27]. Deregulation of AtBAM7 and AtBAM8 (the *bam7bam8* double mutant) causes altered leaf growth and development. These unique features suggest a regulatory role of AtBAM7 and AtBAM8 in controlling plant growth and development through crosstalk with brassinosteroid signaling and starch metabolism [28]. Besides the above, AtBAM4 has no catalytic capacity but facilitates starch breakdown independently of AtBAM1 and AtBAM3 [22]. Silencing of *StBAM1* and collective silencing of *StBAM1* and *StBAM9* in potato (*Solanum tuberosum* L.) demonstrate decreased β-amylase activity in cold-stored tubers. Meanwhile, soluble starch content increases in the RNA interference (RNAi) line of *StBAM1*, but decreases in the RNAi line of *StBAM9*, suggesting that StBAM1 may regulate cold-induced sweetening in potato tubers by hydrolyzing soluble starch, whereas StBAM9 directly acts on starch granules [29]. These observations raise the possibility that BAMs perform regulatory roles within starch metabolic pathways.

In *Arabidopsis*, BAMs act at the non-reducing ends of α−1, 4-linked glucan chains to produce maltose, which are the main hydrolytic enzymes during the night when leaf starch is broken down [22]. Maltose is the central product of starch degradation [30] and is well documented for its protective role upon cold stress [11,31]. *AtBAM1* could be transcriptionally induced by heat stress at 40 °C, while *AtBAM3* could be influenced by cold stress at 5 °C, consistent with the known daytime role of BAM1 in the guard cell stroma or the nighttime role of AtBAM3 in the mesophyll cells, respectively [2]. These changes correlate with maltose accumulation; the in vitro assays demonstrate that maltose functions as a compatible-solute stabilizing factor to protect proteins, membranes, and the photosynthetic electron transport chain in the chloroplast stroma under acute temperature stresses [11]. Similarly, the RNAi line of *AtBAM3* prevents maltose accumulation upon cold shock, which accordingly increases the sensitivity of PSII photochemical efficiency of *Arabidopsis* under freezing stress conditions [17]. Additionally, diurnal starch degradation is triggered by thioredoxin-regulated AtBAM1 in osmotically stressed mesophyll cells, and the produced osmolytes (maltose) reveals an active role in protection against osmotic stress [9]. Current studies on the upstream activators or repressors of BAMs are still limited. In rice (*Oryza sativa* L.), three OsBAMs function downstream of a negative regulator OsMYB30 under cold stress conditions, a further study on OsmiR528a has established a model that OsmiR528a targets OsMYB30-BAMs to enhance cold stress tolerance of rice [31,32]. *PtrBAM1* is a chloroplast-localizing *BAM* gene of *Poncirus trifoliate*, which has been reported to function in cold tolerance by modulating starch degradation. The C-repeat binding factor (CBF), termed *PtrCBF* interacts with the promoter of *PtrBAM1*, and regulates *PtrBAM1* in cold stress response by modulating soluble sugar levels [33]. Nevertheless, more regulation machinery of BAMs in stress signaling and response remains to be determined.

B. napus L. belongs to *Brassica* genus (*B. genus*), which is a major oil crop widely distributed worldwide. The allotetraploid *B. napus* (AACC, 2n = 38, http://www.genoscope.cns.fr) is generated by the crossing of *Brassica rapa* (AA, 2n = 20) and *Brassica oleracea* (CC, 2n = 18) [34]. Throughout the life cycle, *B. napus* experiences adverse environments, like extreme temperatures, drought, salinity or flood [35–40]. With the publishing of *B. napus* genome, genome-wide characterization of functional genes has been accelerated. The *BAM* gene family has been recognized across the plant kingdom, including 9 *AtBAMs* in *Arabidopsis* [41], 17 *PbBAMs* in pear (*Pyrus bretschneideri* Rehd.) [19], 10 *OsBAMs* in rice [42]. Nevertheless, the BAM members have not been reported in *B. genus*. The present study aims to characterize the BAM multi-gene family in *B. napus*, *B. oleracea* and *B. rapa*. The comprehensive analyses involving a phylogenetic tree, gene structure, chromosomal distribution, homologs and collinearity were conducted. Furthermore, the expression patterns under various abiotic stress conditions were evaluated to determine the function of *BAM* genes in *B. napus*. The information will provide a platform for future functional study of the molecular regulatory mechanisms of the *BAM* gene family in *Brassica* crops under abiotic stress conditions.

2. Materials and Methods

2.1. Identification of BAM Genes in B. napus, B.oleracea and B. rapa

The protein sequences of *Arabidopsis BAM* genes were obtained from Tair website (Tair10, http://www.arabidopsis.org) [43]. These *AtBAMs* were used as queries to search against the genomes of *Brassica* crops to characterize the candidate BAMs. The genome-related data from *B. napus* (Darmor-bzh), *B. rapa* (Chiifu), *B. oleracea* (var.capitata line 02–12) were derived from http://www.genoscope.cns.fr, http://brassicadb.org, https://plants.ensembl.org, respectively. The potential BAMs were checked in order to verify the existence of the conserved glycoside hydrolyase 14 domain (PF01373) by the Pfam (http://pfam.sanger.ac.uk) database [44]. The molecular weight and isoelectric point of each BAM protein were retrieved from the ExPASy program (http://web.expasy.org/protparam/).

2.2. Phylogenetic Analysis

The phylogenetic tree was constructed for BAM members of *Brassica* crops and *Arabidopsis* using the MEGA 7.0 software based on the Neighbour–Joining method. The evolutionary distances were evaluated using the Poisson correction method based on the units of the number of amino acid substitutions per site. The nodes of the tree were evaluated by bootstrap analysis with 1000 replicates [45]. Multiple alignments were carried out based on the full-length protein sequences of *BAM* gene family via ClustalX software (ver.1.83) with default settings, and Jalview [46].

2.3. Gene Structure, Chromosomal Location and Synteny Analysis

The exon/intron structural features of the corresponding BAM genes were explored using the Gene Structure Display Server (GSDS) database (http://gsds.cbi.pku.edu.cn). The start and stop locations of all *BAM* gene members were obtained from the *Brassica* database, MapChart software was used to visualize the distribution of each gene as described previously [47]. BLASTP program was employed to identify the tandem duplicated genes using BAM protein sequences of four dicots (i.e., *Arabidopsis*, *B. napus*, *B. rapa*, *B. oleracea*) with E-value cutoff $\leq 1 \times 10^{-10}$. These BAM proteins were employed to identify orthologous regions with the parameters ($e = 1 \times 10^{-20}$). The syntenic analysis maps of four plant species were constructed using MCSCanX and linearized on the chromosomes using the Circos program [45].

2.4. Materials and Treatments

The semi-winter type of *B. napus*, named Zhongshuang 6 (ZS6), was used for gene expression analysis and physiological measurements under abiotic stress conditions. The germinated seeds of ZS6 were transplanted to 10 cm × 10 cm pots with uniform growth, each pot containing five seedlings. The ZS6 variety was planted in three pots as three biological repeats and grown in a chamber (MLR-35IH, Panasonic, Japan) with a 16-h-light/8-h-dark cycle at 21 °C, the light intensity was 250 μ mol m^{-2} s^{-1}, and the humidity was controlled at 55% (±5%). To performed the short-term abiotic stress treatment, the four-leaf seedlings were treated with cold (continuous 4 °C in the chamber), dehydration (stopping water supply in the chamber), flooding (submerging completely with water in the chamber), salinity (irrigation with 100 mM/L NaCl solution in the chamber), heat (continuous 37 °C in the chamber). The leaves were sampled for three biological replications before stress treatments (denoted as 0 h), and 1 h, 6 h, 12 h as well as 24 h after stress treatments according to previous studies [11,19,33,39]. To perform the long-term cold stress treatment, the four-leaf seedlings were treated with a 16-h-light (8 °C)/8-h-dark (4 °C) cycle for 21 days. The leaves of seedlings were sampled for three biological replications before stress treatments (denoted as 0 d), and 1 d, 7 d, 14 d as well as 21 d after stress treatments.

2.5. Expression Analysis

Total RNA was extracted using the TransZol Up Plus RNA Kit (TransGen Biotech Co., LTD, Beijing, China) according to the manual instructions. Three micrograms total RNA was synthesized to the first-strand cDNA using the EasyScript®One-Step cDNA Synthesis SuperMix (TransGen Biotech Co., LTD, Beijing, China) according to the manufacturer's instructions. The Quantitative Real-time PCR (qRT-PCR) was performed using the SYBR®Green Premix kit according to the manufacturer's instructions on a StepOnePlusReal-Time PCR System (Applied Biosystems, Waltham, MA, USA). The relative expression level was determined by the $2^{-\Delta\Delta Ct}$ method based on three biological repetitions [48]. The primers used in this study are listed in an Additional file (Additional file 1: Table S1).

2.6. Starch Content and β-Amylase Activity Determination

To reflect the changes at the physiological level of *B. napus* L. under abiotic stress conditions, the starch and β-amylase activity of ZS6 seedlings were measured. The leaves of ZS6 were collected during the drought, flooding, heat, cold and salt treatments at a designed time and stored in −80 °C. The frozen leaves were grounded to powder in liquid nitrogen, and 0.2 g sample was added to 1 mL of 0.1 mol/L citric acid buffer, the extraction buffer was centrifuged for 15 min at 4 °C at 6000× *g*. The extraction was used to determine the β-amylase activity using a β-amylase kit (G0511W, Suzhou Grace Bio-technology Co., LTD, Suzhou, China) as previously described [29]. Meanwhile, the frozen leaves were grounded to powder in liquid nitrogen, and 0.1 g sample was added to 1 mL of 80% ethanol and extracted at 80 °C water bath for 30 min, the extraction buffer was centrifuged for 5 min at room temperature at 3000× *g*. The extraction was used to determine the starch content using the kit (G0507W, Suzhou Grace Bio-technology Co., LTD, Suzhou, China) [17].

2.7. Statistical Analysis

All experiments were performed with three biological replicates. Values are presented as mean ± SD. The significance of the data was evaluated using the Least-Significant Difference (LSD) test with Statistix 8.1 software. The significance level was expressed in lower case letters at $p < 0.05$.

3. Results

3.1. Characterization of BAM Genes in Brassica Crops

The full-length protein sequences of 9 *AtBAMs* in *Arabidopsis* were used as BLAST queries against the *Brassica* crops genomics database. A total of 30 *BAM* genes in *B. napus*, 11 *BAM* genes in *B. rapa* and 20 *BAM* genes in B. *oleracea* were obtained, respectively. The newly identified genes were named according to the closest homologs in *Arabidopsis* (*Arabidopsis thaliana* (L.) Heynh.) (Table 1). In Table 1, "random" means that the gene was not assembled into the specific location of the genome. *AtBAM1, AtBAM2, AtBAM6, AtBAM7, AtBAM8* presented the same number of homologs in the tetraploid *Brassica napus* L. (*B. napus* L.) with total homologs in *Brassica rapa* L. (*B. rapa* L.) and *Brassica oleracea* L. (*B. oleracea* L.). However, *AtBAM3, AtBAM4* had more homologs in *B. napus* than that in the parental diploid species; *AtBAM5, AtBAM9* had fewer homologs in *B. napus* than that in *B. rapa* and *B. oleracea.* The newly identified *BAM* genes exhibited coding potentials from 181 to 700 amino acids, and the protein molecular weights varied from 20.86 kDa to 79.06 kDa. Among the 30 *B. napus* genes, *BnaBAM8.2* showed the longest open reading frame (ORF) of 2103 bp; *BnaBAM3.6* showed the shortest ORF of 546 bp. To get insights into the evolutionary relationship of different BAM members, the phylogenetic tree was generated using the MEGA 7.0 software based on the BAM proteins of *Arabidopsis* (*Arabidopsis thaliana* (L.) Heynh.) and the *Brassica* L. species (Figure 1). BAM proteins were grouped into four clades, which was consistent with previous results [22]. Six BAM proteins of *Brassica* species (i.e., BnaBAM5.4, BraBAM9.2, BoBAM5.5, BoBAM5.6, BoBAM9.3, BoBAM9.4) differed highly from the other members and can not fall into any subfamilies. Subfamily IV contained the largest family members (18 BAM proteins) in three *Brassica* species, followed by 15 BAM proteins in Subfamily III, 12 BAM proteins in Subfamily II, and 10 BAM proteins in Subfamily I.

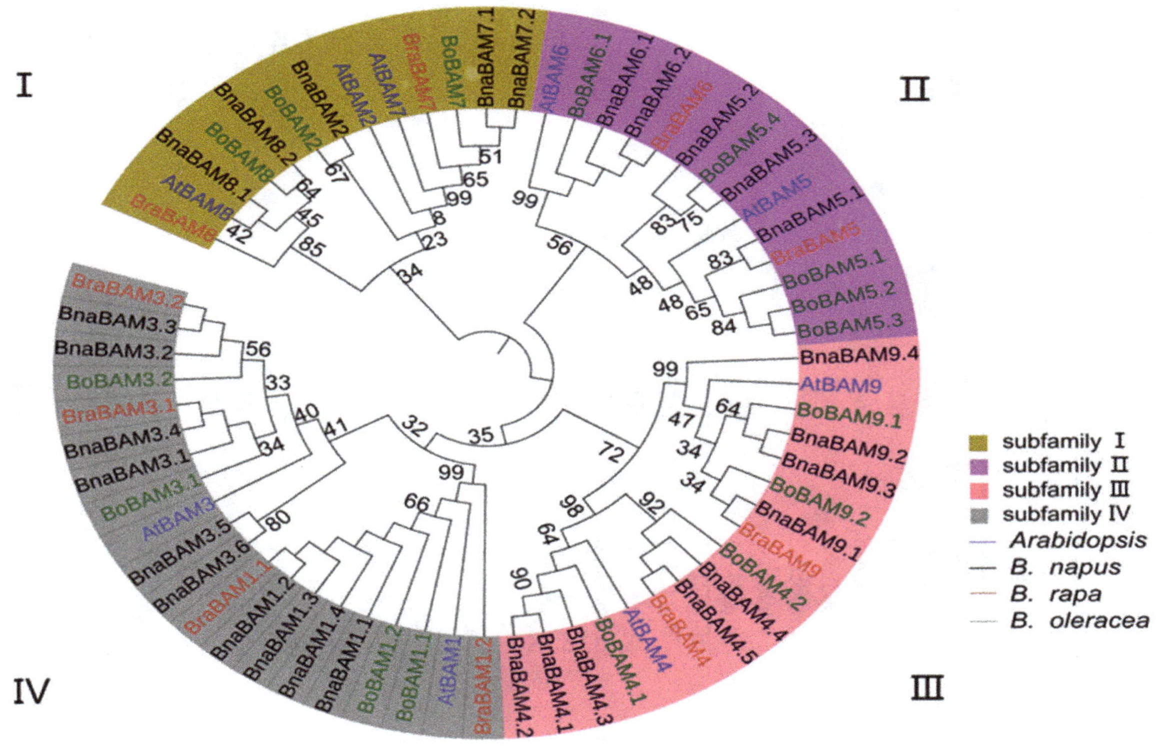

Figure 1. Phylogenetic analysis of β-amylase (BAM) proteins in *Arabidopsis, B. napus, B. rapa* and *B. oleracea*. Bootstrap values are shown near the nodes. The different colored arcs represents different subfamilies of the BAM proteins. BAM proteins in *Arabidopsis, B. napus, B. rapa* and *B. oleracea* were colored in blue, black, red and green.

Table 1. List of identified BAM gene family of *B. napus*, *B. rapa*, *B. oleracea* exhibiting accession number, opening reading frame (ORF), intron number, genomic location, length of amino acid, molecular weight (MW), and iso-electric point (pI).

Serial No.	Gene Name	Accession Numbers	ORF (bp)	Introns No.	Genomic Location	Protein		PI
						Length (a.a)	MW (kDa)	
1	BraBAM1.1	Bra015025	1713	3	A07: 4200669-4203428	570	63.38	5.31
2	BraBAM1.2	Bra001937	1695	3	A03: 19461784-19463944	564	62.99	5.68
3	BraBAM3.1	Bra026230	1650	3	A01: 10498540-10500425	549	61.72	6.98
4	BraBAM3.2	Bra012676	1647	3	A03: 22804184-22806081	548	61.45	6.28
5	BraBAM4	Bra035586	1587	9	A02: 7408259-7410757	528	59.80	7.54
6	BraBAM5	Bra038088	1497	6	A08: 6619783-6622218	498	56.18	5.07
7	BraBAM6	Bra005581	1746	6	A05: 6338859-6341224	581	66.53	5.58
8	BraBAM7	Bra004944	2019	8	A05: 2587413-2590431	672	75.08	5.64
9	BraBAM8	Bra021962	1306	8	A02: 17928394-17931297	653	73.63	5.60
10	BraBAM9.1	Bra002178	1614	2	A10: 11069000-11070867	537	58.59	5.82
11	BraBAM9.2	Bra006473	2112	0	A03: 3657746-3659857	703	79.70	6.03
12	BoBAM1.1	Bo3g082110.1	1638	3	C3: 30584504-30586921	545	61.16	5.62
13	BoBAM1.2	Bo7g034320.1	1713	3	C7: 13225248-13228060	570	63.41	5.31
14	BoBAM2	Bo9g002380.1	1623	7	C9: 245084-247243	540	61.17	5.20
15	BoBAM3.1	Bo1g052940.1	1647	3	C1: 15029435-15031326	548	61.86	7.18
16	BoBAM3.2	Bo7g104960.1	1644	3	C7: 40419711-40421605	547	61.37	6.28
17	BoBAM4.1	Bo3g021390.1	1656	9	C3: 7167639-7170389	551	62.65	7.91
18	BoBAM4.2	Bo2g037740.1	1917	10	C2: 10399330-10402605	638	71.78	5.76
19	BoBAM5.1	Bo8g040610.1	1179	4	C8: 13860892-13862943	392	44.89	5.59
20	BoBAM5.2	Bo8g040630.1	915	3	C8: 13904477-13905924	304	34.18	4.77
21	BoBAM5.3	Bo02108s010.1	915	3	Unknown chromosome	304	34.18	4.77
22	BoBAM5.4	Bo3g185170.1	1005	4	C3: 64572299-64575054	334	38.65	5.57
23	BoBAM5.5	Bo03100s010.1	498	2	Unknown chromosome	165	18.60	5.20
24	BoBAM5.6	Bo3g185180.1	645	3	C3:64580047-64581495	214	24.43	6.11
25	BoBAM6	Bo4g045870.1	1818	7	C4: 10696860-10699390	605	69.45	5.66
26	BoBAM7	Bo4g025050.1	2037	9	C4: 4131171-4134476	678	75.88	5.62
27	BoBAM8	Bo2g127050.1	2031	9	C2: 39136275-39139346	676	76.20	5.58
28	BoBAM9.1	Bo3g013270.1	1596	2	C3: 4592661-4594899	531	57.85	5.59
29	BoBAM9.2	Bo9g154650.1	1749	2	C9: 46305431-46307444	582	64.25	6.34
30	BoBAM9.3	Bo08131s010.1	243	0	Unknown chromosome	81	8.98	9.01

Table 1. *Cont.*

Serial No.	Gene Name	Accession Numbers	ORF (bp)	Introns No.	Genomic Location	Protein		PI
						Length (a.a)	MW (kDa)	
31	BoBAM9.4	Bo8g046130.1	609	4	C8: 15583635-15585005	202	23.04	8.45
32	BnaBAM1.1	BnaA03g37260D	1668	3	chrA03: 18437399-18439520	555	61.91	6.36
33	BnaBAM1.2	BnaC03g43570D	1638	3	chrC03: 28618604-28621024	545	61.16	5.88
34	BnaBAM1.3	BnaC09g21440D	1695	3	chrC09: 18644151-18647237	564	62.62	5.53
35	BnaBAM1.4	BnaA07g05790D	1713	3	chrA07: 6121654-6124373	570	63.31	5.55
36	BnaBAM2	BnaA09g51890D	1620	7	chrA09_random: 126993-129149	539	61.12	5.22
37	BnaBAM3.1	BnaC01g21190D	1647	3	chrC01: 14819094-14820988	548	61.76	7.65
38	BnaBAM3.2	BnaA03g42940D	1626	4	chrA03: 21561950-21563838	541	60.72	6.46
39	BnaBAM3.3	BnaC07g34180D	1647	3	chrC07: 37099448-37101348	548	61.40	6.59
40	BnaBAM3.4	BnaA01g17940D	1650	3	chrA01: 9483862-9485751	549	61.75	7.64
41	BnaBAM3.5	BnaA08g08230D	774	5	chrA08: 8068950-8076869	257	29.80	8.20
42	BnaBAM3.6	BnaC08g10900D	546	4	chrC08: 16342627-16343635	181	20.86	8.73
43	BnaBAM4.1	BnaC03g14120D	1602	9	chrC03: 6768572-6771333	533	60.49	8.74
44	BnaBAM4.2	BnaC03g71580D	897	4	chrC03_random: 222415-223959	298	33.68	9.32
45	BnaBAM4.3	BnaA03g11260D	1599	9	chrA03: 5050886-5053547	532	60.21	8.57
46	BnaBAM4.4	BnaA02g08900D	1572	9	chrA02: 4429617-4432105	523	59.15	8.89
47	BnaBAM4.5	BnaC02g12830D	1587	9	chrC02: 8167152-8169713	528	59.65	8.51
48	BnaBAM5.1	BnaA08g05660D	1497	6	chrA08: 5553480-5555924	498	56.22	5.20
49	BnaBAM5.2	BnaA08g00480D	1389	6	chrA08: 296747-299203	462	53.3	5.84
50	BnaBAM5.3	BnaC03g78310D	1005	4	chrC03_random: 6484548-6487309	334	38.65	5.57
51	BnaBAM5.4	BnaC03g78320D	645	3	chrC03_random: 6492299-6493747	214	24.43	6.11
52	BnaBAM6.1	BnaC04g12480D	1749	6	chrC04: 9717378-9719767	582	66.78	5.77
53	BnaBAM6.2	BnaA05g10780D	1746	6	chrA05: 5901832-5904179	581	66.75	6.20
54	BnaBAM7.1	BnaA05g05170D	2031	8	chrA05: 2697113-2700136	676	75.57	5.86
55	BnaBAM7.2	BnaC04g04570D	2019	9	chrC04: 3393159-3396247	672	75.19	5.71
56	BnaBAM8.1	BnaA02g36650D	1899	9	chrA02_random: 1168405-1171865	632	71.06	5.62
57	BnaBAM8.2	BnaC02g31510D	2103	8	chrC02: 33718069-33721130	700	79.06	5.99
58	BnaBAM9.1	BnaA10g16290D	1611	2	chrA10: 12426993-12428859	536	58.52	6.03
59	BnaBAM9.2	BnaA03g07240D	1593	2	chrA03: 3214063-3216446	530	57.83	5.73
60	BnaBAM9.3	BnaC03g09200D	1596	2	chrC03: 4394436-4396680	531	57.95	5.76
61	BnaBAM9.4	BnaC09g39140D	1614	2	chrC09: 41822923-41824840	537	58.86	6.50

3.2. Identification of the Gene Structure and Conserved Regions of BAM Genes

To analyze the structural diversification of *BnaBAMs* genes, gene structure display server software was used to map the intron–exon structures. The evolutionary analysis revealed the diversity of the structural compositions of *BnaBAMs* genes (Figure 2), which was consistent with the phylogenetic analyses. It also showed that 30 *BnaBAM* genes have different exon number varying from 3 to 10. The maximum number of exons was found in *BnaBAM4.1-4.5* with an exception of *BnaBAM4.2*. The structures and sizes were found to be well conserved among subfamily IV (*BnaBAM1.1-1.4, BnaBAM3.1-3.4*), subfamily I (*BnaBAM7.1-7.2, BnaBAM8.1-8.2*). However, subfamily II (*BnaBAM5.1-5.4, BnaBAM6.1-6.2*), and subfamily III (*BnaBAM4.1-4.5, BnaBAM9.1-9.4*) showed a very different intron–exon distribution.

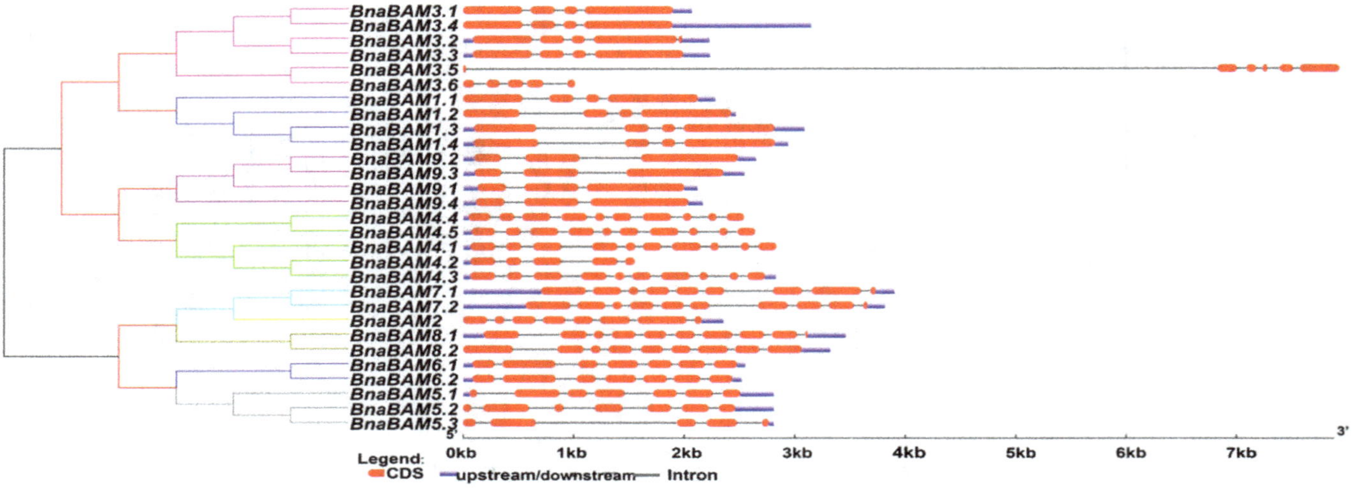

Figure 2. Gene structures of *BnaBAM* genes family. Gene structures of *BnaBAMs* based upon the number and position of exons (red boxes), introns (grey lines) and untranslated regions (blue lines).

The substrate-binding pocket and the active site of the active β-amylase GmBMY1 have been identified in soybean (*Glycine max* (L.) Merr.) [49,50]. The catalytic-related regions include two catalytic residues Glu-186 and Glu-380, the flexible loop, the inner loop [50,51]. In the present study, conserved amino acid residues in BnaBAM proteins were characterized via a multiple sequence alignment (Figure 3). Glu-186 was well conserved in 28 BnaBAM proteins, while notable divergence was observed within the other three catalytic regions. In the active enzyme AtBAM5, all of the 22 amino acids lining the active site were similar to those of GmBMY1. BnaBAM3.1-3.4 of subfamily IV presented the highest similarity to AtBAM5, and BnaBAM1.1-1.4 substituted at two places of the conserved active sites. Besides above, BnaBAM5.1 of subfamily II showed only one substitution. These observations were similar with the demonstrated enzyme activity of AtBAM1, AtBAM3 and AtBAM5, suggesting a potential catalytic role of BnaBAM1, BnaBAM3s and BnaBAM5, while the genes are encoding β-amylase-like proteins.

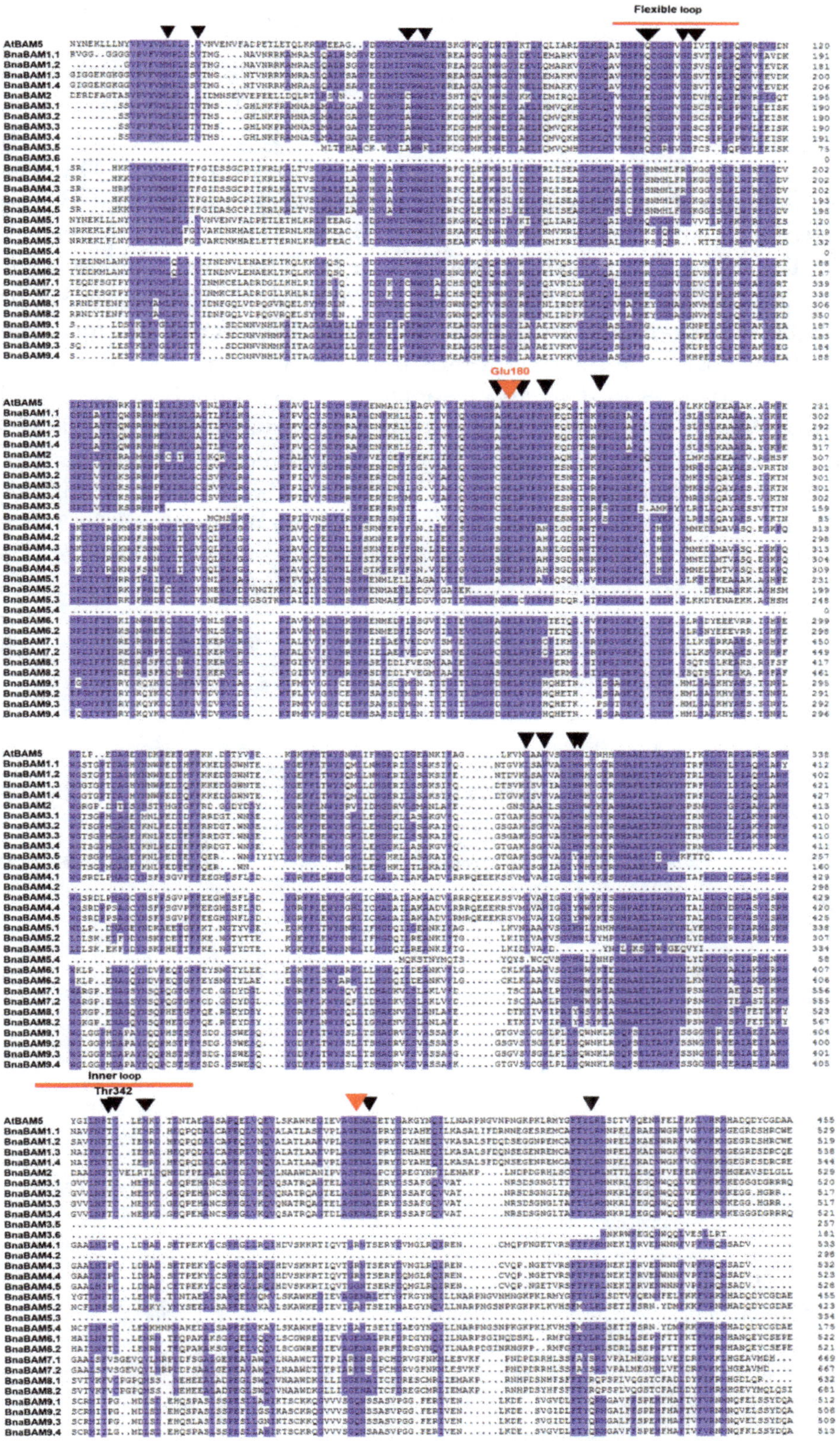

Figure 3. Alignment of the *B. napus* BAM proteins and the AtBAM5 protein. The blue shading indicates highly conservative substitutions. Unshaded residues are not conserved. Black arrowheads indicate substrate-binding residues. Red arrowheads indicate the two catalytic residues. Red lines indicate the flexible loop structure. The alignment was made using ClustalX software program.

3.3. Chromosomal Localization Analysis of the BAM Gene Family in Brassica Crops

In order to explore the chromosomal arrangement of 30 *BnaBAM* genes, each *BnaBAM* gene was mapped to *B. napus* chromosomes according to their physical locations (Figure 4A). The *BnaBAM* genes were widely distributed in 15 of 19 chromosomes of *B. napus*. However, the location of *BnaBAM* genes on chromosomes was relatively dispersed and uneven. The number of *BnaBAM* genes mapped on each chromosome varied from 1 to 6. *B. napus* was generated from the interspecific hybridization between A and C-genome ancestors (*B.rapa* and *B. oleracea*), which lead to genome duplication [52]. Therefore, the physical locations of *BraBAM* and *BoBAM* genes in the parental species was also presented (Figure 4B,C). *BoBAM5.3*, *BoBAM5.5* and *BoBAM9.3* were distributed to the unassembled genomic scaffolds, and could not be mapped to the chromosome. Overall, half of the *BnaBAM* genes were correlated with chromosomes inherited from parental-related species. A01, A05, A07 and A10 chromosomes of *B. napus* presented one or two *BnaBAM* genes, which was uniform with the distributions of *BraBAMs* in *B. rapa*. Similarly, the distributions of *BnaBAMs* on C01, C02, C04 chromosomes of *B. napus* were uniform with that in *B. oleracea*. Curiously, some *BnaBAMs* shared different distributions with that from *B. rapa* and *B. oleracea*. *BnaBAM8.1*, *BnaBAM4.4* were clustered on certain fragments of A02 chromosomes and separated by approximately 3 Mb, while *BraBAM4* and *BraBAM8* were located in a wider interval of 10 Mb on C02 chromosomes of *B. rapa*. Additionally, 5 *BnaBAM* genes were densely located on C03 chromosome of *B. napus* while 5 *BoBAM* genes were distributing throughout the C03 chromosome of *B. oleracea*.

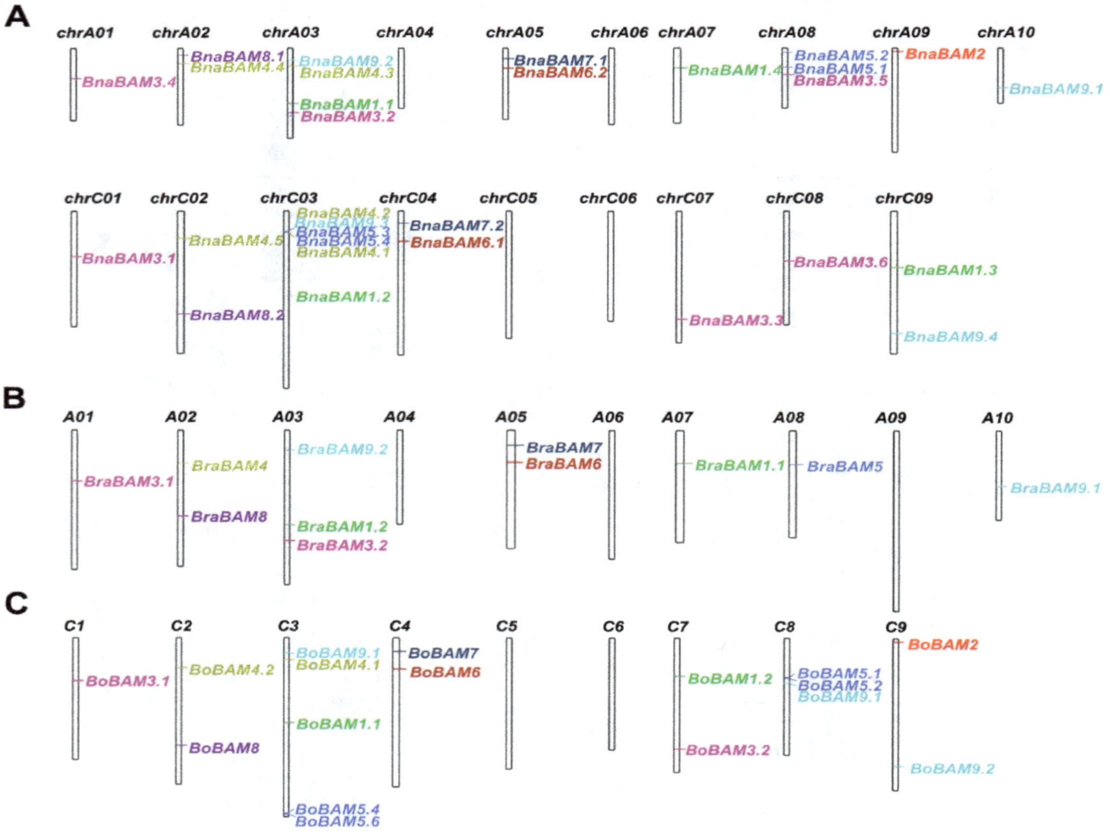

Figure 4. Chromosomal distribution of *BAM* gene family in *B. napus*, *B. rapa* and *B. oleracea*. The arrangement of 30 *BnaBAM* genes on 19 chromosomes of *B. napus* (**A**), 11 *BraBAM* genes on 10 chromosomes of *B. rapa* (**B**), and 17 *BoBAM* genes on 9 chromosomes of *B. oleracea* (**C**).

3.4. *Collinearity Analysis of Detected BAM Genes*

Arabidopsis genome has experienced two recent whole-genome duplications (WGD) of the crucifer (*Brassicaceae* Burnett) lineage, and a triplication event which is probably shared by most dicots [53]. To investigate the genetic divergence and gene duplications of *BAM* genes within *Arabidopsis* and *Brassica* species (*B. napus, B. rapa, B. oleracea*), the syntenic relationships of *BAM* genes were investigated (Figure 5). Overall, the distribution of collinear genes on chromosomes was relatively uneven. Among 30 *BnaBAM* genes, 21 members were found to be collinear with 9 *AtBAM* genes; 25 members were collinear with 10 *BraBAM* genes and 15 *BoBAM* genes (Additional file 2: Table S2). The results indicated that allotetraploid was the primary driving force for the rapid expansion of the *BnaBAM* gene family in *B. napus*. Because of the triplication event from their common ancestor with *Arabidopsis*, one *Arabidopsis BAM* should theoretically correspond to three orthologs in *B. rapa* and *B. oleracea*. However, only 9 orthologous gene pairs were obtained between *Arabidopsis* and *B. rapa*, followed by 12 orthologous gene pairs between *Arabidopsis* and *B. oleracea* (Additional file 2: Table S3). The synteny between *BAM* genes of parental-related species (*B. rapa* and *B. oleracea*) and their *Arabidopsis* homologs was less than expected, suggesting that duplicated genes might be lost during evolution. Eight *AtBAM* genes were found to be collinear with no fewer than one *BAM* gene of the three *Brassica* crops, indicating that these *BAM* genes may play vital roles during the evolution processes.

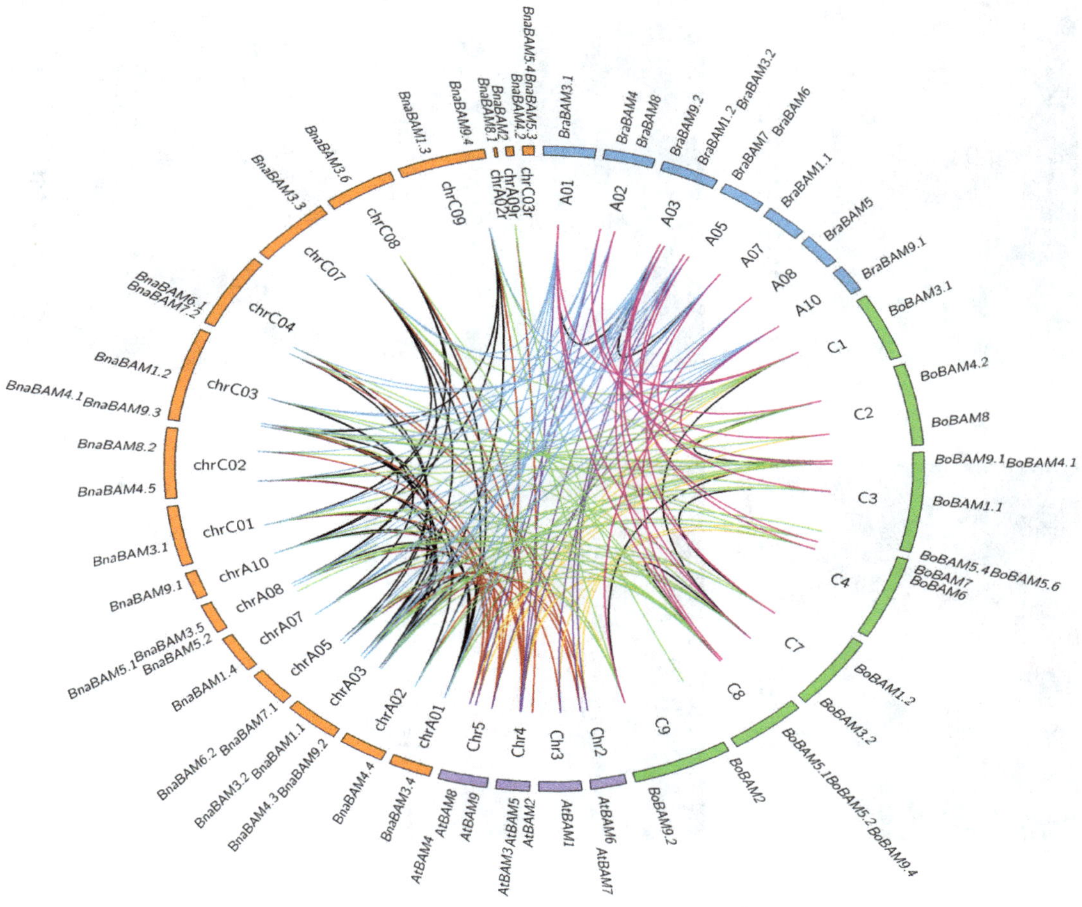

Figure 5. Collinearity analysis of the *BAM* genes between *Arabidopsis* and three *Brassica* crops. The synteny relationship between each pair of *BAM* genes was detected using linear regression. Genes with the syntenic relationship are linked by lines of the same color. The inner-circle indicates the chromosome numbers and the outer circle indicates the location of *AtBAMs*, *BnaBAMs*, *BraBAMs* and *BoBAMs* on chromosomes of *Arabidopsis*, *B. napus*, *B. rapa* and *B. oleracea*.

3.5. BnaBAM Genes Response to Various Abiotic Stresses

It has been well-documented that *AtBAMs* help plants adapt to unfavourable conditions [2], but how the environmental factors affect the *BnaBAM* gene family is still unclear. To further explore the potential roles of *BnaBAM* genes in response to different abiotic stresses, an extensive-expression analysis was performed in *B. napus* under drought, high temperature, low temperature, flooding and salt stress conditions. The results showed strong induction of *BnaBAM* genes at the transcript level, and the expression patterns of all *BnaBAM* genes varied greatly among different treatment groups. Sixteen *BnaBAM* genes were found to be continuously up-regulated by drought stress, whereas 14 genes were down-regulated (Figure 6A). Subfamily III included the homologs of *AtBAM4* and *AtBAM9*, and showed a systemic up-regulation after drought treatment. On the contrary, subfamily II (homologs of *AtBAM5* and *AtBAM6*), and subfamily I (homologs of *AtBAM2*, *AtBAM7*, *AtBAM8*) were generally down-regulated by drought stress. However, *BnaBAM* genes (homologs of *AtBAM1* and *AtBAM3*) belonging to subfamily IV performed distinct expression patterns with each other. Nonetheless, flooding stress demonstrated a negative modulation on most *BnaBAM* genes, with exceptions of the slight induction during the initial period of flooding treatment. *BnaBAM5.3* and *BnaBAM3.6* differed from others, which were greatly up-regulated by flooding with time changing (Figure 6B).

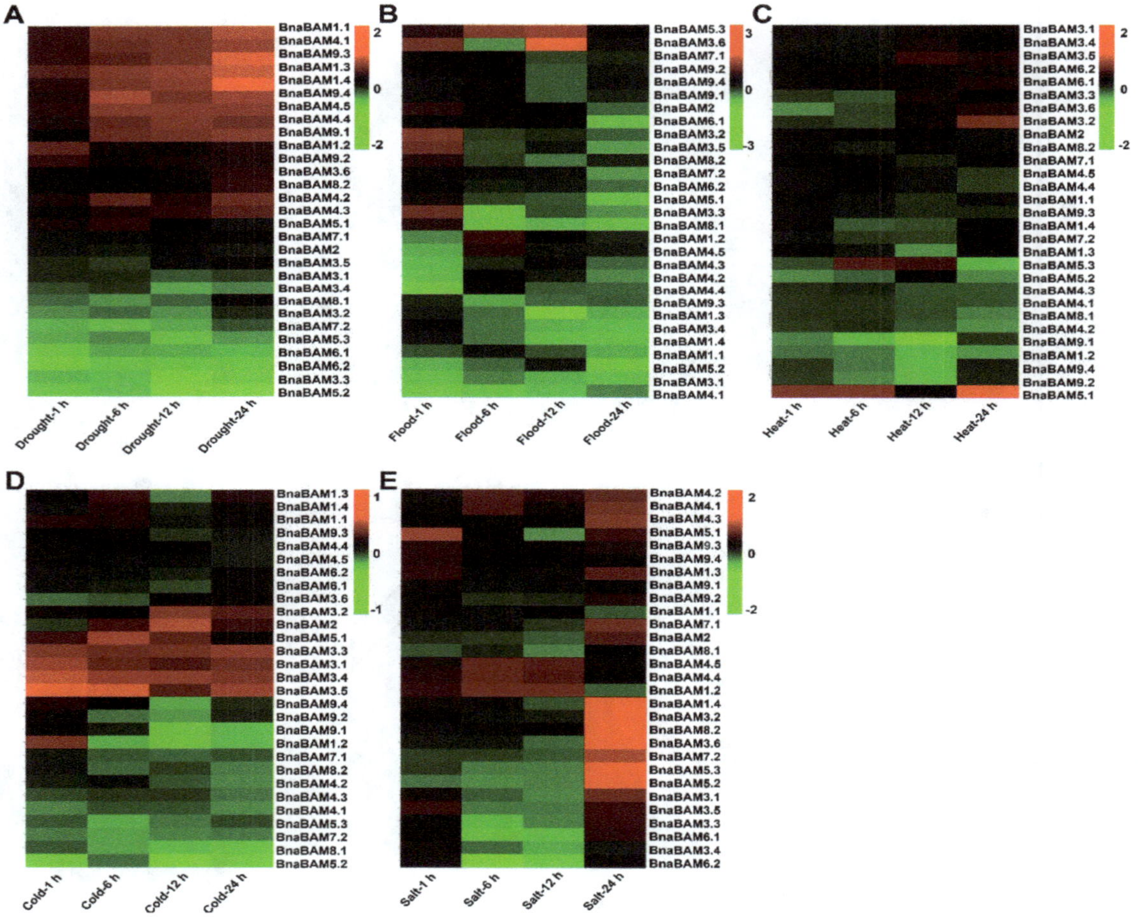

Figure 6. Expression profile of *BnaBAM* genes under different abiotic stress conditions. The transcripts of *BnaBAM* genes under drought (**A**), flooding (**B**), heat (**C**), cold (**D**) and salt (**E**) conditions are investigated and depicted as heat maps. The color scale represents $\log_2$ (expression values), with red denoting high level of transcription and green denoting a low level of transcription.

BAM genes are widely known for their involvement in extreme temperatures, such as cold and heat stresses [2,11,17,29,31]. *BnaBAM3.2, BnaBAM5.1* were significantly up-regulated by heat stress, whereas subfamily III was shown to be repressed (Figure 6C). When seedlings of *B. napus* were exposed to cold stimulation, subfamily IV including *BnaBAM1.1-1.4, BnaBAM3.1-3.5* were highly induced, and the other three subfamilies with an exception of *BnaBAM2* and *BnaBAM5* exhibited down-regulated patterns (Figure 6D), which supported the critical roles of AtBAM1 and AtBAM3 in the response of temperature stresses [2]. Considering the well-described function for *BAM* genes in the cold adaption of plants, 16 *BnaBAMs* which significantly responded to cold shock stress, were selected to further evaluate the expression levels under lone-term cold stress (4–8 °C). As observed in Figure S1, *BnaBAM3.2-3.3, BnaBAM2* markedly increased about 3- to 36- fold at 1 d-14 d of cold treatment, in contrast, the repressed expression pattern was observed for subfamily II and III, which exhibited agreement with the qRT-PCR investigations under cold shock stress condition (Figure 6D).

Unlike the responses mentioned above, *BnaBAMs* showed more dynamic expression patterns under salt stress conditions (Figure 6E). Specifically, *BnaBAM3.1-3.6, BnaBAM9.1-9.4* were markedly induced at 1 h and 24 h after exposure to salt stress, but repressed at 6 h and 12 h after salt treatment. Meanwhile, *BnaBAM4.1-4.5, BnaBAM1.2* were found to be increased over time under salt stress conditions. These differential expression patterns of *BnaBAM* genes under various abiotic stresses indicated that *BnaBAM* genes of *B. napus* may participate in stress response processes, which provides information for guiding future study on *BnaBAM* gene family in response to stress stimulus.

3.6. Abiotic Stresses Affect Starch Content and β-Amylase Activity in B. napus

Overall, most *BnaBAM* genes were responsive to abiotic stresses. Therefore, to determine how abiotic stresses affect starch metabolism of *B. napus*, the current study investigated the starch content and β-amylase activity in a widely cultivated variety ZS6, which was exposed to different abiotic stresses. It was found that total starch content was significantly reduced, and then increased to the original level after drought treatment, whereas starch content continuously declined after flood and cold treatment. A puzzling example was heat treatment, which resulted in increased starch content after 1 h of treatment, followed by a decrease during 6 h to 12 h, and finally a rise to the original level. However, salt stress had no significant effect on the starch content of rapeseed (Figure 7). Notably, there was no direct association between starch accumulation and β-amylase activity. The β-amylase activity showed approximately 20% to 80% reduction at different time points after drought, flood, heat, cold and salt treatments when compared with that under normal conditions (Figure 8). It was noted that salt stress exhibited severe influence on the enzyme activity than other treatments, and the β-amylase activity raised again at 24 h after salt treatment, and flooding caused a slight fluctuation in the β-amylase activity. These results reflected that abiotic stresses exhibited stimulatory impacts on β-amylase activity and the related starch content, however, the impacts on β-amylase activity were not correlated with the stress-induced alterations in starch metabolism.

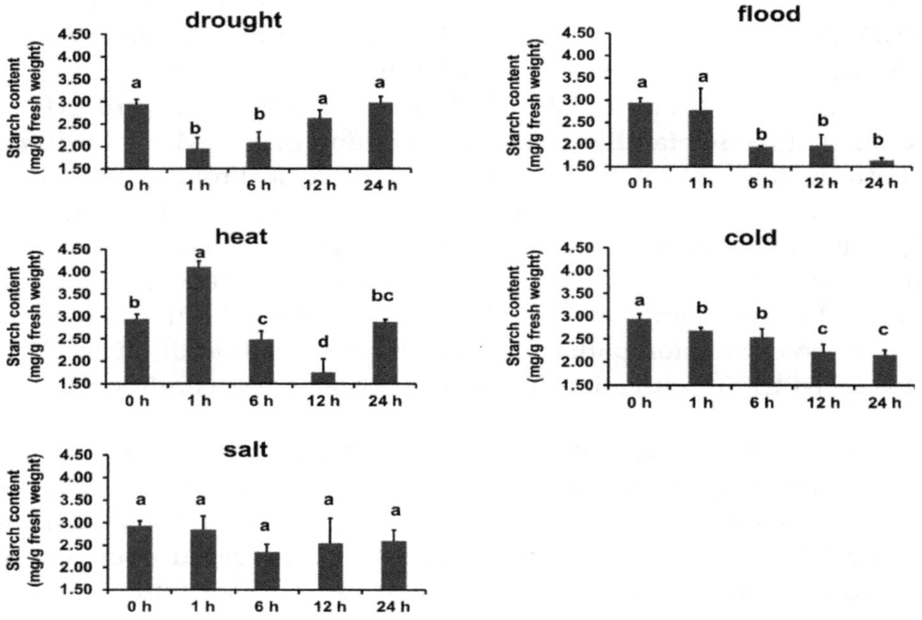

Figure 7. The effect of various abiotic stresses on starch content of *B. napus*. Investigations of starch content in *B. napus* seedling under drought, flooding, heat, cold and salt conditions. Bars indicate the SE of three biological replicates. Statistical analysis is determined by least-squares difference (LSD) test ($p < 0.05$).

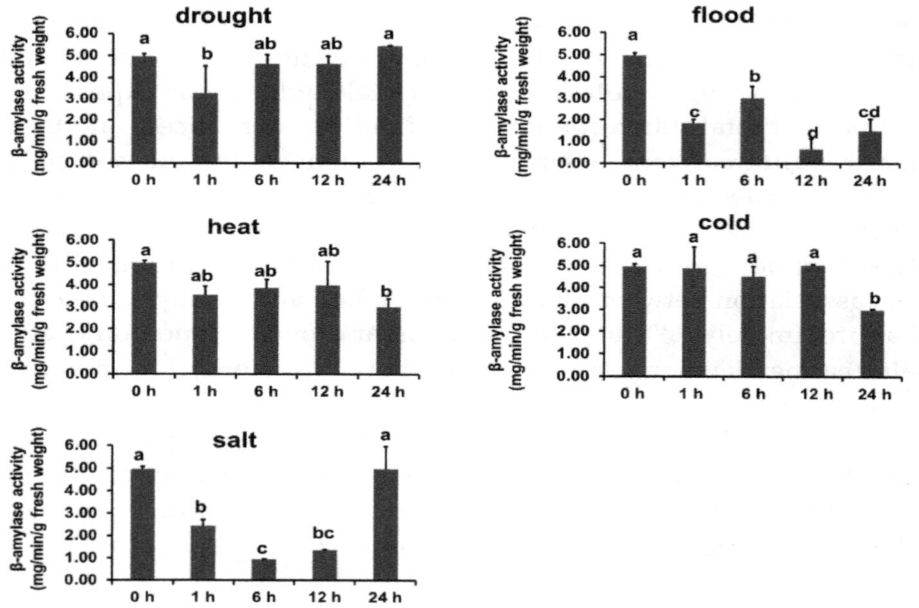

Figure 8. The effect of various abiotic stresses on β-amylase activity of *B. napus*. Investigations of β-amylase activity in *B. napus* seedling under drought, flooding, heat, cold and salt conditions. Bars indicate the SE of three biological replicates. Statistical analysis is determined by LSD test ($p < 0.05$).

4. Discussion

B. napus is the most productive and widely planted species among three oilseed rapes (*B. napus*, *B. rapa*, and *B. juncea*). To date, sets of genes in *B. napus*, including *BnaCBF17*, *LEA3*, *BnSIP1-1*, *BnNAC485*, *BnCOL2* have been reported to be associated with responses of abiotic stresses [54–58]. *BAM* genes hold great potential for modulating the sugar homeostasis of plants under abiotic stress conditions [28,43]. Therefore, the current study systematically analyzed the *BAM* gene family of *Brassica* crops, especially in *B. napus*. The *BAM* genes of three *Brassica* species could be categorized into four subfamilies (Figures 1 and 2) based on amino acid sequence alignments, which was in accordance with that of *Arabidopsis*, pear, banana (*Musa balbisiana* Colla), soybean and *Triticeae* Dumort. [19,59–62], suggesting that *BAM* genes were evolutionarily conserved among different plants. However, based on the conservation of intron positions in the BAM family in *Arabidopsis*, the recent studies have divided *BAM* genes into two subfamilies, which help to trace the origin of each *Arabidopsis* BAM gene back to the early land plant ancestors [23,43]. Most genome-wide replication events, including whole-genome duplication and whole-genome triplication (WGT), are accompanied by loss of genes [63]. The number of *BAM* genes in *B. napus* is almost the sum of that in *B. rapa* and *B. oleracea* with only one gene lost (Table 1), suggesting the gene loss might occur in the *B. napus* lineage process. A notable location was found on A03 and C03 chromosomes of *B. napus*, several *BnaBAM* genes were densely located on chromosomes and possessed higher accumulation, this suggested that tandem duplications and chromosomes segmental duplications might contribute to the expansion of the *BAM* gene family. In total, 25 of 30 *BnaBAM* genes exhibited the syntenic relationships between those in parental-related species (*B. rapa* and *B. oleracea*), followed by 24 *BnaBAM* genes that were found to be collinear with *BAM* genes of *Arabidopsis*. The above results may be contributable to the comparative study with the function-known *BAM* genes from *Arabidopsis*.

BAM genes have been reported to involve multiple abiotic stress responses of plants [3]. As expected, *BnaBAMs* generally responded to drought, flooding, heat, cold and salt treatments (Figure 6). A good example was the ubiquitous up-regulation of subfamily IV, *AtBAM3* of *Arabidopsis* corresponded to 6 homologs in *B. napus*, most of *BnaBAM3s* were induced by flooding, cold, heat, and salt treatments; meanwhile, 4 homologs (*BnaBAM1s*) of *AtBAM1* were up-regulated by drought stress. The greatest induction of *BnaBAM1s* and *BnaBAM3s* from subfamily IV was observed under cold stress conditions, which was consistent with the predominant role of AtBAM1 and AtBAM3 in cold tolerance [2,17]. In addition, the present study found evidence that two *BnaBAM3s* responded to long-term cold stress in *B. napus* (Additional file 3: Figure S1), therefore, *BnaBAM3s* were assumed to be responsible for cold stress adaption in *B. napus*. By contrast, subfamily I was widely down-regulated under drought, flooding, heat and cold stresses over time; and a puzzling result was found for subfamily II, *BnaBAM5.1* and *BAM6.1-6.2* showed opposite expression patterns under cold stress conditions, demonstrating distinct roles of *BnaBAM2s*, *BnaBAM5s*, *BnaBAM6s*, *BnaBAM7s*, *BnaBAM8s* in stress responses. Taken together, *BnaBAMs* may play a unique role in connecting starch metabolism and responses to abiotic stresses, while it is difficult to draw conclusions at this stage. However, the *BnaBAM* gene family would be a subject of considerable interest in understanding the genetic basis of stress resistance of *B. napus*.

Over the past few decades, progress has been achieved on the mechanisms of responding and adapting to environmental factors in plants, which composes of signal perception, signal transduction, transcriptional regulation, physiological and metabolic responses [64]. Precisely, plant cells sense abiotic signals, leading to the changes in membrane fluidity and cytoskeleton organization, which accordingly activate the second signal messengers (Ca^{2+}, reactive oxygen species, etc.), thus the downstream signal transduction cascades (kinases, transcription factors, etc.) are activated, that finally trigger the intricate responsive reactions [65]. Osmotic adjustment is one of these mechanisms, which is generated by the accumulation of a large number of osmolytes or compatible constitutes. Starch, as well as the derived metabolites, functions as osmoprotectants to mitigate the negative effects on plants that are caused by challenging environments [8,66–69]. In the majority of current literature, starch content has been reported to be decreased during abiotic stress treatments [8,10,18,22,70,71]. Similarly, reduced starch content was observed under abiotic stress conditions in the present study (Figure 7), suggesting that starch degradation could be triggered by abiotic stresses in *B. napus*, and was independent of the analyzed plant species. However, β-amylase activity showed a decreased trend after exposure to abiotic stresses in *B. napus*, and it was not necessarily correlated with the alterations in starch content (Figure 8). The enzyme activity of tobacco (*Nicotiana nudicaulis* Watson) slightly accumulates at 1 d and then maintains decreasing until the end of cold treatment [33]. While the total β-amylase activity of leaves remains constant under heat and cold shock conditions in *Arabidopsis* and pea plants, even the accumulation of maltose is observed [11]. One possible explanation may be the intricate process of starch breakdown, which consists of multiple synergistic actions [21]. β-amylase acts as the major hydrolytic enzyme during the nighttime in the leaf chloroplasts [22], but the *Arabidopsis* genome encodes additional enzymes for starch degradation, such as phosphorylases, α-amylases and limits dextrinase [3]. For instance, AtBAM1 works synergistically with the α-amylase 3 (AtAMY3) to efficiently degrade starch in the guard cells [72]. Thus, there are likely to be more enzymes involved in stress-induced starch degradation than those identified so far [3]. To date, mechanisms controlling starch degradation are largely unknown in *B. napus*. Studying the knockout mutants of the key *BnaBAMs* may help to determine whether a correlation exists between starch content and β-amylase activity in *B. napus*.

5. Conclusions

In conclusion, the data presented here identified 61 *BAM* genes in three *Brassica* species, including 30 *BnaBAM* genes, 11 *BraBAM* genes and 20 *BoBAM genes*. All *BnaBAMs* were highly conserved by the presence of the BAM domain, and clustered into four subfamilies based on gene evolution analysis. The exon–intron structures were prone to be similar within the same subfamily. A total of 25 *BnaBAM* genes demonstrated collinearity with 10 *BraBAM* genes and 15 *BoBAM genes*. The extensive examination at the transcript level and physiology level here provided a responsive overview of β-amylase under abiotic stress conditions, indicating their critical roles in diverse stress responses. The results presented in this report provide strong support for a functional study on the *BAM* gene family of *Brassica* crops, furthermore, facilitate our understanding of the molecular basis of the defenses against abiotic stresses in *B. napus*.

Author Contributions: D.L. and Y.L. conceived and designed the experiments, D.L., Z.J., X.Z., Y.C. performed the experiments and analyzed the data. D.L., Y.L. wrote the manuscript. All authors have read and agreed to the published version of the manuscript.

References

1. Gong, Z.Z.; Xiong, L.M.; Shi, H.Z.; Yang, S.H.; Herrera-Estrella, L.R.; Xu, G.H.; Chao, D.Y.; Li, J.R.; Wang, P.Y.; Qin, F.; et al. Plant abiotic stress response and nutrient use efficiency. *Sci. China Life Sci.* **2020**, *63*, 635–674. [CrossRef]
2. Monroe, J.D.; Storm, A.R.; Badley, E.M.; Lehman, M.D.; Platt, S.M.; Saunders, L.K.; Schmitz, J.M.; Torres, C.E. β-Amylase1 and β-Amylase3 are Plastidic Starch Hydrolases in *Arabidopsis* That Seem to Be Adapted for Different Thermal, pH, and Stress Conditions. *Plant Physiol.* **2014**, *166*, 1748–1763. [CrossRef]
3. Thalmann, M.; Santelia, D. Starch as a determinant of plant fitness under abiotic stress. *New Phytol.* **2017**, *214*, 943–951. [CrossRef]
4. Watkinson, J.I.; Hendricks, L.; Sioson, A.A.; Heath, L.S.; Bohnert, H.J.; Grene, R. Tuber development phenotypes in adapted and acclimated, drought-stressed *Solanum tuberosum* ssp. *andigena* have distinct expression profiles of genes associated with carbon metabolism. *Plant Physiol. Biochem.* **2008**, *46*, 34–45. [CrossRef] [PubMed]
5. Hummel, I.; Pantin, F.; Sulpice, R.; Piques, M.; Rolland, G.; Dauzat, M.; Christophe, A.; Pervent, M.; Bouteille, M.; Stitt, M.; et al. *Arabidopsis* plants acclimate to water deficit at low cost through changes of carbon usage: An integrated perspective using growth, metabolite, enzyme, and gene expression analysis. *Plant Physiol.* **2010**, *154*, 357–372. [CrossRef]

6. Yang, J.C.; Zhang, J.H.; Wang, Z.Q.; Zhu, Q.S. Activities of starch hydrolytic enzymes and sucrose-phosphate synthase in the stems of rice subjected to water stress during grain filling. *J. Exp. Bot.* **2001**, *52*, 2169–2179. [CrossRef]
7. Harb, A.; Krishnan, A.; Ambavaram, M.M.; Pereira, A. Molecular and physiological analysis of drought stress in *Arabidopsis* reveals early responses leading to acclimation in plant growth. *Plant Physiol.* **2010**, *154*, 1254–1271. [CrossRef] [PubMed]
8. Thalmann, M.; Pazmino, D.; Seung, D.; Horrer, D.; Nigro, A.; Meier, T.; Kolling, K.; Pfeifhofer, H.W.; Zeeman, S.C.; Santelia, D. Regulation of leaf starch degradation by abscisic acid is important for osmotic stress tolerance in plants. *Plant Cell* **2016**, *28*, 1860–1878. [CrossRef] [PubMed]
9. Valerio, C.; Costa, A.; Marri, L.; Issakidis-Bourguet, E.; Pupillo, P.; Trost, P.; Sparla, F. Thioredoxin-regulated β-amylase (BAM1) triggers diurnal starch degradation guard cells, and in mesophyll cells under osmotic stress. *J. Exp. Bot.* **2011**, *62*, 545–555. [CrossRef]
10. Vasseur, F.; Pantin, F.; Vile, D. Changes in light intensity reveal a major role for carbon balance in *Arabidopsis* responses to high temperature. *Plant Cell Environ.* **2011**, *34*, 1563–1576. [CrossRef]
11. Kaplan, F.; Guy, C.L. β-amylase induction and the protective role of maltose during temperature shock. *Plant Physiol.* **2004**, *135*, 1674–1684. [CrossRef] [PubMed]
12. Kempa, S.; Krasensky, J.; Dal, S.S.; Kopka, J.; Jonak, C. A central role of abscisic acid in stress-regulated carbohydrate metabolism. *PLoS ONE* **2008**, *3*, e3935. [CrossRef] [PubMed]
13. Chen, H.J.; Chen, J.Y.; Wang, S.J. Molecular regulation of starch accumulation in rice seedling leaves in response to salt stress. *Acta Physiol. Plant.* **2008**, *30*, 135–142. [CrossRef]
14. Scarpeci, T.E.; Valle, E.M. Rearrangement of carbon metabolism in *Arabidopsis thaliana* subjected to oxidative stress condition: An emergency survival strategy. *Plant Growth Regul.* **2008**, *54*, 133–142. [CrossRef]
15. Krasensky, J.; Jonak, C. Drought, salt, and temperature stress-induced metabolic rearrangements and regulatory networks. *J. Exp. Bot.* **2012**, *63*, 1593–1608. [CrossRef]
16. Hoermiller, I.I.; Naegele, T.; Augustin, H.; Stutz, S.; Weckwerth, W.; Heyer, A.G. Subcellular reprogramming of metabolism during cold acclimation in *Arabidopsis thaliana*. *Plant Cell Environ.* **2016**, *40*, 602–610. [CrossRef]
17. Kaplan, F.; Guy, C.L. RNA interference of *Arabidopsis* beta-amylase8 prevents maltose accumulation upon cold shock and increases sensitivity of PSII photochemical efficiency to freezing stress. *Plant J.* **2005**, *44*, 730–743. [CrossRef]
18. Zanella, M.; Borghi, G.L.; Pirone, C.; Thalmann, M.; Pazmino, D.; Costa, A.; Santelia, D.; Trost, P.; Sparla, F. β-amylase1 (BAM1) degrades transitory starch to sustain proline biosynthesis during drought stress. *J. Exp. Bot.* **2016**, *67*, 1819–1826. [CrossRef]
19. Zhao, L.Y.; Gong, X.; Gao, J.Z.; Dong, H.Z.; Zhang, S.L.; Tao, S.T.; Huang, X.S. Transcriptomic and evolutionary analyses of white pear (*Pyrusbretschneideri*) β-amylase genes reveals their importance for cold and drought stress responses. *Gene* **2019**, *20*, 102–113. [CrossRef]
20. Stitt, M.; Zeeman, S.C. Starch turnover: Pathways, regulation and role in growth. *Curr. Opin. Plant Biol.* **2012**, *15*, 282–292. [CrossRef]
21. Streb, S.; Zeeman, S.C. Starch metabolism in *Arabidopsis*. *Arab. Book* **2012**, *10*, 73–98. [CrossRef] [PubMed]
22. Fulton, D.C.; Stettler, M.; Mettler, T.; Vaughan, C.K.; Li, J.; Francisco, P.; Gil, M.; Reinhold, H.; Eicke, S.; Messerli, G.; et al. β-AMYLASE4, a noncatalytic protein required for starch breakdown, acts upstream of three active β-amylases in *Arabidopsis* chloroplasts. *Plant Cell* **2008**, *20*, 1040–1058. [CrossRef] [PubMed]
23. Monroe, J.D.; Breault, J.S.; Pope, L.E.; Torres, C.E.; Gebrejesus, T.B.; Berndsen, C.E.; Storm, A.R. *Arabidopsis* β-amylase2 is a K^+-requiring, catalytic tetramer with sigmoidal kinetics. *Plant Physiol.* **2017**, *175*, 1525–1535. [CrossRef] [PubMed]
24. Lao, N.T.; Schoneveld, O.; Mould, R.M.; Hibberd, J.M.; Gray, J.C.; Kavanagh, T.A. An *Arabidopsis* gene encoding a chloroplast-targeted β-amylase. *Plant J.* **1999**, *20*, 519–527. [CrossRef] [PubMed]
25. Li, J.; Francisco, P.; Zhou, W.X.; Edner, C.; Steup, M.; Ritte, G.; Bond, C.S.; Smith, S.M. Catalytically-inactive β-amylase BAM4 required for starch breakdown in *Arabidopsis* leaves is a starch-binding-protein. *Arch. Biochem. Biophys.* **2009**, *489*, 92–98. [CrossRef] [PubMed]
26. Steidle, E.A. Investigation of the Role of BAM9 in Starch Metabolism in *Arabidopsis thaliana*. Master's Thesis, James Madison University, Harrisonburg, VA, USA, 2010.

27. Soyk, S.; Simkova, K.; Zurcher, E.; Luginbuhl, L.; Brand, L.H.; Vaughan, C.K.; Wanke, D.; Zeeman, S.C. The Enzyme-Like Domain of *Arabidopsis* Nuclear β-Amylases Is Critical for DNA Sequence Recognition and Transcriptional Activation. *Plant Cell* **2014**, *26*, 1746–1763. [CrossRef] [PubMed]
28. Reinhold, H.; Soyk, S.; Simkova, K.; Hostettler, C.; Marafino, J.; Mainiero, S.; Vaughan, C.K.; Monroe, J.D.; Zeeman, S.C. β-Amylase-Like Proteins Function as Transcription Factors in *Arabidopsis*, Controlling Shoot Growth and Development. *Plant Cell* **2011**, *23*, 1391–1403. [CrossRef] [PubMed]
29. Hou, J.; Zhang, H.L.; Liu, J.; Reid, S.; Liu, T.F.; Xu, S.J.; Tian, Z.D.; Sonnewald, U.; Song, B.; Xie, C.H. Amylases *StAmy23*, *StBAM1* and *StBAM9* regulate cold-induced sweetening of potato tubers in distinct ways. *J. Exp. Bot.* **2017**, *68*, 2317–2331. [CrossRef]
30. Niittyla, T.; Messerli, G.; Trevisan, M.; Chen, J.; Smith, A.M.; Zeeman, S.C. A Previously Unknown Maltose Transporter Essential for Starch Degradation in Leaves. *Science* **2004**, *303*, 87–89. [CrossRef]
31. Lv, Y.; Yang, M.; Hu, D.; Yang, Z.Y.; Ma, S.Q.; Li, X.H.; Xiong, L.Z. The *OsMYB30* Transcription Factor Suppresses Cold Tolerance by Interacting with a JAZ Protein and Suppressing β-Amylase Expression. *Plant Physiol.* **2017**, *173*, 1475–1491. [CrossRef]
32. Tang, W.; Thompson, W.A. OsmiR528 Enhances Cold Stress Tolerance by Repressing Expression of Stress Response-related Transcription Factor Genes in Plant Cells. *Curr. Genom.* **2019**, *20*, 100–114. [CrossRef] [PubMed]
33. Peng, T.; Zhu, X.F.; Duan, N.; Liu, J.H. *PtrBAM1*, a β-amylase-coding gene of *Poncirus trifoliata*, is a CBF regulon member with function in cold tolerance by modulating soluble sugar levels. *Plant Cell Environ.* **2014**, *37*, 2754–2767. [CrossRef] [PubMed]
34. Chalhoub, B.; Denoeud, F.; Liu, S.Y.; Parkin, I.A.P.; Tang, H.B.; Wang, X.Y.; Chiquet, J.; Belcram, H.; Tong, C.B.; Samans, B.; et al. Early allopolyploid evolution in the post-Neolithic *Brassica napus* oilseed genome. *Science* **2014**, *345*, 950–953. [CrossRef] [PubMed]
35. Georges, F.; Das, S.; Ray, H.; Bock, C.; Nokhrina, K.; Kolla, V.A.; Keller, W. Over-expression of *Brassica napus* phosphatidylinositol-phospholipase C2 in canola induces significant changes in gene expression and phytohormone distribution patterns, enhances drought tolerance and promotes early flowering and maturation. *Plant Cell Environ.* **2009**, *32*, 1664–1681. [CrossRef] [PubMed]
36. Diksaityte, A.; Virsile, A.; Zaltauskaite, J.; Januskaitiene, I.; Juozapaitiene, G. Growth and photosynthetic responses in *Brassica napus* differ during stress and recovery periods when exposed to combined heat, drought and elevated CO_2. *Plant Physiol. Biochem.* **2019**, *142*, 59–72. [CrossRef]
37. Wollmer, A.C.; Pitann, B.; Muhling, K.H. Timing of Waterlogging Is Crucial for the Development of Micronutrient Deficiencies or Toxicities in Winter Wheat and Rapeseed. *J. Plant Growth Regul.* **2019**, *38*, 824–830. [CrossRef]
38. Cui, J.Q.; Hua, Y.P.; Zhou, T.; Liu, Y.; Huang, J.Y.; Yue, C.P. Global Landscapes of the *Na+/H+ Antiporter (NHX)* Family Members Uncover their Potential Roles in Regulating the Rapeseed Resistance to Salt Stress. *Int. J. Mol. Sci.* **2020**, *21*, 3429. [CrossRef]
39. Yan, L.; Shah, T.; Cheng, Y.; Lv, Y.; Zhang, X.K.; Zou, X.L. Physiological and molecular responses to cold stress in rapeseed (*Brassica napus* L.). *J. Integr. Agric.* **2019**, *18*, 2742–2752. [CrossRef]
40. Huang, R.Z.; Liu, Z.H.; Xing, M.Q.; Yang, Y.; Wu, X.L.; Liu, H.Q.; Liang, W.F. Heat Stress Suppresses *Brassica napus* Seed Oil Accumulation by Inhibition of Photosynthesis and *BnWRI1* Pathway. *Plant Cell Physiol.* **2019**, *60*, 1457–1470. [CrossRef]
41. Smith, S.M.; Fulton, D.C.; Chia, T.; Thorneycroft, D.; Chapple, A.; Dunstan, H.; Hylton, C.; Zeeman, S.C.; Smith, A.M. Diurnal Changes in the Transcriptome Encoding Enzymes of Starch Metabolism Provide Evidence for Both Transcriptional and Posttranscriptional Regulation of Starch metabolism in *Arabidopsis* leaves. *Plant Physiol.* **2004**, *136*, 2687–2699. [CrossRef]
42. Koide, T.; Ohnishi, Y.; Horinouchi, S. Characterization of recombinant β-amylases from *Oryza sativa*. *Bios. Biot. Bioch.* **2011**, *75*, 793–796. [CrossRef]
43. Monroe, J.D.; Storm, A.R. Review: The *Arabidopsis* β-amylase (BAM) gene family: Diversity of form and function. *Plant Sci.* **2018**, *276*, 163–170. [CrossRef] [PubMed]
44. Totsuka, A.; Nong, H.V.; Kadokawa, H.; Kim, C.S.; Itoh, Y.; Fukazawa, C. Residues essential for catalytic activity of soybean β-amylase. *Eur. J. Biochem.* **1994**, *221*, 649–654. [CrossRef] [PubMed]

45. Wang, M.M.; Liu, M.M.; Ran, F.; Guo, P.C.; Ke, Y.Z.; Wu, Y.W.; Wen, J.; Li, P.F.; Li, J.N.; Du, H. Global Analysis of *WOX* Transcription Factor Gene Family in *Brassica napus* Reveals Their Stress- and Hormone-Responsive Patterns. *Int. J. Mol. Sci.* **2018**, *19*, 3470. [CrossRef] [PubMed]
46. Clamp, M.; Cuff, J.; Searle, S.M.; Barton, G.J. The Jalview Java alignment editor. *Bioinformatics* **2004**, *20*, 426–427. [CrossRef] [PubMed]
47. Miao, L.M.; Gao, Y.Y.; Zhao, K.; Kong, L.J.; Yu, S.B.; Li, R.R.; Liu, K.W.; Yu, X.L. Comparative analysis of basic helix-loop-helix gene family among *Brassica oleracea, Brassica rapa,* and *Brassica napus*. *BMC Genom.* **2020**, *21*, 178. [CrossRef] [PubMed]
48. Wang, Z.; Chen, Y.; Fang, H.D.; Shi, H.F.; Chen, K.P.; Zhang, Z.Y.; Tan, X.L. Selection of reference genes for quantitative reverse-transcription polymerase chain reaction normalization in *Brassica napus* under various stress conditions. *Mol. Genet. Genom.* **2014**, *289*, 1023–1035. [CrossRef] [PubMed]
49. Kang, Y.N.; Adachi, M.; Utsumi, S.; Mikami, B. The roles of Glu186 and Glu380 in the catalytic reaction of soybean beta-amylase. *J. Mol. Biol.* **2004**, *339*, 1129–1140. [CrossRef]
50. Kang, Y.N.; Tanabe, A.; Adachi, M.; Utsumi, S.; Mikami, B. Structural Analysis of Threonine 342 Mutants of Soybean β-Amylase: Role of a Conformational Change of the Inner Loop in the Catalytic Mechanism. *Biochemistry* **2005**, *44*, 5106–5116. [CrossRef] [PubMed]
51. Thalmann, M.; Coiro, M.; Meier, T.; Wicker, T.; Zeeman, S.C.; Santelia, D. The evolution of functional complexity within the β-amylase gene family in land plants. *BMC Evol. Biol.* **2019**, *19*, 66. [CrossRef]
52. Cheung, F.; Trick, M.; Drou, N.; Lim, Y.P.; Park, J.Y.; Kwon, S.; Kim, J.A.; Scott, R.; Pires, J.C.; Paterson, A.H.; et al. Comparative Analysis between Homoeologous Genome Segments of *Brassica napus* and Its Progenitor Species Reveals Extensive Sequence-Level Divergence. *Plant Cell* **2009**, *21*, 1912–1928. [CrossRef] [PubMed]
53. Yu, J.Y.; Tehrim, S.; Zhang, F.Q.; Tong, C.B.; Huang, J.Y.; Cheng, X.H.; Dong, C.H.; Zhou, Y.Q.; Qin, R.; Hua, W.; et al. Genome-wide comparative analysis of NBS-encoding genes between Brassica species and *Arabidopsis thaliana*. *BMC Genom.* **2014**, *15*, 3. [CrossRef] [PubMed]
54. Liang, Y.; Kang, K.; Gan, L.; Ning, S.B.; Xiong, J.Y.; Song, S.Y.; Xi, L.Z.; Lai, S.Y.; Yin, Y.G.; Gu, J.W.; et al. Drought-responsive genes, late embryogenesis abundant group3 (*LEA3*) and vicinal oxygen chelate, function in lipid accumulation in *Brassica napus* and *Arabidopsis* mainly via enhancing photosynthetic efficiency and reducing ROS. *Plant Biotechnol. J.* **2019**, *17*, 2123–2142. [CrossRef] [PubMed]
55. Dahal, K.; Gadapati, W.; Savitch, L.V.; Singh, J.; Huner, N.P.; Norman, P.A. Cold acclimation and *BnCBF17*-over-expression enhance photosynthetic performance and energy conversion efficiency during long-term growth of *Brassica napus* under elevated CO_2 conditions. *Planta* **2012**, *236*, 1639–1652. [CrossRef] [PubMed]
56. Luo, J.L.; Tang, S.H.; Mei, F.L.; Peng, X.J.; Li, J.; Li, X.F.; Yan, X.H.; Zeng, X.H.; Liu, F.; Wu, Y.H.; et al. *BnSIP1-1*, a Trihelix Family Gene, Mediates Abiotic Stress Tolerance and ABA Signaling in *Brassica napus*. *Front. Plant Sci.* **2017**, *8*, e71136. [CrossRef] [PubMed]
57. Lu, Y.; Chen, H.Y.; Cai, W.M. *BnNAC485* is involved in abiotic stress responses and flowering time in *Brassica Napus*. *Plant Physiol. Biochem.* **2014**, *79*, 77–78.
58. Liu, L.D.; Ding, Q.Y.; Liu, J.; Yang, C.L.; Chen, H.; Zhang, S.F.; Zhu, J.C.; Wang, D.J. *Brassica napus* COL transcription factor *BnCOL2* negatively affects the tolerance of transgenic *Arabidopsis* to drought stress. *Environ. Exp. Bot.* **2020**, *178*, 104171. [CrossRef]
59. Mita, S.; Hirano, H.; Murano, N.; Nakamura, K. Genetic analysis on sugar-induced expression of a gene for β-amylase in *Arabidopsis thaliana*. *Plant Cell Physiol.* **1995**, *36*, S83.
60. Bang, C.B.; Huyen, T.T.T. Analysis of β-amylase gene family in soybean (*Glycine max*). *Tap. Chi. Sinh. Hoc.* **2015**, *37*, 165–176. [CrossRef]
61. Miao, H.X.; Sun, P.G.; Miao, Y.L.; Liu, J.H.; Zhang, J.B.; Jia, C.H.; Wang, J.K.; Wang, Z.; Jin, Z.Q.; Xu, B.Y. Genome-wide identification and expression analysis of the β-amylase genes strongly associated with fruit development, ripening, and abiotic stress response in two banana cultivars. *Front. Agric. Sci. Eng.* **2016**, *3*, 346–356. [CrossRef]
62. Mason-Gamer, R.J. The beta-amylase genes of grasses and a phylogenetic analysis of the Triticeae (*Poaceae*). *Am. J. Bot.* **2005**, *92*, 1045–1058. [CrossRef] [PubMed]
63. Bowers, J.E.; Chapman, B.A.; Rong, J.; Paterson, A.H. Unravelling angiosperm genome evolution by phylogenetic analysis of chromosomal duplication events. *Nature* **2003**, *422*, 433–438. [CrossRef] [PubMed]

64. Zhu, J.K. Abiotic Stress Signaling and Responses in Plants. *Cell* **2016**, *167*, 313–324. [CrossRef] [PubMed]
65. Guo, X.Y.; Liu, D.F.; Chong, K. Cold signaling in plants: Insights into mechanisms and regulation. *J. Integr. Plant Biol.* **2018**, *60*, 745–756. [CrossRef]
66. Balibrea, M.E.; Amico, J.D.; Bolarin, M.C.; Perez-Alfocea, F. Carbon partitioning and sucrose metabolism in tomato plants growing under salinity. *Physiol. Plant.* **2000**, *110*, 503–511. [CrossRef]
67. Kerepesi, I.; Galiba, G. Osmotic and salt stress-induced alteration in soluble carbohydrate content in wheat seedlings. *Crop Sci.* **2000**, *40*, 482–487. [CrossRef]
68. Liu, T.; Staden, J.V. Partitioning of carbohydrates in salt-sensitive and salt-tolerant soybean callus cultures under salinity stress and its subsequent relief. *Plant Growth Regul.* **2001**, *33*, 13–17. [CrossRef]
69. Brumos, J.; Colmenero-Flores, J.M.; Conesa, A.; Izquierdo, P.; Sanchez, G.; Iglesias, D.J.; Lopez-Climent, M.F.; Gomez-Cadenas, A.; Talon, M. Membrane transporters and carbon metabolism implicated in chloride homeostasis differentiate salt stress responses in tolerant and sensitive Citrus rootstocks. *Funct. Integr. Genom.* **2009**, *9*, 293–309. [CrossRef]
70. Gonzalez-Cruz, J.; Pastenes, C. Water-stress-induced thermotolerance of photosynthesis in bean (*Phaseolus vulgaris* L.) plants: The possible involvement of lipid composition and xanthophyll cycle pigments. *Environ. Exp. Bot.* **2012**, *77*, 127–140. [CrossRef]
71. Goyal, A. Osmoregulation in *Dunaliella*, Part II: Photosynthesis and starch contribute carbon for glycerol synthesis during a salt stress in *Dunaliella tertiolecta*. *Plant Physiol. Biochem.* **2007**, *45*, 705–710. [CrossRef]
72. Horrer, D.; Flütsch, S.; Pazmino, D.; Matthews, J.S.A.; Thalmann, M.; Nigro, A.; Leonhardt, N.; Lawson, T.; Santelia, D. Blue Light Induces a Distinct Starch Degradation Pathway in Guard Cells for Stomatal Opening. *Curr. Biol.* **2016**, *26*, 362–370. [CrossRef] [PubMed]

4

Performance and Stability of Commercial Wheat Cultivars under Terminal Heat Stress

Ibrahim S. Elbasyoni

Crop Science Department, Faculty of Agriculture, Damanhour University, Damanhour 22516, Egypt; ibrahim.salah@agr.dmu.edu.eg or ielbasyoni2@unl.edu

Abstract: Egypt, the fifteenth most populated country and the largest wheat importer worldwide, is vulnerable to global warming. Ten of the commercial and widely grown wheat cultivars were planted in two locations, i.e., Elbostan and Elkhazan for three successive seasons 2014/2015, 2015/2016, and 2016/2017 under two sowing dates (recommended and late). Elbostan and Elkhazan are the two locations used in this study because they represent newly reclaimed sandy soil and the Nile delta soil (clay), respectively. A split-plot, with main plots arranged as a randomized complete block design and three replicates, was used. The overall objective of this study was to identify the ideal cultivar for recommended conditions and heat stressed conditions. The results revealed that heat stress had a significant adverse impact on all traits while it raised the prevalence and severity of leaf and stem rust which contributed to overall yield losses of about 40%. Stability measurements, the additive main effects and multiplicative interaction model (AMMI) and genotype main effect plus genotype × environment interaction (GGE), were useful to determine the ideal genotypes for recommended and late sowing conditions (heat stressed). However, inconsistency was observed among some of these measurements. Cultivar "Sids12" was stable and outperformed other tested cultivars under combined sowing dates across environments. However, cultivar "Gemmeiza9" was more stable and outperformed other cultivars across environments under the recommended sowing date. Moreover, cultivar "Gemmeiza12" was the ideal cultivar for the late sown condition. Based on our findings, importing and evaluating heat stress tolerant wheat genotypes under late sown conditions or heat stressed conditions in Egypt is required to boost heat stress tolerance in the adapted wheat cultivars.

Keywords: wheat; heat stress; stability

---

1. Introduction

Wheat (*Triticum aestivum* L.) accounts for 30% of the cereal grains production while providing 55% of the carbohydrates and 20% of the food calories consumed globally [1]. Thus, wheat is considered a strategic cereal crop for several countries around the globe including Egypt, in which wheat production became one of the crucial elements of food security [2]. Furthermore, wheat is cultivated and grown in a wide range of environmental and climatic conditions [3]. Therefore, the impact of climate change is expected to affect wheat production in several regions around the globe. The Mediterranean basin is one of the regions (hot spots) that is expected to have an annual mean temperature increase of 3 to 4 °C, which might lead to total grain yield reduction of about 18% to 24% [4]. Heat stress was defined as the rise in temperature for a period and beyond the point that causes irreversible damage to the plant growth and development [5].

The impact of heat stress on several aspects of wheat phenology and physiology during the reproductive stage was studied by several researchers, in which they reported that heat stress trigger senescence-related metabolic changes in wheat [6]. Heat stress decreases photosynthesis as a result of photosystem II (PSII) inhibition [7–10]. Moreover, during the reproductive stage, heat stress decreased the grain-filling duration significantly [11], increased floral abortion, and decreased the number of seeds [12–14]. Nevertheless, heat stress significantly increased protein concentration, but with lower end-use quality because the functionality of protein was reduced by the high temperature [15]. The increased protein concentration under terminal heat stress could be due to upregulation of heat shock proteins (HSPs) which is a plant mechanism to alleviate the effect of heat stress [16]. In wheat, 6560 probe sets for HSPs displayed expression upregulation under heat stress treatment of 34 °C and 40 °C [16].

Conceptually, the phenotype (P) of any plant is a result of the genotype (G), the environment (E), and the genotype–environment interaction (G × E). Based on this concept, one can cope with the negative impact of the heat stress either by altering the environment or using heat stress tolerant genotypes. Exploring highly adapted local genotypes to identify heat stress tolerant wheat genotypes is the first step to start a wheat breeding program for heat stress [5,17,18]. Whereas, if the breeder was able to find high yielding and stress tolerant genotype, then years of evaluation and crosses to develop such a cultivar can be saved. Plant breeders always look for genotypes that perform better across environments with minimal G × E interaction, but that seldom occurs, especially under the dynamic weather conditions and the fluctuations in the environmental conditions from year to year and location to location. Thus, measuring the stability of a given cultivar is an essential criterion before releasing new cultivars. Selecting superior genotypes using stability measurements instead of average performance is highly recommended because genotypes selected using stability measures are more reliable across environments with a minimized G × E interaction, or the provide a predictable response across environments. Studies have shown that stability analyses according to various measures can result in better identification of stable genotypes, even when there were no interactions among the measures [19]. Stability and G × E measurements can be classified into parametric and non-parametric measurements [20]. The most frequently utilized two parametric stability methods are partitioning of G × E interaction [21] and the regression model [22]. The additive main effects and multiplicative interaction model (AMMI) [23] and the genotype main effect plus G × E interaction (GGE) [24] are the most frequently utilized non-parametric methods. Both AMMI and GGE biplot analyses are based on the principal component analysis (PCA). However, GGE biplot is based on environment-centered principal component (PCA), whereas AMMI analysis is a double centered PCA method [25].

In the current study, ten wheat cultivars that represent around 60% of the commercially grown wheat cultivars in Egypt (based on 2017/2018 grown wheat cultivars) were evaluated under recommended and late sowing conditions (heat stressed). These cultivars were used to study the effect of heat stress (late sown) on several physiological and morphological traits. Egypt, the fifteenth most populated country worldwide, is one of the Mediterranean basin countries that is vulnerable to the effects of global warming [26]. In addition to the impact of the global warming, the Egyptian population is projected to be 125,870,736 inhabitants in 2030, which will require the production of more wheat grain. The recommended sowing date for wheat in Egypt is around mid-November. However, due to limited availability of land and water, several growers in Egypt tend to grow wheat after sugar beet (*Beta vulgaris* L.), carrot (*Daucus carota* L.), or pea (*Pisum sativum* L.), which results in sowing wheat around Mid-January. These later sowing dates expose wheat plants to terminal heat stress during the reproductive stage.

Therefore, evaluating wheat cultivars for heat tolerance has environmental and socioeconomic importance in this region of the world. Although screening for heat stress tolerance in wheat has been done in the past, limited research was conducted to study the effect of heat stress on the grain yield stability or wheat resistance to leaf and stem rust using commercial newly developed wheat cultivars under clay and sandy soils across three successive growing seasons. Thus, the primary objectives of this study were to; 1—evaluate ten of the commercially and widely distributed wheat cultivars under heat stressed and recommended (control) conditions across several environments.

2—Study the relationships among relevant phenological and physiological traits under normal and heat stress conditions.

2. Materials and Methods

2.1. Plant Materials and Field Conditions

Ten of the commercially widely distributed Egyptian wheat cultivars were used in this study (Table 1). The studied materials were planted in three successive growing seasons, i.e., 2014/2015, 2015/2016, and 2016/2017, in two locations; Elkhazan (31°05′35.2″ N, 30°30′10.4″ E) and Elbostan (30°45′19.4″ N, 30°29′04.8″ E), Behera governorate, Egypt. In each growing season for both locations, two sowing dates were used, i.e., recommended and late sown dates. The recommended sowing dates were 19, 12, 14 of November for 2014/2015, 2015/2016, and 2016/2017, respectively. While the late-sown dates were 7, 9, 13 of January, for 2014/2015, 2015/2016, and 2016/2017, respectively. A Split-plot, with main plots arranged as a randomized complete block design and three blocks, was used. The two sowing dates were randomly assigned to the main plots within each of the three blocks. The ten wheat cultivars were assigned randomly to the subplots within each main plot (sowing date). The experimental units (plots) were two meters long and four rows wide by 25 cm between rows. The surrounding border of the experimental areas of one meter wide, planted with the wheat cultivar "Morocco", a "spreader" cultivar so named because it is susceptible to currently prevalent races of leaf and stem rust. Standard agronomic practices including recommended fertilization and irrigation schedules were followed.

Table 1. Name, pedigree, and year of release of the ten wheat cultivars used in this study.

Cultivar	Pedigree	Year of Release
Giza168	MRL/BUC//SERI	1995
Gemmeiza7	7CMH74A-630/SX//SERI82/AGEN	1999
Gemmeiza9	ALD"s"/HUAC//CMH74A-630/SX	1999
Gemmeiza10	Maya74"S"/ON/1160-147/3/Bb/G11/4/chat"S"/5/crow"S"CGM5820-3GM-1GM-2GM-0GM	2004
Gemmeiza11	BOW "S"/KVZ "S"//7C/SERI82/3/GIZA168/SKHA61	2011
Gemmeiza12	OTUS/3/SARA/THB//VEECMSS97Y00227S-5Y-	2011
Sakha94	OPATA/RAYON//KAUZ	2004
Sids12	BUC//7C/ALD/5/MAYA74/ON//1160-147/3/BB/GLL/4/CHAT"S"/6/MAYA/VUL-4SD-1SD-1SD-0SD.	2007
Sids13	KAUZ "S"//TSI/SNB"S". ICW94-0375-4AP-2AP-030AP-0APS-3AP-0APS-050AP-0AP-0SD.	2010
Misr2	SKAUZ/BAV92. CMSS96M0361S-1M-010SY-010M-010SY-8M-0Y-0S.	2011

2.2. Phenotypic Measurements

The number of days to flowering date was recorded visually as the number of days to anther exertion from 50% of the main spikes (days). Leaf area (LA) was estimated on three samples according to the following equation [25]:

$$\text{Leaf area (LA)} = L \times W \times 0.75 \quad (1)$$

where L and W are the length and width, respectively, of the flag leaf.

Plant height was measured on a random sample of five plants in each plot as the distance from the soil surface to the tip of the spike awns excluded at harvest time (cm). Grain yield was measured by harvesting the four rows of each plot (tons/ha). Total leaf chlorophyll content (SPAD index) estimated using spad-502 chlorophyll meter (spad-502 plus, Konica Minolta, Kearney, NE, USA), during the flowering stage. Canopy temperatures (Tc) were measured using a handheld infrared thermometer (KM 843, Comark Ltd., Hertfordshire, UK) with a field view of 100 mm to 1000 mm. Canopy temperatures (Tc) data were taken from the same side of each plot at 1m distance from the edge and approximately 50 cm above the canopy at an angle of 30° to the horizontal. Readings were made between 1300 and 1500 h on sunny days. Grain filling duration (GFD) was measured from flowering to physiological maturity (when the peduncle changed color). Leaf and stem rust screening under recommended and late sown conditions were conducted using the modified Cobb's scale described by Peterson et al. [26]. The infection type was expressed in the following types, i.e., Immune = 0,

R = resistant, small uredinia surrounded by necrosis; MR = Moderately resistant, medium to large uredinia surrounded by necrosis; MS = moderately susceptible, medium to large uredinia surrounded by chlorosis; S = susceptible, large uredinia without necrosis or chlorosis [27]. The statistical analysis was conducted on the infection type after replacing the infection types with 0, 0.2, 0.4, 0.80, and 1 scores for immune, resistant, moderately resistant, moderately susceptible, and susceptible, respectively.

2.3. Statistical Analysis

Analysis of variance was carried out using SAS 9.2 (SAS v9.2; SAS Institute Inc., Cary, NC, USA), by fitting the following linear model [27]:

$$Y_{ijlm} = \mu + E_i + EB_{(i)j} + T_l + ET_{il} + TBE_{(i)jl} + G_m + EG_{im} + ETG_{ilm} + \varepsilon_{ijlm} \quad (2)$$

where Y_{ijlm} is the response measured on the $_{ijlm}$ plot, μ is the overall mean, E_i is the effect of ith environment (three seasons and two locations which compose six environments), $EB_{(i)j}$ is j^{th} block nested within i^{th} environment, T_l is the effect of the lth sowing date, ET_{il} is the interaction between i^{th} environment and l^{th} sowing date, $TBE_{(i)jl}$ *is* interaction between l^{th} sowing date and j^{th} replicates within i^{th} environment as an error term for environment, sowing date, and environment × sowing date. G_m is the effect of mth cultivar, ETG_{ilm} is the interaction effect among i^{th} environment, l^{th} sowing date, and G^{th} cultivar, and ε_{ijlm} is the experimental error.

Means were compared using the least significance difference (*LSD*) test at p-value < 0.05, according to Gomez and Gomez [28]. Homogeneity of the variance in different environments was tested following Bartlett's Test [29]. Combined analyses of variance were performed among the different environments with homogeneous variance, as outlined by Cochran and Cox [30]. Correlation coefficients were conducted using Pearson correlation coefficient.

2.4. Stability Analysis and Genotype × Interaction (G × E)

The following stability measurements were performed on grain yield under the six environments (three seasons and two locations which compose six environments), i.e., coefficient of variability (CV_i) [31], regression coefficient (b_i) [22], Wricke's ecovalance (W_i) [32], superiority measure (P_i) [21], Perkins and Jinks (D_i) [33], and average absolute rank difference of genotype on the environment ($S_i(1)$) [24] . Moreover, the additive main effects and multiplicative interaction model (AMMI) [23] was applied on the grain yield variable for each sowing date separately, and after combing them. Then the genotype main effect plus G × E interaction (GGE biplot) [24] was used to visualize the G × E interaction. The stability and G × E analysis was conducted using R (software) package GEA-R (Version 4.0, 2017, CIMMYT, El Batán, Mexico) [34].

3. Results

3.1. Analysis of Variance

The analysis of variance for total chlorophyll content, canopy temperature, leaf area, grain filling duration, plant height, grain yield, leaf rust scores, stem rust scores, and the number of days to flowering are presented in Table 2. The results indicated highly significant effect (p-value < 0.01) for the six environments (three growing seasons and two locations which compose six environments) on all traits except for leaf rust scores and stem rust scores, in which the environmental effect was not found to be statistically significant. Furthermore, sowing dates (Sd) had a highly significant effect on all traits. More importantly, the analysis of variance revealed highly significant variance among the studied cultivars. Our results suggested that the magnitude of differences among cultivars was sufficient to provide a scope to characterize the effect of terminal heat stress (late sown condition). All traits except leaf and stem rust scores, had significant two-way and three-way interactions (environments × cultivars, environments × sowing dates, sowing dates × cultivars and environments × sowing dates × cultivars) effects. As for leaf and stem rust scores, the interaction effect of environments ×

cultivars and sowing dates × cultivars found to be highly significant, while the interaction effect of environments × sowing dates × cultivars was not statistically significant.

Table 2. Analysis of variance for total chlorophyll content (CHLOR), canopy temperature (CANO), leaf area (LA), grain filling duration (GFD), plant height (PH), grain yield (YIELD), leaf rust (LR), stem rust (SR), and number of days to flowering (NDF) under two sowing date (SD) and different environments (ENV).

SOURCE	DF	Mean Squares								
		CHLOR	CANO	LA	GFD	PH	YIELD	LR	SR	NDF
ENV	5	72.6 **	1316.1 **	212.7 **	492.5 **	785.5 **	34.02 **	0.025	0.016	2263.5 **
Replication(ENV)	12	0.60	2.27	1.44	3.36	37.73	0.10	0.017	0.017	67.00 **
SD	1	1055.4 **	61871.3 **	11969.29 **	6002.5 **	63374.9 **	836.1 **	14.88 **	16.72 **	85069.9 **
ENV * SD	5	119.0 **	463.2 **	115.33 **	16.3 *	393.3 **	13.0 **	0.02	0.036	360.9 **
Main plot Error	12	0.59	1.36	2.10	6.83	37.55	0.12	0.042	0.011	68.9
CULTIVAR	9	15.1 **	182.5 **	142.8 **	116.7 **	1141.2 **	6.6 **	1.46 **	1.77 **	3500.5 **
SD * CULTIVAR	9	6.4 **	84.3 **	20.2 **	43.8 **	761.1 **	2.9 **	1.47 **	0.89 **	2003.76 **
ENV * CULTIVAR	45	9.08 **	109.7 **	53.4 **	38.04 **	371.1 **	2.2 **	0.02	0.035 *	600.3 **
ENV * SD * CULTIVAR	45	6.8 **	193.18 **	28.59 **	44.9 **	423.2 **	2.8 **	0.019	0.031	682.9 **
ERROR	216	0.14	2.71	0.69	1.4	14.95	0.1	0.019	0.02	22.2

*, **: Significant at the 0.05 and 0.01 probability levels, respectively. SD: Sowing date; ENV: Environment.

The late sown condition had a significant adverse effect on the total chlorophyll content, leaf area, grain filling duration, plant height, and grain yield. The average of the total chlorophyll content measured for all cultivars under recommended and late sown condition was 31.9 and 28.5, respectively. Moreover, the late sown condition decreased leaf area from 35.15 to 23.5 cm^2. In the same manner, the late sown condition shortened the grain filling duration from 32.5 days to 25.5 days. The late sown condition had an adverse effect on plant height in which mean plant height across cultivars was dropped from 83.00 cm to 56.77 cm. Moreover, the late sown condition increased canopy temperature from 32.76 °C to 56.16 °C. Furthermore, the late sown condition decreased overall resistance to leaf rust (leaf rust scores increased from 0.34 to 0.73) and stem rust (stem rust scores increased from 0.36 to 0.79).

Results in Table 3 illustrate the effect of sowing date on the response of the studied cultivars for the total chlorophyll content, canopy temperature, leaf area, grain filling duration, plant height, grain yield, and leaf and stem rust scores across environments. Cultivar "Giza168" exhibited the tallest plants (95.8 and 64.9 cm), while the shortest plants were for cultivar "Gemmeiza10" (65.7 and 52.2 cm) obtained from recommended and late sown conditions, respectively. Furthermore, cultivar "Gemmeiza12" produced the highest grain yield (8.8 ton/hectare) under the recommended sowing date. While cultivar "Gemmeiza9" produced the highest grain yield (4.87 ton/hectare) under the late sown condition. Results in Table 3 indicated that cultivars Gemmeiza10 and Gemmeiza12 flowered earlier than other cultivars under the recommended and late sown conditions. Recommended sowing date extended the number of days to flowering for all cultivars across environments. Results of the adult plant resistance to leaf and stem rust indicated a negative impact of late sowing date on wheat resistance to both stem and leaf rust. Among the ten wheat cultivars tested based on the infection type (IT) and under the recommended sowing date, five cultivars were resistant to leaf rust, i.e., Gemmeiza9, Gemmeiza10, Gemmeiza12, Giza168, and Sids13. Furthermore, Sids12, Gemmeiza7, and Gemmeiza11 were moderately resistant to leaf rust. Whereas, Sakha94 was susceptible to leaf rust under the recommended sowing date. Under late sowing date, Gemmeiza9, Gemmeiza11, and Sids12 showed moderate resistance (MR) to leaf rust across all environments, while the rest of the cultivars were moderately susceptible or susceptible to leaf rust. Moreover, the stem rust results under recommended sowing date indicated that five wheat cultivars were stem rust resistant, i.e., Sids12, Sakha94, Gemmeiza10, Gemmeiza11, and Gemmeiza12, across all environments. Furthermore, Gemmeiza7, Gemmeiza9, and Misr2 were moderately resistant to stem rust, but Giza168 and Sids13 were susceptible. Nevertheless, stem rust results obtained from the late sowing date indicated that Sids12, Gemmeiza9, and Gemmeiza11 were moderately resistant, but Gemmeiza7, Gemmeiza10, Gemmeiza 12, Misr2, Sakha94, Giza168, and Sids13 were susceptible.

Table 3. Response of the ten wheat cultivars to recommended (R) and late sowing (L) dates for total chlorophyll content (CHLOR), canopy temperature (CANO), leaf area (LA), grain filling duration (GFD), plant height(PH), grain yield (**GY**), leaf rust scores (LR), stem rust scores(SR), and number of days to flowering (NDF).

	PH		GY		NDF		LR		SR		CHLOR		CANO		LA		GFD	
Sowing date	**R**	**L**	**R**	**L**	**R**	**L**	**R**	**L**	**R**	**L**	**R**	**L**	**R**	**L**	**R**	**L**	**R**	**L**
Sids12	83.5	62.5	8.08	4.27	139.5	112.1	0.40 (40.7)	0.36 (49.9)	0.09 (37.52)	0.46 (53.2)	32.6	29.0	33.9	46.9	36.1	24.6	31.32	28.9
Gemmeiza10	65.7	52.2	7.65	3.12	116.2	103.4	0.06 (40.5)	0.85 (57.1)	0.14 (32.6)	0.90 (46.8)	31.3	28.0	32.2	44.4	33.8	22.4	32.2	27.9
Gemmeiza7	88.3	58.3	4.67	4.15	132.5	124.4	0.41 (30.4)	0.94 (54.1)	0.40 (39.2)	0.95 (43.2)	32.6	28.6	36.6	42.6	38.36	25.2	34.8	25.25
Gemmeiza9	89.3	58.3	7.6	4.87	133.7	119.8	0.11 (38.8)	0.41 (68.1)	0.40 (32.8)	0.46 (55.5)	32.4	28.5	37.1	44.1	38.5	25.1	34.5	29.3
Gemmeiza11	88.4	52.3	5.0	2.9	126	109.6	0.32 (20.6)	0.36 (43.9)	0.16 (26.3)	0.41 (54.6)	30.7	28.4	34.3	41.9	34.4	24.6	30.2	24.2
Gemmeiza12	76.3	60.1	8.8	3.12	117.5	95.2	0.05 (36.6)	0.84 (47.7)	0.16 (26.3)	0.93 (54.6)	30.8	28.3	33.1	47.6	33.4	23.1	33.3	22.3
Misr2	90.6	61.5	5.0	3.4	128.4	110.7	0.78 (38.4)	0.83 (56.7)	0.35 (38.0)	0.91 (55.1)	31.3	28.7	32.4	60.4	36.25	24.2	34.9	25.3
Sakha94	75.5	63.9	5.0	3.3	127.0	121.4	0.94 (46.5)	0.82 (44.3)	0.08 (39.4)	0.95 (56.8)	30.8	27.8	31.1	47.5	29.9	20.1	28.3	21.8
Giza168	95.8	64.9	5.4	4.0	121.8	114.0	0.06 (39.4)	0.93 (59.8)	0.91 (32.9)	0.97 (45.5)	33.5	28.5	29.6	42.5	34.5	22.9	32.5	25.9
Sids13	79.9	58.3	5.0	3.7	126.7	110.3	0.11 (29.7)	0.92 (52.2)	0.88 (45.6)	0.85 (52.5)	32.7	29.5	33.5	48.7	36.5	23.2	32.7	24.2
Lsd	6.0		0.5		1.8		0.071		0.065		0.6		2.6		1.3		1.8	
Sowing date mean	83.3	59.2	6.2	3.7	126.9	112.1	0.32	0.73	0.36	0.78	31.9	28.5	33.4	46.7	35.2	23.5	32.5	25.5

Lsd: least significant difference for the interaction between sowing dates and genotypes at $\alpha = 0.05$. Numbers in brackets refer to the severity of leaf and stem rust diseases at the adult plant stage.

Even though wheat cultivars showed highly significant variance across environments for total chlorophyll content, the range of difference among values was rather narrow (Table 3). Cultivar Giza168 exhibited the highest values for total chlorophyll content (33.5 SPAD units) while the lowest values were for cultivar "Gemmeiza11" (30.7 SPAD) obtained from the recommended sowing date. Furthermore, under the late sown condition, cultivar "Sids13" has the highest total chlorophyll content value (29.5 SPAD). Besides, canopy temperature measurement indicated higher values on all cultivars under the recommended sowing date compared to the late sown condition. As shown in Table 3, cultivar "Gemmeiza9" had the highest canopy temperature (37.1) under the recommended sowing date. Cultivar "Misr2" had the highest canopy temperature (60.4) under the late sown condition. Significant reduction in leaf area due to the late sown condition was also detected. Leaf areas under the recommended sowing date ranged from 38.6 to 29.9 cm^2 for cultivars Gemmeiza7 and Sakha94, respectively. However, leaf area ranges from 25.2 to 20.1 cm^2 for the same cultivars under late sown condition. Under the recommended sowing date, grain filling duration ranged from 34.8 to 28.3 days for cultivars Gemmeiza7 and Sakha94, respectively. Furthermore, under the recommended sown date, cultivar "Misr2" had the most extended grain filling duration (34.9 days). Nevertheless, under the late sown condition cultivar, Gemmeiza9 had the most extended grain filling duration (29.3 days).

3.2. *Interrelationships among the Studied Traits under Recommended Sown Condition*

Pearson correlation coefficients among the studied traits under the recommended sowing date (normal) are presented in Table 4 (above diagonal). Results in Table 4 indicated significant positive correlation (p-value < 0.05) among total chlorophyll content, leaf area, and plant height. Significant negative correlation was detected for the relationship among total chlorophyll content, the number of days to flowering and leaf rust. However, non-significant correlation (p-value > 0.05) of total chlorophyll content with canopy temperature, grain filling duration, grain yield, and stem rust was detected (Table 4, above diagonal). There was a significant positive correlation between canopy temperature and leaf area. Furthermore, canopy temperature was negatively correlated with the number of days to flowering. A non-significant correlation was detected for the canopy temperature with grain filling duration, plant height, grain yield, and leaf and stem rust scores. Correlations of leaf area with grain filling duration, plant height, and grain yield were significant and positive. Whereas, the correlation of leaf area with stem rust and number of days to flowering were significant but negative. Grain filling duration was positive and significantly correlated with grain yield, but it was negative and significantly correlated with stem rust and the number of days to flowering. No significance was detected for the correlation among grain filling duration, plant height, and leaf rust. Plant height was significantly and positively correlated with grain yield but significantly and negatively correlated with both stem rust and number of days to flowering. Additionally, non-significant correlation was detected for plant height with stem and leaf rust scores. Grain yield was significantly and negatively correlated with the number of days to flowering and stem rust, but it was not significantly correlated with leaf rust scores. The correlation between leaf rust and number of days to flowering was positive and significant. The correlation between stem and leaf rust was significant. Finally, the correlation between stem rust and number of days to flowering was not significant.

3.3. *Interrelationships among the Studied Traits under the Late Sown Condition*

The Pearson correlations coefficients among the studied traits under the late sown condition are presented in Table 4 (below diagonal). The correlation of total chlorophyll content with leaf area, grain filling duration, plant height, and grain yield were significant and positive, but it was not significantly correlated with canopy temperature, leaf rust, stem rust, and the number of days to flowering. Canopy temperature was significantly and positively correlated with leaf and stem rust scores but negatively correlated with grain filling duration, grain yield, and the number of days to flowering. Non-significant correlation for canopy temperature with plant height and leaf area was detected. Leaf area was significantly and negatively correlated with leaf and stem rust, but it was significantly and positively correlated with grain filling duration, plant height, and grain yield. Furthermore, Leaf area was not significantly correlated with the number of days to flowering. Grain filling duration was significant and positively correlated with plant height and grain yield, but significant and negatively correlated with leaf and stem rust scores. However, non-significant correlation for grain filling duration with the number of days to flowering was detected. The correlation of plant height with grain yield was significant and positive. Plant height was significantly and negatively correlated with leaf rust scores, stem rust scores, and the number of days to flowering. Grain yield was significantly and negatively correlated with stem and leaf rust, but non-significant correlation was detected between grain yield and number of days to flowering. The correlation between leaf rust and stem rust scores was significant and positive. Furthermore, leaf rust was significant and negatively correlated with the number of days to flowering. Nevertheless, non-significant correlation between stem rust and the number of days to flowering was detected.

Table 4. Pearson correlation coefficients among total chlorophyll content (CHLOR), canopy temperature (CANO), leaf area (LA), grain filling duration (GFD), plant height (PH), grain yield (YIELD), leaf rust (LR), stem rust (SR) scores, and the number of days to flowering (NDF) under recommended sowing date (above diagonal), and correlation coefficients among same traits under late sown condition (below diagonal).

	CHLOR	CANO	LA	GFD	PH	YIELD	LR	SR	NDF
CHLOR		−0.17	0.55 **	0.27	0.43 *	0.07	−0.33 *	0.07	−0.47 **
CANO	0.00		0.44 **	−0.11	0.10	−0.20	−0.05	−0.25	−0.31 *
LA	0.42 *	−0.21		0.61 **	0.50 **	0.43 *	0.06	−0.63 **	−0.79 **
GFD	0.40 *	−0.57 **	0.56 **		0.31	0.60 **	0.17	−0.69 **	−0.56 **
PH	0.55 **	−0.08	0.60 **	0.39 *		0.45 *	0.01	−0.06	−0.90 **
YIELD	0.39 *	−0.55 **	0.55 **	0.77 **	0.43 *		0.01	−0.67 **	−0.59 **
LR	−0.33	0.37 *	−0.71 **	−0.81 **	−0.53 **	−0.82 **		−0.38	0.42 *
SR	−0.28	0.41 *	−0.63 **	−0.84 **	−0.42 *	−0.85 **	0.97 **		−0.06
NDF	−0.30	−0.80 **	−0.30	0.25	−0.46 **	0.23	−0.53 **	−0.1	

*, **: Significant at the 0.05, 0.01 probability levels, respectively.

3.4. Genotype × Environment Interaction (G × E) for Grain Yield

Grain yield is a quantitative and complex trait that was found to be responsive to genotype by environment interaction (G × E). Additionally, grain yield is the most critical parameter that determines a cultivar's acceptance by growers. Thus, in this part of the study, we performed stability analysis on the grain yield. The results of the stability parameters used in this study are presented in Table 5. Under the combined sowing dates (12 environments) model, i.e., two sowing dates, two locations and three years, Gemmeiza9 was the most stable cultivar in several measurements such as coefficient of variation (C.V%), Superiority measure (P_i) and Wrike's ecovalence (W_i).Gemmeiza10 was the most stable cultivar for other measurements such as Regression coefficient (b_i), and Perkins and Jinks (D_i). However, Gemmeiza12 was the most stable genotype for the average absolute rank difference of genotype on the environment ($S_i(1)$). Moreover, under the recommended sowing date, i.e., two locations and three years (six environments), coefficient of variation (C.V%), and Superiority measure (P_i) identified Sids12 to be the most stable cultivar. Wrike's ecovalence (W_i) and average absolute rank difference of genotype on environment ($S_i(1)$) identified Gemmeiza12 to be the most stable cultivar. Moreover, regression coefficient (b_i) and Perkins and Jinks (D_i) identified Gemmeiza10 to be the most stable cultivar. Furthermore, under the late sown condition, coefficient of variation (C.V%), Wrike's ecovalence (W_i), Superiority measure (P_i), and the average absolute rank difference of genotype on the environment ($S_i(1)$) identified Sids12 to be the most stable genotype. Furthermore, Giza168 was the most stable genotype under regression coefficient (b_i) and Perkins and Jinks (D_i).

Table 5. Stability parameters estimates for grain yield and ten cultivars, under combined, early, and late sown conditions.

Sowing Date	Genotypes	Coefficient of Variation C.V%	Regression Coefficient (b_i)	Perkins and Jinks (D_i)	Wrike's Ecovalence (W_i)	Superiority Measure (P_i)	Average Absolute Rank ($S_i(1)$)
Recommended & late combined	Sids12	49.03	1.60	0.60	28.42	0.78	0.82
	Gemmeiza10	32.43	0.60	−0.40	12.53	6.35	0.32
	Gemmeiza7	31.70	0.88	−0.12	10.38	2.29	0.61
	Giza168	40.77	1.10	0.10	8.82	2.66	0.41
	Gemmeiza11	47.59	1.46	0.46	13.67	1.34	0.68
	Gemmeiza12	32.28	0.61	−0.39	12.83	5.75	0.21
	Misr2	33.89	0.75	−0.25	8.11	4.58	0.45
	Sakha94	47.22	1.11	0.11	18.29	3.56	0.52
	Gemmeiza9	30.03	1.06	0.06	3.91	0.54	0.24
	Sids13	35.91	0.83	−0.17	7.49	4.39	0.48
Recommended	Sids12	6.14	0.32	−0.68	4.15	0.16	0.53
	Misr2	31.73	1.47	0.47	2.52	7.75	0.53
	Gemmeiza7	24.62	0.62	−0.38	5.79	11.01	0.47
	Gemmeiza9	31.65	1.73	0.73	6.18	4.50	1.00
	Gemmeiza11	30.89	1.54	0.54	7.61	4.44	1.20
	Gemmeiza12	13.53	0.86	−0.14	1.09	0.91	0.33
	Gemmeiza10	19.67	0.18	−0.82	9.98	9.74	0.60
	Sakha94	23.90	0.83	−0.17	4.52	7.84	1.00
	Giza168	34.97	1.47	0.47	13.11	5.37	1.47
	Sids13	17.07	0.98	−0.02	3.31	1.08	0.60
late	Sids13	24.03	2.32	1.32	1.30	1.40	0.67
	Gemmeiza9	22.99	2.48	1.48	1.35	1.03	0.73
	Gemmeiza7	25.24	2.26	1.26	1.36	1.69	0.67
	Gemmeiza10	6.74	−0.33	−1.33	1.47	0.09	0.60
	Gemmeiza11	8.37	−0.12	−1.12	1.21	0.88	0.87
	Gemmeiza12	30.57	2.56	1.56	2.74	1.77	0.73
	Misr2	11.30	0.50	−0.51	0.63	1.77	0.80
	Sakha94	17.77	0.68	−0.32	1.65	1.32	0.83
	Giza168	24.80	−0.84	−1.84	5.09	1.75	0.97
	Sids12	5.27	0.49	−0.51	0.36	0.00	0.07

The results of stability measures used in the current study indicated inconsistency among some of the stability measures used. Thus, to complement the results of the previous stability measures, Genotype by environment (G × E) was further investigated using the additive main effect and multiplicative interaction (AMMI) analysis (Table 6) and genotype main effect plus genotype × environment interaction (GGE). Additionally, two models were fitted in the AMMI and GGE biplot; the first was by considering sowing dates as part of the environments, i.e., 12 environments, while the second was by running the AMMI analysis across years and locations within each sowing date, i.e., six environments. In the first model (12 environments), the analysis of variance for AMMI model indicated significant effect of the environments, genotypes, and genotype × environment interaction. Whereas the variance of the environment was 63.2%, while the variance due to genotypes was 14.6% and that for genotype × environment interaction was 22.2%.

In the second model (6 environments within each sowing date), the variance of the AMMI model for the recommended sowing date was 29.52%, 46%, and 24.48% for the environment, genotypes, and genotypes–environment interaction, respectively. Moreover, the variance of the AMMI model for the late sown condition was 12.99%, 48.57%, and 38.44% for the environment, genotypes, and genotypes–environment interaction, respectively. In both models, the genotype–environment interaction was highly significant (p-value < 0.01) implying differential response of genotypes to environments. Substantial variance for the environment in the first model compare to the second model was detected, which indicates an amplification effect of sowing dates on the environmental effect.

Table 6. Summary of the analysis of variance and partitioning of the G × E interaction by , the additive main effects and multiplicative interaction model (AMMI) for grain yield.

Sowing Date	Source	DF	MS	% of Variance Explained
Recommend and late combined	Environments (E)	11	96.80 **	63.20
	Genotypes (G)	9	27.40 **	14.60
	G × E	99	3.80 **	22.20
	PC1	19	8.90 **	45.30
	PC2	17	6.00 **	27.10
	PC3	15	3.70 **	14.80
	Residuals	240	0.10	
Recommend	Environments (E)	5	42.15 **	29.52
	Genotypes (G)	9	36.50 **	46.00
	G × E	45	3.88 **	24.48
	PC1	13	6.74 **	50.13
	PC2	11	4.82 **	30.36
	PC3	9	2.25 **	11.59
	Residuals	120	0.19	
late	Environments (E)	5	3.48 **	12.99
	Genotypes (G)	9	7.23 **	48.57
	G × E	45	1.14 **	38.44
	PC1	13	2.42 **	61.16
	PC2	11	1.20 **	25.59
	PC3	9	0.54 **	9.41
	Residuals	120	0.03	

**: Significant at 0.01 probability level.

Based on AMMI analysis, the genotype–environments interaction was divided into three main principal components that explain 87.2% of the total variance under the first model (combined sowing dates). Furthermore, the first three principal components explained 92.08% and 96.16%of the interaction between genotype and environment under recommended and late sown conditions, respectively. A graphical representation of the relationship between cultivars and sowing dates across environments regarding grain yield is shown in a GGE biplot (Figure 1A,B).

The previous biplot and the AMMI analysis of variance indicated a variable response of the genotypes under the two sowing dates. Therefore, in addition to running the biplot and stability analysis on the combined sowing dates, it was refitted within each sowing date. Thus, three biplots were generated; for combined, early, and late sown conditions (Figures 1–3). For the three biplots, a polygon was formed by connecting the genotypes that were further away from the biplot origin, such that all other genotypes were contained in the polygon. Genotypes located on the vertices of the polygon performed either the best or the poorest in one or more locations since they had the longest distance from the origin of biplot. The vertex cultivars in the first GGE biplot (combined sowing dates) were Giza168, G7 (Gemmeiza7), Misr2, Sids13, Sakha94, G9 (Gemmeiza9), Sids12, and G10 (Gemmeiza10) (Figure 1B). Under the recommended sowing date (Figure 2B) the vertex genotypes were Giza168, G7 (Gemmeiza7), Sids13, Sakha94, and G10 (Gemmeiza10).

Moreover, the vertex genotypes for the late sown condition were G9 (Gemmeiza9), Sids12, G12 (Gemmeiza12), Giza168, and Sakha94 (Figure 3B). The best genotype for combined sowing dates was Sids12 (Figure 4A). However, the best genotype for recommended sowing date was G12 (Gemmeiza12 (Figure 4B). Furthermore, G9 (Gemmeiza9) followed by Sids12 were the better genotypes for the late sown condition (Figure 4C).

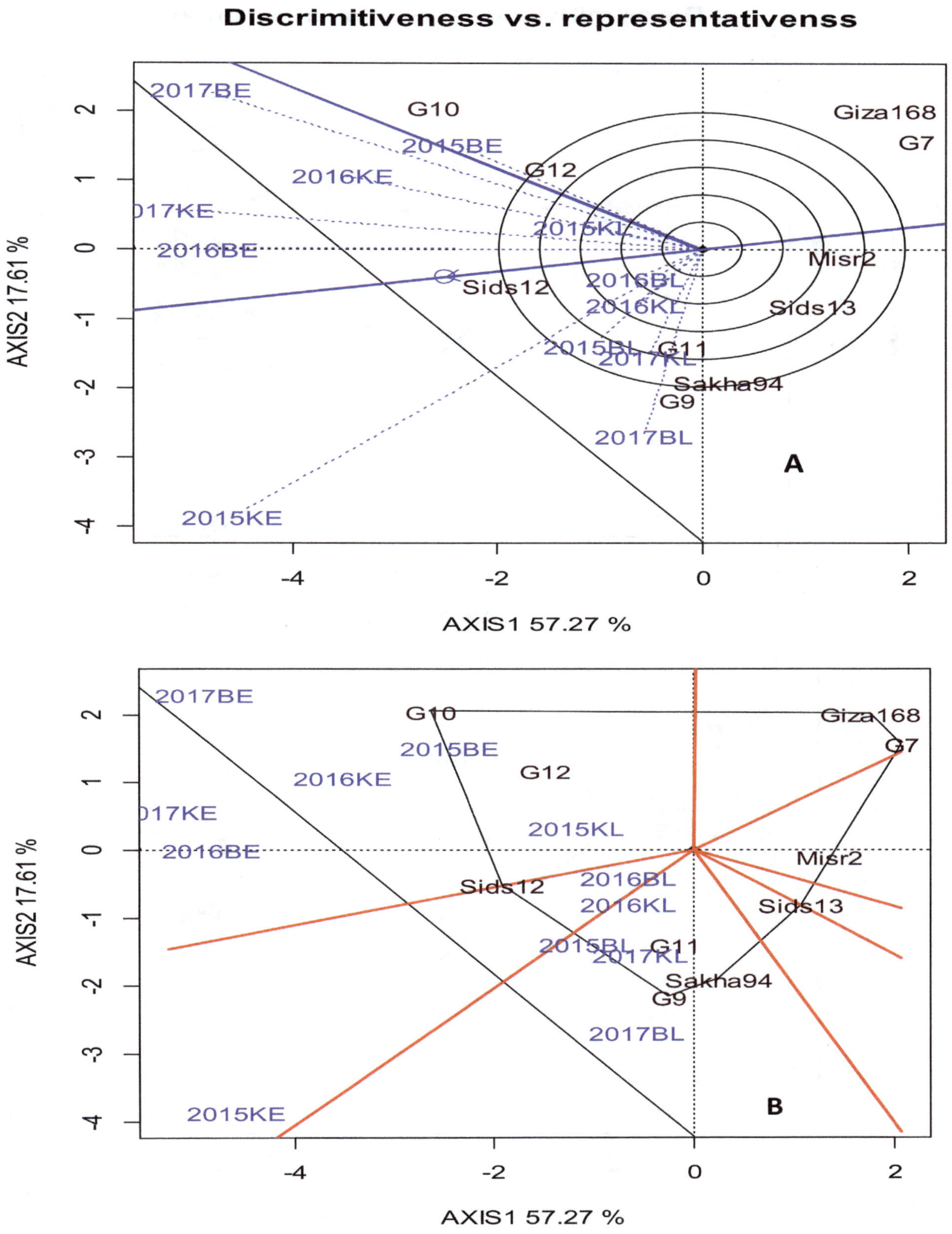

Figure 1. Vector view of GGE biplot for relationships among environments (**A**), and GGE biplot identification of winning cultivars across environments (two years i.e., 2015 and 2016, and two locations i.e., Elbostan (B), and Elkhazan (K)) under recommended (E) and late (L) sowing dates (**B**). AXIS1 and 2 refer to Principal component 1 (PC1) and Principal component 2 (PC2), respectively.

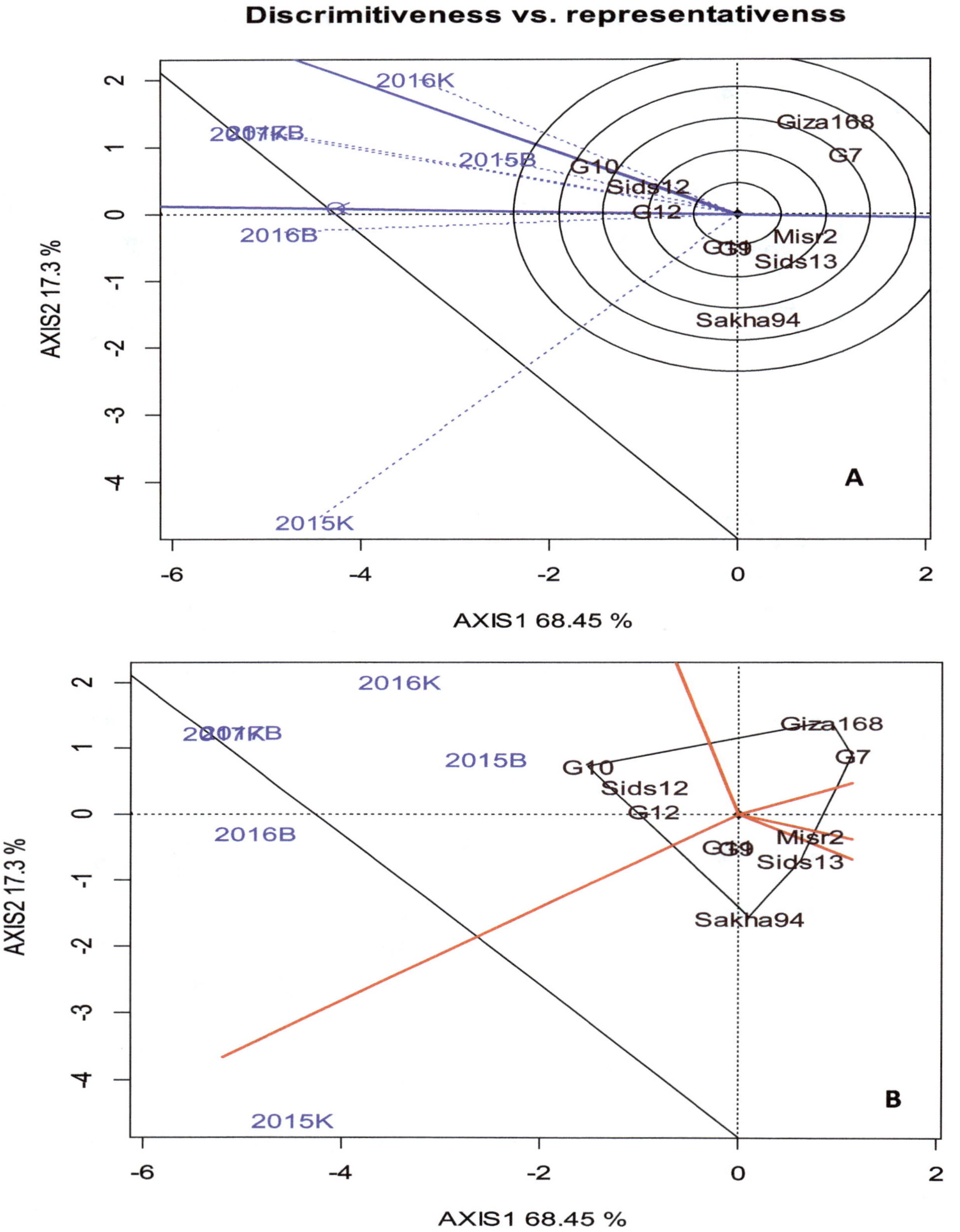

Figure 2. Vector view of GGE biplot for relationships among environments (**A**), and GGE biplot identification of winning cultivars across environments (two years, i.e., 2015 and 2016, and two locations, i.e., Elbostan (B), and Elkhazan (K)) under early sowing date (**B**). AXIS1 and 2 refer to Principal component 1 (PC1) and Principal component 2 (PC2), respectively.

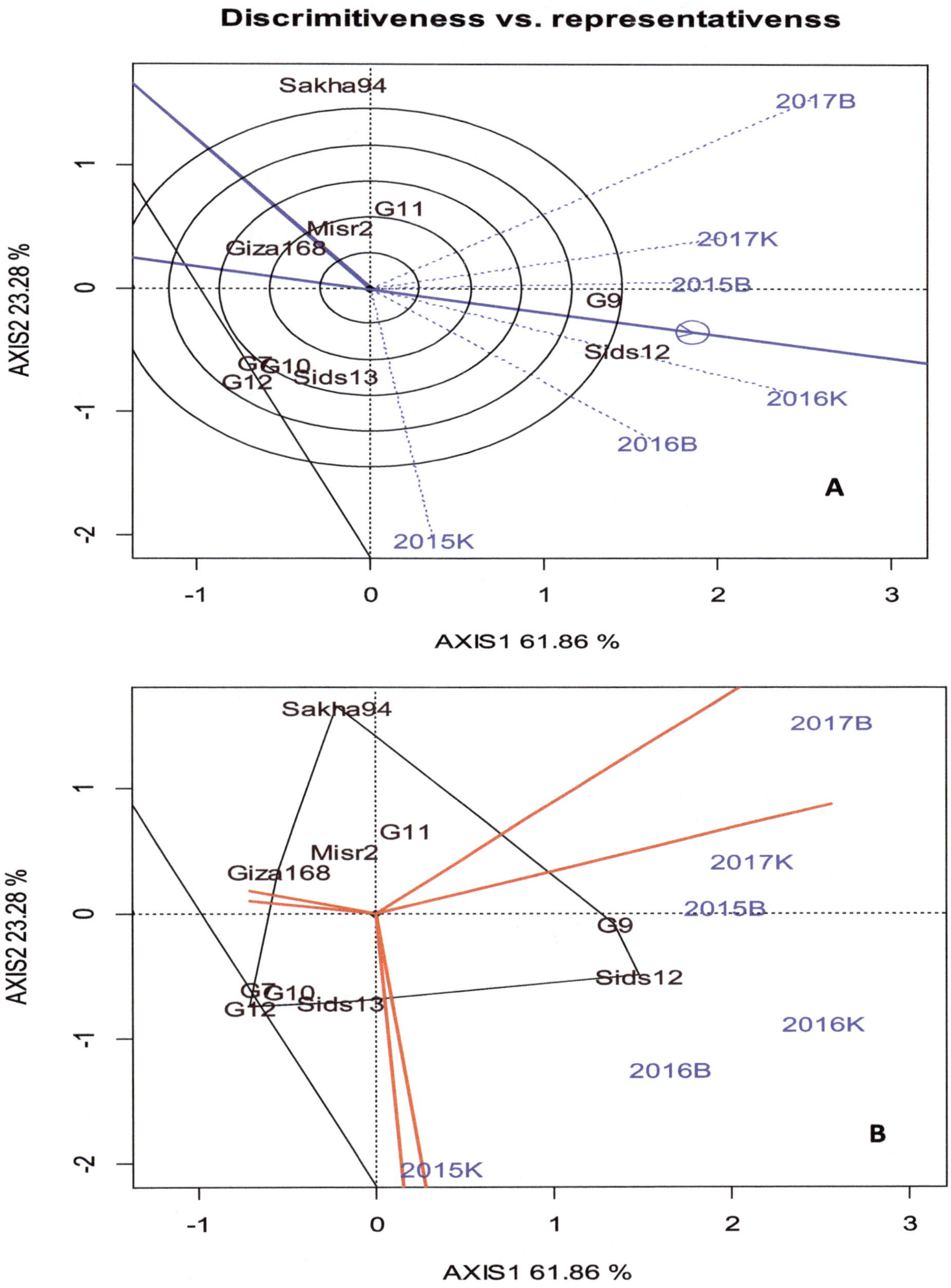

Figure 3. Vector view of GGE biplot for relationships among environments (**A**), and GGE biplot identification of winning cultivars across environments (two years, i.e., 2015 and 2016, and two locations, i.e., Elbostan (B), and Elkhazan (K)) under late sown condition (**B**). AXIS1 and 2 refer to Principal component 1 (PC1) and Principal component 2 (PC2), respectively.

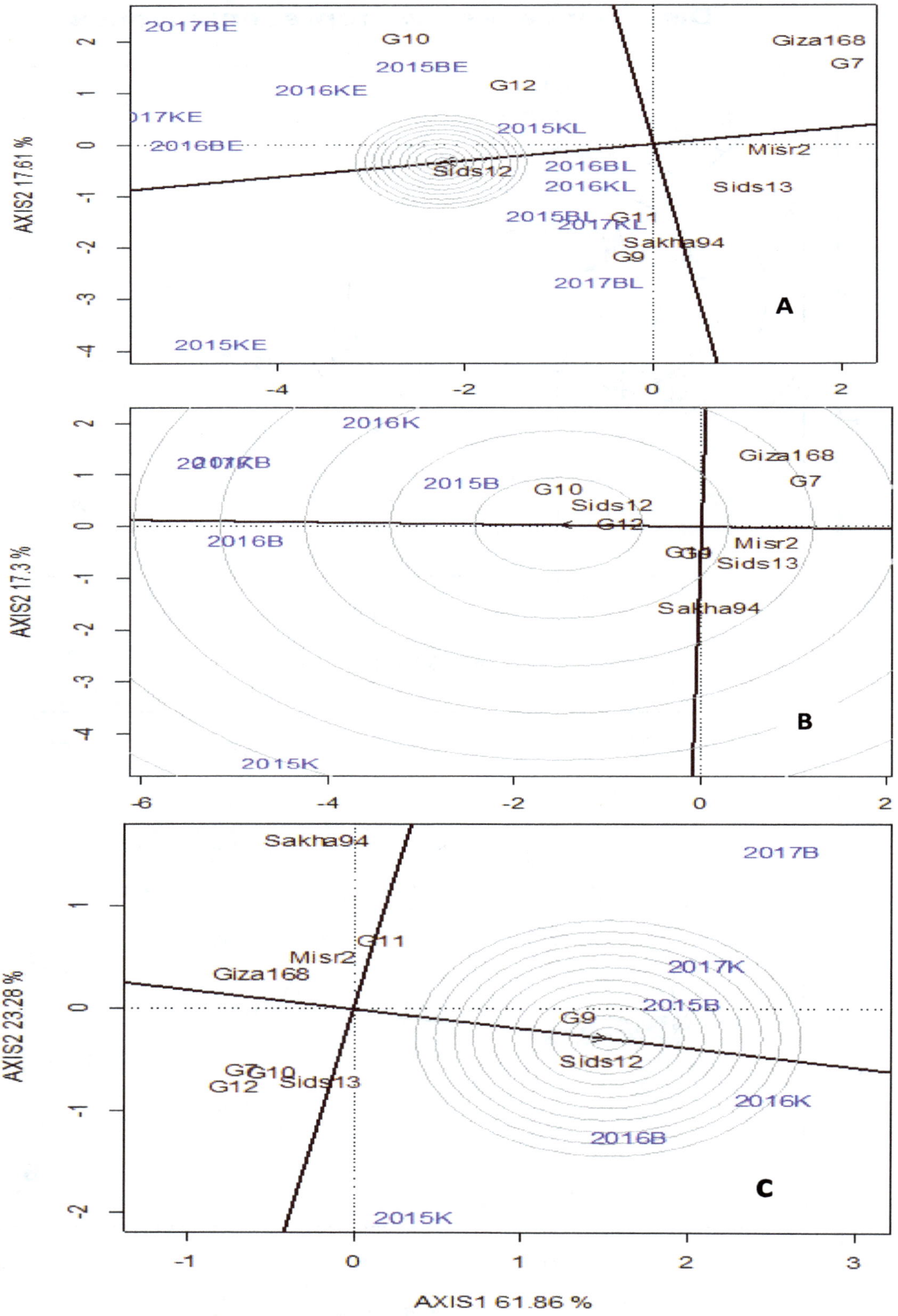

Figure 4. GGE biplot for the ideal genotype under both sowing dates (**A**), recommended sowing date (**B**) and late sown condition (**C**), across environments (two years, i.e., 2015 and 2016, and two locations, i.e., Elbostan (B), and Elkhazan (K)). AXIS1 and 2 refer to Principal component 1 (PC1) and Principal component 2 (PC2), respectively.

4. Discussion

The world bank identified Egypt as one of the potentially vulnerable countries in the Mediterranean basin that might be affected by global warming [35]. Wheat is the backbone of food security in Egypt, where it provides more than 30% of the population's calorie intake [36]. Wheat is a cool season crop that found to be sensitive to heat stress during the reproductive stage [37]. Heat stress tolerance is a complicated process controlled by several small effect genes or QTLs and is often confounded by differences in plant morphology and physiology under different environments [38–40]. Thus, to understand the complexity of plant responses to heat stress, it is vital to account for the morphological, physiological, and genetic basis of this response under the field conditions.

The current study intended to investigate the effect of heat stress on several morphological traits in addition to the grain yield using commercial wheat cultivars that are widely grown in Egypt [41]. Two locations in Egypt were used; Elbostan and Elkhazan, the first represents newly reclaimed sandy soil while the former represents the Nile delta soil (clay). In Elbostan the mean temperature during flowering under the late sown condition was higher than the recommended sowing conditions by 3.8, 2.3, and 3.2 °C for the 2015, 2016, 2017 growing seasons, respectively. Moreover, the mean temperature during flowering for Elkhazan was higher than the recommended sowing conditions by 3.8, 2.2, and 3.4 °C for the 2015, 2016, 2017 growing seasons, respectively.

The increased temperature during the late sown condition decreased total chlorophyll content, leaf area, grain filling duration, plant height, and grain yield. Moreover, the increased terminal temperature during flowering increased canopy temperature and leaf and stem rust susceptibility. Total chlorophyll content declined significantly under the late sown condition, suggesting structural damage to the chloroplast due to heat stress [7–9,42]. The decrease in the total chlorophyll content in response to induced heat stress has also been reported previously in several crops [43–47]. However, in the current study, total chlorophyll content was not significantly correlated with grain yield in either the timely-sown or the late-sown plots. The reasons for the weak correlation between total chlorophyll content and yield might be due to the growth stage in which we measured the total chlorophyll content. Several researchers indicated that plants start losing chlorophyll when the grain filling stage starts [48].

Heat stress (late sown condition) increased average canopy temperature across all cultivars. The correlation between canopy temperature and grain yield was negative and significant which agrees with previously reported results [49–51]. The prevalence and severity of stem and leaf rust were increased due to the late sown condition. A possible explanation is that the late sown condition creates favorable environmental conditions for both stem and leaf rust [52–58].

Overall, the late sown condition decreased the period of grain filling duration, while the sudden rise in the temperature during the reproductive stage decreased the amount of the assimilates [33]. High temperature increases the rate of seed filling, but the increase in the seed filling did not compensate for the loss in the grain filling duration [39]. Our results indicated variable response of the studied cultivars to heat stress across all traits under different environmental conditions which agrees with previous studies [40,59]. Moreover, the high temperature increased stem and leaf rust prevalence and severity during flowering stage under the late sown condition. Failure to understand the mechanisms of grain yield stability might impact both traditional breeding and the use of modern genetics in improving yield production [60]. Heat stress tolerance and maintaining high grain yield under heat stress is considered as one of the most critical aspects of wheat improvement [61–64]. Genotype–environment (G × E) interaction makes genotypic evaluation a complicated process because of the different response of genotypes under different locations or years. Obviously, grain yield is the most important trait that might solely determine the success of a plant breeder. At the same time, grain yield is a complex and quantitative trait that has strong G × E interactions. Thus, measuring the stability of a giving cultivar became one of the plant breeding routines before releasing the cultivar. The advantage of selecting superior genotypes using stability analysis instead of average performance is that stable genotypes are reliable across environments which reduce G × E interaction. Studies have shown that stability analyses according to various measures can result in better identification of stable genotypes, even when there were no interactions among the measures. In the current study, seven stability parameters were used, which were inconsistent in identifying ideal genotypes for environments or sowing dates; this result agrees with previous results [65–67]. The inconsistency could be attributed to the difference in statistical and mathematical methods that the stability parameters rely on [68]. AMMI and GGE biplot were also applied to identify the most stable genotypes. In our study, the ideal genotype was defined agronomically as the genotype that performs well, as "High yielding" and stable across a wide range of environments. Furthermore, in our study, we defined the ideal genotype statistically as the genotype that was stable in at least one of the seven traditional stability measures in addition to AMMI.

In the current study, some genotypes performed well under recommended sowing date but not under the late sown condition (heat stressed conditions) and vice-versa, while some of the studied materials performed well under both sowing dates. Previous results imply that the genotypes used in this study contain different combinations of genes governing their response to sowing dates and tolerance to heat stress, and this result agrees with previous findings [59,69–73]. The ultimate goal of this study was to estimate the effect of global warming and heat stress on grain yield using a representative sample of commercially distributed and widely grown wheat cultivars in Egypt. Thus, wheat breeders and decision makers in this region might have an idea about the effect of heat stress on wheat production.

5. Conclusions

The results revealed that wheat performance was significantly influenced by environment, genotype, and their interaction. Heat stress had a negative impact on all traits. The late sown condition increased the prevalence and severity of leaf and stem rust which most likely contributed to the overall yield losses. The late sown condition decreased the overall yield production for the studied cultivars by 45%. The cultivars showed high G $\times$ E interaction, and the sowing date increased that interaction. Stability measurements were useful in determining the most stable genotypes. However, inconsistency was observed among some measurements. AMMI and GGE biplots were adequate for analyzing and visualizing the patterns of G $\times$ E. Sids12 was stable and outperformed the tested materials under both the recommended and late sown conditions. Gemmeiza12 was more stable and outperformed the tested materials under the recommended sown conditions. Gemmeiza9 followed by Sids12 were stable and performed better than the rest of the tested cultivars under the late sown conditions. However, late sown conditions reduced yield by 47.15%, 64.5%, and 59.2% for cultivars Sids12, Gemmeiza9, and Gemmeiza10, respectively. As a result of this work, we recommend importing and evaluating wheat accessions known to be tolerant to heat stress and cross them with the adapted wheat cultivars in this region to boost heat stress tolerance. Even though this study was conducted using cultivars grown mainly in Egypt; we expect that heat stress will have a similar effect on the spring wheat cultivars grown in the Mediterranean region because most of the wheat breeding programs in the Mediterranean region share lines from each other and international wheat breeding organizations.

Acknowledgments: The author sincerely thanks the science and technology development fund (STDF), grant number 14935 and Damanhour University for funding and supporting this study. I would also like to thank Stephen Baenziger for reviewing, editing and helpful suggestions while preparing this manuscript.

References

1. Enghiad, A.; Ufer, D.; Countryman, A.M.; Thilmany, D.D. An Overview of Global Wheat Market Fundamentals in an Era of Climate Concerns. *Int. J. Agron.* **2017**, *2017*, 3931897. [CrossRef]
2. Curtis, T.; Halford, N.G. Food security: The challenge of increasing wheat yield and the importance of not compromising food safety. *Ann. Appl. Biol.* **2014**, *164*, 354–372. [CrossRef] [PubMed]
3. Olmstead, A.L.; Rhode, P.W. Adapting North American wheat production to climatic challenges, 1839–2009. *Proc. Natl. Acad. Sci. USA* **2011**, *108*, 480–485. [CrossRef]

4. Asseng, S.; Ewert, F.; Martre, P.; Rötter, R.P.; Lobell, D.B.; Cammarano, D.; Kimball, B.A.; Ottman, M.J.; Wall, G.W.; White, J.W.; et al. Rising temperatures reduce global wheat production. *Nat. Clim. Chang.* **2015**, *5*, 143–147. [CrossRef]
5. Bita, C.E.; Gerats, T. Plant tolerance to high temperature in a changing environment: Scientific fundamentals and production of heat stress-tolerant crops. *Front. Plant Sci.* **2013**, *4*, 273. [CrossRef] [PubMed]
6. Farooq, M.; Bramley, H.; Palta, J.A.; Siddique, K.H.M. Heat Stress in Wheat during Reproductive and Grain-Filling Phases. *Crit. Rev. Plant Sci.* **2011**, *30*, 491–507. [CrossRef]
7. Jat, R.K.; Singh, P.; Jat, M.L.; Dia, M.; Sidhu, H.S.; Jat, S.L.; Bijarniya, D.; Jat, H.S.; Parihar, C.M.; Kumar, U.; et al. Heat stress and yield stability of wheat genotypes under different sowing dates across agro-ecosystems in India. *Field Crops Res.* **2018**, *218*, 33–50. [CrossRef]
8. Akter, N.; Rafiqul Islam, M. Heat stress effects and management in wheat. A review. *Agron. Sustain. Dev.* **2017**, *37*, 37. [CrossRef]
9. Spiertz, J.H.J.; Hamer, R.J.; Xu, H.; Primo-Martin, C.; Don, C.; van der Putten, P.E.L. Heat stress in wheat (*Triticum aestivum* L.): Effects on grain growth and quality traits. *Eur. J. Agron.* **2006**, *25*, 89–95. [CrossRef]
10. Feng, B.; Liu, P.; Li, G.; Dong, S.T.; Wang, F.H.; Kong, L.A.; Zhang, J.W. Effect of Heat Stress on the Photosynthetic Characteristics in Flag Leaves at the Grain-Filling Stage of Different Heat-Resistant Winter Wheat Varieties. *J. Agron. Crop Sci.* **2014**, *200*, 143–155. [CrossRef]
11. Gonzalez, A.; Bermejo, V.; Gimeno, B.S. Effect of different physiological traits on grain yield in barley grown under irrigated and terminal water deficit conditions. *J. Agric. Sci.* **2010**, *148*, 319–328. [CrossRef]
12. Matsui, T.; Omasa, K.; Horie, T. The Difference in Sterility due to High Temperatures during the Flowering Period among Japonica-Rice Varieties. *Plant Prod. Sci.* **2001**, *4*, 90–93. [CrossRef]
13. Vara Prasad, P.V.; Craufurd, P.Q.; Summerfield, R.J. Sensitivity of peanut to timing of heat stress during reproductive development. *Crop Sci.* **1999**, *39*, 1352–1357. [CrossRef]
14. Batts, G.R.; Ellis, R.H.; Morison, J.I.L.; Nkemka, P.N.; Gregory, P.J.; Hadley, P. Yield and partitioning in crops of contrasting cultivars of winter wheat in response to CO_2 and temperature in field studies using temperature gradient tunnels. *J. Agric. Sci.* **1998**, *130*, 17–27. [CrossRef]
15. Wardlaw, F.; Sofieldb, I.; Cartwrightc, P.M. Factors Limiting the Rate of Dry Matter Accumulation in the Grain of Wheat Grown at High Temperature. *Aust. J. Plant Physiol.* **1980**, *7*, 387–400. [CrossRef]
16. Iqbal, M.; Raja, N.I.; Yasmeen, F.; Hussain, M.; Ejaz, M.; Shah, M.A. Impacts of Heat Stress on Wheat: A Critical Review. *Adv. Crop Sci. Technol.* **2017**, *5*, 251. [CrossRef]
17. Adams, S. Effect of Temperature on the Growth and Development of Tomato Fruits. *Ann. Bot.* **2001**, *88*, 869–877. [CrossRef]
18. Abiko, M.; Akibayashi, K.; Sakata, T.; Kimura, M.; Kihara, M.; Itoh, K.; Asamizu, E.; Sato, S.; Takahashi, H.; Higashitani, A. High-temperature induction of male sterility during barley (*Hordeum vulgare* L.) anther development is mediated by transcriptional inhibition. *Sex. Plant Reprod.* **2005**, *18*, 91–100. [CrossRef]
19. Al-Otayk, S.M. Performance of Yield and Stability of Wheat Genotypes under High Stress Environments of the Central Region of Saudi Arabia. *Meteorol. Environ. Arid Land Agric. Sci.* **2010**, *21*, 81–92. [CrossRef]
20. Farshadfar, E.; Sabaghpour, S.H.; Zali, H. Comparison of parametric and non-parametric stability statistics for selecting stable chickpea (*Cicer arietinum* L.) genotypes under diverse environments. *Aust. J. Crop Sci.* **2012**, *6*, 514–524.
21. Lin, C.S.; Binns, M.R. A superiority measure of cultivar performance for cultivar × location data. *Can. J. Plant Sci.* **1988**, *68*, 193–198. [CrossRef]
22. Eberhart, S.A.; Russell, W.A. Stability Parameters for Comparing Varieties. *Crop Sci.* **1966**, *6*, 36–40. [CrossRef]
23. Romagosa, I.; Fox, P.N. Genotype × environment interaction and adaptation. In *Plant Breeding*; Springer: Dordrecht, The Netherlands, 1993; pp. 373–390, ISBN 9401046654.
24. Akcura, M.; Kaya, Y. Nonparametric stability methods for interpreting genotype by environment interaction of bread wheat genotypes (*Triticum aestivum* L.). *Genet. Mol. Biol.* **2008**, *31*, 906–913. [CrossRef]
25. Bavec, M.; Vuković, K.; Grobelnik Mlakar, S.; Rozman, Č.; Bavec, F. Leaf area index in winter wheat: response on seed rate and nitrogen application by different varieties. *J. Cent. Eur. Agric.* **2007**, *8*, 337–342. [CrossRef]
26. Peterson, R.F.; Campbell, A.B.; Hannah, A.E. A diagrammatic scale for estimating rust intensity on leaves and stems of cereals. *Can. J. Res.* **1948**, *26*, 496–500. [CrossRef]

27. Roelfs, A.P.; Singh, R.P.; Saari, E.E. *Rust Diseases of Wheat: Concepts and Methods of Disease Management*; CIMMYT (International Maize and Wheat Improvement Center): Mexico, D.F., Mexico, 1992; ISBN 968612747X.
28. Gomez, K.A.; Gomez, A.A.; Gomez, K.A. *Statistical Procedures for Agricultural Research*; John Wiley & Sons: New York, NY, USA, 1984; ISBN 9780471870920.
29. Steel, R.G.D.; Torrie, J.H. *Principles and Procedures of Statistics: A Biometrical Approach*, 2nd ed.; McGraw-Hill Publishing Co.: New York, NY, USA, 1980; 631p.
30. Mellenbergh, G.J.; Arce, C. Experimental design and analysis. *J. Mark. Res. (JMR)* **1992**, *29*, 155–156. [CrossRef]
31. Pinthus, M.J. Estimate of genotypic value: A proposed method. *Euphytica* **1973**, *22*, 121–123. [CrossRef]
32. Mulusew, F.; Bing, D.J.; Tadele, T.; Amsalu, A. Comparison of biometrical methods to describe yield stability in field pea (*Pisum sativum* L.) under south eastern Ethiopian conditions. *Afr. J. Agric. Res.* **2014**, *9*, 2574–2583. [CrossRef]
33. Blanco, A.; Mangini, G.; Giancaspro, A.; Giove, S.; Colasuonno, P.; Simeone, R.; Signorile, A.; De Vita, P.; Mastrangelo, A.M.; Cattivelli, L.; et al. Relationships between grain protein content and grain yield components through quantitative trait locus analyses in a recombinant inbred line population derived from two elite durum wheat cultivars. *Mol. Breed.* **2012**, *30*, 79–92. [CrossRef]
34. Pacheco, Á.; Vargas, M.; Alvarado, G.; Rodríguez, F.; Crossa, J.; Burgueño, J. GEA-R (Genotype × Environment Analysis with R for Windows) Version 4.0. Available online: http://hdl.handle.net/11529/10203 (accessed on 29 January 2018).
35. Hereher, M.E. Time series trends of land surface temperatures in Egypt: a signal for global warming. *Environ. Earth Sci.* **2016**, *75*, 1218. [CrossRef]
36. Valizadeh, J.; Ziaei, S.M.; Mazloumzadeh, S.M. Assessing climate change impacts on wheat production (a case study). *J. Saudi Soc. Agric. Sci.* **2014**, *13*, 107–115. [CrossRef]
37. Castro, M.; Peterson, C.J.; Dalla Rizza, M.; Díaz Dellavalle, P.; Vázquez, D.; Ibañez, V.; Ross, A. Influence of heat stress on wheat grain characteristics and protein molecular weight distribution. In *Wheat Production in Stressed Environments*; Springer: Dordrecht, The Netherlands, 2007; pp. 365–371, ISBN 978-1-4020-5496-9.
38. Mohammed, A.R.; Tarpley, L. Impact of high nighttime temperature on respiration, membrane stability, antioxidant capacity, and yield of rice plants. *Crop Sci.* **2009**, *49*, 313–322. [CrossRef]
39. Ayeneh, A.; Van Ginkel, M.; Reynolds, M.P.; Ammar, K. Comparison of leaf, spike, peduncle and canopy temperature depression in wheat under heat stress. *Field Crops Res.* **2002**, *79*, 173–184. [CrossRef]
40. Fu, G.; Feng, B.; Zhang, C.; Yang, Y.; Yang, X.; Chen, T.; Zhao, X.; Zhang, X.; Jin, Q.; Tao, L. Heat Stress Is More Damaging to Superior Spikelets than Inferiors of Rice (*Oryza sativa* L.) due to Their Different Organ Temperatures. *Front. Plant Sci.* **2016**, *7*, 1637. [CrossRef] [PubMed]
41. Gadallah, A.; Milad, I.; Yossef, Y.A.; Gouda, M.A. Evaluation of Some Egyptian Bread Wheat (*Triticum aestivum*) Cultivars under Salinity Stress. *Alex. Sci. Exch.* **2017**, *38*, 260.
42. Chen, W.R.; Zheng, J.S.; Li, Y.Q.; Guo, W.D. Effects of high temperature on photosynthesis, chlorophyll fluorescence, chloroplast ultrastructure, and antioxidant activities in fingered citron. *Russ. J. Plant Physiol.* **2012**, *59*, 732–740. [CrossRef]
43. Kreslavski, V.D.; Lyubimov, V.Y.; Shabnova, N.I.; Balakhnina, T.I.; Kosobryukhov, A.A. Heat-induced impairments and recovery of photosynthetic machinery in wheat seedlings. Role of light and prooxidant-antioxidant balance. *Physiol. Mol. Biol. Plants* **2009**, *15*, 115–122. [CrossRef] [PubMed]
44. Kreslavski, V.; Tatarinzev, N.; Shabnova, N.; Semenova, G.; Kosobryukhov, A. Characterization of the nature of photosynthetic recovery of wheat seedlings from short-term dark heat exposures and analysis of the mode of acclimation to different light intensities. *J. Plant Physiol.* **2008**, *165*, 1592–1600. [CrossRef] [PubMed]
45. Xu, S.; Li, J.; Zhang, X.; Wei, H.; Cui, L. Effects of heat acclimation pretreatment on changes of membrane lipid peroxidation, antioxidant metabolites, and ultrastructure of chloroplasts in two cool-season turfgrass species under heat stress. *Environ. Exp. Bot.* **2006**, *56*, 274–285. [CrossRef]
46. Guo, Y.-P.; Zhou, H.-F.; Zhang, L.-C. Photosynthetic characteristics and protective mechanisms against photooxidation during high temperature stress in two citrus species. *Sci. Hortic.* **2006**, *108*, 260–267. [CrossRef]
47. Havaux, M. Characterization of thermal damage to the photosynthetic electron transport system in potato leaves. *Plant Sci.* **1993**, *94*, 19–33. [CrossRef]

48. Del Río, L.A.; Pastori, G.M.; Palma, J.M.; Sandalio, L.M.; Sevilla, F.; Corpas, F.J.; Jiménez, A.; López-Huertas, E.; Hernández, J.A. The Activated Oxygen Role of Peroxisomes in Senescence. *Plant Physiol.* **1998**, *116*, 1195–1200. [CrossRef] [PubMed]
49. Asthir, B. Protective mechanisms of heat tolerance in crop plants. *J. Plant Interact.* **2015**, *10*, 202–210. [CrossRef]
50. Hasanuzzaman, M.; Nahar, K.; Alam, M.M.; Roychowdhury, R.; Fujita, M. Physiological, biochemical, and molecular mechanisms of heat stress tolerance in plants. *Int. J. Mol. Sci.* **2013**, *14*, 9643–9684. [CrossRef] [PubMed]
51. Hemantaranjan, A. Heat Stress Responses and Thermotolerance. *Adv. Plants Agric. Res.* **2014**, *1*. [CrossRef]
52. Liu, J.Q.; Kolmer, J.A. Genetics of Leaf Rust Resistance in Canadian Spring Wheats AC Domain and AC Taber. *Plant Dis.* **1997**, *81*, 757–760. [CrossRef]
53. Huerta-Espino, J.; Singh, R.P.; Germán, S.; McCallum, B.D.; Park, R.F.; Chen, W.Q.; Bhardwaj, S.C.; Goyeau, H. Global status of wheat leaf rust caused by Puccinia triticina. *Euphytica* **2011**, *179*, 143–160. [CrossRef]
54. Liu, J.Q.; Kolmer, J.A. Inheritance of Leaf Rust Resistance in Wheat Cultivars Grandin and CDC Teal. *Plant Dis.* **1997**, *81*, 505–508. [CrossRef]
55. Herrera-Foessel, S.A.; Singh, R.P.; Huerta-Espino, J.; Crossa, J.; Yuen, J.; Djurle, A. Effect of Leaf Rust on Grain Yield and Yield Traits of Durum Wheats with Race-Specific and Slow-Rusting Resistance to Leaf Rust. *Plant Dis.* **2006**, *90*, 1065–1072. [CrossRef]
56. Broers, L.H.M. Partial resistance to wheat leaf rust in 18 spring wheat cultivars. *Euphytica* **1989**, *44*, 247–258. [CrossRef]
57. Broers, L.H.M. Influence of development stage and host genotype on three components of partial resistance to leaf rust in spring wheat. *Euphytica* **1989**, *44*, 187–195. [CrossRef]
58. Draz, I.S.; Abou-Elseoud, M.S.; Kamara, A.-E.M.; Alaa-Eldein, O.A.-E.; El-Bebany, A.F. Screening of wheat genotypes for leaf rust resistance along with grain yield. *Ann. Agric. Sci.* **2015**, *60*, 29–39. [CrossRef]
59. Khan, A.A.; Kabir, M.R. Evaluation of Spring Wheat Genotypes (*Triticum Aestivum* L.) for Heat Stress Tolerance Using Different Stress Tolerance Indices. *Cercet. Agron. Mold.* **2015**, *47*, 49–63. [CrossRef]
60. Mir, R.R.; Zaman-Allah, M.; Sreenivasulu, N.; Trethowan, R.; Varshney, R.K. Integrated genomics, physiology and breeding approaches for improving drought tolerance in crops. *Theor. Appl. Genet.* **2012**, *125*, 625–645. [CrossRef] [PubMed]
61. Cossani, C.M.; Reynolds, M.P. Physiological Traits for Improving Heat Tolerance in Wheat. *Plant Physiol.* **2012**, *160*, 1710–1718. [CrossRef] [PubMed]
62. Aziz, A.; Mahmood, T.; Mahmood, Z.; Shazadi, K.; Mujeeb-Kazi, A.; Rasheed, A. Genotypic Variation and Genotype × Environment Interaction for Yield-Related Traits in Synthetic Hexaploid Wheats under a Range of Optimal and Heat-Stressed Environments. *Crop Sci.* **2018**, *58*, 295–303. [CrossRef]
63. Acevedo, E.; Silva, P.; Silva, H. *Wheat Growth and Physiology*; FAO Plant Production and Protection Series (FAO): Rome, Italy, 2002.
64. Altenbach, S.B.; Dupont, F.M.; Kothari, K.M.; Chan, R.; Johnson, E.L.; Lieu, D. Temperature, Water and Fertilizer Influence the Timing of Key Events During Grain Development in a US Spring Wheat. *J. Cereal Sci.* **2003**, *37*, 9–20. [CrossRef]
65. Chamekh, Z.; Karmous, C.; Ayadi, S.; Sahli, A.; Hammami, Z.; Fraj, M.B.; Benaissa, N.; Trifa, Y.; Slim-Amara, H. Stability analysis of yield component traits in 25 durum wheat (*Triticum durum* Desf.) genotypes under contrasting irrigation water salinity. *Agric. Water Manag.* **2015**, *152*, 1–6. [CrossRef]
66. Mohamed, N.E.M.; Said, A.A. 7 Stability Parameters for Comparing Bread Wheat Genotypes under Combined Heat and Drought Stress. *Egypt. J. Agron.* **2014**, *36*, 123–146.
67. Abbas Mosavi, A.; Babaiean Jelodar, N.; Kazemitabar, K. Environmental Responses and Stability Analysis for Grain Yield of Some Rice Genotypes. *World Appl. Sci. J.* **2013**, *21*, 105–108. [CrossRef]
68. Witcombe, J.R. Estimates of stability for comparing varieties. *Euphytica* **1988**, *39*, 11–18. [CrossRef]
69. Lopes, M.S.; El-Basyoni, I.; Baenziger, P.S.; Singh, S.; Royo, C.; Ozbek, K.; Aktas, H.; Ozer, E.; Ozdemir, F.; Manickavelu, A.; et al. Exploiting genetic diversity from landraces in wheat breeding for adaptation to climate change. *J. Exp. Bot.* **2015**, *66*, 3477–3486. [CrossRef] [PubMed]

70. Omae, H.; Kumar, A.; Shono, M. Adaptation to High Temperature and Water Deficit in the Common Bean (*Phaseolus vulgaris* L.) during the Reproductive Period. *J. Bot.* **2012**, *2012*, 803413. [CrossRef]
71. Qu, A.-L.; Ding, Y.-F.; Jiang, Q.; Zhu, C. Molecular mechanisms of the plant heat stress response. *Biochem. Biophys. Res. Commun.* **2013**, *432*, 203–207. [CrossRef] [PubMed]
72. Lobell, D.B.; Field, C.B. Global scale climate–crop yield relationships and the impacts of recent warming. *Environ. Res. Lett.* **2007**, *2*, 14002. [CrossRef]
73. Krasensky, J.; Jonak, C. Drought, salt, and temperature stress-induced metabolic rearrangements and regulatory networks. *J. Exp. Bot.* **2012**, *63*, 1593–1608. [CrossRef] [PubMed]

5

Halotolerant Bacterial Diversity Associated with *Suaeda fruticosa* (L.) Forssk. Improved Growth of Maize under Salinity Stress

Faiza Aslam and Basharat Ali *

Department of Microbiology and Molecular Genetics, University of the Punjab, Lahore-54590, Pakistan; faiza.mmg@gmail.com

* Correspondence: basharat.ali.mmg@pu.edu.pk

Abstract: Halotolerant bacterial strains associated with the rhizosphere and phytoplane of *Suaeda fruticosa* (L.) Forssk. growing in saline habitats were isolated to mitigate the salinity stress of *Zea mays* L. 16S rRNA gene sequencing confirmed the presence of strains that belong to *Gracilibacillus, Staphylococcus, Virgibacillus, Salinicoccus, Bacillus, Zhihengliuella, Brevibacterium, Oceanobacillus, Exiguobacterium, Pseudomonas, Arthrobacter*, and *Halomonas* genera. Strains were screened for auxin production, 1-aminocyclopropane-1-carboxylate (ACC)-deaminase, and biofilm formation. Bacterial auxin production ranged from 14 to 215 μg mL^{-1}. Moreover, several bacterial isolates were also recorded as positive for ACC-deaminase activity, phosphate solubilization, and biofilm formation. In pot trials, bacterial strains significantly mitigated the salinity stress of *Z. mays* seedlings. For instance, at 200 and 400 mM NaCl, a significant increase of shoot and root length (up to onefold) was recorded for *Staphylococcus jettensis* F-11. At 200 mM, *Zhihengliuella flava* F-9 (45%) and *Bacillus megaterium* F-58 (42%) exhibited significant improvements for fresh weight. For dry weight, *S. jettensis* F-11 and *S. arlettae* F-71 recorded up to a threefold increase at 200 mM over the respective control. The results of this study suggest that natural plant settings of saline habitats are a good source for the isolation of beneficial salt-tolerant bacteria to grow crops under saline conditions.

Keywords: antioxidant enzymes; bacterial auxin production; halotolerant bacteria; halophytes; maize; plant-growth-promoting rhizobacteria; salinity stress; *Suaeda fruticose* (L.) Forssk.

---

1. Introduction

Soil salinity is one of the most severe environmental factors limiting agricultural productivity in arid or semiarid regions around the world. Salinity adversely affects the physical and chemical parameters of the soil as well as crop production to a significant extent [1]. Halophytes are the vegetation of saline habitats. Halophytes possess special morphological, anatomical, and physiological characteristics that are well suited to cope with saline habitats. *Suaeda fruticosa* (L.) Forssk. is a plant species that belongs to the family Amaranthaceae. It is a halophyte that occurs in arid and semiarid saline habitats or salt marches. *S. fruticosa* has a strong ability to accumulate and sequester NaCl [2]. It can be used for soil reclamation to reduce salinity and contamination of heavy metals. Halophytes have a number of adaptations to mitigate salinity stress, including salt discharge from roots, accumulation of organic acids, reduced stomatal conductance, lower water potential, and uptake of inorganic ions [3]. Moreover, specific types of amino acids, carbohydrates, and glycine betaine are accumulated as compatible solutes under abiotic stress [4].

High salt content in soil and irrigation water is a major threat to the sustainability of agriculture around the world. The presence of excess salts in the rhizosphere can cause severe injury to the root system followed by their gradual accumulation in other plant tissues. It can cause heavy damage to plant metabolism, which can lead to stunted plant growth and reduced yield [5]. Salinity can disrupt the water uptake and ion equilibrium of plants. It can lead to oxidative damage to plants due to the production of reactive oxygen species (ROS). This includes the production of superoxide anion (O_2^-), hydrogen peroxide (H_2O_2), and hydroxyl radical (OH) etc. Halophytes have the ability to keep the level of these ROS at minimal levels due to the presence of an antioxidant system that consists of enzymes like catalase (CAT), peroxidase (POD), and superoxide dismutase (SOD) [6,7].

Rhizobacteria associated with the root system of halophytes can enhance the growth, development, and stress tolerance of plants growing in saline soils [8]. It has been demonstrated that beneficial microorganisms play a significant role in alleviating salt stress in plants, which results in increased crop yield. Plant-growth-promoting rhizobacteria (PGPR) are a group of microorganisms that colonize the root system of plants and trigger plant growth and development under stress conditions. PGPR can employ several direct or indirect mechanisms to enhance plant growth and productivity. Auxin production, 1-aminocyclopropane-1-carboxylate (ACC)-deaminase, siderophores production, and phosphate solubilization are well documented in the literature [9,10]. Indole-3-acetic acid (IAA) represents one of the most extensively studied auxins in plants. It mediates multidimensional developmental processes including apical dominance, tropic responses, cellular differentiation, and pattern formation [11]. IAA is also quantitatively the most abundant phytohormones secreted by rhizobacteria as secondary metabolites. Auxin production has been reported for several halotolerant bacterial genera including *Bacillus, Enterobacter, Arthrocnemum, Agrobacterium, Ochrobactrum, Pseudomonas*, and *Pantoea* in NaCl-amended medium [8,9]. Moreover, soil microorganisms possess the enzyme ACC-deaminase that converts ACC (precursor of ethylene) to α-ketobutyrate and ammonia. When plants are exposed to stress conditions, ethylene is synthesized, which results in retardation of root growth and senescence [12,13]. Although ethylene mediates growth responses in plants, it can inhibit plant growth at elevated levels (under stress), leading to physiological changes in plant tissues. Tolerance to biotic or abiotic stresses can be increased by treating plant roots with ACC-deaminase-producing rhizobacteria [14]. Halotolerant PGPR have been reported to confer upon plants the ability to mitigate environmental or abiotic stresses [7]. For instance, *Bacillus aquimaris, Bacillus thuringiensis, Enterobacter cloacae, Enterobacter asburiae, Ochrobactrum anthropi*, and *Pseudomonas stutzeri* significantly increased vegetative and yield parameters under saline conditions [9,10,15]. Maize (*Zea mays* L.) is the third most important cereal crop that is grown under a wide range of climatic conditions. Salt-tolerant PGPR isolated from halophytes have been reported to enhance the vegetative growth parameters of maize under induced salinity [16].

Soil salinity affects approximately 6.3 million hectors of irrigated land in Pakistan and has caused considerable losses to the agriculture sector [17]. There is a need to devise environmentally friendly strategies that can ameliorate soil salinity by the application of resident halotolerant bacteria. In the present study, salt-tolerant bacteria associated with halophytic plants have been evaluated to mitigate the salinity stress of maize. In Pakistan, *S. fruticosa* is distributed in coastal and inland saline habitats [6]. Little is known about the association of halotolerant bacterial communities with this plant. Therefore, the purpose of this study was to evaluate the bacterial diversity of the rhizosphere and phytoplane of *S. fruticosa* growing in saline habitats of the Photohar Plateau, a region of northeastern Pakistan. Bacterial communities associated with the surfaces of *S. fruticosa* may have naturally adapted to tolerate high salt concentrations. Therefore, screening of resident bacterial flora may indicate potential candidates for agricultural applications under saline habitats.

2. Materials and Methods

2.1. Isolation of Bacterial Strains

Suaeda fruticosa (L.) Forssk. is a halophytic plant that grows in the saline habitats of the Photohar Plateau, a region of northeastern Pakistan. Halotolerant bacterial strains were isolated from the rhizosphere and phytoplane of *S. fruticosa*. For rhizospheric samples, 1 g of soil was dissolved in 99 mL of sterilized distilled water. Then, the serial dilutions of the samples were prepared successively as described in Cappuccino and Sherman [18]. For endophytes, leaves were sterilized by soaking in 70% ethanol for 2–3 min and then washed with autoclaved distilled water. One gram of leaves was grounded to paste and transferred to a test tube containing 9 mL of autoclaved distilled water. About 50 μL from soil and leaf suspensions was plated on L-agar supplemented with 0.5 M and 1 M NaCl. Plates were incubated at 30 °C for 24 h. Discrete bacterial colonies were selected and purified by many rounds of streaking.

2.2. 16S rRNA Gene Sequencing

Bacterial strains were identified by 16S rRNA gene sequencing. Total genomic DNA from overnight grown bacterial cultures was extracted using the FavorPrep™ Tissue Genomic DNA Extraction Mini Kit (Favorgen Biotech Corporation, Ping-Tung, Taiwan). PCR amplification of 16S rRNA gene sequencing was accomplished by using forward primer 27f (5′-AGAGTTTGATCCTGGCTCAG-3′) and reverse primer 1522r (5′-AAGGAGGTGATCCA (AG)CCGCA-3′) [19]. PCR amplification was performed in 50 μL of Dream Taq™ Green PCR Master Mix (Fermentas, Waltham, MA, USA) with 0.5 μg of chromosomal DNA template and 0.5 μM of each primer. The reaction mixtures were incubated in a thermocycler Primus 96 (PeQLab, Erlangen, Germany) at 94 °C for 5 min and passed through 30 cycles: Denaturation for 20 s at 94 °C, primer annealing for 20 s at 50 °C, and extension at 72 °C for 2 min. The amplified products were purified by using the FavorPrep™ Gel Purification Mini Kit (Favorgen Biotech Corporation, Ping-Tung, Taiwan). Purified PCR products were sent to First Base Laboratories, Singapore for sequencing. Sequences were aligned with a multiple sequence alignment program (Clustal W) by using Molecular Evolutionary Genetics Analysis (MEGA) 6. software) [20]. The phylogenetic relationships among halotolerant bacterial strains were studied by constructing phylogenetic trees by the neighbor-joining method [21].

2.3. Halophility Assay

L-broth medium supplemented with 0, 0.5, 1, and 1.5 M NaCl was used to evaluate the salt tolerance of bacterial strains. After inoculation, strains were incubated overnight at 37 °C for 24 h at 130 rpm/min. For all treatments, three replicates were placed for comparison. The optical density of the cultures was recorded at 600 nm by a spectrophotometer (CECIL CE 7200, Cecil Instruments Limited, Cambridge, UK). Proline analysis of bacterial strains grown under salt stress was also accomplished by following the method of Cha-Um and Kirdmanee [22].

2.4. Plant-Growth-Promoting Traits

For auxin quantification, strains were inoculated in triplicate in L-broth supplemented with 1 M NaCl and 500 μg mL^{-1} filter sterilized solution of L-tryptophan. After incubation at 37 °C for 72 h, cells were removed from stationary phase cultures by centrifugation. One milliliter of culture supernatant was mixed with 2 mL of Salkowski reagent [23]. Afterwards, samples were incubated in the dark for the development of pink color. Optical density was recorded at 535 nm with a spectrophotometer (CECIL CE 7200). The standard curve was constructed by using different concentrations of authentic auxins to calculate the auxin production by different bacterial strains. The phosphate solubilization potential of bacterial isolates was evaluated by streaking on Pikovskaya's agar plates [24].

The 1-aminocyclopropane-1-carboxylate (ACC)-utilization ability of the strains was evaluated by following the method described in [25]. Bacterial strains were inoculated in 5 mL of L-broth in triplicate and incubated overnight at 28 °C on a shaker at 200 rpm/min. Two milliliters of each culture were harvested in a 2-mL Eppendorf tube by centrifugation at 8000× *g* for 5 min. The cell pellet was washed twice with 1 mL of liquid Dworkin-Foster(DF) medium. Afterwards, the pellet was suspended in 2 mL of DF-ACC medium in a 12-mL culture tube. The tubes were incubated at 28 °C on a shaker at 200 rpm/min for 24 h. A 2-mL sample of DF-ACC medium without inoculation was incubated in parallel. One milliliter of each culture was centrifuged in a 1.5 mL Eppendorf tube as mentioned above. One hundred microliters of each supernatant were diluted to 1 mL with liquid DF medium in a 1.5-mL Eppendorf tube. Sixty microliters of each tenfold-diluted supernatant was used for the 96-well PCR-plate ninhydrin-ACC assay and mixed with 120 μL of ninhydrin reagent. The PCR plate was then heated on a boiling water bath for 30 min. The DF medium was used as a blank. Each diluted supernatant was run in triplicate. After transfer of 100 μL of the remaining reaction solution, the absorbance of the samples was measured at 570 nm with a microplate spectrophotometer (Epoch, BioTek, Winooski, VT, USA).

2.5. *Biofilm Formation*

The biofilm-forming ability of bacterial strains was evaluated by following the method of [26]. Bacterial strains were inoculated in triplicate in Tryptic Soy Broth (TSB) supplemented with 0, 200, and 400 mM NaCl and incubated at 37 °C for 24 h. Following incubation, 20 μL of the cultures was transferred to the wells of a 96-well flat bottom microtiter plate that contained 180 μL of TSB supplemented with the abovementioned NaCl concentrations. Negative controls contained only 200 μL of the TSB medium and the assay was performed in triplicate. To promote biofilm formation, the plates were incubated on a shaker at 37 °C for 72 h. After incubation, the well contents were discarded and washed thrice with 250 μL of sterile distilled water to remove any nonadherent and weakly adherent cells. Plates were air dried for 30 min. The biofilm formed in the wells was fixed with 250 μL/well 98% methanol for 15 min. After air drying, the fixed bacterial cells were stained with 200 μL of 0.1% *v*/*v* crystal violet solution for 5 min. Excessive stain was removed by placing the plate under low-running tap water and the plate was air dried. Resolubilization of crystal violet with the adherent cells was done by adding 200 μL/well of 33% *v*/*v* glacial acetic acid. The optical density was measured at 570 nm using a microplate spectrophotometer (Epoch, BioTek, Winooski, VT, USA).

2.6. *In Vitro Plant Bioassay*

Seeds of maize (*Zea mays* L. Var. Sahiwal-2002) were obtained from Punjab Seed Corporation (Lahore, Pakistan). Seeds were surface sterilized by treatment with 0.1% $HgCl_2$ for 2–3 min followed by repeated washing with sterilized distilled water. Rooting assay was performed with single or mixed bacterial cultures. Single bacterial cultures included F-9, F-11, F-12, F-35, F-37, F-58, F-71, F-72, F-81, F-83, F-84, and F-87. Mixed culture combinations C-1 (F-9, F-11, F-12, F-87), C-2 (F-35, F-71, F-81, F-83), and C-3 (F-37, F-58, F-72, F-84) were also included for seed inoculation. For inoculum preparation, strains were cultivated in L-broth for 24 h. Then, cultures were harvested by centrifugation and the cell pellet was washed with autoclaved distilled water. The optical densities of the cultures were adjusted to a final concentration of 10^7 colony forming units (CFU)/mL in sterilized distilled water. Sterilized seeds were soaked in a bacterial suspension for 30 min and placed in Petri plates containing two sterilized filter papers. Water-treated seeds were used for comparison. For each strain and combination, five seeds were inoculated in duplicate. Petri plates were incubated in an Environmental Test Chamber (MRL-350H; Sanyo, Osaka, Japan) at 25 °C. After 7 days, shoot length, root length, number of roots, and fresh and dry weights of seedlings were recorded.

2.7. *Pot Trials with Salt Stress*

Pot experiments were conducted to evaluate the growth-promoting effect of bacterial strains on maize under different salt-stress conditions. Seeds were surface sterilized as mentioned earlier and placed on moistened autoclaved double-filter paper in Petri plates. Healthy germinated seeds were selected for bacterization with single and mixed cultures of bacterial strains as mentioned above. Germinated seeds were sown in pots containing 100 g of autoclaved soil. Five seeds were sown in each pot and experiments were carried out in triplicate. Three salt-stress conditions, i.e., 0, 200, and 400 mM, were applied up to the field capacity of soil. Salt stress was applied twice—at the time of sowing and to 7-day-old seedlings. Water-treated seeds were used as a control. Pots were kept under conditions of 12 h of photoperiod and a temperature of 25 °C in the Environmental Test Chamber (MRL-350H;

Sanyo, Osaka, Japan). After two weeks, plant-growth parameters such as shoot length, root length, number of roots, and fresh weight of seedlings were recorded. For dry weight, plants were incubated in an oven at 80 °C for 24 h.

2.8. Biochemical Analysis of Plants

The proline content of plants grown under 0, 200, and 400 mM NaCl was measured as described by Cha-Um and Kirdmanee [22]. Similarly, the method of Racusen and Foote [27] was used for the quantitative estimation of peroxidases from fresh leaves. Acid phosphatase was extracted from plant leaves following the method of Iqbal and Rafique [28].

2.9. Statistical Analysis

Data for bacterial auxin production and plant-growth parameters were subjected to statistical analysis by using SPSS 20 software (IBM Corporation, New York, NY, USA). The data were also subjected to analysis of variance (ANOVA). Finally, mean values were separated by using Duncan's multiple range test ($p \leq 0.05$). Correlation analysis between bacterial growth and NaCl concentrations was also performed ($p = 0.05$).

3. Results

3.1. Strains Identification

One hundred salt-tolerant bacterial strains were isolated from the rhizosphere and phytoplane of *Suaeda fruticosa* (L.) Forssk. *fruticosa* growing in saline habitats. 16S rRNA gene sequences were compared with already-deposited sequences in GenBank. The comparison indicated around 99% similarity with respective identified species. Analysis showed that the strains belong to the genera *Gracilibacillus, Staphylococcus, Virgibacillus, Salinicoccus, Bacillus, Zhihengliuella, Brevibacterium, Oceanobacillus, Exiguobacterium, Pseudomonas, Arthrobacter*, and *Halomonas*. Sequences were submitted to GenBank under accession numbers KT027652–KT027742. Rhizospheric soil samples recorded the presence of 45 salt-tolerant rhizobacteria and were represented by 9 bacterial genera (Table 1). Similarly, the phytoplane recorded the presence of 46 endophytic bacterial strains represented by 8 bacterial genera (Table 2). Figure 1 shows the phylogenetic relationships among different salt-tolerant bacteria isolated from the rhizosphere and phytoplane of *S. fruticosa*. Analysis showed that the majority of the Gram-positive strains (firmicutes) clustered in one major group. This included strains from the genera *Bacillus* and *Staphylococcus*. A few Gram-positive strains were also represented by the genera *Oceanobacillus, Salinococcus, Gracilibacillus, Brevibacterium*, and *Exiguobacterium*. In addition to that, Gram-positive *Arthrobacter* and *Zhihengliuella* clustered in a separate group. Gram-negative strains were represented by the genera *Pseudomonas* and *Halomonas* and occupied a separate cluster in the phylogenetic tree.

Table 1. 16S rRNA gene sequencing of halotolerant bacterial strains isolated from the rhizosphere of *Suaeda fruticose* (L.) Forssk.

S. No.	Strains	Identified as	Accessions
1	F-1	*Gracilibacillus saliphilus* F-1	KT027652
2	F-2	*Staphylococcus petrasii* F-2	KT027653
3	F-3	*Virgibacillus salarius* F-3	KT027654
4	F-4	*G. saliphilus* F-4	KT027655
5	F-5	*Salinicoccus sesuvii* F-5	KT027656
6	F-6	*Bacillus licheniformis* F-6	KT027657
7	F-7	*B. subtilis* F-7	KT027658
8	F-8	*B. mojavensis* F-8	KT027659

Table 1. *Cont.*

S. No.	Strains	Identified as	Accessions
9	F-9	*Zhihengliuella flava* F-9	KT027660
10	F-10	*B. licheniformis* F-10	KT027661
11	F-11	*S. jettensis* F-11	KT027662
12	F-12	*S. arlettae* F-12	KT027663
13	F-13	*B. sonorensis* F-13	KT027664
14	F-14	*B. subtilis* F-14	KT027665
15	F-15	*B. aerius* F-15	KT027666
16	F-16	*B. licheniformis* F-16	KT027667
17	F-17	*B. subtilis* F-17	KT027668
18	F-18	*B. atrophaeus* F-18	KT027669
19	F-19	*Brevibacterium halotolerans* F-19	KT027670
20	F-20	*B. subtilis* F-20	KT027671
21	F-21	*B. licheniformis* F-21	KT027672
22	F-22	*B. axarquiensis* F-22	KT027673
23	F-23	*B. amyloliquefaciens* F-23	KT027674
24	F-24	*B. subtilis* F-24	KT027675
25	F-25	*B. subtilis* F-25	KT027676
26	F-26	*G. thailandensis* F-26	KT027677
27	F-27	*B. subtilis* F-27	KT027678
28	F-28	*Oceanobacillus picturae* F-28	KT027679
29	F-29	*S. devriesei* F-29	KT027680
30	F-30	*S. jettensis* F-30	KT027681
31	F-31	*S. petrasii* F-31	KT027682
32	F-32	*B. subtilis* F-32	KT027683
33	F-33	*B. infantis* F-33	KT027684
34	F-34	*B. subtilis* F-32	KT027685
35	F-35	*Exiguobacterium mexicanum* F-35	KT027686
36	F-36	*B. subtilis* F-36	KT027687
37	F-37	*S. caprae* F-37	KT027688
38	F-38	*S. devriesei* F-38	KT027689
39	F-39	*S. hominis* F-39	KT027690
40	F-83	*B. stratosphericus* F-83	KT027734
41	F-84	*B. pumilus* F-84	KT027735
42	F-85	*B. pumilus* F-85	KT027736
43	F-86	*B. marisflavi* F-86	KT027737
44	F-87	*Pseudomonas japonica* F-87	KT027738
45	F-89	*Sal. sesuvii* F-89	KT027740

Table 2. 16S rRNA gene sequencing of halotolerant bacterial strains isolated from the phytoplane of *S. fruticosa*.

S. No.	Strains	Identified as	Accessions
1	F-40	*Bacillus subtilis* F-40	KT027691
2	F-41	*Exiguobacterium* sp. F-41	KT027692
3	F-42	*Staphylococcus jettensis* F-42	KT027693
4	F-43	*Salinicoccus sesuvii* F-43	KT027694
5	F-44	*Oceanobacillus kapialis* F-44	KT027695
6	F-45	*B. flexus* F-45	KT027696
7	F-46	*B. pumilus* F-46	KT027697
8	F-47	*S. devriesei* F-47	KT027698
9	F-48	*B. malacitensis* F-48	KT027699
10	F-49	*B. malacitensis* F-49	KT027700
11	F-50	*E. aquaticum* F-50	KT027701
12	F-51	*O. polygoni* F-51	KT027702
13	F-52	*O. kimchii* F-52	KT027703

Table 2. *Cont.*

S. No.	Strains	Identified as	Accessions
14	F-53	*B. mojavensis* F-53	KT027704
15	F-54	*B. megaterium* F-54	KT027705
16	F-55	*B. pumilus* F-55	KT027706
17	F-56	*Sal. roseus* F-56	KT027707
18	F-57	*B. subtilis* F-57	KT027708
19	F-58	*B. megaterium* F-58	KT027709
20	F-59	*B. aryabhattai* F-59	KT027710
21	F-60	*B. tequilensis* F-60	KT027711
22	F-61	*B. malacitensis* F-61	KT027712
23	F-62	*B. subtilis* F-62	KT027713
24	F-63	*S. warneri* F-63	KT027714
25	F-64	*Arthrobacter bergerei* F-64	KT027715
26	F-65	*A. ardleyensis* F-65	KT027716
27	F-66	*A. arilaitensis* F-66	KT027717
28	F-67	*B. subtilis* F-67	KT027718
29	F-68	*B. cereus* F-68	KT027719
30	F-69	*S. arlettae* F-69	KT027720
31	F-70	*S. cohnii* F-70	KT027721
32	F-71	*S. arlettae* F-71	KT027722
33	F-72	*S. jettensis* F-72	KT027723
34	F-73	*Pseudomonas rhizosphaerae* F-73	KT027724
35	F-74	*S. gallinarum* F-74	KT027725
36	F-75	*S. petrasii* F-75	KT027726
37	F-76	*S. lugdunensis* F-76	KT027727
38	F-77	*S. capitis* F-77	KT027728
39	F-78	*S. pasteuri* F-78	KT027729
40	F-79	*S. jettensis* F-79	KT027730
41	F-80	*S. equorum* F-80	KT027731
42	F-81	*Halomonas nanhaiensis* F-81	KT027732
43	F-82	*B. safensis* F-82	KT027733
44	F-88	*O. kimchii* F-88	KT027739
45	F-90	*E. mexicanum* F-90	KT027741
46	F-91	*Exiguobacterium* sp. F-91	KT027742

3.2. Halophility Assay

Bacterial isolates showed variable growth responses at different NaCl concentrations (Figure 2). The majority of the isolates that included *S. jettensis* F-11, *S. arlettae* F-12, *S. caprae* F-37, *S. arlettae* F-71, *S. jettensis* F-72, and *B. safensis* F-83 showed good growth up to 1.5 M NaCl. A few strains recorded sensitivity against increasing levels of NaCl. For example, *Z. flava* F-9, *B. megaterium* F-58, and *B. pumilus* F-84 recorded poor growth as compared to other halotolerant bacterial strains. Proline is produced as an osmoprotectant under osmotic stress induced at high salinity. It accumulates as a compatible solute in living cells to counter the effects of NaCl toxicity. Bacterial proline content did not show a significant difference with the increasing salt content from 0 to 1.5 M NaCl. The highest concentration of proline was observed with *S. jettensis* F-11 (228 $\mu g\ g^{-1}$) and *S. jettensis* F-72 (216 $\mu g\ g^{-1}$) at 1.5 M NaCl (Figure 3).

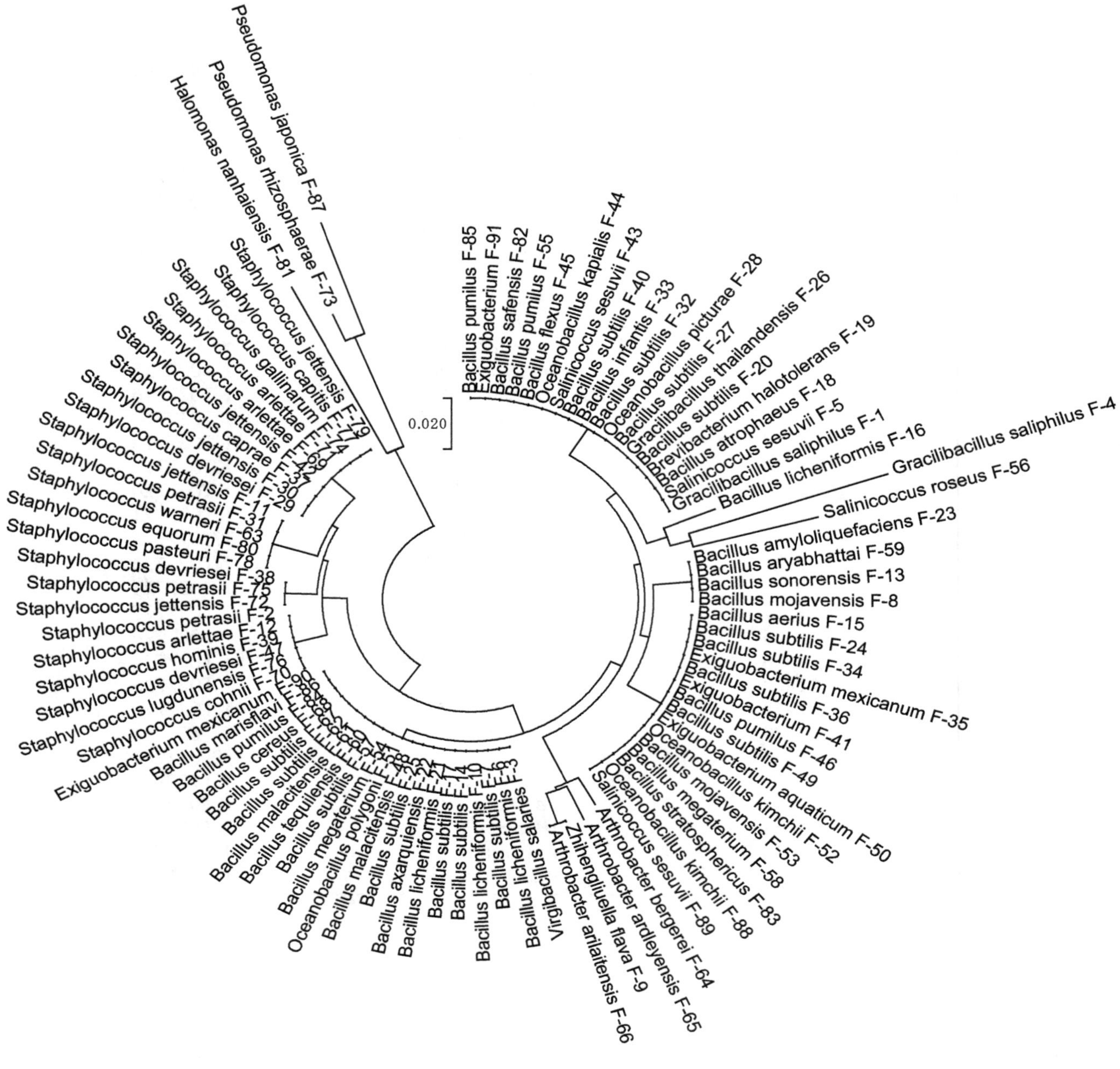

Figure 1. Phylogenetic tree showing the relationships among different halotolerant bacterial strains isolated from rhizosphere and phytoplane of *S. fruticosa*. (L.) Forssk. Scale bar represents mutations per nucleotide position.

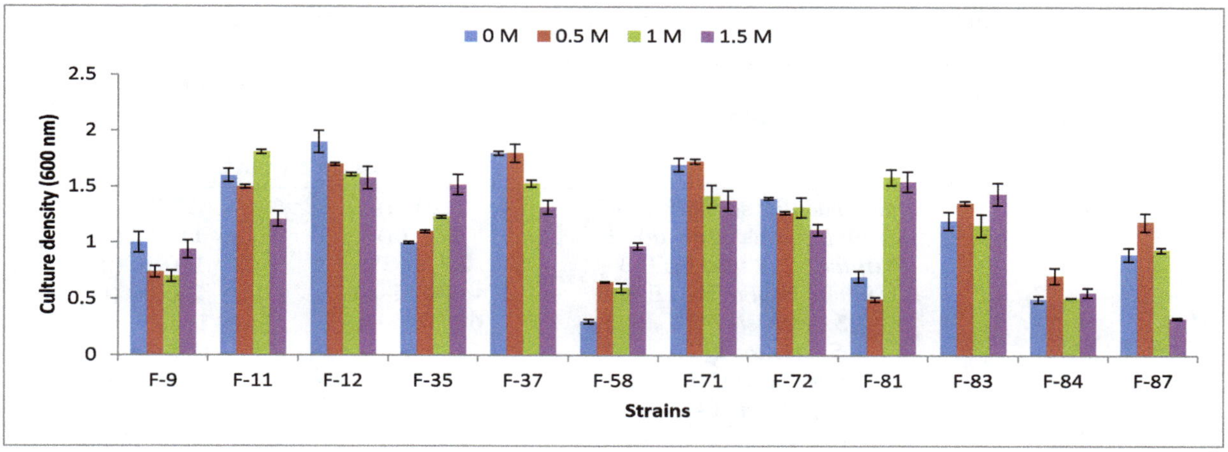

Figure 2. Halophility assay of salt-tolerant rhizobacteria in the presence of different concentrations of NaCl. Bar represents mean $\pm$ SE of three replicates.

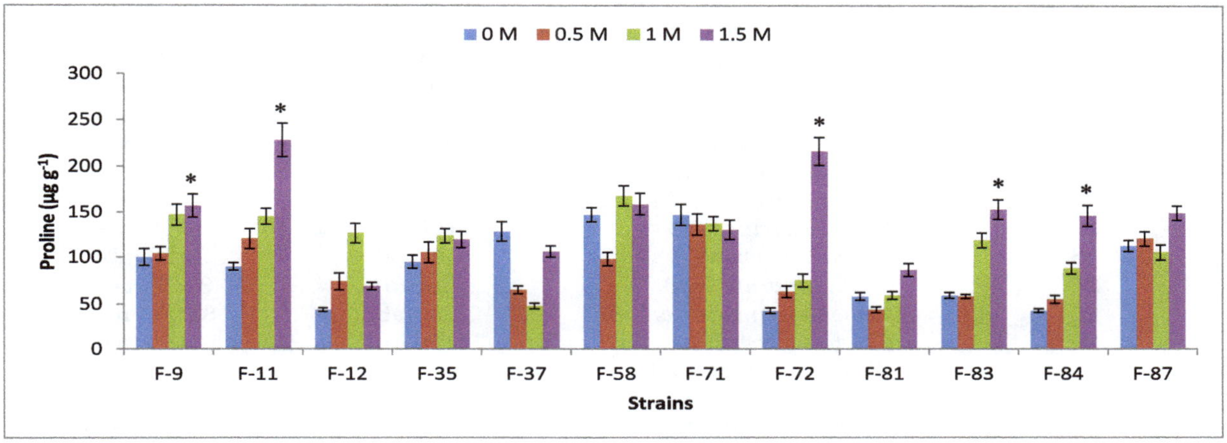

Figure 3. Effect of NaCl concentrations on proline content of halotolerant bacteria. Bars represent mean $\pm$ SE of three replicates. Treatments followed by * indicate significant difference over control using Duncan's multiple range test ($p \leq 0.05$).

3.3. Plant-Growth-Promoting Traits

One hundred bacterial strains were screened for in vitro auxin production by colorimetric analysis. Screening indicated significant auxin production by 30 bacterial strains (Table 3). In the absence of L-tryptophan, bacterial auxin production ranged from 5.70 to 87.90 μg mL^{-1}. Bacterial strains produced significant levels of auxin in culture supernatant in the presence of 500 μg mL^{-1} L-tryptophan as compared to the nonamended control. For instance, *S. petrasii* F-2, *V. salarius* F-3, *Z. flava* F-9, *S. jettensis* F-11, *S. arlettae* F-12, and *S. arlettae* F-71 recorded 10-, 2-, 5-, 3-, 3-, and 2-fold increases in auxin production over the control. Overall, auxin production ranged from 14 to 215 μg mL^{-1} in L-tryptophan-amended medium. On the other hand, 14 isolates were recorded as positive for phosphate solubilization. *S. jettensis* F-11, *S. arlettae* F-12, and *S. jettensis* F-72 were the most efficient as compared to other strains.

Bacterial isolates showed variable potential for ACC-deaminase activity. Highly significant activity was observed for *S. caprae* F-37, *S. arlettae* F-71, *B. subtilis* F-14, *E. mexicanum* F-35, and *O. kapialis* F-44. For biofilm formation, *S. caprae* F-37 and *B. pumilus* F-84 recorded good production at 200 mM NaCl. However, at higher NaCl levels, poor biofilm formation was recorded for the majority of the strains (Figure 4).

Table 3. L-tryptophan-dependent bacterial auxin production in the presence of 1 M NaCl.

S. No.	Strains	L-tryptophan (μg mL^{-1})	
		0	500
		Auxin (μg mL^{-1})	
1	*Gracilibacillus saliphilus* F-1	6.20 (a)	15.70 (a)
2	*Staphylococcus petrasii* F-2	9.30 (a)	113.50 (o)
3	*Virgibacillus salarius* F-3	87.90 (i)	150.00 (p)
4	*Zhihengliuella flava* F-9	44.70 (d–h)	215.00 (r)
5	*S. jettensis* F-11	63.20 (g–i)	185.70 (qr)
6	*S. arlettae* F-12	51.40 (e–i)	174.30 (pq)
7	*Exiguobacterium* sp. F-41	6.40 (a)	43.50 (a–f)
8	*S. jettensis* F-42	33.60 (d–g)	78.70 (l–o)
9	*Salinicoccus sesuvii* F-43	16.50 (abc)	28.40 (abc)
10	*S. devriesei* F-47	78.30 (hi)	28.00 (abc)
11	*Bacillus megaterium* F-58	20.80 (abc)	48.70 (a–g)
12	*B. aryabhattai* F-59	25.90 (d–g)	61.20 (i–n)
13	*S. warneri* F-63	6.40 (a)	14.00 (a)
14	*Arthrobacter bergerei* F-64	13.40 (ab)	25.50 (abc)
15	*A. ardleyensis* F-65	9.10 (a)	29.60 (abc)
16	*S. arlettae* F-69	14.30 (abc)	54.20 (h–n)
17	*S. cohnii* F-70	15.70 (abc)	42.10 (a–f)
18	*S. arlettae* F-71	53.60 (f–i)	88.60 (no)
19	*S. jettensis* F-72	30.60 (d–g)	59.80 (i–n)
20	*Pseudomonas rhizosphaerae* F-73	36.80 (d–g)	67.70 (g–n)
21	*S. gallinarum* F-74	15.40 (abc)	23.00 (abc)
22	*S. pasteuri* F-78	10.40 (a)	36.90 (a–c)
23	*S. jettensis* F-79	5.70 (a)	24.50 (abc)
24	*S. equorum* F-80	7.60 (a)	32.20 (a–d)
25	*Halomonas nanhaiensis* F-81	21.10 (abc)	83.30 (mno)
26	*B. safensis* F-82	9.50 (a)	43.90 (a–f)
27	*B. stratosphericus* F-83	23.60 (def)	72.70 (k–n)
28	*B. pumilus* F-84	7.90 (a)	18.10 (ab)
29	*B. pumilus* F-85	32.60 (a–d)	59.80 (i–n)
30	*B. marisflavi* F-86	22.60 (abc)	38.60 (a–e)

Mean of three replicates. Different letters followed by numerical values within same column indicate significant difference between treatments using Duncan's multiple range test ($p \leq 0.05$).

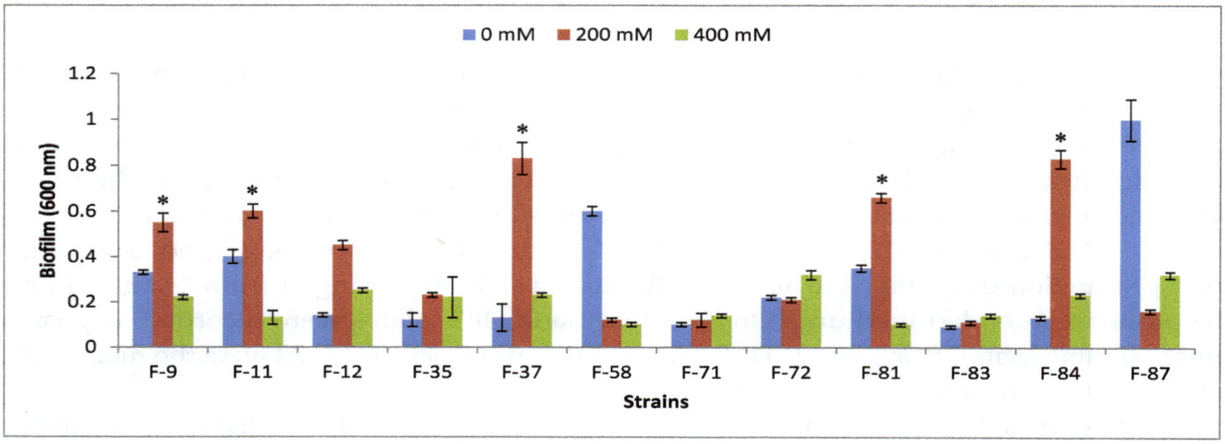

Figure 4. Effect of different NaCl concentrations on biofilm-forming ability of bacterial strains. Bars represent mean ± SE of three replicates. Treatments followed by * indicate significant difference over control using Duncan's multiple range test ($p \leq 0.05$).

3.4. In Vitro Plant Bioassay

Bacterial strains as single or mixed cultures significantly enhanced the growth of *Z. mays* under in vitro conditions. Maximum increases in shoot length over the water-treated control were observed for *S. jettensis* F-72 (61%), *S. arlettae* F-71 (48%), and *S. arlettae* F-12 (48%). Similarly, for root length, *S. arlettae* F-12, *S. jettensis* F-72, and *B. safensis* F-83 showed 92%, 85%, and 78% increases, respectively, over the control. A statistically significant increase in the number of roots per plant was recorded for *S. jettensis* F-72 and mixed culture C-2. For fresh and dry weights, *S. arlettae* F-12 and mixed culture combination C-1 were the most promising (Table 4).

Table 4. Effect of bacterial strains on growth parameters of *Zea mays* under in vitro conditions.

Strains	Shoot Length (cm)	Root Length (cm)	Roots/Plant	Fresh Weight (g)	Dry Weight (g)
Control	23 (a)	13 (b)	6.4 (de)	4.6 (a)	0.7 (ab)
F-9	28 (bcd)	20 (cd)	4.8 (a)	5.3 (ab)	0.9 (abc)
F-11	32 (defg)	20 (cde)	5.6 (abcd)	6.8 (a–e)	1.1 (bc)
F-12	34 (fgh)	25 (f)	6.2 (bcde)	8.8 (ef)	1.3 (c)
F-35	31 (c–g)	21 (de)	5.7 (abcd)	5.7 (abc)	0.8 (ab)
F-37	29 (cd)	20 (cde)	5.5 (abcd)	10.2 (f)	0.9 (abc)
F-58	33 (efgh)	20 (cde)	6.0 (bcd)	7.4 (b–f)	0.9 (abc)
F-71	34 (gh)	22 (def)	5.7 (abcd)	8.1 (cdef)	1.1 (bc)
F-72	37 (h)	24 (f)	7.1 (e)	7.8 (b–f)	1.0 (abc)
F-81	30 (cdef)	20 (cd)	5.0 (ab)	7.5 (b–f)	1.3 (c)
F-83	29 (cde)	23 (ef)	6.2 (cde)	4.2 (a)	0.6 (a)
F-84	31 (c–g)	22 (def)	5.1 (abc)	6.0 (abcd)	1 (abc)
F-87	28 (bc)	19 (cd)	5.4 (abcd)	8.6 (def)	1.1 (bc)
* C-1	22 (a)	10 (a)	5.6 (abcd)	9.1 (ef)	1.1 (bc)
* C-2	24 (ab)	14 (b)	6.5 (de)	7.8 (b–f)	1.3 (c)
* C-3	23 (a)	17 (c)	5.6 (abcd)	10.1 (f)	1.4 (ab)

Mean of 10 plants. Different letters within same column indicate significant difference between treatments using Duncan's multiple range rest ($p \leq 0.05$); * Mixed culture combinations: C-1 (F-9, F-11, F-12, F-87), C-2 (F-35, F-71, F-81, F-83), C-3 (F-37, F-58, F-72, F-84).

3.5. Pot Trials under Salt Stress

Halotolerant bacterial strains were evaluated to mitigate the effects of salinity on the plants of *Z. mays*. Bacterial strains were used as single or mixed cultures as mentioned above. Bacterization of seeds significantly enhanced growth parameters at 0 mM NaCl. A highly significant response for shoot and root length was shown by *S. arlettae* F-12 and *S. arlettae* F-71 over the water-treated control. For fresh weight, *Z. flava* F-9 was the most effective. For dry weight, *S. jettensis* F-11, *S. caprae* F-37, and *H. nanhaiensis* F-81 recorded up to threefold increases. Increases in salt content (0–400 mM NaCl) negatively affected the growth of plants in control treatments (Table 5). However, inoculations by halotolerant rhizobacteria significantly improved different growth parameters under saline conditions. At 200 mM NaCl, significant increases in shoot length were observed with *S. jettensis* F-11 (59%), *S. arlettae* F-12 (55%), and *S. arlettae* F-71 (50%) over the respective control. At higher stress (400 mM NaCl), *S. jettensis* F-11, *S. arlettae* F-12, and *B. marisflavi* F-87 recorded around a onefold increase for shoot elongation over the respective control. For root length, *S. jettensis* F-11, *S. arlettae* F-12, *S. arlettae* F-71, and *H. nanhaiensis* F-81 gave a promising response at 200 mM NaCl. Similarly, at 400 mM salt stress, the highest response for root elongation was observed with *S. jettensis* F-11 (85%) and *S. arlettae* F-12 (69%). For number of roots, *S. jettensis* F-72 (98%) and *E. mexicanum* F-35 (52%) recorded significant improvements over the control. At 400 mM stress, a 46% increase in root number was observed with *S. arlettae* F-12. Seed bacterization also significantly influenced plant weight under saline conditions. For instance, at 200 mM salt stress, *Z. flava* F-9 (45%) and *B. megaterium* F-58 (42%) demonstrated statistically significant improvements for fresh weight. Similarly, at higher NaCl content (400 mM), *S. jettensis* F-11 (44%) was the most effective (Table 5). On the other hand, for dry weight, *S. jettensis*

F-11 and *S. arlettae* F-71 recorded two- to threefold increases over the 200 mM control. At 400 mM salt stress, the mixed culture C-1 gave good results for dry weight.

Table 5. Effect of halotolerant bacterial strains on the growth of *Z. mays* in salt amended soils.

Strains	NaCl (mM)	Shoot Length (cm)	Root Length (cm)	Roots/Plant	Fresh Weight (g)	Dry Weight (g)
Control	0	22 (b–g)	13 (abc)	5.0 (jklm)	0.9 (abcd)	0.1 (a)
	200	22 (bcde)	16 (bcde)	4.2 (i–m)	1.2 (c–j)	0.1 (abcd)
	400	14 (a)	13 (abc)	4.3 (a–e)	1.1 (c–g)	0.1 (ab)
F-9	0	29 (i–o)	19 (e–k)	5.0 (b–j)	2.0 (r)	0.3 (e–l)
	200	29 (i–p)	21 (g–n)	5.2 (b–l)	1.7 (pqr)	0.2 (abcd)
	400	25 (d–i)	21 (f–m)	4.6 (a–f)	1.6 (l–q)	0.2 (c–h)
F-11	0	32 (n–s)	20 (e–k)	6.0 (e–l)	1.8 (qr)	0.4 (nop)
	200	35 (rs)	25 (m)	6.0 (f–l)	1.6 (m–r)	0.4 (op)
	400	32 (n–r)	24 (klmn)	5.0 (b–k)	1.6 (n–r)	0.2 (e–j)
F-12	0	34 (pqrs)	25 (mn)	6.2 (h–m)	1.6 (n–r)	0.3 (j–p)
	200	34 (pqrs)	23 (j–n)	6.0 (f–m)	1.4 (g–n)	0.3 (d–j)
	400	30 (k–r)	22 (h–n)	6.3 (jklm)	1.3 (f–l)	0.3 (e–k)
F-35	0	31 (l–r)	21 (g–n)	6.0 (f–l)	1.5 (j–p)	0.2 (c–h)
	200	26 (f–l)	21 (g–n)	6.4 (klm)	1.2 (c–i)	0.2 (c–g)
	400	25 (e–i)	21 (g–m)	5.0 (b–j)	1.0 (b–f)	0.2 (e–j)
F-37	0	29 (i–o)	20 (f–l)	5.5 (e–l)	1.0 (a–e)	0.4 (k–p)
	200	27 (g–m)	21 (g–n)	5.3 (b–l)	0.7 (ab)	0.1 (abc)
	400	23 (b–g)	16 (bcde)	5.1 (b–k)	1.4 (g–n)	0.3 (i–o)
F-58	0	33 (o–s)	21 (f–m)	6.0 (g–m)	1.2 (d–k)	0.3 (e–j)
	200	29 (i–o)	21 (f–m)	5.0 (b–i)	1.7 (opqr)	0.2 (c–i)
	400	23 (b–g)	21 (f–m)	3.4 (a)	1.1 (c–g)	0.4 (l–p)
F-71	0	34 (qrs)	22 (h–n)	6.0 (f–l)	1.4 (h–p)	0.3 (f–m)
	200	33 (o–s)	23 (j–n)	6.0 (g–m)	1.5 (l–q)	0.3 (e–k)
	400	23 (c–h)	18 (d–i)	5.0 (b–i)	1.5 (k–p)	0.3 (g–m)
F-72	0	37 (s)	24 (lmn)	7.1 (m)	1.7 (opqr)	0.3 (f–m)
	200	32 (n–r)	18 (d–h)	8.6 (n)	1.6 (m–r)	0.2 (c–h)
	400	25 (e–i)	21 (g–n)	5.0 (b–h)	1.3 (g–m)	0.3 (i–o)
F-81	0	30 (i–q)	19 (e–j)	5.0 (b–k)	1.4 (i–p)	0.4 (p)
	200	28 (i–o)	20 (f–l)	5.0 (b–g)	1.5 (j–p)	0.2 (bcde)
	400	20 (bcd)	16 (cdef)	4.6 (a–g)	1.0 (b–f)	0.1 (abc)
F-83	0	29 (i–o)	23 (i–n)	6.2 (i–m)	1.0 (a–f)	0.2 (c–g)
	200	29 (i–o)	22 (g–n)	5.5 (e–l)	1.0 (abcd)	0.2 (c–h)
	400	21 (bcde)	19 (e–k)	4.6 (a–f)	0.7 (a)	0.2 (b–f)
F-84	0	31 (m–r)	22 (h–n)	5.1 (b–k)	1.1 (c–g)	0.2 (d–j)
	200	28 (i–n)	18 (d–h)	5.0 (b–i)	1.1 (c–g)	0.2 (b–f)
	400	20 (bcd)	18 (d–h)	4.1 (abcd)	1.4 (g–o)	0.3 (h–n)
F-87	0	28 (i–n)	19 (e–j)	5.4 (d–l)	1.2 (c–h)	0.2 (d–j)
	200	32 (n–s)	22 (g–n)	6.0 (f–m)	1.2 (e–l)	0.3 (e–l)
	400	28 (h–n)	20 (f–l)	5.0 (b–k)	1.4 (g–n)	0.3 (e–j)
* C-1	0	22 (b–f)	10 (a)	6.0 (e–l)	1.3 (f–l)	0.2 (c–h)
	200	19 (bc)	12 (ab)	5.2 (b–l)	1.4 (g–n)	0.2 (b–f)
	400	25 (e–i)	18 (d–h)	4.0 (ab)	1.6 (m–r)	0.4 (mnop)
* C-2	0	24 (d–i)	14 (abcd)	6.5 (lm)	1.4 (g–n)	0.2 (c–h)
	200	19 (b)	15 (bcde)	5.4 (c–l)	1.4 (g–n)	0.2 (c–g)
	400	26 (e–k)	15 (bcde)	4.6 (a–f)	1.4 (g–o)	0.2 (c–h)
* C-3	0	23 (c–h)	17 (defg)	5.6 (e–l)	1.1 (c–g)	0.2 (abcd)
	200	22 (bcde)	18 (d–h)	4.6 (a–g)	1.0 (abc)	0.2 (c–g)
	400	19 (bc)	21 (f–m)	4.0 (abc)	1.2 (c–i)	0.3 (i–o)

Mean of 15 plants. Different letters within same column indicates significant difference between treatments using Duncan's multiple range rest ($p \leq 0.05$); * Mixed culture combinations.

3.6. Antioxidant Analysis

Bacterial inoculations recorded high proline accumulation in plants. At 200 mM NaCl, *S. arlettae* F-12, *B. safensis* F-83, and C-1 (mixed culture) recorded the highest accumulation of proline. Similarly, at 400 mM salt stress, *S. arlettae* F-12 and *S. arlettae* F-71 resulted in considerable increases in proline concentrations over water-treated control plants (Figure 5a). Regarding peroxidases, plants produced a very low content in water-treated seeds. However, with increasing salinity (200–400 mM), considerable enhancements in peroxidase activity was observed (Figure 5b). Acid phosphatase production significantly increased at 200 mM NaCl with *Z. flava* F-9, *S. arlettae* F-12, and *S. jettensis* F-72. At 400 mM salt stress, *Z. flava* F-9, *S. jettensis* F-72, and *B. safensis* F-83 recorded up to twofold increases for acid phosphatase content over the control (Figure 5c).

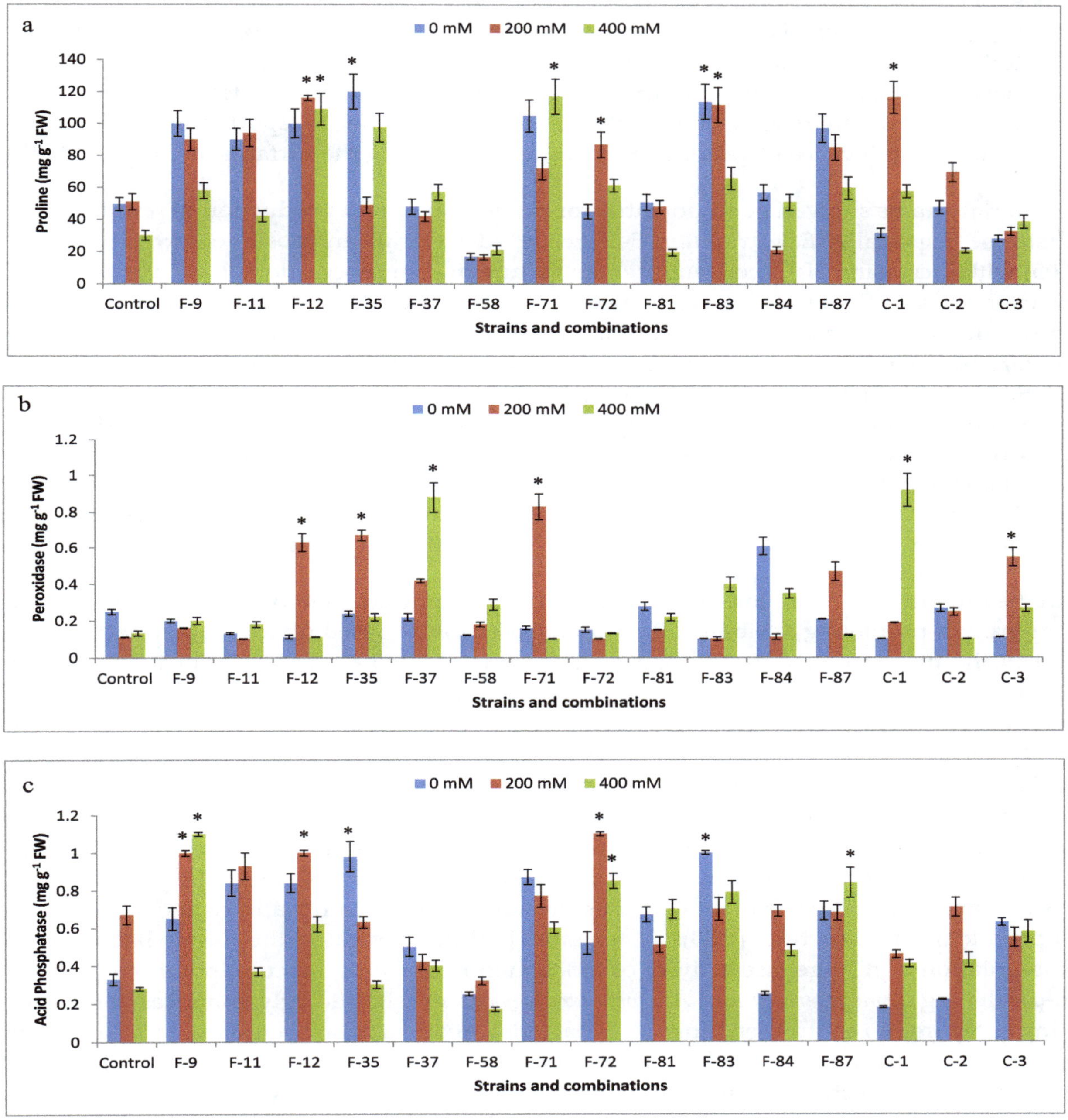

Figure 5. Effect of single and mixed bacterial inoculations and different salinity levels on the activities of antioxidant metabolites or enzymes: (**a**) Proline; (**b**) Peroxidase; (**c**) Acid phosphatase. Bars represent mean ± SE of three replicates. Treatments followed by * indicate significant difference over respective control using Duncan's multiple range test ($p \leq 0.05$).

4. Discussion

Soil salinity severely affects crop productivity and resident soil microbial communities. In Pakistan, soil salinity results in approximately 26% losses for the agroeconomy because it affects the germination, growth, and respiration of seeds [17]. The present study was carried out to examine the bacterial diversity of *S. fruticosa* growing in saline habitats of northeastern Pakistan. Extreme habitats have been reported to be colonized by a variety of microorganisms that are well adapted to these environments [29–31]. Natural plant settings growing in these localities may harbor beneficial microbes for agricultural applications under saline conditions [9]. *S. fruticosa* is a halophytic plant that mainly colonizes saline habitats. Therefore, we targeted the isolation of plant-growth-promoting rhizobacteria (PGPR) associated with the surfaces and rhizosphere of this plant. Analysis of 16S rRNA gene sequences confirmed the presence of several strains of halotolerant bacteria. Since little is known about the bacterial diversity associated with *S. fruticosa*, we therefore reported in detail the diversity and agricultural significance of bacterial communities colonizing the surfaces of this halophytic plant.

Bacterial strains showed good potential for salt tolerance and production of proline up to 1.5 M NaCl stress. Especially, *E. mexicanum* F-35 recorded a significant positive correlation ($r = 0.967$; $p = 0.05$) with increasing NaCl concentration in culture media. Inoculation of *Z. mays* by halotolerant rhizobacteria also influenced the production of antioxidant enzymes. For instance, plants grown under salinity stress exhibited the accumulation of proline as a compatible solute to retain water that otherwise might be lost due the surrounding hypertonic environment. Similarly, bacterial inoculations also influenced the peroxidase and acid phosphatase contents of plants. Peroxidases and acid phosphatases are considered very important enzymes to mitigate salinity-induced stresses. Exposure of plants to environmental stresses can lead to the production of reactive oxygen species (ROS) that can damage plant macromolecules. In addition to plants, microbes can augment the supply of enzymatic and nonenzymatic metabolites to detoxify the impact of ROS [32,33]. Previously, bioassays for the antioxidant system have also been reported with different plants and microorganisms [34,35].

In the present study, halotolerant rhizobacteria also showed good potential to produce a variety of plant-growth-promoting attributes. For example, strains recorded up to a 10-fold increase in auxin content in the presence of L-tryptophan over the unamended control. Under in vitro conditions, strains of genus *Staphylococcus* (F-12, F-71, F-72) were the most promising to enhance the rooting or shooting of inoculated maize plants. In pot trials, strains belonging to *Bacillus*, *Staphylococcus*, *Zhihengliuella*, and *Halomonas* genera were very effective at enhancing plant growth in salt-amended soils. However, strains of *S. arlettae* (F-12, F-71) recorded consistent results under both sets of experiments. Auxin-producing PGPR have been reported to improve plant growth under different sets of conditions. Charcoal-based formulations of rhizobacteria with auxin-production ability have been shown to enhance the vegetative and yield parameters of wheat [36]. Previously, we also reported the growth and yield improvement of plants grown under different salinity and water stresses by auxin-producing rhizobacteria [9,33]. Defez et al. [37] demonstrated the role of bacterial auxin in the upregulation of nitrogenase activity of inoculated rice plants. Inoculation of maize seeds with auxin-producing *B. megaterium* and *Azotobacter chroococcum* significantly enhanced the rooting and shooting compared to control seedlings [38]. Bacterial auxin has also been shown to alleviate the salt stress of plants grown under saline conditions. For instance, halotolerant bacteria from saline habitats stimulated plant growth by mitigating osmotic stress [9,39]. Similarly, salt-tolerant rhizobacteria exhibiting multiple plant-growth-promoting traits stimulated the growth of wheat in salt-affected soils [40]. In another study, strains of *Bacillus* facilitated the growth of maize under water-stress conditions [41].

High salinity levels in soil or in irrigation water are a major threat to the sustainability of agricultural production. The accumulation of salts in the vicinity of a plant's rhizosphere can cause severe injury to the root system. High concentrations of salts can result in the production of a gaseous hormone ethylene after seed germination. Elevated levels of ethylene can suppress the root system, which can result in compromised plant growth and development. Salt-tolerant rhizobacteria have the ability to produce the enzyme ACC-deaminase, which can degrade the substrate (ACC) of ethylene into α-ketobutyrate and ammonia [13,32]. Hence, the ACC-deaminase activity of rhizobacteria mitigates deleterious levels of stress-induced ethylene. In the present study, highly significant ACC-deaminase activity was exhibited by *S. caprae* F-37, *S. arlettae* F-71, and *B. subtilis* F-14. PGPR with ACC-deaminase activity can protect inoculated plants from the deleterious effects of abiotic stressors. Orhan et al. [42] reported the salt-stress alleviation of wheat by halotolerant and halophilic PGPR that exhibited ACC-deaminase activity. Halotolerant *Klebsiella* with ACC-deaminase activity has also been shown to promote the vegetative growth parameters and chlorophyll content of plants under salt stress [43]. Similarly, Tank and Saraf [44] reported the positive effects of ACC-deaminase-containing bacteria on plant growth against salinity stress.

Biofilm formation is another important trait of PGPR that can help plants tolerate abiotic stress. It is a complex association of bacterial cells attached to the plant's root system and plays a very critical role in retaining moisture and protecting against different biotic or abiotic stresses. In this study, *S. caprae* F-37 and *B. pumilus* F-84 demonstrated good production of biofilm at 200 mM NaCl. The biofilm-forming ability of halotolerant rhizobacteria has been shown to enhance plant biomass under saline conditions [45]. Inoculation of barley plants with biofilm-forming *B. amyloliquefaciens* ameliorated salt stress and also stimulated different plant-growth parameters [46].

5. Conclusions

Finally, it can be concluded that *S. fruticosa* growing in saline areas harbor a range of beneficial plant-growth-promoting rhizobacteria. Bacterial strains showed a good ability to grow in up to 1.5 M NaCl. Halotolerant bacterial strains showed good auxin-production potential, ACC-deaminase activity, and biofilm formation. In pot trials, strains recorded good results for the amelioration of salt stress of maize plants. Especially, *S. jettensis* F-11, *S. arlettae* F-12, *B. marisflavi* F-87, *Z. flava* F-9, and *H. nanhaiensis* F-81 were shown to exhibit multiple plant-growth-promoting traits. In pot trials, strains were also very effective at mitigating the salinity stress of maize plants. For instance, *S. jettensis* F-11, F-12, *S. arlettae* F-71, *B. marisflavi* F-87, *H. nanhaiensis* F-81, *Z. flava* F-9, and *E. mexicanum* F-35 enhanced plant-growth parameters under salt stress. The data presented in this work are very encouraging for the use halotolerant rhizobacteria to enhance plant growth under salinity stress. The results of this study also indicated that the natural plant settings of a saline habitat are a good source for isolating beneficial PGPR to grow crops under saline conditions.

Author Contributions: F.A. conducted the lab experiments related to the isolation, identification of bacterial strains and pot trials. B.A. developed the experimental layout, performed statistical analysis and wrote this manuscript.

References

1. Singh, K. Microbial and enzyme activities of saline and sodic Soils. *Land Degrad. Dev.* **2016**, *27*, 706–718. [CrossRef]
2. Labidi, N.; Ammari, M.; Mssedi, D.; Benzerti, M.; Snoussi, S.; Abdelly, C. Salt excretion in *Suaeda fruticosa*. *Acta Biol. Hung.* **2010**, *61*, 299–312. [CrossRef] [PubMed]
3. Khan, M.A.; Aziz, I. Salinity tolerance in some mangrove species from Pakistan. *Wetl. Ecol. Manag.* **2001**, *9*, 229–233. [CrossRef]
4. Mahajan, S.; Tuteja, N. Cold, salinity and drought stresses: An overview. *Arch. Biochem. Biophys.* **2005**, *444*, 139–158. [CrossRef] [PubMed]
5. Shrivastava, P.; Kumar, R. Soil salinity: A serious environmental issue and plant growth promoting bacteria as one of the tools for its alleviation. *Saudi J. Biol. Sci.* **2015**, *22*, 123–131. [CrossRef] [PubMed]

6. Hameed, A.; Hussain, T.; Gulzar, S.; Aziz, I.; Gul, B.; Khan, M.A. Salt tolerance of a cash crop halophyte *Suaeda fruticosa*: Biochemical response to salt and exogenous chemical treatments. *Acta Physiol. Plant.* **2012**, *34*, 2331–2340. [CrossRef]
7. Ngumbi, E.; Kloepper, J. Bacterial-mediated drought tolerance: Current and future prospects. *Appl. Soil Ecol.* **2016**, *105*, 109–125. [CrossRef]
8. Sharma, S.; Kulkarni, J.; Jha, B. Halotolerant rhizobacteria promote growth and enhance salinity tolerance in peanut. *Front. Microbiol.* **2016**, *7*, 1600. [CrossRef] [PubMed]
9. Raheem, A.; Ali, B. Halotolerant rhizobacteria: Beneficial plant metabolites and growth enhancement of *Triticum aestivum* L. in salt amended soils. *Arch. Agron. Soil Sci.* **2015**, *61*, 1691–1705. [CrossRef]
10. Bhise, K.K.; Bhagwat, P.K.; Dandge, P.B. Plant growth-promoting characteristics of salt tolerant *Enterobacter cloacae* strain KBPD and its efficacy in amelioration of salt stress in *Vigna radiata* L. *J. Plant Growth Regul.* **2017**, *36*, 215–226. [CrossRef]
11. Woodward, A.W.; Bartel, B. Auxin: Regulation, action and interaction. *Ann. Bot.* **2005**, *95*, 707–735. [CrossRef] [PubMed]
12. Nadeem, S.M.; Zahir, Z.A.; Naveed, M.; Asghar, H.N.; Arshad, M. Rhizobacteria capable of producing ACC-deaminase may mitigate salt stress in wheat. *Soil Sci. Soc. Am. J.* **2010**, *74*, 533–542. [CrossRef]
13. Glick, B.R. Bacteria with ACC deaminase can promote plant growth and help to feed the world. *Microbiol. Res.* **2014**, *169*, 30–39. [CrossRef] [PubMed]
14. Zhang, F.; Zhang, J.; Chen, L.; Shi, X.; Lui, Z.; Li, C. Heterologous expression of ACC deaminase from *Trichoderma asperellum* improves the growth performance of *Arabidopsis thaliana* under normal and salt stress conditions. *Plant Physiol. Biochem.* **2015**, *94*, 41–47. [CrossRef] [PubMed]
15. Li, H.Q.; Jiang, Z.W. Inoculation with plant growth-promoting bacteria (PGPB) improves salt tolerance of maize seedling. *Russ. J. Plant Physiol.* **2017**, *64*, 235–241. [CrossRef]
16. Ullah, S.; Bano, A. Isolation of plant-growth-promoting rhizobacteria from rhizospheric soil of halophytes and their impact on maize (*Zea mays* L.) under induced salinity. *Can. J. Microbiol.* **2015**, *61*, 307–313. [CrossRef] [PubMed]
17. Murtaza, G.; Murtaza, B.; Usman, H.M.; Ghafoor, A. Amelioration of saline-sodic soil using gypsum and low quality water in following sorghum-berseem crop rotation. *Int. J. Agric. Biol.* **2013**, *15*, 640–648.
18. Cappuccino, J.G.; Sherman, N. Microbiology: A Laboratory. In *Microbial Populations in Soil: Enumeration*, 6th ed.; Pearson Education: Singapore, 2002; pp. 349–354.
19. Johnson, J.L. Methods for general and molecular bacteriology. In *Similarity Analysis of rRNAs*; Gerhardt, P., Murray, R.G.E., Wood, W.A., Krieg, N.R., Eds.; American Society for Microbiology: Washington, DC, USA, 1994; pp. 625–700.
20. Tamura, K.; Stecher, G.; Peterson, D.; Filipski, A.; Kumar, S. MEGA6: Molecular evolutionary genetics analysis version 6.0. *Mol. Biol. Evol.* **2013**, *30*, 2725–2729. [CrossRef] [PubMed]
21. Saitou, N.; Nei, M. The neighbor-joining method: A new method for reconstructing phylogenetic trees. *Mol. Biol. Evol.* **1987**, *4*, 406–425. [PubMed]
22. Cha-Um, S.; Kirdmanee, C. Effect of salt stress on proline accumulation, photosynthetic ability and growth characters in two maize cultivars. *Pak. J. Bot.* **2009**, *41*, 87–98.
23. Tang, W.; Borner, J. Enzymes involved in synthesis and breakdown of indoleacetic acid. *Mod. Method Plant Anal.* **1979**, *7*, 238–241.
24. Pikovskaya, R.I. Mobilization of phosphorus in soil in connection with viral activity of some microbial species. *Mikrobiologiya* **1948**, *17*, 362–370.
25. Li, Z.; Chang, S.; Lin, L.; Li, Y.; An, Q. A colorimetric assay of 1-aminocyclopropane-1-carboxylate (ACC) based on ninhydrin reaction for rapid screening of bacteria containing ACC deaminase. *Lett. Appl. Microbiol.* **2011**, *53*, 178–185. [CrossRef] [PubMed]
26. Christensen, G.D.; Simpson, W.; Younger, J.; Baddour, L.; Barrett, F.; Melton, D.; Beachey, E. Adherence of coagulase-negative staphylococci to plastic tissue culture plates: A quantitative model for the adherence of staphylococci to medical devices. *J. Clin. Microbiol.* **1985**, *22*, 996–1006. [PubMed]
27. Racusen, D.; Foote, M. Protein synthesis in dark-grown bean leaves. *Can. J. Bot.* **1965**, *43*, 817–824. [CrossRef]
28. Iqbal, J.; Rafique, N. Toxic effects of $BaCl_2$ on germination, early seedling growth, soluble-proteins and acid-phosphatases in *Zea mays* L. *Pak. J. Bot.* **1987**, *19*, 1–8.

29. Demergasso, C.; Escudero, L.; Casamayor, E.O.; Chong, G.; Balagué, V.; Pedrós-Alió, C. Novelty and spatio-temporal heterogeneity in the bacterial diversity of hypersaline lake Tebenquiche (Salar de Atacama). *Extremophiles* **2008**, *12*, 491–504. [CrossRef] [PubMed]
30. Roohi, A.; Ahmed, I.; Iqbal, M.; Jamil, M. Preliminary isolation and characterization of halotolerant and halophilic bacteria from salt mines of Karak, Pakistan. *Pak. J. Bot.* **2012**, *44*, 365–370.
31. Goswami, D.; Dhandhukia, P.; Patel, P.; Thakker, J.N. Screening of PGPR from saline desert of Kutch: Growth promotion in Arachis hypogea by Bacillus licheniformis A2. *Microbiol. Res.* **2014**, *169*, 66–75. [CrossRef] [PubMed]
32. Forni, C.; Duca, D.; Glick, B.R. Mechanisms of plant response to salt and drought stress and their alteration by rhizobacteria. *Plant Soil* **2017**, *410*, 335–356. [CrossRef]
33. Raheem, A.; Shaposhnikov, A.; Belimov, A.A.; Dodd, I.C.; Ali, B. Auxin production by rhizobacteria was associated with improved yield of wheat (*Triticum aestivum* L.) under drought stress. *Arch. Agron. Soil Sci.* **2018**, *64*, 574–587. [CrossRef]
34. Chakraborty, U.; Chakraborty, B.; Chakraborty, A.; Dey, P. Water stress amelioration and plant growth promotion in wheat plants by osmotic stress tolerant bacteria. *World J. Microbiol. Biotechnol.* **2013**, *29*, 789–803. [CrossRef] [PubMed]
35. Estrada, B.; Aroca, R.; Barea, J.M.; Ruiz-Lozano, J.M. Native arbuscular mycorrhizal fungi isolated from a saline habitat improved maize antioxidant systems and plant tolerance to salinity. *Plant Sci.* **2013**, *201*, 42–51. [CrossRef] [PubMed]
36. Aslam, F.; Ali, B. Efficacy of charcoal based formulations of *Bacillus* and *Escherichia coli* to enhance the growth and yield of *Triticum aestivum* L. *Res. J. Biotechnol.* **2015**, *10*, 81–88.
37. Defez, R.; Andreozzi, A.; Bianco, C. The overproduction of indole-3-acetic acid (IAA) in endophytes upregulates nitrogen fixation in both bacterial cultures and inoculated rice plants. *Microb. Ecol.* **2017**, *74*, 441–452. [CrossRef] [PubMed]
38. Đorđević, S.; Stanojević, D.; Vidović, M.; Mandić, V.; Trajković, I. The use of bacterial indole-3-acetic acid (IAA) for reduce of chemical fertilizers doses. *Hem. Ind.* **2017**, *71*, 195–200.
39. Kim, K.; Jang, Y.; Lee, S.; Oh, B.; Chae, J.; Lee, K. Alleviation of salt stress by *Enterobacter* sp. EJ01 in tomato and *Arabidopsis* is accompanied by up-regulation of conserved salinity responsive factors in plants. *Mol. Cells* **2014**, *37*, 109–117. [CrossRef] [PubMed]
40. Tiwari, S.; Singh, P.; Tiwari, R.; Meena, K.K.; Yandigeri, M.; Singh, D.P.; Arora, D.K. Salt-tolerant rhizobacteria-mediated induced tolerance in wheat (*Triticum aestivum*) and chemical diversity in rhizosphere enhance plant growth. *Biol. Fertil. Soils* **2011**, *47*, 907. [CrossRef]
41. Sajid, M.; Raheem, A.; Ali, B. Phylogenetic diversity of drought tolerant *Bacillus* spp. and their growth stimulation of *Zea mays* L. under different water regimes. *Res. J. Biotechnol.* **2017**, *12*, 38–46.
42. Orhan, F. Alleviation of salt stress by halotolerant and halophilic plant growth-promotingbacteria in wheat (*Triticum aestivum*). *Braz. J. Microbiol.* **2016**, *47*, 621–627. [CrossRef] [PubMed]
43. Singh, R.P.; Jha, P.; Jha, P.N. The plant-growth-promoting bacterium *Klebsiella* sp. SBP-8 confers induced systemic tolerance in wheat (*Triticum aestivum*) under salt stress. *J. Plant Physiol.* **2015**, *184*, 57–67. [CrossRef] [PubMed]
44. Tank, N.; Saraf, M. Salinity-resistant plant growth promoting rhizobacteria ameliorates sodium chloride stress on tomato plants. *J. Plant Interact.* **2010**, *5*, 51–58. [CrossRef]
45. Qurashi, A.W.; Sabri, A.N. Bacterial exopolysaccharide and biofilm formation stimulate chickpea growth and soil aggregation under salt stress. *Braz. J. Microbiol.* **2012**, *43*, 1183–1191. [CrossRef] [PubMed]
46. Kasim, W.A.; Gaafar, R.M.; Abou-Ali, R.M.; Omar, M.N.; Hewait, H. Effect of biofilm forming growth promoting rhizobacteria on salinity tolerance in barley. *Ann. Agric. Sci.* **2016**, *61*, 217–227. [CrossRef]

6

Functional Metabolomics—A Useful Tool to Characterize Stress-Induced Metabolome Alterations Opening New Avenues towards Tailoring Food Crop Quality

Corinna Dawid * and Karina Hille

Chair of Food Chemistry and Molecular Sensory Science, Technical University of Munich, Lise-Meitner-Strasse 34, 85354 Freising, Germany; karina.hille@tum.de

* Correspondence: corinna.dawid@tum.de

Abstract: The breeding of stress-tolerant cultivated plants that would allow for a reduction in harvest losses and undesirable decrease in quality attributes requires a new quality of knowledge on molecular markers associated with relevant agronomic traits, on quantitative metabolic responses of plants to stress challenges, and on the mechanisms controlling the biosynthesis of these molecules. By combining metabolomics with genomics, transcriptomics and proteomics datasets a more comprehensive knowledge of the composition of crop plants used for food or animal feed is possible. In order to optimize crop trait developments, to enhance crop yields and quality, as well as to guarantee nutritional and health factors that provide the possibility to create functional food or feedstuffs, knowledge about the plants' metabolome is crucial. Next to classical metabolomics studies, this review focuses on several metabolomics-based working techniques, such as sensomics, lipidomics, hormonomics and phytometabolomics, which were used to characterize metabolome alterations during abiotic and biotic stress in order to find resistant food crops with a preferred quality or at least to produce functional food crops.

Keywords: plant stress; abiotic stress; biotic stress; metabolomics; phytometabolomics; sensomics; phytohormonics; liquid chromatography-mass spectrometry (LC-MS/MS); nuclear magnetic resonance spectroscopy (NMR); targeted metabolomics; untargeted metabolomics; functional food

---

1. Importance of Metabolomics for Agricultural Research

In the environment, plants are often exposed to an enormous number of biotic and abiotic stress factors, such as pathogen and insect infestation as well as extreme temperatures, drought, salinity, pollutants, heavy metals or nutritional deficiencies. That leads to harvest or quality losses, such as the formation of a pronounced bitter off-flavor and sometimes causes toxicological problems as well as huge global economic losses [1]. The breeding of stress-tolerant cultivated plants that would allow for a reduction in harvest losses and undesirable decrease in quality attributes requires a new quality of knowledge on molecular markers associated with relevant agronomic traits, on quantitative metabolic responses of plants on stress challenges, and on the mechanisms controlling the biosynthesis of these molecules [2]. Thereby, knowledge of the biologically active metabolites, especially secondary metabolites as well as metabolic networks of primary and secondary metabolites affected by plant stress conditions, is essential.

The metabolome of plants represents the complete set of low-molecular weight metabolites (such as primary metabolites including all necessary metabolic intermediates, hormones and other signaling molecules, as well as secondary metabolites) in a given organism, a biological cell, tissue, or organ at a certain point in time and development. For a long time, primary and also secondary metabolites were simply considered as one of the end-products of gene expression and protein activity. Nowadays, it is increasingly accepted that low-molecular weight molecules modulate macromolecular processes through, e.g., feedback inhibition and by signaling phytohormons [3]. Therefore, Dixon et al. [3] conclude that, metabolomic studies are "intended to provide an integrated view of the functional status of an organism". But monitoring the whole metabolome with exclusively one technique is not possible at the moment [4]. Although for a single plant like *Arabidosis thaliana* no more than 5000 metabolites have been assumed, the plant kingdom is reported to contain between 200,000 and 1,000,000 metabolites, all of them having different structural features and polarities [5–9]. This wide structural diversity with different chemical properties is a challenging task during sample extraction and analysis [4]. In addition, the knowledge about functional, secondary metabolites, on having an antifungal, toxic, off-flavor activity or promising human health-promoting activities, is not yet sufficient.

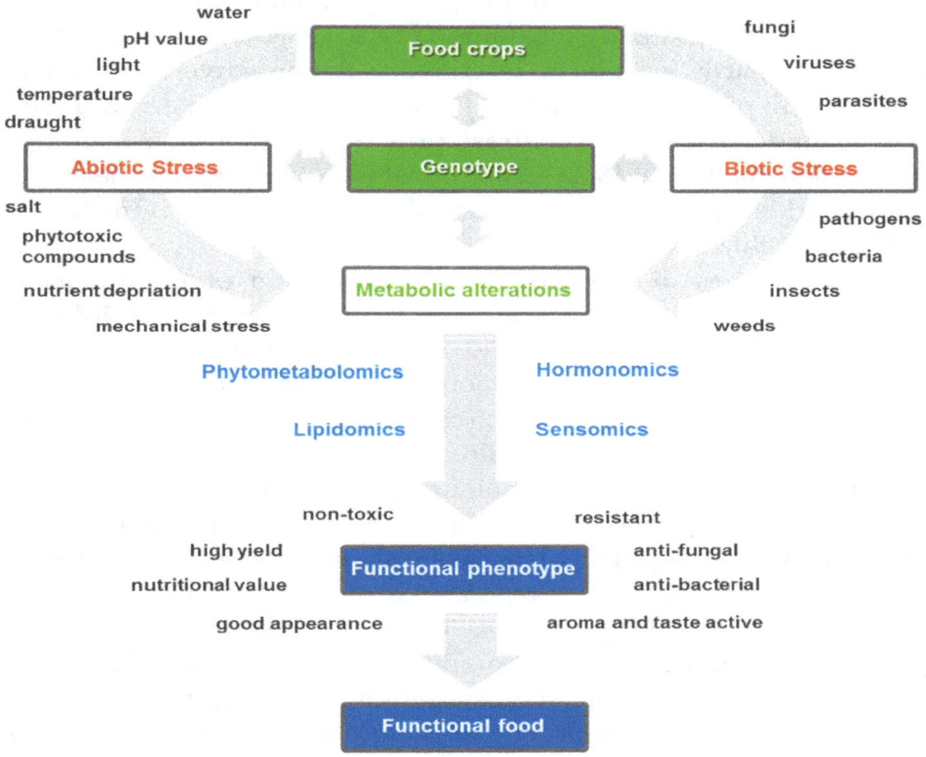

Figure 1. Functional metabolomics—a useful tool to discover metabolome alterations during abiotic and biotic stress to find the perfect functional pheno-/genotype to produce functional food.

Therefore, an important aspect of the use of crop plants for human consumption is the characterization and determination of their nutritional value as well as their individual flavor-active, health-promoting, non-toxic, functional constituents. The aim of crop plant research should be to find an optimal functional pheno-/genotype as a golden standard with high resistance and best possible quality criteria for future breeding experiments (cf. Figure 1). Analytical marker compounds can help to support molecular breeding efforts to obtain phenotype knowledge-based plant populations, no matter which analysis method is used, bulk segregant analysis, mapping by sequencing or quantitative trait loci (QTL) analysis [10,11]. In sum, metabolomics, especially the characterization of functional relevant metabolites, also called functional metabolomics, can be used as a tool for functional genomics [2,9,12,13]. For example, integration of metabolomic data with genomics and transcriptomics, especially that obtained from transcriptome coexpression network analysis, helped to decode several functions of genes and to gain a deep insight into the biological processes of the model plant *Arabaidopsis* [2,9]. Although more studies are desirable, at the moment only manuscripts are available in which secondary, functional metabolites of non-model plants are characterized [2,9,12,13]. Next to studies using potatoes, strawberries, carrots or *Brassica napus* [9,14–18], Rambla et al. [18] performed a QTL analysis using a cross between *Solanum lycopersicum* and a red-fruited wild tomato species *Solanum pimpinellifolium* to relate 102 QTLs to 39 corresponding volatile organic compounds, which partly contribute to the typical flavor impression of tomatoes. In addition, by means of genome-wide association studies 323 associations among 143 single nucleotide polymorphisms and 89 metabolites were identified, including non-volatile, functional, secondary metabolites, which can be used for new human drug production [19].

Next to the genetic code, environmental stress especially influences the metabolome composition of plants. Stress conditions are sensed by the plant, which then activates a network of signaling pathways, including the participation of phytohormons, and leads to changes, on the one hand, in the primary metabolism and, on the other hand, in the up-regulation of phytochemicals [20].

For example, during fungi infestations, antifungal phytochemicals, especially metabolic pathways leading to isoprenoids, phenylpropanoids, alkaloids, fatty acids, and polyketides are up-regulated [21]. Many phytoalexins, which partly have been described as flavor-active, nutritional components and as a source for development of health-promoting food products, have been well documented in crop plants in the field of plant defense [21]. In contrast, primary metabolism is documented to be mainly influenced by abiotic stress challenges [5,20–22]. For instance, in *Arabidopsis* leaves, next to the amino acid profiles, oligosaccharides, γ-amino butyrate and the metabolites from the tricarboxylic acid cycle are influenced during drought conditions [23].

Analysis of those changes in the molecular composition of plant material by means of mass spectrometry (MS) or nuclear magnetic resonance spectroscopy (NMR) based techniques is called metabolomics, which can profile the impact of time, stress, nutritional status and environmental perturbation on hundreds of metabolites simultaneously. It is an essential technology for functional genomics and systems biology which can visualize and answer questions about biological systems [3,24]. Kushalappa and Gunnaiah [25] already predicted that while in the past, genetic tools were mainly used for crop improvement on yield, flagged as the green revolution, in the next few decades functional genomics, which allows the identification of genes that enhance crop yields without losses in nutritional value while minimizing loss due to stress, will be the main topic. Metabolomics is especially helpful to discover metabolites up-regulated during stress conditions [2]. These metabolites then can be tested for their biological functions. Thereby, the combination and understanding of information received from genomics, transcriptomics, proteomics and metabolomics studies is a complementary tool for functional genomics and systems biology investigations. Although nowadays, next to the human genome [26], the genomes of several plants such as *Arabidopsis thaliana* [27], rice [28,29], tomato [30], and barley [31] have been decoded by means of automated nucleotide sequencing, knowledge about the regulation of gene expression, about the action of these genetic products, and especially about the metabolic networks resulting from catalytic proteins, is rather fragmentary [3].

Besides aiming to avoid quality or yield losses induced by abiotic or biotic stress conditions, there are exciting new strategies of using plant stress as a way of producing phytoalexin-enriched functional foods [32] or to enhance the flavor quality of crop plants by stress or elicitor challenges to produce new functional foodstuffs with high quality [33]. Dixon et al. [3] additionally suggest that the long-term goal should be the application of new knowledge of functional metabolites for "rational custom-designed breeding by classical methods as well the application of genetic engineering techniques to improve and develop new aromatic plants". Therefore, publicly open databases, containing all spectroscopic, spectrometric and bioactivity data as well as the biosynthetic pathways of worthy metabolites are necessary in the future [3].

Targeted as well as non-targeted metabolomic-based working techniques, such as phytometabolomics, sensomics, lipidomics or hormonomics can help to unravel metabolic pathways, signal transmissions as well as metabolites with different bioactivity which can be simultaneously up- or down-regulated by stress conditions. After decoding marker compounds by means of these non-targeted approaches, marker metabolites can be isolated from the plant material and characterized by means of 1/2D-NMR or MS based structure identification techniques as well as by biological assays (e.g., using antifungal susceptibility test systems). Although the last-named structure identification techniques and the testing of the metabolite's bioactivity do not belong to the metabolomics approach, its application is mandatory to understand biological processes.

2. Analytical Techniques Used to Characterize Stress-Induced Metabolome Alterations in Plants

Enormous technological progress has been achieved since the early days of plant metabolome characterization. Thin-layer chromatography (TLC) together with gas chromatography coupled to a flame ionization detector (GC-FID) have so far been applied to characterize primary and secondary metabolites in plant materials. Nowadays, besides NMR, high-throughput, sensitive techniques using gas chromatography, liquid chromatography, ultra-high performance liquid chromatography/time-of-flight as well as capillary electrophoresis coupled to mass spectrometry (GC-MS, LC-MS/MS, UHPLC-TOF-MS or CE-MS), together with bioinformatics tools, are state-of the art working techniques [3,13]. With the development of mass spectrometry imaging approaches, including matrix-assisted laser desorption (MALDI) or desorption electrospray (DESI) ionization procedures coupled to high resolution mass spectrometers, it is possible to perform metabolomics analysis in situ, as well [34]. Although NMR-based techniques are used in plant metabolomics, it is less common in plant-stress metabolomics. However, new working techniques including NMR-technology, such as differential off-line LC-NMR (DOLC-NMR), show promise for characterizing metabolites up- or down-regulated during biotic or abiotic stress influences on plants [35].

Although the term plant metabolomics is defined as the identification and quantification of all low-molecular weight substances in an organism, at a defined point of harvest time and development stage and in a given organ, tissue or cell type, there is no single work-up protocol or standard technique available to detect all metabolites using only one platform. Therefore, several analytical separation and detection techniques are usually combined to visualize abiotic and biotic stress induced metabolome alterations of different plants induced by abiotic and biotic stress [4].

In the last four decades, different methods have been established for the targeted and non-targeted characterization of plant metabolites (cf. the following reviews: Shualev et al. [20]; Urano et al. [22]; Obata and Fernie, [5]; Arbona et al. [4]; Kumar et al. [2]). Nowadays, we differentiate between the non-targeted, sometimes also called untargeted, and the targeted metabolomics approach. While the non-targeted approach represents a comprehensive analysis of all measurable compounds, including unknown metabolites, targeted metabolomics is the measurement of defined groups of chemically characterized and biochemically annotated substances [36]. For both, targeted and non-targeted analyses GC-MS, UHPLC-TOF-MS, as well as LC-MS/MS are the most common techniques to characterize stress-induced metabolome alterations in plants [5].

In particular, primary metabolites, such as carbohydrates, amino acids or fatty acids and all volatile compounds are mainly characterized by means of targeted GC-MS with or without derivatization [37]. Due to the fact that standard compounds are commercially available and spectrometric data is published in several public libraries, such as NIST (National Institute of Standards and Technology), the identification and exact quantification of primary metabolites by means of GC-MS analysis is straightforward. In contrast, structure identification of secondary metabolites exclusively by means of MS data is rather challenging [4,38]. Metabolomics approaches mainly identify secondary non-volatile compounds by means of non-targeted UHPLC-TOF-MS techniques. In contrast to GC-MS, LC-MS libraries for structure identification are less developed, as instrument-type dependent mass spectra and MS fragmentation patterns, as well as retention time shifts depending on used LC columns complicate the comparison of structure identification results using different instruments [39]. In the future, a worldwide concept of raw data processing, extensive mass spectral libraries, and powerful database management systems that can store and provide both raw and meta data is desirable.

At the moment, biological active compounds are unequivocally identified by means of MS and 1/2D NMR experiments after their isolation by means of preparative HPLC techniques. But the MS spectra obtained has mostly not been shared with the metabolomics community due to the absence of a standardized data-handling concept. Once identified, sensitive quantitation methods using stable isotope dilution assays (SIDA) are carried out by means of LC-MS/MS or GC-MS techniques. Therefore, nowadays, determination of the exact concentration of the isotopically labelled internal standards can best be achieved by means of quantitative NMR spectroscopy (q NMR) [40].

Although the aforementioned analytical metabolomics techniques nowadays are reliable and very sensitive, they still have their limitations regarding small sample volumes or low quantities of metabolites (partly in the nanomolar range). This is especially so when metabolomics is used to answer specific questions in the field of functional genomics or systems biology to get a comprehensive understanding of plant growth, development, defense or productivity, and knowledge about metabolic pathways and networks at cellular levels is required [9,12,41–43]. In addition to single cell-cell genomics, transcriptomics or proteomics, several groups developed plant single-cell and single-cell type metabolomics during the last decade [43–46]. While using single-cell metabolomics, the metabolome of one individual cell, such as a single epidermal cell, is analyzed, while in single-cell-type metabolomics the metabolome of a population of cells, such as guard cells or trichomes, is characterized [43]. Whereas in studies using single animal cells, more than 300 compounds could, nowadays, be visualized by means of capillary electrophoresis-electrospray ionization-time-of-flight-mass spectrometry (CE-ESI-TOF-MS) methods. The amounts of metabolites characterized in smaller plant-based single-cells is still much lower [42,43]. To gain a better understanding of biological systems and characterize metabolic pathways in single cells or in situ, special attention should, in the future, be given to the development of more sensitive MS apparatuses. This is especially the case in the development of more sensitive visualizing MS techniques, such as MALDI or DESI imaging experiments, and in the improvement of extraction protocols [13,43]. Newly developed derivatisation strategies are also promising strategies to improve the MS sensitivities in future. Next to limitations in measuring low sample amounts, data interpretation sometimes is challenging.

A major challenge for combining all omics-techniques is the integration of detailed metabolomics data into transcriptomic and proteomic profiling data sets [3,9,12,13,20]. Although Arbona et al. [4] only recently reviewed the role of metabolomics as a tool to investigate abiotic plant stress in the context of systems biology, and how metabolomics can be integrated into other omics data sets, there is still a lack of knowledge. Furthermore, Urano et al. [22] summarized arguments that holistic omics analysis and their data interpretation are essential to identify the broad function of metabolite regulatory networks during the responses to plant stress. Therefore, bioinformatics knowledge and tools should be used more often in the future to combine the information from both research fields. Although databases and pathway viewers (e.g., an overview is given by Bino et al. [12] or de Souoza et al. [47]) have been further developed in recent years, inter alia Sumner et al. [13] also emphasized that standardized databases integrating metabolomics with other global-omics data are required.

3. Phytometabolomics—From Plant Stress to Metabolic Response

Plants continuously encounter various biotic and abiotic environmental stresses during their growth and development phases. While biotic stress is caused by pathogens, parasites, predators, and other competing organisms, abiotic stress arises from inappropriate levels of physical components in the environment, such as temperature or water extremes [21]. Both stress factors lead likewise to yield and quality losses in plant crops used for human nutrition. The aim is, on the one hand, to avoid those immense quality and yield losses to ensure feeding the world's growing population and on the other hand, to develop new functional foods and to reduce insecticides and pesticides used in agriculture. Therefore, knowledge of molecular networks and pathway activation of plants during their stress response is essential.

Only recently Mithöfer and Boland [48] asked the question "Do you speak chemistry?". They highlighted that plants always respond by emitting chemical compounds to signal their environment. Moreover, they pointed out that plants use metabolites as chemical sensing and communication systems. On the one hand, signaling molecules and phytoalexins are formed, on the other hand it is suggested that plants "cry for help" by producing secondary metabolites as indirect defense system. Those cry-for-help compounds, usually belonging to the group of volatile organic compounds, attract predators or herbivores helping the plant to get rid of other stress triggers [48,49]. Therefore, it is important to notice that stress-specific as well as common metabolites, so-called generalist and specialist, can either be formed or their amounts up-regulated [5].

The understanding of metabolic stress pathways is just possible if comparative metabolite characterization by means of metabolomics analysis of stressed and non-stressed plants followed by structure identification experiments is performed. In the following three sub-sections an initial overview of the literature about biotic and abiotic stress studies is undertaken.

3.1. From Abiotic Plant Stress to Metabolic Response

It is well expected that climatic changes due to the global warming will lead to massive abiotic stress factors influencing the metabolome spectra of all plants in the future. Next to water stress, including soil flooding and drought stress, salt stress, temperature stress, light stress, sulfur and phosphorus stress, oxidative stress, as well as heavy metal stress will play important roles for food crop producers. To understand and avoid quality and yield losses induced by abiotic stress factors, several metabolomics studies have been carried out (cf. the following reviews: Shulaev et al. [20]; Kaplan et al. [21]; Urano et al. [22]; Obata and Fernie [5]; Arbona et al. [4]).

Among them, Arbona et al. [4] already summarized in their review "Metabolomics as a tool to investigate abiotic stress tolerance in plants" that the plants' metabolic responses to abiotic stress can mainly be found in the primary metabolism. Next to metabolic responses in mono- and disaccharides, sugar alcohols, amino acids, especially proline, polyamines and members of the tricarboxylic acid cycle were found to be influenced by abiotic factors. Thereby, the carbohydrate metabolism in particular is directly influenced by stress conditions, because during stress instead of glucose, plants use fructanes and starch as an energy source [50].

In their review, Obata and Fernie [5] compared several metabolic fingerprints of *Arabidopsis* leaves influenced by dehydration, salt, heat and cold, high light and sulfur limitation, ultraviolet (UV) light quality change, low nitrogen amounts, and potassium limitation with each other. Thereby, they highlighted that specific compounds and compound classes are generally accumulated during abiotic stress factors (except of light treatment), such as sucrose, raffinose, proline, other branched chain amino acids or γ-aminobutyric acid (GABA). However, the amounts of such up-regulated "generalists" varies between stress factors. They concluded, that the stress-specific plant response is the result of an inhibition or activation of a defined metabolic pathway. In particular, enzymes activities are known to be influenced by temperature or ion concentrations in plants. By contrast, other metabolites are accumulated only under specific conditions, such as trehalose.

Next to the carbohydrates, the lipid composition of plants is additionally influenced by abiotic stress factors. Lipids in plants are a crucial and diverse class of biomolecules that forms the so called plant lipidome. Lipidomics is defined as the identification and quantification of all lipids within a biological system at a defined stage of development [51]. Within the field of omics-analysis lipidomics represents a sub-unit of targeted and non-targeted metabolomics approaches, mainly using MS- and NMR-based techniques. The development of lipidomics and its increased promise in systems biology over the past two decades has been the subject of several reviews (e.g., Blanksby and Mitchell [34]). Changes in the lipidome, induced by modifications of its biosynthesis, regulation, adaption, remodeling, function, role, and interaction, as well as membrane lipid remodeling are relevant responses from plant cells to counteract biotic and abiotic stress challenges [51]. In particular, the oxidation of polyunsaturated fatty acids is one of the most fundamental reactions in lipid chemistry, which for example leads to off-flavor (perceived rancidity) or to phytohormon precursor formations [52]. Different groups of lipids, such as fatty acids, phosphatidic acids, inositol phosphates, diacylglycerols, oxylipins, shingolipids, and *N*-acetylethanolamine are involved in signaling systems during stress conditions as well [53–58]. Those compounds are shown to be directly synthesized after abiotic stress influences by a wide range of enzymes, such as fatty acid amide hydrolases, phospholipases, acyl hydrolases, diacylglycero kinases or phytoshingosine kinases [58]. In addition, it is known that in response to abiotic stress factors, lipids migrate into cell walls to repair damage through membrane remodeling [58–60]. Moreover, a combination of transcriptomics and lipidomic profiling showed that cold stress induces the prokaryotic pathway and suppresses the eukaryotic pathway for glycerolipid biosynthesis [61]. Next to cold stress, light and temperature stress, as well as nutrient starvation as typical abiotic stress factors were studied by means of lipidomic approaches in *Arabidopsis* (for details cf. the following review: Tenenboim et al. [51]). Only recently, lipidomics studies led to the clarification of the role of leaf lipids in thyme plant response to drought stress [58].

In the future, more stress-related lipidomic studies are expected, due to the fact that in nutritional research lipid profiling also plays a crucial role [51]. The food industry, for example, shows great interest in the use of lipidomics to characterize defined crop plant genotypes. As a replacement of palm kernel fat by a natural alternative, new plant genotypes having the same triglyceride pattern are needed. Furthermore, although most genes annotated as lipid-related are functionally characterized, their corresponding metabolites are not assigned. That would be an interesting field of research in the future, as well.

Next to numerous papers and reviews which highlight metabolome alterations induced by one single stress factor, Karin Köhl's review [62] gives a very nice overview of the metabolites influenced by combined abiotic stress effects in crops. Due to the fact that plants often have to deal with more than one stress factor, further metabolomics studies in crops which take several stress factors into account would be useful in the future [62].

In sum, although integrated omics data of model plants such as *Arabidopsis* have markedly increased our knowledge of understanding the mechanisms involved in plants' response to various abiotic stress factors [22,63,64], knowledge about secondary metabolome alterations in abiotic stressed food crops is rather fragmentary. Only few studies indicate abiotic stress influences the amounts of phenolic compounds, glucosinolates, carotinoides terpene derivatives and phytohormons in plants [4,51]. Urano et al. [22] emphasize again and again that phytohormon levels, especially those of abscisic acid, vary a lot in different plant compartments. Next to knowledge gaps in the stress influence on the secondary metabolome, more studies are needed to characterize stress combinations as well [5].

3.2. From Biotic Stress Metabolomics to Metabolic Response

During biotic stress exposure, plants use qualitative and quantitative measures to resist pathogen attacks [65]. While, in the past, the qualitative resistance, based on monogenetic inheritance, has been successfully used to elite cultivars to improve resistance, the quantitative resistance, which is presumed to be durable, non-race-specific, and effective against a large bouquet of pathogens, is mainly unknown [25]. However, a huge number of quantitative trait loci (QTLs) associated with qualitative resistance against pathogen-associated molecular patterns (PAMs), found for example in soy beans (*Glycine max*), and with quantitative resistance, found for blast in barley (*Hordeum vulgare*), or powdery mildew in wheat, have already been identified for biotic stress challenges [25,66–70]. However, both the mechanisms of resistance and the metabolites up-regulated during plant stress controlled by the QTLs are mainly unexplored [66]. In particular, targeted metabolomics approaches enable the possibility to monitor marker compounds up-regulated during plant stress. They can then be isolated, identified and biologically characterized, for example by application of an antifungal test.

Non-targeted metabolomics in comparison with bioinformatics and database research can also give an idea of which metabolic pathways are activated during a plant's stress response. Moreover, targeted analysis of known bioactive, e.g., anti-microbial or anti-fungal, compounds can help to identify QTLs, by using targeted metabolomics techniques for phenotyping experiments in breeding crossing lines. Thereby, alongside breeding improvement, metabolomics helps in the understanding of biological functions and the study of host–pathogen interactions [25].

It is already known from plant–pathogen interaction studies that signaling molecules, such as ethylene, salicilic acid, jasmonic acid or inositol, as well as antifungal/antimicrobial phytochemicals and cell wall compartments are formed or their concentrations up-regulated during plant stress [25]. Numerous studies already indicate that a complex network of secondary metabolites is influenced by biotic stress challenges. Next to phenylpropanoids, including flavonoids [34,50,51,71], alkaloids [72,73], terpenoids [33], and fatty acids [74] are also known to be up-regulated [25]. In addition, lipids act under pathogen or herbivore attack as mechanical barriers. The plants' wax layers form the first line of defense against pathogens and herbivores [51,75]. Additionally, some lipids act as signaling molecules, namely oxylipins and jasmonates, which participate in immune response cascades. Phytoalexins, a group of fatty acid degradation products, have antimicrobial and antifungal activities [51]. In addition, only recently Caroline Gutjahr's lab could show that fatty acids, like carbohydrates, are also transported from the plant host to fungi [76].

Moreover, as plants produce a diverse array of more than 100,000 low molecular weight natural products [32], countless studies are available using targeted quantifications of phytoalaxins in plants.

In 2009, Boue et al. [32] published a forward-looking paper called "phytoalexin-enriched functional foods" in which they proposed to use stress conditions to enhance the amounts of phytoalexins in food plants to produce functional foods. Several phytoalaxins are known for their health beneficial properties, including antioxidant activities, anti-inflammatory activities, cholesterol-lowering abilities, and anticancer activities [32]. In particular, flavonoids, which are ubiquitous in many food plants, have been linked to important health-promoting activities. For example, consumption of legumes, especially soy, have been linked to the reduction of cancer risks and coronary heart diseases simply due to its high flavonoid content [32]. Therefore, besides resistance and yield stability, functional food ingredients from the phytoalexin family could also be a possible target for plant breeders. In addition, mild stress conditions could possibly help to enhance the amounts of bioactive secondary metabolites, thereby producing functional foods [32].

Next to aspects in nutrition, diet and health, food and environmental safety can especially be monitored by means of metabolomics [3]. Toxic compounds can, for example, be formed as a stress response by plants, such as furanocoumarins in celery and glycoalcaloides in potatoes, or by fungi contamination, which can produce human toxic mycotoxins [77]. Therefore, versatile, sensitive, reliable and fast-targeted LC-MS/MS and GC-MS methods using stable isotope dilution techniques have been developed in the past to monitor food quality [78,79].

In sum, in the future more versatile and sensitive multiple-targeted MS quantification methods to evaluate food and crop quality are required.

3.3. *Functional Phytometabolomics—Characterization Approach of Plant Stress Metabolites*

In conclusion, the science-driven breeding of stress-tolerant cultivated plants that would allow for a reduction in harvest losses and undesirable decrease in quality attributes requires a new quality of knowledge on molecular markers associated with relevant agronomic traits, on quantitative metabolic stress responses of plants, and the mechanisms controlling their biosynthesis. The field of "functional phytometabolomics", using targeted and non-targeted MS or NMR techniques to quantitatively assess key metabolome alterations in plant-derived crops and foods induced by biotic stress challenges as well as abiotic stress conditions, is, therefore, a promising field of research.

In the phytometabolomics approach, metabolites up-regulated during stress challenges are visualized by means of working techniques used in the field of metabolomics. Markers previously not published in literature could e.g., be visualized by retention time and mass to charge ratio in UHPLC-TOF-MS analysis. They are isolated in purities higher than 98% from the plant material using medium pressure liquid chromatography (MPLC) and preparative HPLC techniques and are identified by means of LC-MS, LC-MS/MS, UHPLC-TOF-MS and 1/2D NMR experiments. Testing of different biological activities (anti-fungal, anti-bacterial, anti-oxidant activities etc.) of those compounds allows initial insights into their biological functions. To translate the knowledge on how stress-resistant traits master their successful defense against stress conditions into breeding programs, genotype-specific metabolome alterations have to be characterized. Subsequently, the gene clusters controlling the biosynthetic pathways of key stress metabolites have to be identified by means of genome-wide association and QTL mapping studies. Just to highlight two examples: Firstly, Matsuda et al. [19] characterized several bioactive flavon glycosides, which can be used for new human drug discovery in rice by means of a metabolom genome-wide association study. Thereby, they highlighted "that one plant species produces more diverse phytochemicals than previously expected, and plants still contain many useful compounds for human applications". Secondly, only recently Rambla et al. [18] used a QTL analysis, to show that 102 QTLs correspond to 39 different volatile organic compounds, including flavor active key metabolites, in tomatoes. This research will help to navigate breeding programs and to optimize post-harvest treatment of plant-derived food products from producer to consumer/processor towards the production of high-quality food products.

4. Phytohormone Profiling by Means of Plant Hormonomics

Phytohormones are a class of low molecular weight, structurally diverse, but highly bioactive compounds in plants. They act as chemical messengers, triggering and controlling physiological processes during plant growth and development (e.g., cell elongation, regulation of apical dominance, vascular differentiation, fruit development, latal and adventitious root formation) as well as in response to abiotic and biotic stress conditions [2,80,81]. Next to ethylene, auxins, cytokinins, brassinosteroids, gibberellins, jasmonates, salicylates, polyamines, abscisates and signal peptides, strigolactones are part of the phytohormone family [81,82]. During stress exposure those phytohormone classes interact with each other by means of synergistic or antagonistic cross-talks, resulting in each other's biosynthesis or up-regulation response [83]). In the past, several studies provided evidence that plant hormones are necessary for plants to adapt during stress conditions, especially, during abiotic stress factors by mediating a wide range of adaptive responses [83]. They play a key role in the plant's intricate signal networks, often immediately altering gene expression by inducing or preventing the degradation of transcriptional regulators via the ubiquitin–proteasome system [83,84]. Thanks to Kumar et al. [2], metabolic engineering of phytohormons can be used to improve quality and stress tolerance of crops. Although the analysis of plant hormones, such as auxins, especially their quantification by means of SIDA-LC-MS/MS revealed significant insights in their tissue- and cell-type-specific analysis, their distribution profiles in plant organs, tissues, and cells still remains elusive [81]. A wide variety of targeted GC-MS and LC-MS/MS methods have been published dealing with the quantification of exact amounts of single members of the phytohormone family (e.g., cf.: Porifiro et al. [85]; Novák et al. [81]). However, in the last two decades, only one method was described to characterize those physiologically important molecules in their network interactions in one single run [86]. Only recently, Ondřej Novák and his team developed a new, versatile and sensitive targeted UHPLC-MS/MS method, which enables the simultaneous quantification of 101 phytohormone-related metabolites (phytohormones and their precursors) in less than 20 mg plant material [86]. This allows the characterization of the majority of known phytohormones as well as their biosynthetic precursors, that regulate diverse processes in plants by intricate signaling networks. With this newly developed, targeted metabolomics approach, so called "plant hormonomics", Šimura et al. [86] were able to detect 45 and quantify a total of 43 endogenous compounds out of the 101 phytohormones in both root and shoot samples, in salt-stressed and non-stressed, 12-day-old *Arabidopsis thaliana* seedlings. Subsequent multivariate statistical analysis cross-compared with data obtained from transcriptomic studies enabled the identification of the main phytohormones involved in the adaption of *Arabidopsis thaliana* to salt stress. The multivariate statistical analysis revealed that 23 of the quantified 43 metabolites significantly differed between the salt-stressed and non-stressed roots. In contrast, the shoots and the roots differed in the concentrations of 15 compounds. Those hormone profiles obtained were cross-compared with transciptomic data. Well in line with findings from Rhy and Cho [87], who found that jasmoic acid and abscisic acid promote salt tolerance in *Arabidopsis*, Šimura et al. [86] observed increased levels of abscisic acid, its oxidation products phaseic acid, dihydrophaseic acid, and jasmonic acid in salt-stressed root samples. In addition, different responses of gibberillinic acid derivatives, especially its active form GA4, and its transcriptomic data, were found by Šimura et al. [86]. In sum, this newly developed multiple parallel analysis of phytohormones, called "plant hormonomics" enables the real-time profiling of hormone networks of large collections of phytohormones, their precursors, transport forms and degradation products in single stressed or non-stressed samples [86,88].

Although, Šimura et al. [86] were already able to quantify the main phytohormones in less than 20 mg of salt stressed *Arabidopsis* samples, special derivatisation reactions of the analytes prior to their analysis will conceivably lead to an increase in sensitivity and selectivity during MS analysis in the future [85,88–90]. For example, the response/sensitivity during LC-ESI-MS/MS analysis of metabolites only present in very low concentrations in plants, like indole-3-acetic acid, a member of the auxin family, can be tremendously increased up to 200-fold after methylation [89]. In addition, several aldehyde trapping derivatisation reagents are known to enhance the sensitivity during ESI-MS analysis of biological mixtures [91]. In future, it could be an extremely important and advantageous step in hormonomics to analyze a combination of derivatised and non-derivatised phytohormone classes, simultaneously in one method.

5. Sensomics—A Phenotyping Tool to Characterize Crops Flavor Impression

In general, more than 83% of consumers say that flavor influences their decision the most when purchasing any kind of food products or beverages [92]. Therefore, it is no surprise that, although aided by visual inspection, the final recognition and quality evaluation of food crops made by consumers are mainly mediated by its flavor perception. Human flavor perception is induced by the interaction of volatile odor-active and non-volatile taste-active molecules with ~390 odorant receptor proteins located in the *regio olfactoria* in the nose and ~40 taste receptor proteins on the tongue [93]. To meet consumers' demand for continuously available, fresh foods with a premium quality, the "flavor blueprint" of a golden standard, which is a combinatorial code of the entire set of odor- and taste-active metabolites in their natural concentrations, has to be known [94]. To decode all flavor-active molecules of a crop, the so-called sensometabolome, several high-end working techniques, including combinations of state-of-the-art metabolomics and chromatography approaches in combination with human sensory science experiments, such as aroma or taste dilution analysis, have been developed in the past. This so called molecular sensory science or sensomics approach has especially been shaped by Peter Schieberle's and Thomas Hofmann's working groups during the last three decades [93,94]. Today, it is well accepted that the presence of certain structural elements, so-called olfactophores and gustophores, as well as specific concentrations exceeding the sensory thresholds are important prerequisites of low molecular weight metabolites to become flavor-active [94]. Although a total of more than 10,000 volatiles occur in crop plants, the use of the so-called sensomics approach gave evidence that the typical aroma impression of a crop-based foodstuff is caused by a limited number of aroma-active volatiles [93]. In conclusion, only a surprisingly small group of so called "key odorants in food" contribute to the typical aroma profile of a crop plant. For example, only two compounds, namely ethyl (2*S*)-2-methylbutanoate and 1-(ethylsulfanyl)ethane-1-thiol, are necessary to explain the typical aroma impression of durian fruits [95], and to mimic the typical aroma impression of mangos only eight compounds are required [96]. While only a small number of aroma-active compounds contribute to the key odorants in food and interact with a huge number of odorant receptors, several thousands of non-volatile taste-active compounds are already known. In particular, the application of taste dilution analysis followed by dose/activity considerations led to the discovery of many bitter, sweet, umami, pungent, astringent, salty, or sour sensing molecules in several plants, such as carrots [97,98], cocoa [99], asparagus [100–102], pepper [103], red currants [104], tea [105], stevia [106] or spinach [107] in the past. It is nowadays well accepted that not only a single flavor-impact molecule, but a combinatorial code of multiple odor- and taste-active key compounds, each in its specific concentration, reflect the chemosensory phenotype and trigger the typical flavor profile of food products. In particular, for flavor improvement of crops, the analysis of aroma- and taste-active compounds presents a major challenge for flavor improvement of crops [94]. It has been shown in the past, that the biosynthesis of several key odorants in food and tastants is controlled by genes, the expression of which is altered or even induced by biotic or abiotic stress challenges. But the primary target of crop production then was field performance, yield, and storage characteristics, while ignoring quality traits, such as the flavor code [33,108].

On the one hand, induced by several biotic and abiotic stress factors during growth in the field as well as during post-harvest storage, the attractive sensory quality of miscellaneous crop plants is hindered by sporadic off-flavors, which is often the reason for consumer complaints and therefore a major problem for plant processors. On the other hand, mild stress factors can lead to an increase in in the concentration of flavor-active constituents and hence to a more intensive desirable aroma or taste impressions [33,108].

Moreover, abiotic factors, such as mechanical stress, are reported to increase the amounts of the key bitter tastants, members of the C17-polycetylenes, present in native carrots (*Daucus carota* L.). The increase causes a perceived bitter off-flavor, occurring especially often during the production process for infant diet carrot products [97,98,109–115]. In addition, a decrease in flavor quality accompanied by an increase in bitter taste has also been reported in raw hazelnuts (*Corylus avellana* L.) upon biotic stress challenges, such as upon infection by bugs, belonging to the hemipteran family, like *Gonocerus acuteangulatus* and *Coreus marginatus* [116].

Besides the non-volatile, taste-active metabolites, aroma-active compounds are also known to be influenced by individual stress factors [33]. In particular, studies dealing with tea (*Camellia sinensis*), which is enjoyed as freshly brewed green, black, oolong, or decaffeinated tea infusion, and its quality changes induced by stress are available in literature. For example, the tea green leafhopper (*Empoasca* (*Matsumurasca*) *onukii* Matsuda) attacks, at pre-harvest stages, can decisively influence the unique aroma quality of tea leaves as a result of the upregulation of the linalool synthase (CsLIS1 and CsLIS2) causing higher concentrations of the key odorant, linalool [33,117]. Besides linalool, the formation of key odorants, such as (*E*)-nerolidol, can also be influenced by a combination of low-temperature stress and mechanical damage [33,118]. Therefore, Wüst [33] concluded that the tea-related findings illustrate how the use of the stress response of plants within the sensometabolome can lead to an improvement of flavor of agricultural products. In addition, he pointed out, that the plants' contact with stress elicitors, such as methyl jasmonate instead of the actual biotic or abiotic stress factors could also lead to "stress induced" flavor improvement [33,118].

In order to gain a more comprehensive knowledge on the chemical mechanisms involved in quality changes of cultivated crop plants in response to biotic or abiotic stress challenges or to improve the flavor quality by the application of moderate, well controlled stress, numerous volatile, aroma-active as well as non-volatile taste-active key metabolites, the so-called sensometabolites, of stressed and non-stressed plant genotypes should be comparatively characterized by means of a fast and robust high-throughput GC-MS systems with high peak separation capacity and sensitivity, such as comprehensive two-dimensional gas chromatography/time-of-flight mass spectrometry (GC $\times$ GC/TOF-MS) or UPLC-TOF-MS metabolic profiling analysis in the future. This strategy aims at reducing the flavor deficiencies in modern commercial varieties as a "green" alternative to genetic engineering. The workflow for a successful implementation of this approach—from the identification of key odorants by molecular science techniques to the investigation of mechanisms controlling their biosynthesis—is complex and calls for interdisciplinary research [33,108].

6. Conclusions

Plants continuously encounter various biotic and abiotic environmental stresses during their growth and development phases, which leads likewise to yield and quality losses. To avoid those previously listed economical losses, to reduce insecticides and pesticides used in agriculture, to ensure feeding the world's growing population, and to develop new functional foods knowledge about molecular networks and the pathway activation of plants during their stress response is essential.

This review has emphasized the importance of metabolomics-based working techniques to discover metabolome alterations during abiotic and biotic stress conditions. Therefore, metabolomics is a promising tool for knowledge-based targeted breeding programs. It also shows that due to missing databases and non-standardized LC-MS conditions, basic metabolomics, lipidomics and phytohormonics strategies, without isolation and unequivocal structure identification experiments, are sometimes insufficient. Therefore, techniques using biological and molecular structural characterizations of marker metabolites in combination with metabolomics techniques, such as phytometabolomics or sensomics approaches, are useful solutions to produce high-quality phytoalexin-enriched functional foods in the future.

Author Contributions: C.D. wrote the manuscript. K.H. critically revised the final version of the manuscript.

References

1. Börner, H. *Pflanzenkrankheiten und Pflanzenschutz*, 8th ed.; Springer-Verlag: Berlin/Heidelberg, Germany, 2009.
2. Kumar, R.; Bohra, A.; Pandey, A.K.; Pandey, M.K.; Kumar, A. Metabolomics for Plant Improvement. Status and Prospects. *Front. Plant Sci.* **2017**, *8*, 1302. [CrossRef] [PubMed]
3. Dixon, R.A.; Gang, D.R.; Charlton, A.J.; Fiehn, O.; Kuiper, H.A.; Reynolds, T.L.; Tjeerdema, R.S.; Jeffery, E.H.; German, J.B.; Ridley, W.P.; et al. Application of metabolomics in agriculture. *J. Agric. Food Chem.* **2006**, *54*, 8984–8994. [CrossRef] [PubMed]
4. Arbona, V.; Manzi, M.; de Ollas, C.; Gómez-Cadenas, A. Metabolomics as a tool to investigate abiotic stress tolerance in plants. *Int. J. Mol. Sci.* **2013**, *14*, 4885–4911. [CrossRef] [PubMed]
5. Obata, T.; Fernie, A. The use of metabolomics to dissect plant responses to abiotic stresses. *Cell. Mol. Life Sci.* **2012**, *69*, 3225–3243. [CrossRef] [PubMed]
6. De Luca, V.; St Pierre, B. The cell and developmental biology of alkaloid biosynthesis. *Trends Plant Sci.* **2000**, *5*, 168–173. [CrossRef]
7. D'Auria, J.C.; Greshenzon, J. The secondary metabolism of *Arabidopsis thalina*: Growing like a weed. *Curr. Opin. Plant Biol.* **2005**, *8*, 308–316. [CrossRef] [PubMed]
8. Davies, H. A role for "omics" technologies in food safety assessment. *Food Control* **2010**, *21*, 1601–1610. [CrossRef]
9. Saito, K.; Matsuda, F. Metabolomics for funtional genomics, systems biology, and biotechnology. *Annu. Rev. Plant Biol.* **2010**, *61*, 463–489. [CrossRef] [PubMed]
10. Cantu, D.; Govindarajulu, M.; Kozik, A.; Wang, M.; Chen, X.; Kojima, K.K.; Dubcovsky, J. Next generation sequencing provides rapid access to the genome of *Puccinia striiformis* f. sp. *tritici*, the causal agent of wheat stripe rust. *PLoS ONE* **2011**, *6*, e24230.
11. Schneeberger, K.; Weigel, D. Fast-forward genetics enabled by new sequencing technologies. *Trends Plant Sci.* **2011**, *16*, 282–288. [CrossRef] [PubMed]
12. Bino, R.J.; Hall, R.D.; Fiehn, O.; Kopka, J.; Saito, K.; Draper, J.; Nikolau, B.J.; Mendes, P.; Roessner-Tunali, U.; Beale, M.H.; et al. Potential of metabolomics as a functional genomics tool. *Trends Plant Sci.* **2004**, *9*, 1360–1385. [CrossRef] [PubMed]
13. Sumner, L.W.; Lei, Z.; Nikolau, B.J.; Saito, K. Modern plant metabolomics. Advanced natural product gene discoveries, improved technologies, and future prospects. *Nat. Prod. Rep.* **2015**, *32*, 212–229. [CrossRef] [PubMed]
14. Aharoni, A.; Keizer, L.C.; Bouwmeester, H.J.; Sun, Z.; Alvarez-Huerta, M.; Verhoeven, H.A.; Blaas, J.; van Houwelingen, A.M.; De Vos, R.C.; van der Voet, H.; et al. Identification of the SAAT gene involved in strawberry flavor biogenesis by use of DNA microarrays. *Plant Cell.* **2000**, *12*, 647–662. [CrossRef] [PubMed]

15. Carreno-Quintero, N.; Acharjee, A.; Maliepaard, C.; Bachem, C.W.B.; Mumm, R.; Bouwmeester, H.; Visser, R.G.F.; Keurentjes, J.J.B. Untargeted metabolic quantitative trait loci analyses reveal a relationship between primary metabolism and potato tuber quality. *Plant Physiol.* **2012**, *158*, 1306–1318. [CrossRef] [PubMed]
16. Feng, J.; Long, Y.; Shi, L.; Shi, J.; Barker, G.; Meng, J. Characterization of metabolite quantitative trait loci and metabolic networks that control glucosinolate concentration in the seeds and leaves of *Brassica napus*. *New Phytol.* **2012**, *193*, 96–108. [CrossRef] [PubMed]
17. Keilwagen, J.; Lehnert, H.; Berner, T.; Budahn, H.; Mothnagel, T.; Ulrich, D.; Dunemann, F. The terpene synthase gene family of carrot (*Daucus carota* L.): Identification of candidate genes associated with terpenoid volatile compounds. *Front. Plant Sci.* **2017**, *8*, 1930. [CrossRef] [PubMed]
18. Rambla, J.L.; Medina, A.; Fernández-Del-Carmen, A.; Barrantes, W.; Grandillo, S.; Cammareri, M.; López-Casado, G.; Rodrigo, G.; Alonso, A.; García-Martínez, S.; et al. Identification, introgression, and validation of fruit volatile QTLs from a red-fruited wild tomato species. *J. Exp. Bot.* **2017**, *68*, 429–442. [CrossRef] [PubMed]
19. Matsuda, F.; Nakabayashi, R.; Yang, Z.; Okazaki, Y.; Yonemaru, J.; Ebana, K.; Yano, M.; Saito, K. Metabolome-genome-wide association study dissects genetic architecture for generating natural variation in rice secondary metabolism. *Plant J. Cell Mol. Boil.* **2015**, *81*, 13–23. [CrossRef] [PubMed]
20. Shulaev, V.; Cortes, D.; Miller, G.; Mittler, R. Metabolomics for plant stress response. *Physiol. Plant.* **2008**, *132*, 199–208. [CrossRef] [PubMed]
21. Kaplan, F.; Kopka, J.; Haskell, D.W.; Zhao, W.; Schiller, K.C.; Gatzke, N.; Sung, D.Y.; Guy, C.L. Exploring the temperature-stress metabolom of Arabidopsis. *Plant Physiol.* **2004**, *136*, 4159–4168. [CrossRef] [PubMed]
22. Urano, K.; Kurihara, Y.; Seki, M.; Shinozaki, K. 'Omics' analyses of regulatory networks in plant abiotic stress responses. *Curr. Opin. Plant Biol.* **2010**, *13*, 132–138. [CrossRef] [PubMed]
23. Urano, K.; Maruyama, K.; Ogata, Y.; Morishita, Y.; Takeda, M.; Sukarai, N.; Suzuki, H.; Saito, K.; Shibata, D.; Kobayashi, M.; et al. Characterization of the ABA-regulated global responses to dehydration in *Arabidopsis* by metabolomics. *Plant J.* **2009**, *57*, 1065–1078. [CrossRef] [PubMed]
24. Ghatak, A.; Chaturvedi, P.; Weckwerth, W. Metabolomics in Plant Stress Physiology. In *Advances in Biochemical Engineering/Biotechnology*; Springer: Berlin/Heidelberg, Germany, 2018.
25. Kushalappa, A.C.; Gunnaiah, R. Metabolo-proteomics to discover plant biotic stress resistance genes. *Trends Plant Sci.* **2013**, *18*, 522–531. [CrossRef] [PubMed]
26. Venter, J.C.; Adams, M.D.; Myers, E.W.; Li, P.W.; Mural, R.J.; Sutton, G.G.; Smith, H.O.; Yandell, M.; Evans, C.A.; Holt, R.A.; et al. The sequence of the human genome. *Science* **2001**, *291*, 1304–1351. [CrossRef] [PubMed]
27. The Arabidopsis Genome Initiative. Analysis of the genome sequence of the flowering plant Arabidopsis thaliana. *Nature* **2000**, *408*, 796–815. [CrossRef] [PubMed]
28. Goff, S.A.; Ricke, D.; Lan, T.H.; Presting, G.; Wang, R.L.; Dunn, M.; Glazebrook, J.; Sessions, A.; Oeller, P.; Varma, H.; et al. A draft sequence of the rice genome (*Oryza sativa* L. ssp. japonica). *Science* **2002**, *296*, 92–100. [CrossRef] [PubMed]
29. Yu, J.; Hu, S.N.; Wang, J.; Wong, G.K.S.; Li, S.G.; Liu, B.; Deng, Y.J.; Dai, L.; Zhou, Y.; Zhang, X.Q.; et al. A draft sequence of the rice genome (*Oryza sativa* L. ssp. indica). *Science* **2002**, *296*, 79–92. [CrossRef] [PubMed]
30. Sato, S.; Tabata, S.; Hirakawa, H.; Asamizu, E.; Shirasawa, K.; Isobe, S.; Kaneko, T.; Nakamura, Y.; Shibata, D.; Aoki, K.; et al. The tomato genome sequence provides insights into fleshy fruit evolution. *Nature* **2012**, *485*, 635–641.
31. Mascher, M.; Gundlach, H.; Himmelbach, A.; Beier, S.; Twardziok, S.O.; Wicker, T.; Radchuk, V.; Dockter, C.; Hedley, P.E.; Russell, J.; et al. A chromosome conformation capture ordered sequence of the barley genome. *Nature* **2017**, *544*, 427–433. [CrossRef] [PubMed]
32. Boue, S.M.; Cleveland, T.E.; Carter-Wientjes, C.; Shih, B.Y.; Bhatnagar, D.; McLachlan, J.M.; Burow, M.E. Phytoalexin-enriched functional foods. *J. Agric. Food Chem.* **2009**, *57*, 2614–2622. [CrossRef] [PubMed]
33. Wüst, M. Smell of stress: Identification of induced biochemical pathways affecting the volatile composition and flavor quality of crops. *J. Agric. Food Chem.* **2018**, *66*, 3616–3618. [CrossRef] [PubMed]
34. Blanksby, S.J.; Mitchell, T.W. Advances in mass spectrometry for lipidomics. *Annu. Rev. Anal. Chem.* **2010**, *3*, 433–465. [CrossRef] [PubMed]

35. Hammerl, R.; Frank, O.; Hofmann, T. Differential off-line LC-NMR (DOLC-NMR) metabolomics to monitor tyrosine-induced metabolome alterations in *Saccharomyces cerevisiae*. *J. Agric. Food Chem.* **2017**, *65*, 3230–3241. [CrossRef] [PubMed]
36. Roberts, L.D.; Souza, A.L.; Gerszten, R.E.; Clish, C.B. Targeted Metabolomics. *Curr. Protoc. Mol. Biol.* **2012**, *30*. [CrossRef] [PubMed]
37. Desbrosses, G.; Steinhauser, D.; Kopka, J. Metabolom analysis using GC-MS. In *Lotus Japonicus Handbook*; Springer: Dordrecht, The Netherlands, 2005; pp. 165–174.
38. Burton, L.; Ivosev, G.; Tate, S.; Impey, G.; Wingate, J.; Bonner, R. Instrumental and experimental effects in LC-MS-based metabolomics. *J. Chromatogr. B* **2008**, *871*, 227–235. [CrossRef] [PubMed]
39. Moco, S.; Bino, R.J.; Vorst, O.; Verhoeven, H.A.; de Groot, J.; van Beek, T.A.; Vervoort, J.; de Vos, C.H. A liquid-chromatography-mass spectrometry based metabolome database for tomato. *Plant Physiol.* **2006**, *141*, 1205–1218. [CrossRef] [PubMed]
40. Frank, O.; Kreißl, J.K.; Daschner, A.; Hofmann, T. Accurate determination of reference materials and natural isolates by means of quantitative 1H NMR spectroscopy. *J. Agric. Food Chem.* **2014**, *62*, 2506–2515. [CrossRef] [PubMed]
41. Sumner, L.W.; Mendes, P.; Dixon, R.A. Plant metabolomics: Large-scale phytochemistry in the functional genomics area. *Phytochemistry* **2003**, *62*, 817–836. [CrossRef]
42. Wang, D.; Bodowitz, S. Single cell analysis: The new fronzier in òmics. *Trends Biotechnol.* **2010**, *28*, 281–290. [CrossRef] [PubMed]
43. Misra, B.B.; Assmann, S.M.; Chen, S. Plant single-cell and single-cell-type metabolomics. *Trends Plant Sci.* **2014**, *19*, 637–646. [CrossRef] [PubMed]
44. Lange, B.M. Single cell genomics. *Curr. Opin. Plant Biol.* **2005**, *8*, 236–241. [CrossRef] [PubMed]
45. Tang, F.; Lao, K.; Surani, M.A. Development and applications of single cell transcriptome analysis. *Nat. Methods* **2011**, *8*, 6–11. [CrossRef] [PubMed]
46. Dai, S.; Chen, S. Single cell-type proteomics: Toward a holistic understanding of plant function. *Mol. Cell. Proteom.* **2012**, *11*, 1622–1630. [CrossRef] [PubMed]
47. De Souza, L.P.; Naake, T.; Tohge, T.; Fernie, A.R. From chromatogram to analyte to metabolite. How to pick horses for courses from the massive web resources for mass spectral plant metabolomics. *GigaScience* **2017**, *6*, 1–20. [CrossRef] [PubMed]
48. Mithöfer, A.; Boland, W. Do you speak chemistry. *EMBO Rep.* **2016**, *17*, 626–629. [CrossRef] [PubMed]
49. Mithöfer, A.; Boland, W. Plant defense against herbivores. Chemical aspects. *Annu. Rev. Plant Biol.* **2012**, *63*, 431–450. [CrossRef] [PubMed]
50. Kaplan, F.; Guy, C.L. β-Amylse induction and the protective role of maltose during temperature shock. *Plant Physiol.* **2004**, *135*, 1674–1684. [CrossRef] [PubMed]
51. Tenenboim, H.; Burgos, A.; Willmitzer, L.; Brotman, Y. Using lipidomics for expanding the knowledge on lipid metabolism in plants. *Biochimie* **2016**, *130*, 91–96. [CrossRef] [PubMed]
52. Frankel, E.N. *Lipid Oxidation*, 2nd ed.; Woodhead Publishing in Food Science, Technology and Nutrition: Philadelphia, PA, USA, 2012.
53. El-Hafid, L.; Pham, T.A.; Zuily-Fodil, Y.; Vieira da Silva, J. Enzymatic Breakdown of Polar Lipids in Cotton Leaves under Water Stress. 1. Degradation of Monogalactosyl-Diacylglycerol. Plant Physiology Biochemistry 1989. Available online: http://agris.fao.org/agris-search/search.do?recordID=FR9001726 (accessed on 29 June 2018).
54. Hubac, C.; Guerrier, D.; Ferran, J.; Tremolieres, A. Change of leaf lipid composition during water stress in two genotypes of lupinus albus resistant or susceptible to drought. *Plant Physiol. Biochem.* **1989**, *27*, 737–744.
55. Pham, T.A.T.; Vieira da Silva, J.; Mazliak, P. The role of membrane lipids in drought resistance of plants. Bulletin de la Société Botanique de France. *Actual. Bot.* **1990**, *137*, 99–114.
56. Quartacci, M.F.; Pinzino, C.; Sgherri, C.L.; Navari-Izzo, F. Lipid composition and protein dynamics in thylakoids of two wheat cultivars differently sensitive to drought. *Plant Physiol.* **1995**, *108*, 191–197. [CrossRef] [PubMed]
57. Kaoua, M.; Serraj, R.; Benichou, M.; Hsissou, D. Comparative sensitivity of two Moroccan wheat varieties to water stress: The relationship between fatty acids and proline accumulation. *Bot. Stud.* **2006**, *47*, 51–60.
58. Moradi, P.; Mahdavi, A.; Khoshkam, M.; Iriti, M. Lipidomics unravels the role of leaf lipids in thyme plant response to drought stress. *Int. J. Mol. Sci.* **2017**, *18*, 2067. [CrossRef] [PubMed]

59. De Paula, F.M.; Thi, A.P.; de Silva, J.V.; Justin, A.; Demandre, C.; Mazliak, P. Effects of water stress on the molecular species composition of polar lipids from *Vigna unguiculata* L. Leaves. *Plant Sci.* **1990**, *66*, 185–193. [CrossRef]
60. Hamrouni, I.; Salah, H.B.; Marzouk, B. Effects of water-deficit on lipids of safflower aerial parts. *Phytochemistry* **2001**, *58*, 277–280. [CrossRef]
61. Li, Q.; Zheng, Q.; Shen, W.; Cram, D.; Fowler, D.B.; Wie, Y. Understanding the biochemical basis of temperature-induced lipid pathway adjustments in plants. *Plant Cell* **2015**, *27*, 86–103. [CrossRef] [PubMed]
62. Köhl, K. Metabolomics on Combined Abiotic Stress Effects in Crops. In *Drought Stress Tolerance in Plants*; Hossain, M., Wani, S., Bhattacharjee, S., Burritt, D., Tran, L.S., Eds.; Springer: Cham, Switzerland, 2016; Volume 2.
63. Nakabayashi, R.; Saito, K. Integrated metabolomics for abiotic stress responses in plants. *Curr. Opin. Plant Boil.* **2015**, *24*, 10–16. [CrossRef] [PubMed]
64. Okazaki, Y.; Saito, K. Integrated metabolomics and phytochemical genomics approaches for studies on rice. *GigaScience* **2016**, *5*, 1–7. [CrossRef] [PubMed]
65. Agrios, G. Genetics of plant disease. In *Pant Pathology*, 5th ed.; Elsvier Academic Press: Cambridge, MA, USA, 2005; pp. 125–174.
66. Aghnoum, R.; Marcel, T.C.; Johrde, A.; Pecchioni, N.; Schweizer, P.; Niks, R.E. Basal host resistance of barley to powdery mildew: Connecting quantitative trait loci and candidate genes. *Mol. Plant Microbe Interact.* **2010**, *23*, 91–102. [CrossRef] [PubMed]
67. Valdés-López, O.; Thibivilliers, S.; Qiu, J.; Xu, W.W.; Nguyen, T.H.; Libault, M.; Le, B.H.; Goldberg, R.B.; Hill, C.B.; Hartman, G.L.; et al. Identification of quantitative trait loci controlling gene expression during the innate immunity response of soybean. *Plant Physiol.* **2011**, *157*, 1975–1986. [CrossRef] [PubMed]
68. Massman, J.; Cooper, B.; Horsley, R.; Neat, S.; Dill-Macky, R.; Chao, S.; Dong, Y.; Schwarz, P.; Muehlbauer, G.J.; Smith, K.P. Genome-wide association mapping of Fusarium head blight resistance in contemporary barley breeding germplasm. *Mol. Breed.* **2011**, *27*, 439–454. [CrossRef]
69. Gunnaiah, R.; Kushalappa, A.C.; Duggavathi, R.; Fox, S.; Somers, D.J. Integrated metabolo-proteomic approach to decipher the mechanisms by which wheat QTL (Fhb1) contributes to resistance against Fusarium graminearum. *PLoS ONE* **2012**, *7*, e40695. [CrossRef] [PubMed]
70. Bollina, V.; Kumaraswamy, G.K.; Kushalappa, A.C.; Choo, T.M.; Dion, Y.; Rioux, S.; Faubert, D.; Hamzehzarghani, H. Mass spectrometry-based metabolomics application to identify quantitative resistance-related metabolites in barley against Fusarium head blight. *Mol. Plant Pathol.* **2010**, *11*, 769–782. [CrossRef] [PubMed]
71. Ballester, A.R.; Lafuente, M.T.; de Vos, R.C.; Bovy, A.G.; González-Candelas, L. Citrus phenylpropanoids and defence against pathogens. Part I: Metabolic profiling in elicited fruits. *Food Chem.* **2013**, *136*, 178–185. [CrossRef] [PubMed]
72. Machado, A.R.T.; Campos, V.A.C.; da Silva, W.J.R.; Campos, V.P.; de Mattos Zeri, A.C.; Olivera, D.F. Metabolic profiling in the roots of coffee plants exposed to the coffee root-knot nematode, *Meloidogyne exigua*. *Eur. J. Plant Pathol.* **2012**, *134*, 431–441. [CrossRef]
73. Sana, T.R.; Fischer, S.; Wohlgemuth, G.; Katrekar, A.; Jung, K.H.; Ronald, P.C.; Fiehn, O. Metabolomic and transcriptomic analysis of the rice response to the bacterial blight pathogen *Xanthomonas oryzae* pv. oryzae. *Metabolomics* **2010**, *6*, 451–465. [CrossRef] [PubMed]
74. Batovska, D.I.; Todorova, I.T.; Nedelcheva, D.V.; Parushev, S.P.; Atanassov, A.I.; Hvarleva, T.D.; Djakova, G.J.; Bankova, V.S.; Popov, S.S. Preliminary study on biomarkers for the fungal resistance in *Vitis vinifera* leaves. *J. Plant Physiol.* **2008**, *165*, 791–795. [CrossRef] [PubMed]
75. Suh, M.C.; Samuels, A.L.; Jetter, R.; Kunst, L.; Pollard, M.; Ohlrogge, J. Cuticular lipid composition, surface structure, and gene expression in *Arabidopsis* stem epidermis. *Plant Physiol.* **2005**, *139*, 1649–1665. [CrossRef] [PubMed]
76. Keymer, A.; Pimprikar, P.; Wewer, V.; Huber, C.; Brands, M.; Bucerius, S.L.; Delaux, P.M.; Klingl, V.; Röpenack-Lahaye, E.V.; Wang, T.L.; et al. Lipid transfer from plants to arbuscular mycorrhiza fungi. *Elife* **2017**, *6*, e29107. [CrossRef] [PubMed]
77. Cellini, F.; Chesson, A.; Colquhoun, I.; Constable, A.; Davies, H.V.; Engel, K.H.; Gatehouse, A.M.R.; Kärenlami, S.; Kok, E.J.; Leguay, J.-J.; et al. Unintended effects and their detection in genetically modified crops. *Food Chem. Toxicol.* **2004**, *24*, 1089–1125. [CrossRef] [PubMed]

78. Pinu, F.R. Metabolomics—The new frontier in food safety and quality research. *Food Res. Int.* **2015**, *72*, 80–81. [CrossRef]
79. Shephard, G.S. Current status of mycotoxin analysis: A critical review. *J. AOAC Int.* **2016**, *99*, 842–848. [CrossRef] [PubMed]
80. Taiz, L.; Zeiger, E. *Plant Physiology*, 5th ed.; Sinauer Associates: Sunderland, MA, USA, 2010.
81. Novák, O.; Napier, R.; Ljung, K. Zooming in on plant hormone analysis: Tissue- and cell-specific approaches. *Annu. Rev. Plant Biol.* **2017**, *68*, 323–348. [CrossRef] [PubMed]
82. Davies, P.J. *Plant Hormones: Biosynthesis, Signal Transduction, Action!* 3rd ed.; Kluwer Academic Publishers: Dordrecht, The Netherlands, 2010.
83. Peleg, Z.; Blumwald, E. Hormone balance and abiotic stress tolerance in crop plants. *Curr. Opin. Plant Biol.* **2011**, *14*, 290–295. [CrossRef] [PubMed]
84. Santner, A.; Estelle, M. The ubiquitin—Proteasome system regulates plant hormone signaling. *Plant J.* **2010**, *61*, 1029–1040. [CrossRef] [PubMed]
85. Porfírio, S.; Gomes da Silva, M.D.R.; Peixe, A.; Cabrita, M.J.; Azadi, P. Current analytical methods for plant auxin quantification—A review. *Anal. Chim. Acta* **2016**, *902*, 8–21. [CrossRef] [PubMed]
86. Šimura, J.; Antoniadi, I.; Široká, J.; Tarkowska, D.; Strnad, M.; Ljung, K.; Novak, O. Plant hormonomics: Multiple phytohormone profiling by targeted metabolomics. *Plant Physiol.* **2018**. [CrossRef] [PubMed]
87. Rhy, H.; Cho, Y. Plant hormons in salt stress tolerance. *J. Plant Biol.* **2015**, *58*, 147–155.
88. Ramos, L. Critical overview of selected contemporary sample preparation techniques. *J. Chromatogr. A* **2012**, *1221*, 84–98. [CrossRef] [PubMed]
89. Prinsen, E.; Van Dongen, W.; Esmans, E.L.; Van Onckelen, H.A. HPLC linked electrospray tandem mass spectrometry: A rapid and reliable method to analyse indole-3-acetic acid metabolism in bacteria. *J. Mass Spectrom.* **1997**, *32*, 12–22. [CrossRef]
90. Waterval, J.; Lingeman, H.; Bult, A.; Underberg, W.J. Derivatization trends in capillary electrophoresis. *Electrophoresis* **2000**, *21*, 4029–4045. [CrossRef]
91. Egging, M.; Wijtmans, M.; Ekkebus, R.; Lingeman, H.; de Esch, I.J.; Kool, J.; Niessen, W.M.A.; Irth, H. Development of a selective ESI-MS derivatization reagent: Synthesis and optimization for the analysis of aldehydes in biological mixtures. *Anal. Chem.* **2008**, *80*, 9042–9051. [CrossRef] [PubMed]
92. International Food Information Council Foundation, Washington, D.C. 2016 Food and Health Survey. Available online: http://www.foodinsight.org/articles/2016-food-and-health-survey-food-decision-2016-impact-growing-national-food-dialogue (accessed on 3 December 2016).
93. Dunkel, A.; Steinhaus, M.; Kotthoff, M.; Nowak, B.; Krautwurst, D.; Schieberle, P.; Hofmann, T. Nature's chemical signatures in human olfaction: A foodborne perspective for future biotechnology. *Angew. Chem. Int. Ed. Engl.* **2014**, *53*, 7124–7143. [CrossRef] [PubMed]
94. Schieberle, P.; Hofmann, T. Mapping the combinatorial code of food flavors by means of molecular sensory science approach. In *Food Flavors—Chemical, Sensory and Technological Properties*; Jelen, H., Ed.; CRC Press: Boca Raton, FL, USA, 2012; pp. 411–437.
95. Li, J.-X.; Schieberle, P.; Steinhaus, M. Insights into the key compounds of durian (*Durio zibethinus* L. 'Monthong') pulp odor by odorant quantitation and aroma simulation experiments. *J. Agric. Food Chem.* **2016**, *65*, 639–647. [CrossRef] [PubMed]
96. Munafo, J.P.; Didzbalis, J.; Schnell, R.J.; Steinhaus, M. Insights into the key aroma compounds in mango (*Mangifera indica* L. 'Haden') fruits by stable isotope dilution quantitation and aroma simulation experiments. *J. Agric. Food Chem.* **2016**, *64*, 4312–4318. [CrossRef] [PubMed]
97. Czepa, A.; Hofmann, T. Quantitative studies and sensory analyses on the influence of cultivar, spatial tissue distribution, and industrial processing on the bitter Off-taste of carrots (*Daucus carota* L.) and carrot products. *J. Agric. Food Chem.* **2004**, *52*, 4508–4514. [CrossRef] [PubMed]
98. Schmiech, L.; Uemra, D.; Hofmann, T. Reinvestigation of the bitter compounds in carrots (*Daucus carota* L.) by using a molecular sensory science approach. *J. Agric. Food Chem.* **2008**, *56*, 10252–10260. [CrossRef] [PubMed]
99. Stark, T.; Hofmann, T. Isolation, structure determination, synthesis, and sensory activity of *N*-phenylpropenoyl-L-amino acids from cocoa (*Theobroma cacao*). *J. Agric. Food Chem.* **2005**, *53*, 5419–5428. [CrossRef] [PubMed]

100. Dawid, C.; Hofmann, T. Structural and sensory characterization of bitter tasting steroidal saponins from asparagus spears (*Asparagus officinalis* L.). *J. Agric. Food Chem.* **2012**, *60*, 11889–11900. [CrossRef] [PubMed]
101. Dawid, C.; Hofmann, T. Identification of sensory-active phytochemicals in asparagus (*Asparagus officinalis* L.). *J. Agric. Food Chem.* **2012**, *60*, 11877–11888. [CrossRef] [PubMed]
102. Dawid, C.; Hofmann, T. Quantitation and bitter taste contribution of saponins in fresh and cooked white asparagus (*Asparagus officinalis* L.). *Food Chem.* **2013**, *145*, 427–436. [CrossRef] [PubMed]
103. Dawid, C.; Henze, A.; Frank, O.; Glabasnia, A.; Rupp, M.; Buening, K.; Orlikowski, D.; Bader, M.; Hofmann, T. Structural and sensory characterization of key pungent and tingling compounds from black pepper (*Piper nigrum* L.). *J. Agric. Food Chem.* **2012**, *60*, 2884–2895. [CrossRef] [PubMed]
104. Schwarz, B.; Hofmann, T. Sensory-Guided Decomposition of Red Current Juice (*Ribes rubrum*) and Structure Determination of Key Astringent Compounds. *J. Agric. Food Chem.* **2007**, *55*, 1394–1404. [CrossRef] [PubMed]
105. Scharbert, S.; Holzmann, N.; Hofmann, T. Identification of the astringent taste compounds in black tea infusions by combining instrumental analysis and human bioresponse. *J. Agric. Food Chem.* **2004**, *52*, 3498–3508. [CrossRef] [PubMed]
106. Hellfritsch, C.; Brockhoff, A.; Stähler, F.; Meyerhof, W.; Hofmann, T. Human psychometric and taste receptor responses to steviol glycosides. *J. Agric. Food Chem.* **2012**, *60*, 6782–6793. [CrossRef] [PubMed]
107. Brock, A.; Hofmann, T. Identificatioion of the Key Astringent Compounds in Spinach (*Spinacia oleracea*) by Means of the Taste Dilution Analysis. *Chem. Percept.* **2008**, *1*, 268–281. [CrossRef]
108. Hofmann, T.; Krautwurst, D.; Schieberle, P. Current status and future perspectives in flavor research: Highlights of the 11th Wartburg Symposium on flavor chemistry & biology. *J. Agric. Food Chem.* **2018**, *66*, 2197–2203. [PubMed]
109. Harding, V.K.; Heale, J.B. The accumulation of inhibitory compounds in the induced resistance response of carrot root slices to *Botrytis cinerea*. *Physiol. Plant Pathol.* **1981**, *18*, 7–15. [CrossRef]
110. Lund, E.D.; White, J.M. Polyacetylenes in Normal and waterstressed 'Orlando Gold' carrots (*Daucus carota*). *J. Sci. Food Agric.* **1990**, *51*, 507–516. [CrossRef]
111. Olsson, K.; Svensson, R. The influence of polyacetylenes on the susceptibility of carrots to storage diseases. *J. Phytopathol. Phytopathol. Z.* **1996**, *144*, 441–447. [CrossRef]
112. Kreutzmann, S.; Christensen, L.P.; Edelenbos, M. Investigation of bitterness in carrots (*Daucus carota* L.) based on quantitative chemical and sensory analyses. *LWT Food Sci. Technol.* **2008**, *41*, 193–205. [CrossRef]
113. Kreutzmann, S.; Svensson, V.T.; Thybo, A.K.; Bro, R.; Petersen, M.A. Prediction of sensory quality in raw carrots (*Daucus carota* L.) using multi-block LS-ParPLS. *Food Qual. Preference* **2008**, *19*, 609–617. [CrossRef]
114. Kidmose, U.; Hansen, S.L.; Christensen, L.P.; Edelenbos, M.; Larsen, E.; Nørbæk, R. Effects of genotype, root size, storage, and processing on bioactive compounds in organically grown carrots (*Daucus carota* L.). *J. Food Sci.* **2004**, *69*, S388–S394. [CrossRef]
115. Lund, E.D.; Bruemmer, J.H. Acetylenic compounds in stored packaged carrots. *J. Sci. Food Agric.* **1991**, *54*, 287–294. [CrossRef]
116. Singldinger, B.; Dunkel, A.; Bahmann, D.; Bahmann, C.; Kadow, D.; Bisping, B.; Hofmann, T. New taste-active 3-(*O*-β-D-glucosyl)-2-oxoindole-3-acetic acids and diarylheptanoids in Cimiciato-infected hazelnuts. *J. Agric. Food Chem.* **2018**, *66*, 4660–4673. [CrossRef] [PubMed]
117. Mei, X.; Liu, X.; Zhou, Y.; Wang, X.; Zeng, L.; Fu, X.; Li, J.; Tang, J.; Dong, F.; Yang, Z. Formation and emission of linalool in tea (*Camellia sinensis*) leaves infested by tea green leafhopper (Empoasca (Matsumurasca) onukii Matsuda). *Food Chem.* **2017**, *237*, 356–363. [CrossRef] [PubMed]
118. Zhou, Y.; Zeng, L.; Liu, X.; Gui, J.; Mei, X.; Fu, X.; Dong, F.; Tang, J.; Zhang, L.; Yang, Z. Formation of (*E*)-nerolidol in tea (*Camellia sinensis*) leaves exposed to multiple stresses during tea manufacturing. *Food Chem.* **2017**, *231*, 78–86. [CrossRef] [PubMed]

7

Improving Flooding Tolerance of Crop Plants

Angelika Mustroph

Plant Physiology, University Bayreuth, Universitaetsstr. 30, 95440 Bayreuth, Germany;
angelika.mustroph@uni-bayreuth.de

Abstract: A major problem of climate change is the increasing duration and frequency of heavy rainfall events. This leads to soil flooding that negatively affects plant growth, eventually leading to death of plants if the flooding persists for several days. Most crop plants are very sensitive to flooding, and dramatic yield losses occur due to flooding each year. This review summarizes recent progress and approaches to enhance crop resistance to flooding. Most experiments have been done on maize, barley, and soybean. Work on other crops such as wheat and rape has only started. The most promising traits that might enhance crop flooding tolerance are anatomical adaptations such as aerenchyma formation, the formation of a barrier against radial oxygen loss, and the growth of adventitious roots. Metabolic adaptations might be able to improve waterlogging tolerance as well, but more studies are needed in this direction. Reasonable approaches for future studies are quantitative trait locus (QTL) analyses or genome-wide association (GWA) studies in combination with specific tolerance traits that can be easily assessed. The usage of flooding-tolerant relatives or ancestral cultivars of the crop of interest in these experiments might enhance the chances of finding useful tolerance traits to be used in breeding.

Keywords: hypoxia; waterlogging; submergence; flooding; maize; soybean; barley; aerenchyma

---

1. Introduction

In times of changing climate, agriculture faces increasing problems with extreme weather events leading to considerable yield losses. In combination with a growing population and higher food demand, this presents a challenge to scientists and breeders to maintain the current food supply. Certainly, more effort is required to develop stress-resistant crops and improve agricultural practices in order to cope with these problems.

Plants are, due to their sessile nature, exposed to all changes in abiotic and biotic factors occurring at their habitat. Water availability, for example, is always problematic and can change from periods of drought to periods of flooding after a heavy rainfall. Plants can adapt to changing environmental conditions, but this comes at the cost of reduced growth and reproduction. If stress duration or severity exceeds the plant's ability to adapt, it will eventually die.

Most crop plants are rather sensitive to stresses since they were selected for high yield. In order to improve crop plant resistance to stresses, and to improve their productivity and survival, two research strategies are required. First, mechanisms of tolerance against stresses have to be understood. This requires analyses at the molecular level (transcriptomics, proteomics, metabolomics) in order not only to get to know adaptive mechanisms but also to understand their activation and regulation. The best approach for these analyses is the usage of a stress-resistant plant species closely related or even ancestral to the crop of interest, that potentially has lost some adaptational responses.

Second, due to the long history of breeding all over the world, a huge number of cultivars has become available for many crop plants. These breeding processes often focused on high yield and food quality, concomitant with a loss of genetic diversity and stress resistance. However, older cultivars with low productivity might still contain tolerance loci for a certain stress condition that could be transferred to modern, highly productive cultivars. Strategies to achieve this goal are (1) screening of a wide range of cultivars under a specific stress condition; (2) selection of cultivars with low and high resistance; (3) understanding the physiological basis for resistance, i.e., the tolerance trait; and (4) the genetic analysis of those cultivars by quantitative trait locus (QTL) analysis and other molecular methods in order to find the genetic locus that underlies the tolerance trait. If a genetic locus has been discovered and characterized, it subsequently can be transferred into modern varieties using marker-assisted breeding technology to achieve stress-tolerant cultivars.

In this review, the current progress in crop resistance to flooding will be presented. First, overall plant responses and survival strategies under flooding conditions will be summarized. Then, breeding approaches in different temperate crops will be presented. The focus of this article will be on temperate plants, rather than rice and other tropical crops like cotton or sorghum. Flooding research on rice and its flooding-tolerance traits has been reviewed and discussed in several publications previously [1–5].

2. Plant Responses to Flooding

Besides the low availability of water leading to drought stress, plants can also be affected by too much water. Flooding primarily restricts gas diffusion between the plant and its surroundings due to physical properties (e.g., [6–8]). Oxygen as well as CO_2 cannot be easily exchanged via stomata and cell walls under water. This leads to a lack of oxygen inside flooded plant parts, and mainly limits heterotrophic energy production in mitochondria. Furthermore, low CO_2 availability in flooded leaves restricts photosynthesis. Therefore, flooding causes an energy crisis within plant cells.

Flooding events can be classified by two versions, (1) waterlogging, where only the root system inside the soil is affected; and (2) submergence, where also parts or the whole shoot are under water [9]. In flooded plant parts without ongoing photosynthesis, the oxygen concentration quickly declines and leads to hypoxic conditions (e.g., [10,11]).

Several plant species have developed mechanisms to cope with flooding stress, which enable them even to grow and reproduce in wet soil or under water. But also non-wetland plants can survive flooding, at least for a short period of time. Survival strategies can be divided into two major forms, (1) avoidance of oxygen deficiency within plant tissues; and (2) adaptation to oxygen deficiency. These strategies are described below.

2.1. Avoidance of Oxygen Deficiency by Morphological Modifications

The first strategy, the avoidance of oxygen deficiency inside the flooded plant parts, mainly involves anatomical and morphological modifications that improve gas exchange with the surroundings [8,12]. These modifications are largely mediated by the gaseous plant hormone ethylene that naturally accumulates in flooded plant parts [13].

One of the most prominent modifications is the increase of intercellular gas spaces, the so-called aerenchyma formation, to improve gas transport and distribution inside submerged plant tissues. Aerenchyma can develop in the root cortex as well as in stems and leaves. They are inducible by flooding conditions in several non-wetland species (e.g., wheat, maize [14,15]), or constitutive in many wetland species (e.g., rice, *Zea nicaraguensis* H.H.Iltis and B.F. Benz 2000 [12,15]). The formation of shoot-born adventitious roots has been observed in some plant species under water, for example in rice [16,17] and *Solanum dulcamara* L. [18,19]. Those roots also contain large aerenchyma. In order to restrict gas loss through the root surface, many roots of wetland plants also develop a barrier against radial oxygen loss (ROL) surrounding the aerenchyma-containing tissue (e.g., rice [20,21]; *Z. nicaraguensis* [12,22]).

When the whole plant is under water, some plants have developed the ability to move their leaves up in order to reach the water surface and to restore contact to air. This is achieved by hyponastic growth, meaning the change of the leaf angle to a more upright position, which can be observed in non-wetland (e.g., *Arabidopsis thaliana* (L.) Heynh. [10]) as well as in wetland plants species (e.g., *Rumex palustris* Sm. [23]). Some plant species can go one step further and enhance the shoot growth under water to get their leaves out of the water. This so-called "escape strategy" can be achieved either by growth acceleration in petioles (e.g., *Rumex palustris* Sm. [24,25]; *Ranunculus sceleratus* L. [26]), or by enhanced growth of stems (e.g., rice [27,28]).

2.2. Adaptation to Oxygen Deficiency by Metabolic Modifications

If a plant species is not able to induce morphological modifications, or if water levels are too high to be outgrown, they have to cope with restricted gas exchange, mainly with low-oxygen concentrations. This adaptation involves metabolic modifications (summarized in [7,29]). A first response of plant cells under oxygen deficiency is the induction of fermentation. Since the mitochondrial ATP production is limited by oxygen availability, plants are dependent on glycolytic ATP production. During glycolysis, NADH accumulates and needs to be re-oxidized to NAD in order to maintain the glycolytic process. This is done by lactic acid fermentation, but mainly by ethanolic fermentation via alcohol dehydrogenase and pyruvate decarboxylase.

The higher transcription of genes encoding fermentative enzymes under oxygen deficiency is largely regulated by a group of oxygen-labile transcription factors, group VII of the ethylene-response factor family (groupVII-ERFs). The Arabidopsis groupVII-ERFs AtRAP2.2, AtRAP2.12, and AtRAP2.3 are constitutively expressed, but the proteins are degraded under normoxia by the Arg-branch of the N-end rule pathway. Under hypoxia, they can accumulate and act as transcriptional activators for example for genes encoding fermentative enzymes, but also for other metabolic and regulatory proteins [30–32].

While all plant species analyzed so far are able to induce fermentative enzymes under oxygen deficiency, the availability of carbohydrates and the efficiency to cope with lower energy production (2 Mol ATP per Mol glucose in glycolysis versus 30–36 Mol ATP per Mol glucose in mitochondrial respiration) restrict plant productivity and survival. Sensitive plant species often die from energy deficiency due to exhaustion of fermentable substrates, before the flooding period ends.

However, certain plant species and organs can consume large amounts of carbohydrates from sources such as starch that are otherwise difficult to access under oxygen deficiency. This is achieved by specific amylases and manipulation of the regulatory pathways (e.g., rice coleoptiles [33,34]; *Potamogeton pectinatus* (L.) Böerner tubers [35]). This high carbohydrate availability enables the strong elongation growth in plants exerting the "escape strategy". Other species, among them certain cultivars of rice, restrict metabolism and growth under water and apply the so-called "quiescence strategy", which enables them to survive for longer times with restricted carbohydrate supply. Some plant species might also make use of alternative energy pathways (e.g., the utilization of pyrophosphate instead of ATP for phosphorylation), but this has still not been fully explored [36–38].

2.3. Tolerance Traits for Flooding Survival and Their Usage in Breeding

The overview presented above of adaptational mechanisms employed by plants in response to flooding conditions makes it obvious that there is not only one mechanism or trait of tolerance in a tolerant plant. Hence, the contribution of individual metabolic pathways and morphological modifications to overall flooding tolerance has to be deciphered. This knowledge should enable scientists to focus on specific adaptational mechanisms, and to discover the underlying genetic basis for tolerance traits, which subsequently can be used for breeders in order to improve a crop's tolerance to flooding.

The most prominent example of successful agronomical application of knowledge on a flooding-tolerance trait and its transfer to crops comes from rice. Although this crop is naturally flooding-tolerant, most rice cultivars cannot survive more than one week of complete submergence [39]. Cultivars with the "quiescence strategy" can survive deep floods for up to 14 days, by restricting growth and carbohydrate consumption. On the other hand, deepwater-rice can outgrow a flood within a short time, using the "escape strategy", and thus restores contact to air, enabling long-term survival [27,28]. For both traits, QTL analyses and subsequent molecular investigations have revealed the underlying genes. In both cases transcription factors related to groupVII-ERFs have been made responsible for either restriction of growth under water (SUB1A-1 [39,40]), or for enhanced growth (SNORKEL1/2 [28]). However, only the first genetic trait, the ability to induce quiescence, has been successfully used for crop improvement [5,41], because the second strategy has negative effects on crop stability once the floods recede.

In the next sections, we discuss what progress has been made in improving the flooding tolerance of major temperate crops, and what strategies are currently being applied in research. Among temperate crops, several species have been used for tolerance screenings and QTL analyses. Among the cereals, maize, wheat, and barley are well studied, and some data exist on the pasture grass *Lolium perenne* L. Hardly any data exist on other cereals such as oat and rye (Table 1). Among dicot plants, soybean and rape have been used in several studies, while others like potato or sugar beet have been seldom analyzed. Here, the model species *Arabidopsis thaliana* (L.) Heynh. is the best studied dicot plant with a wealth of expression and metabolic data, and an extended analysis comparing ecotype performances under submergence [11]. However, even here, the underlying mechanisms and genes responsible for tolerance are only now starting to emerge [42–44].

Table 1. Overview over crop species referred to in this review article.

Species	Cultivar Differences in Flooding Tolerance	Quantitative Trait Loci (QTL) Associated with Flooding Tolerance	Genome Sequence	-Omics Data on Flooding/Low-Oxygen Response
Monocots				
Zea mays	yes	yes (Table 2)	[45,46]	available
Triticum aestivum	yes	yes	[47]	not available
Hordeum vulgare	yes	yes (Table 3)	[48]	not available
Avena sativa	unknown	no	not available	not available
Secale cereale	unknown	no	[49]	not available
Lolium perenne	yes	yes	[50]	not available
Dicots				
Glycine max	yes	yes (Table 4)	[51]	available
Brassica napus	yes	yes	[52]	available
Helianthus annuus	unknown	no	[53]	not available
Beta vulgaris	unknown	no	[54]	not available
Solanum tuberosum	unknown	no	[55]	not available

3. Waterlogging Tolerance of Maize and Teosinte

The field crop maize (*Zea mays* L.) is not only a major human food source, but can also be used for animal feed as well as bioethanol production. However, it is relatively flooding-sensitive. Interestingly, maize has several close relatives with higher flooding tolerance, among them the teosinte species *Z. nicaraguensis*, *Z. luxurians*, and *Z. mays* ssp. *huehuetenangensis*. These species have been employed in the past as a basis to improve flooding tolerance of maize, similarly as a huge panel of maize cultivars, as described in detail below.

3.1. Morphological Adaptations of Teosinte as Tolerance Traits

The flooding tolerance of teosinte species has been largely associated with morphological modifications, namely the growth of adventitious roots, the formation of aerenchyma under non-flooded conditions as well as the establishment of a barrier against ROL [22,56,57]. Even under non-waterlogged conditions, *Z. nicaraguensis* and *Z. luxurians* are able to form large aerenchyma in the root cortex. These mechanisms improve aeration of flooded roots and thus enhance waterlogging tolerance [22,56,58]. Multiple QTL analyses revealed several loci associated with the constitutive aerenchyma formation ([59–62], summarized in [63], see also Table 2). Similar studies have been done with another maize relative, *Z. luxurians* [64]. However, the gene(s) that are responsible for constitutive aerenchyma formation have not yet been identified.

Table 2. Overview over promising QTLs to improve flooding tolerance of maize (*Zea mays* L.).

Crossed Cultivars	Treatment	Trait for Tolerance	QTL	Position (Chr, cM)	Reference
		Leaf traits			
Z. mays cv. F1649 × cv. H84	14 days waterlogging with starch solution	Leaf chlorosis	1.03-4	Chr 1	[68]
Z. nicaraguensis CIMMYT 13,451 × *Z. mays* Mi29	16 days waterlogging with starch solution	Leaf chlorosis	Qft-rd4.07-4.11	Chr 4	[69]
Z. mays cv. Mo18W × B73	48 h submergence	Leaf senescence	Subtol6	Chr 6 (162 Mb)	[70]
		Adventitious root formation			
Z. mays ssp. *huehuetenangensis* × *Z. mays* B64	14 days waterlogging	Adventitious root formation	Qarf8.05 Qarf8.03 Qarf5.03	Chr 8 Chr 8 Chr 5	[66]
Z. mays cv. Na4 × B64	14 days waterlogging	Adventitious root formation	Qarf8.05	Chr 8	[71]
Z. mays ssp. *huehuetenangensis* × *Z. mays* Mi29	14 days waterlogging	Adventitious root formation	Qarf8.05 Qarf5.03	Chr 8 Chr 5	[67]
Z. nicaraguensis CIMMYT 13,451 × *Z. mays* Mi29	14 days waterlogging	Adventitious root formation	Qarf3.04 Qarf8.03	Chr 3 Chr 8	[67]
		Constitutive aerenchyma formation			
Z. nicaraguensis CIMMYT 13,451 × *Z. mays* B64	none	Constitutive aerenchyma formation	Qaer1.07 Qaer1.02-3 Qaer5.09 Qaer8.06-7	Chr 1, 144 Chr 1, 35 Chr 5, 138 Chr 8, 97-101	[59]
Z. nicaraguensis CIMMYT 13,451 × *Z. mays* Mi29	none	Constitutive aerenchyma formation	Qaer1.06 Qaer1.11 Qaer5.09n	Chr 1 Chr 1 Chr 5	[60]
Z. nicaraguensis CIMMYT 13,451 × *Z. mays* Mi29	none	Constitutive aerenchyma formation	Qaer1.05-6 Qaer8.05	Chr 1, 45 Chr 8, 0	[61]
Z. luxurians × *Z. mays* B73	none	Constitutive aerenchyma formation	Qaer2.06 Qaer5.05-6	Chr 2, 88 Chr 5, 96	[64]
		Formation of a barrier against radial oxygen loss (ROL)			
Z. nicaraguensis CIMMYT 13,451 × *Z. mays* Mi29	14 days stagnant nutrient solution	ROL formation		Chr 3	[65]

The formation of a barrier against ROL has been observed in *Z. nicaraguensis* under stagnant conditions (i.e., hypoxic nutrient solution), but not in *Z. mays* [22]. Even though maize can form aerenchyma under waterlogging, and therefore transport oxygen-rich air into the roots, root tips usually remain hypoxic. This is due to leaking of oxygen to the outside medium along the whole root. This radial oxygen loss is prevented in teosinte species by a tight barrier in the outer root layers. A locus on chromosome 3 of teosinte was involved in this formation and was sufficient to facilitate barrier formation in maize [65], but more studies need to be done to reveal the responsible gene.

Another close relative, *Z. mays* ssp. *huehuetenangensis*, has a larger potential to form adventitious roots than *Z. mays* cultivars [66,67] (see also Table 2). A QTL analysis suggested loci on chromosomes 4, 5, and 8, but the underlying gene(s) have not been identified yet. Also in *Z. nicaraguensis*, QTLs associated with adventitious root formation have been discovered on chromosomes 3 and 8 [67].

These studies suggest that there is a potential to improve maize flooding tolerance by manipulating its anatomical and morphological responses. It is now important to transfer more of these traits from teosinte into elite maize cultivars in order to improve their flooding tolerance, without negatively affecting yield and food quality. However, it remains to be elucidated, which and how many genes are required for this approach.

3.2. QTL Analyses of Maize Cultivars with Contrasting Flooding Tolerance

Several studies have compared waterlogging or submergence tolerance of different maize cultivars, for example of Chinese origin [72,73], tropic cultivars [74–76], or of a wide range of lines [70]. Thereby, analyses have either focused on metabolic changes, or on anatomical differences associated with tolerance.

Many of these experiments have been complemented with subsequent QTL analyses. For example, Mano et al. [71] determined QTLs associated with adventitious root formation between cultivars B64 and Na4 on chromosomes 3, 7, and 8, the latter potentially linked to a locus identified previously during a species comparison between maize and teosinte ([66], Table 2). Another QTL on chromosome 1 from the tolerant inbred line F1649 compared to sensitive H84 was associated with waterlogging tolerance under reducing conditions that often occur in flooded soils [68]. A cross between the tolerant cultivar HZ32 and the intolerant cultivar K12 [72] identified several gene loci associated with waterlogging tolerance [77–79]. Another cross of the tolerant HZ32 with sensitive Mo17 suggested the gene Cyp51 (of the Cytochrome P450 family) as a potential hypoxia tolerance gene [80]. Among tropical cultivars, tolerant CAWL-46-3-1 was compared to the sensitive line CML311-2-1-3, which again revealed several QTLs [81]. All of these QTLs await further characterization and a link to specific tolerance traits.

Growing knowledge of the molecular response of maize and teosinte to flooding might help during the identification of responsible genes for sensitivity and tolerance traits. The transcriptomic response of maize has been studied under several circumstances and with different goals. Laser microdissection in combination with microarray analysis has been used on root sections to study aerenchyma formation mechanisms [82,83], which could be used in future to improve elite cultivars, in combination with QTL analyses and the comparisons to teosinte as described above.

A comparative study of four maize cultivars with contrasting tolerance (Mo18W & M162W as tolerant and B97 & B73 as sensitive lines) analyzed the transcriptional response to submergence [70]. A major QTL on chromosome 6 was associated with submergence tolerance, Subtol6, but the underlying gene remains to be characterized. Another RNAseq analysis of several tropical maize lines with contrasting waterlogging tolerance also revealed many candidate genes that might be associated with flooding tolerance [76], among them one gene, GRMZM2G055704, that lies in a region on chromosome 1 that had been previously identified by other screens [78].

A comparison of the tolerant maize cultivar HKI1105 and the sensitive cultivar V372 under waterlogging observed the differential expression of many genes [84]. Subsequently, Arora et al. [85] studied the root transcriptional response under waterlogging of this tolerant cultivar and found metabolism-associated genes as well as genes related to aerenchyma formation that could be important for tolerance. Especially cell-wall-related genes could be important for maize tolerance [86], which is again associated with anatomical properties rather than with metabolic adaptations. Another set of cultivars, sensitive Mo17 and tolerant Hz32, was also studied at the transcript level [87–89]. In these experiments, many differentially expressed genes were observed, yet no gene was clearly associable with tolerance.

In summary, so far no gene or gene variant has been verified to be important for maize waterlogging tolerance. The complex regulatory network and multiple responses in morphology as well as metabolism make it unlikely to find the one gene that determines tolerance. Recent progress in sequencing technologies might help to speed up this process. Rather than laborious crosses between two genotypes with contrasting tolerance and subsequent screening of the progeny, information of many genotypes can be included into one analysis. For such genome-wide association studies (GWAS), tolerance traits are correlated with single nucleotid polymorphisms (SNPs). This approach requires knowledge on the genomic sequence of the species (for an overview, see Table 1) as well as genotype-specific sequence information. This technique is now also used in order to improve waterlogging tolerance of maize [76,90].

4. Waterlogging Tolerance of Barley and Other Hordeum Species

Another major cereal for human food production is barley (*Hordeum vulgare* L.), used for brewery and animal feed. Barley is more flooding-sensitive than other cereals [91]. As maize, barley has flooding-tolerant relatives such as *H. marinum* and *H. spontaneum*, which have been utilized in studying flooding-tolerance mechanisms.

Several screens for waterlogging tolerance have been performed, for example with large cultivar collections [92,93] at the germination stage. However, smaller screens at later developmental stages might be more effective and practicable [91,94]. In such screens, scientists analyzed selections of Chinese cultivars [95,96], Australian cultivars [91], Nordic cultivars [97], or selections from bigger collections [98]. Thereby, screening methods and parameters observed differed considerably, leading to very different results.

Subsequent QTL analyses have been performed, but in many cases the association with a specific tolerance trait is still missing, making it hard to functionally study them. From the Nordic cultivars, several major QTLs were revealed by different crosses [99], but the underlying mechanisms or genes have not yet been identified. After the screening of the Chinese cultivars [95,96,100], several crosses were performed to do QTL analyses for tolerance traits. Crosses were done between tolerant TX9425 and sensitive Franklin [94], or with sensitive Naso Nijo [96]. Another QTL analysis was done between tolerant Yerong and sensitive Franklin [94,101,102]. Next, the tolerant line YYXT was used for a cross with the cultivar Franklin [103]. These studies revealed several major and minor QTLs that might be used in breeding (for an overview, see Table 3). Three examples that went further and focused on specific tolerance traits are described below.

Table 3. Promising QTLs from barley (*Hordeum vulgare*) and related species associated with waterlogging tolerance.

Tolerant Cultivar	Treatment	Parameters Analyzed	Name of QTL	Location of QTL (Chr, cM)	References
		***Hordeum vulgare* L.**			
cv. Yerong	9 weeks waterlogging	Survival rate	QWL.YeFr.4H QWL.YeFr.2H.2	4H, 108–117 2H, 113–118	[101]
cv. Yerong	7 days waterlogging	Aerenchyma formation		4H, 80.95–99.08	[104]
cv. YYXT	9 weeks waterlogging	Survival rate	QWl.YyFr.2H QWl.YyFr.3H QWl.YyFr.4H QWl.YyFr.6H	2H, 76.1 3H, 5.2 4H, 121.1 6H, 78.4	[103]
cv. YYXT	21 days stagnant solution	Root porosity		4H, 116	[105]
cv. Psaknon	18 days waterlogging	Chlorophyll fluorescence (ΦPSII)	QY1 QY2	6H, 114 7H, 59	[99]
cv. TX9425	2 days stagnant	Root membrane potential	QMP.TxNn.2H	2H, 8.85	[106]
		H. spontaneum			
cv. TAM407227	7 days waterlogging	Aerenchyma formation	AER.4H	4H, 98.8	[107]

4.1. *Morphological Adaptations in Hordeum Genotypes*

Waterlogging tolerance of *H. marinum* is mediated by anatomical properties, namely high root porosity of adventitious roots as well as a barrier against ROL [108,109]. A study on 35 Hordeum species and genotypes revealed strong variation in the ability to anatomically adapt to waterlogging, by formation of aerenchyma or a barrier against ROL [110]. However, so far *H. marinum* has not been used in breeding processes to improve barley flooding tolerance, probably due to significant genetic variation between both species [110].

Also in some *H. vulgare* cultivar screens, scientists focused on anatomical differences between accessions. A cross between tolerant YYXT and sensitive Franklin was analyzed in respect of aerenchyma formation in adventitious roots leading to higher root porosity and therefore better survival [111]. The first study, still focusing on overall waterlogging tolerance, revealed four major QTLs [103], while a subsequent analysis exposed a major QTL on chromosome 4H associated with root porosity [105]. This QTL was confirmed again in another cross between tolerant Yerong and sensitive Franklin [104]. Moreover, in a cross between sensitive Franklin and tolerant *H. spontaneum*, this QTL was discovered besides several others [107]. Fine mapping narrowed down a region of 58 genes that are candidates underlying this waterlogging-tolerance trait [112]. Further studies are required to identify the responsible gene.

4.2. *Root Ion Transport as a Tolerance Trait*

Another comparison between two barley varieties focused on differences in root ion transport, namely the function of K^+ channels [113]. Root K^+ content was negatively affected under waterlogging in the sensitive variety Naso Nijo, but remained stable in the tolerant variety TX9425. Uptake of ions required for growth and metabolism is energy-dependent, and an energy deficiency under oxygen deficiency should negatively influence ion uptake processes [114]. Recently, a higher K^+ loss through the membrane under oxygen deficiency was associated with lower viability of the root cells [115]. A subsequent QTL analysis of a cross between the two cultivars revealed a major QTL on chromosome 2H underlying this tolerance trait [106]. However, it is currently not clear which gene is responsible for this trait, and whether proton pumps or K^+ channels are involved in the observed tolerance.

Although QTL analyses of barley under waterlogging have been extensively done, hardly anything is known on the transcriptional response to waterlogging as well as on proteomic and metabolomic changes. Very recently, proteomic changes under waterlogging were studied in different tolerant and sensitive barley cultivars, revealing more protection against ROS and higher fermentation capacity in the tolerant varieties [116]. More work is needed here in order to understand and interpret flooding responses in barley.

5. Analysis of Waterlogging Tolerance in Wheat

Wheat is one of the major cereals in Europe, and it is rather waterlogging-sensitive (e.g., [117]). Since wheat is a hexaploid species, the genetic analysis of this cereal is difficult. Furthermore, spring and winter wheat cultivars are available, making this species even more complex.

5.1. *Variation in Wheat Waterlogging Tolerance*

A number of waterlogging tolerance screens has been performed over many years. For example, van Ginkel et al. [118] tested 1344 lines of spring wheat from the Mexican CIMMYT (Centro Internacional de Mejoramiento de Maiz y Trigo) germplasm collection. A subsequent study re-analyzed six of them and included more lines, confirming the high waterlogging tolerance of the cultivar Ducula [119]. Some of these lines were further examined by crossing tolerant and sensitive lines, and suggested at least four genes to be involved in the tolerance mechanism(s) [120]. One mechanism could be the higher root porosity in waterlogged roots of tolerant lines [121]. This observation was also made during a screen of Australian cultivars [122]. Additionally, a small analysis of only three lines revealed the importance of adventitious (seminal) roots and their porosity for waterlogging survival [117].

A screen of 34 winter wheat cultivars not only considered flooding, but also winter hardiness [123]. On a smaller scale, Huang et al. [124] identified Bayles as a sensitive and Savannah as a tolerant genotype out of six winter wheat cultivars. The usage of UK cultivars, however, did not result in superior genotypes [125]. Other screens tested further winter wheat cultivars, with different physiological parameters analyzed (e.g., mineral content [126], grain yield [127], root length [128,129]).

After identification of cultivars with different flooding tolerance, subsequent physiological studies tried to link the tolerance with certain morphological or metabolic traits. In one analysis, Savannah was more tolerant than Bayles partially due to higher root porosity [130]. A recent evaluation of Norwegian genotypes identified tolerant ones that showed specific root traits (e.g., stele and aerenchyma area) in comparison to sensitive cultivars [131,132]. Another experiment demonstrated that the tolerant cultivar Jackson was more tolerant than Coker 9835 probably because of a lower respiration rate [133]. Recently, the same cultivar Jackson was compared to sensitive Frument, and several metabolic differences were reported in leaves between both cultivars, but no clear tolerance mechanism has been found yet [134]. Another experimental set-up also considered different temperatures during a rather artificial anoxia treatment, demonstrating that genotypic differences were more pronounced at higher temperatures, but also here a single tolerance trait was not discovered [135].

Scientists have started to reveal underlying genes responsible for waterlogging tolerance by QTL analyses. A cross between two winter wheat cultivars, USG3209 and Jaypee, exposed two major QTLs on chromosomes 1B and 6D, to be used in further experiments [136]. The implication of the synthetic genotype W7984 together with the cultivated genotype Opata85 described 32 QTLs associated with waterlogging tolerance, which need to be studied further [137]. Another QTL analysis between SHWL1 and Chuanmai 32 discovered ten QTLs [138].

However, these diverse screens with different genotypes and at different locations also revealed a low reproducibility of tolerant and sensitive cultivars, as summarized in Setter et al. [139], pointing to multiple tolerance mechanisms that could be superior at one site, but not at other locations [91]. This could also be the reason for the lack of confirmed QTLs associated with waterlogging tolerance from different studies. The best strategy to continue would be the selection of one specific tolerance trait and its genetic analysis, as has already been done for barley varieties (see above).

5.2. Can Related Species be Used to Improve Wheat Waterlogging Tolerance?

So far, the classical QTL approach has not revealed single tolerance genes or loci in wheat. A different interesting approach was developed in Australia: An amphiploid from wheat with the tolerant grass *H. marinum* was created in order to produce flooding-tolerant wheat cultivars [140]. This approach resulted in some lines with a stronger barrier against ROL [141]. However, those lines show lower growth and grain yield, and are therefore not yet suitable for agriculture. So far, this morphological trait could not be transferred to wheat by use of disomic chromosome addition lines [142]. More analyses are required in order to be successful with this approach.

Potentially, also in wheat there are more flooding-tolerant relatives such as *Triticum macha* L. or *T. dicoccum* cv. Pontus [143], or *T. spelta* [144] that could be used in improving flooding tolerance of wheat. The latter species was included in a QTL analysis that revealed several loci associated with flooding tolerance at the germination state, among them five that were related to enhanced coleoptile growth [145]. However, this developmental stage does not necessarily help in improving waterlogging tolerance in the field.

Despite the wealth of greenhouse and field trials, there is not much progress yet in understanding wheat molecular responses to waterlogging, or in improving its waterlogging tolerance. Studies at the transcriptomic, proteomic, and metabolomic level are required to first understand the reason for the sensitivity of wheat cultivars in order to use this as a base for breeding. Probably the most promising direction would be the introduction of morphological changes that could be transferred from related species. Recent experiments have begun to understand how aerenchyma formation in wheat is regulated [146,147], but also knowledge from other grass species such as rice, maize, and barley should be used.

6. Flooding Tolerance of Ryegrass

One of the most important pasture grasses, *Lolium perenne* L., is often grown on soils not suitable for cereals, for example due to poor drainage. Therefore, it is also a target species to improve its waterlogging tolerance.

One comparison of two genotypes, Aurora6 and Nth African6, together with two F1 lines, was done after four weeks of waterlogging, and Aurora6 was more tolerant than the other genotypes [148]. This cross was subsequently used for a QTL analysis, and 37 loci were identified that were associated with waterlogging tolerance [149]. Another four varieties were studied after one week of waterlogging [150]. Here, antioxidant activity was correlated with waterlogging tolerance. However, more work needs to be done to find loci or genes that are associated with *Lolium perenne* L. waterlogging tolerance.

Another set of studies analyzed the submergence tolerance of 94 to 99 genotypes. In one publication, the behavior of the genotypes was evaluated after seven days of submergence and seven days of recovery, and differential responses of genotypes were observed, ranging from sensitive cultivars to tolerant cultivars with either the quiescence or the escape strategy [151]. This study is an exciting start point for further experiments. In another study, submergence tolerance was correlated with simple sequence repeat (SSR) markers across all genotypes [152]. Finally, a targeted approach was used, and candidate genes were selected from previous physiological experiments and analyzed for SNPs to be related to submergence tolerance [153]. However, verification of these candidate genes is still required.

In summary, work on *Lolium perenne* L. waterlogging and submergence tolerance has only just started. More cultivar screens and QTL analyses are needed, preferentially with specific traits in morphology and metabolism. Furthermore, little is known on the molecular response of ryegrass to waterlogging and submergence, and transcriptomic as well as metabolomic studies need to be done in order to build a basis for breeders and scientists.

7. Soybean Tolerance under Waterlogging

Soybean is a very important crop that can be used as protein-rich food for humans, but is mostly utilized as animal feed. As most other crop species, it is very waterlogging sensitive. In some regions of the world, for example in the mid-south of the US or in Asia, it is grown in rotation with rice on fields that are often flood-prone [154,155]. An improvement of its waterlogging tolerance is therefore of great importance.

Soybean plants are of special importance for food production since they are able to fix nitrogen from air in their nodules with the help of rhizobia, and can facilitate soil enrichment with organic nitrogen compounds. Although nitrogenase is sensitive to oxygen and is therefore protected inside the bacteroids in the nodules, the nitrogen fixation process is very energy-demanding. Therefore, nitrogen fixation quickly stops after waterlogging of soybean roots and nodules, even before roots become fully hypoxic [156].

7.1. Screening for Waterlogging Tolerance in the US and Asia

A first screen on 84 Northern soybean cultivars revealed great variation in waterlogging tolerance [157]. Using one selected tolerant cultivar, Archer, two recombinant inbred line (RIL) populations were created with the sensitive northern cultivars Noir 1 and Minsoy. These were used in a QTL analysis, which exposed one major locus, Sat_064, located on chromosome 18, to be involved in the tolerance [158]. This locus was crossed into two southern genotypes to create near isogenic lines (NILs), but their waterlogging tolerance could not be related to the presence or absence of this locus [154]. Therefore, further analyses tried to identify other QTLs related to waterlogging tolerance in Archer in RILs emerging from crosses with two southern soybean cultivars, A5403 and P9641. These efforts revealed at least five more markers, for example on chromosomes 5 and 13, pointing to several genes involved in stress tolerance [159,160] (see also Table 4). Further crosses of these genotypes produced some RILs and NILs with improved waterlogging tolerance [161], but no distinct gene has yet been associated with the waterlogging tolerance of the cultivar Archer.

Another line combination was used to study flooding tolerance and resistance to the pathogen *Phytophthora sojae*, namely the susceptible elite cultivar S99-2281 and the tolerant exotic cultivar PI 408105A [162]. This analysis identified four QTLs associated with flooding tolerance on chromosomes 11 and 13, of which one overlapped with a QTL for resistance to *Phytophthora sojae* (see also Table 4). Subsequently, physiological features between these two lines were compared that might underlay differences in tolerance [163–166]. Multiple differences were observed between the two genotypes studied, for example in respect to adventitious root and aerenchyma formation (more in the tolerant genotype), gene expression of SUB1-like transcription factors, as well as abscisic acid networks. However, the major contributing factor remains to be determined. Interestingly, differences in adventitious root formation were also observed between Vietnamese genotypes with contrasting tolerance [167], hinting at an importance of morphological traits, as also seen for cereals (see above). Crosses of the susceptible line S99-2281 and another tolerant line, PI 561271, revealed two more QTLs on chromosomes 3 and 10 [160,168] (see also Table 4). The QTL at chromosome 3 was narrowed down to a region of 23 genes, and research is ongoing to identify the responsible tolerance gene [168].

Marker-assisted selection was used to transfer those QTLs into high-yield cultivars, resulting in three new flooding-tolerant germplasm lines for application in breeding programs [160]. Still, further screens with up to 722 cultivars are ongoing, in order to further improve und understand soybean waterlogging tolerance [169,170].

Also in Asia, soybean is increasingly grown on rice paddy fields. Therefore, also Japanese soybean cultivars were used for tolerance screens. A cross between the tolerant cultivar Misuzudaizu and the sensitive cultivar Moshidou Gong 503 was evaluated for waterlogging tolerance of young soybean plants, revealing several QTLs [155]. Other screens were performed with 92 Japanese cultivars [171], 400 Korean cultivars [172], 21 Vietnamese cultivars [173], and 16 Indonesian cultivars [174], or mixtures of different origin [175,176].

Table 4. QTLs associated with waterlogging tolerance in soybean (*Glycine max* (L.) Merr.).

Crossed Cultivars	Treatment	Parameters Analyzed	Name of QTL	Location of QTL	References
cv. Archer × northers lines	14 days waterlogging	Plant growth, seed yield	Gm18	Chr 18, Sat_064	[158]
cv. Archer × southern lines	14 days waterlogging	Damages and survival	Gm5 Gm13	Chr 5, Satt385 Chr 13, Satt269	[159,160]
cv. Misuzudaizu × cv. Moshidou Gong 503	21 days waterlogging	Seed yield	ft1	Chr 6, Satt100	[155]
cv. S99-2281 × cv. PI 408105A	14 days waterlogging	Damages and survival	FTS11 FTS13	Chr 11 Chr 13	[162]
cv. S99-2281 × cv. PI 561271	4–6 days waterlogging	Damages and survival	qWT_Gm03 qWT_Gm10	Chr 3 Chr 10	[160,168]
cv. Iyodaizu × cv. Tachinagaha	7 days 0.1% stagnant agar solution	Root traits	Qhti-12-2	Chr 12, Satt052-Satt302	[177]

In all of those screens, differences in waterlogging tolerance were observed, but physiological and genetic factors underlying these tolerance differences could be studied further only in a few cases. For example, the tolerant Japanese variety Iyodaizu was used for a QTL analysis with sensitive Tachinagaha. Scientists identified 11 QTLs, of which a QTL region on chromosome 12 was most promising, as shown also with NILs [177]. Next, as for other species, soybean seed germination is likewise oxygen dependent. Therefore, seed germination was also studied in several Asian cultivars, revealing four QTLs related to a high germination rate under water [178].

Recently, GWAS have emerged as another strategy to identify genomic loci associated with waterlogging tolerance, and has been used to study another legume, *Phaseolus vulgaris* L. (e.g., [179,180]). These experiments discovered an interesting overlap between *Phaseolus vulgaris* L. flooding tolerance loci with soybean QTLs [179], namely with Sat_064 [158] and Satt187 [178]. Furthermore, Soybean (*agronomy-328455*) has also a close relative that is more waterlogging-tolerant, namely *G. sojae*, which could be included in breeding programs for higher tolerance [160].

7.2. Physiological and Molecular Responses of Soybean to Waterlogging

During one screen of soybean cultivars, scientists discovered spongy white roots in the cultivar Manokin, pointing to an aerenchyma-like structure, but this was not observed in the most tolerant cultivar Delsoy 4710 [181]. Such secondary aerenchyma around roots, stems, and nodules, emerging from phellem, has been observed before in waterlogged soybean, and might improve aeration of the flooded tissues [182,183]. In contrast to several grasses, soybean develops only small primary aerenchyma in its root cortex [184]. Whether secondary aerenchyma formation or their extension is generally related to waterlogging tolerance, remains to be studied. Some experiments suggest that morphological parameters such as adventitious root formation or root porosity might be associated with higher tolerance also in soybean [163,164,167].

The molecular response of soybean to flooding has been well studied already, especially at the proteome level, using different organs, developmental stages, and treatment conditions (summarized in [185]). In one study, several tolerant and sensitive Asian lines, classified at the seedling stage, were analyzed at the proteome level [175]. These experiments revealed multiple differences in protein expression among genotypes but no obvious trend, pointing to multiple tolerance factors.

Also the transcriptional response has been studied in several experiments, providing a rich base for future functional analyses. For example, the response of leaves to seven days of waterlogging treatment of the roots was explored by RNAseq, showing a negative impact of root stress on leaf photosynthesis [186]. Also roots were investigated directly after root hypoxia [187,188]. The same group also developed a flooding-tolerant soybean mutant, by gamma-irradiation [189] which was compared to wildtype soybean. This study revealed many genes that were differentially expressed between both genotypes [190], but a subsequent analysis of candidate genes responsible for tolerance is required, as well as the identification of the mutation.

8. Waterlogging Tolerance of *Brassica napus* L. and Relatives

Rape (*Brassica napus* L.) is an important oil crop and can also be used as animal feed. It has a complex genetic structure since it is an allotetraploid species that originated from the two diploid species *B. rapa* and *B. oleracea*. Despite its rather recent origin, a wide range of cultivars exists ranging from winter to spring types, but also semi-winter types can be found for example in China (e.g., [191]). Rape is very sensitive to waterlogging, even if compared to other Brassica species [192]. This is at least in part due to its inability to form aerenchyma [193], similar to its parent *B. rapa* [194].

8.1. Chinese Semi-Winter Rape Cultivars Show Contrasting Waterlogging Tolerance

So far, most data available for rape waterlogging tolerance come from Chinese semi-winter cultivars. The interest in flooding tolerance of Chinese cultivars is due to the fact that rape in China is often used as a rotation crop on rice paddy fields, and therefore it is often affected by waterlogging.

A comparison of 18 cultivars, with different seed coat color, tested germination ability after 24 h of submergence. Interestingly, yellow-colored cultivars were mainly sensitive to submergence at the seed stage, including GH01, and dark-colored cultivars were mainly tolerant, including Zhongshuang 9 and 10 [195]. These cultivars also came up in other screens for waterlogging tolerance [196]. Further screens with more Chinese lines were performed at later developmental stages [197–199], confirmed previous tolerant lines (Zhongshuang 9 and 10), and identified further lines to be tolerant (Xiangyou 13, Huayouza 9, Ningyou 12) and sensitive (Yuhuang 1, Zhongyouza 3, Zhongshuang 8).

A subsequent genetic analysis suggested two genes to be involved in waterlogging tolerance of Zhongshuang 9, but no QTL has been described yet underlying this tolerance [200]. A cross between six cultivars differing in waterlogging tolerance revealed additive and non-additive effects pointing to several genes or alleles involved, with Zhongshuang 9 being the most potent cultivar [201]. An independent group used a double haploid population between two lines different from the ones mentioned above with high and low waterlogging and drought tolerance [202]. They identified at least 11 QTL for waterlogging tolerance, suggesting a complex regulation also in this species.

An attempt to associate waterlogging tolerance with physiological traits suggested the importance of the antioxidant system [196]. Other analyses observed a difference in nitrate metabolism between two Chinese cultivars under waterlogging [203]. Furthermore, a correlation of low ethanolic fermentation and higher waterlogging tolerance was described [204].

So far, there is little knowledge on the molecular response of rape to waterlogging or submergence, making it hard to find a basis to select candidate genes in the QTL regions. A first attempt to solve this problem was done on two Chinese cultivars with contrasting waterlogging tolerance, GH01 and Zhongshuang 9 [205,206]. The transcriptional response of roots to 12 h of waterlogging was studied by use of RNAseq, and many genes responded to the stress treatment, mainly similar in both genotypes. However, there were some differences between the two cultivars, possibly related to the plant hormone abscisic acid, but this has to be explored further. Unfortunately, these two lines have not yet been subjected to a full QTL analysis. Another transcriptomic study focused on shoot responses after 36 and 72 h of waterlogging, revealing downregulation of photosynthesis [207], as was also shown for soybean [186].

8.2. Can Relatives of Brassica napus Help to Enhance Its Flooding Tolerance?

The genus Brassica contains, among other species, the diploid species *B. rapa*, *B. nigra*, and *B. oleracea*, and the allotretraploid species *B. juncea*, *B. napus*, and *B. carinata*, which often contain multiple subspecies and cultivars (e.g., [208]). Little is known on waterlogging tolerance of other Brassica species, besides one study that described *B. juncea* and *B. carinata* as more waterlogging tolerant than rape [192]. A direct comparison of rape with diploid Brassica species under waterlogging would be most helpful.

So far, some other diploid Brassica species have been studied on their own, revealing also cultivar differences in flooding tolerance. An analysis with two populations of *B. rapa* with different waterlogging tolerance suggested the importance of carbohydrate supply to roots as a potential parameter for tolerance [209]. Second, *B. oleracea* was studied at the seed stage. As reported for other species above, seed germination of Brassica species is particularly strongly dependent on oxygen. A QTL analysis between two *B. oleracea* cultivars, a sensitive Chinese cultivar (A12DHd) and a more tolerant calabrese cultivar (GDDH33), revealed three QTLs that are associated with germination ability under low-oxygen concentrations [210], but associated genes have not yet been identified.

The relative high relatedness of rape to the well-studied model Brassicaceae *Arabidopsis thaliana* (L.) Heynh. might be another way to use existing knowledge in improving rape waterlogging tolerance. First ecotype screens on Arabidopsis have demonstrated differences in submergence tolerance [11, 42–44]. However, since Arabidopsis is also rather flooding-sensitive, the analysis of flooding-tolerant Brassicaceae, for example from the *Rorippa* genus [211,212], might be more suitable.

9. Conclusions

Over recent years, many studies have been published on flooding-tolerant cultivars of temperate crop species. In several cases, QTLs could be discovered, and certain tolerance-related traits were described. However, in most studies, the underlying mechanism(s) or the responsible gene(s) have not yet been identified. It is therefore of great importance to continue with well-designed QTL analyses in order to truly improve crop resistance to flooding. All studies so far have demonstrated that a QTL analysis is only promising when it is associated with a specific, well-defined tolerance trait. The most promising traits so far have been related to morphological adaptations that appear to be similar across genera, such as enhanced root porosity, a barrier against ROL, and the formation of adventitious roots. Metabolic adaptations could be related to the antioxidant system, and to primary metabolism, mainly to carbohydrate availability. However, metabolic traits are rather hard to compare in a large collection of cultivars. The knowledge on the overall crop response to the stress at the transcriptomic, proteomic, and metabolomic level will certainly help to understand the molecular mechanisms that provide the basis for underlying tolerance traits.

Acknowledgments: The author would like to thank Maria Klecker, Bettina Bammer, Jana Müller, and Judith Bäumler for critical reading of the manuscript.

References

1. Voesenek, L.A.C.J.; Bailey-Serres, J. Genetics of high-rise rice: Plant biology. *Nature* **2009**, *460*, 959–960. [CrossRef] [PubMed]
2. Bailey-Serres, J.; Voesenek, L.A.C.J. Life in the balance: A signaling network controlling survival of flooding. *Curr. Opin. Plant Biol.* **2010**, *13*, 489–494. [CrossRef] [PubMed]
3. Bailey-Serres, J.; Fukao, T.; Ronald, P.; Ismail, A.; Heuer, S.; Mackill, D. Submergence Tolerant Rice: SUB1's Journey from Landrace to Modern Cultivar. *Rice* **2010**, *3*, 138–147. [CrossRef]
4. Nagai, K.; Hattori, Y.; Ashikari, M. Stunt or elongate? Two opposite strategies by which rice adapts to floods. *J. Plant Res.* **2010**, *123*, 303–309. [CrossRef] [PubMed]
5. Singh, A.; Septiningsih, E.M.; Balyan, H.S.; Singh, N.K.; Rai, V. Genetics, Physiological Mechanisms and Breeding of Flood-Tolerant Rice (*Oryza sativa* L.). *Plant Cell Physiol.* **2017**, *58*, 185–197. [CrossRef] [PubMed]
6. Armstrong, W. Aeration in Higher Plants. In *Advances in Botanical Research*; Elsevier: London, UK, 1980; Volume 7, pp. 225–332. ISBN 978-0-12-005907-2.
7. Van Dongen, J.T.; Licausi, F. Oxygen Sensing and Signaling. *Annu. Rev. Plant Biol.* **2015**, *66*, 345–367. [CrossRef] [PubMed]
8. Voesenek, L.A.C.J.; Bailey-Serres, J. Flood adaptive traits and processes: An overview. *New Phytol.* **2015**, *206*, 57–73. [CrossRef] [PubMed]
9. Sasidharan, R.; Bailey-Serres, J.; Ashikari, M.; Atwell, B.J.; Colmer, T.D.; Fagerstedt, K.; Fukao, T.; Geigenberger, P.; Hebelstrup, K.H.; Hill, R.D.; et al. Community recommendations on terminology and procedures used in flooding and low oxygen stress research. *New Phytol.* **2017**, *214*, 1403–1407. [CrossRef] [PubMed]
10. Lee, S.C.; Mustroph, A.; Sasidharan, R.; Vashisht, D.; Pedersen, O.; Oosumi, T.; Voesenek, L.A.C.J.; Bailey-Serres, J. Molecular characterization of the submergence response of the Arabidopsis thaliana ecotype Columbia. *New Phytol.* **2011**, *190*, 457–471. [CrossRef] [PubMed]
11. Vashisht, D.; Hesselink, A.; Pierik, R.; Ammerlaan, J.M.H.; Bailey-Serres, J.; Visser, E.J.W.; Pedersen, O.; van Zanten, M.; Vreugdenhil, D.; Jamar, D.C.L.; et al. Natural variation of submergence tolerance among Arabidopsis thaliana accessions. *New Phytol.* **2011**, *190*, 299–310. [CrossRef] [PubMed]
12. Yamauchi, T.; Colmer, T.D.; Pedersen, O.; Nakazono, M. Regulation of Root Traits for Internal Aeration and Tolerance to Soil Waterlogging-Flooding Stress. *Plant Physiol.* **2018**, *176*, 1118–1130. [CrossRef] [PubMed]
13. Sasidharan, R.; Voesenek, L.A.C.J. Ethylene-Mediated Acclimations to Flooding Stress. *Plant Physiol.* **2015**, *169*, 3–12. [CrossRef] [PubMed]
14. Colmer, T.D.; Voesenek, L.A.C.J. Flooding tolerance: Suites of plant traits in variable environments. *Funct. Plant Biol.* **2009**, *36*, 665–681. [CrossRef]

15. Colmer, T.D. Long-distance transport of gases in plants: A perspective on internal aeration and radial oxygen loss from roots. *Plant Cell Environ.* **2003**, *26*, 17–36. [CrossRef]
16. Lorbiecke, R.; Sauter, M. Adventitious Root Growth and Cell-Cycle Induction in Deepwater Rice. *Plant Physiol.* **1999**, *119*, 21–30. [CrossRef] [PubMed]
17. Steffens, B.; Kovalev, A.; Gorb, S.N.; Sauter, M. Emerging Roots Alter Epidermal Cell Fate through Mechanical and Reactive Oxygen Species Signaling. *Plant Cell* **2012**, *24*, 3296–3306. [CrossRef] [PubMed]
18. Dawood, T.; Rieu, I.; Wolters-Arts, M.; Derksen, E.B.; Mariani, C.; Visser, E.J.W. Rapid flooding-induced adventitious root development from preformed primordia in Solanum dulcamara. *AoB Plants* **2014**, *6*. [CrossRef] [PubMed]
19. Dawood, T.; Yang, X.; Visser, E.J.W.; te Beek, T.A.H.; Kensche, P.R.; Cristescu, S.M.; Lee, S.; Floková, K.; Nguyen, D.; Mariani, C.; et al. A Co-Opted Hormonal Cascade Activates Dormant Adventitious Root Primordia upon Flooding in Solanum dulcamara. *Plant Physiol.* **2016**, *170*, 2351–2364. [CrossRef] [PubMed]
20. Kulichikhin, K.; Yamauchi, T.; Watanabe, K.; Nakazono, M. Biochemical and molecular characterization of rice (*Oryza sativa* L.) roots forming a barrier to radial oxygen loss: Metabolic profiles of rice root under waterlogging. *Plant Cell Environ.* **2014**, *37*, 2406–2420. [CrossRef] [PubMed]
21. Shiono, K.; Yamauchi, T.; Yamazaki, S.; Mohanty, B.; Malik, A.I.; Nagamura, Y.; Nishizawa, N.K.; Tsutsumi, N.; Colmer, T.D.; Nakazono, M. Microarray analysis of laser-microdissected tissues indicates the biosynthesis of suberin in the outer part of roots during formation of a barrier to radial oxygen loss in rice (*Oryza sativa*). *J. Exp. Bot.* **2014**, *65*, 4795–4806. [CrossRef] [PubMed]
22. Abiko, T.; Kotula, L.; Shiono, K.; Malik, A.I.; Colmer, T.D.; Nakazono, M. Enhanced formation of aerenchyma and induction of a barrier to radial oxygen loss in adventitious roots of *Zea nicaraguensis* contribute to its waterlogging tolerance as compared with maize (*Zea mays* ssp. *mays*): Aerenchyma and ROL barrier in *Zea nicaraguensis*. *Plant Cell Environ.* **2012**, *35*, 1618–1630. [CrossRef] [PubMed]
23. Cox, M.C.H. Plant Movement. Submergence-Induced Petiole Elongation in *Rumex palustris* Depends on Hyponastic Growth. *Plant Physiol.* **2003**, *132*, 282–291. [CrossRef] [PubMed]
24. Voesenek, L.A.C.J.; Rijnders, J.H.G.M.; Peeters, A.J.M.; van de Steeg, H.M.; de Kroon, H. Plant hormones regulate fast shoot elongation under water: From genes to communities. *Ecology* **2004**, *85*, 16–27. [CrossRef]
25. Van Veen, H.; Mustroph, A.; Barding, G.A.; Vergeer-van Eijk, M.; Welschen-Evertman, R.A.M.; Pedersen, O.; Visser, E.J.W.; Larive, C.K.; Pierik, R.; Bailey-Serres, J.; et al. Two Rumex Species from Contrasting Hydrological Niches Regulate Flooding Tolerance through Distinct Mechanisms. *Plant Cell* **2013**, *25*, 4691–4707. [CrossRef] [PubMed]
26. He, J.B.; Bögemann, G.M.; van de Steeg, H.M.; Rijnders, J.G.H.M.; Voesenek, L.A.C.J.; Blom, C.W.P.M. Survival tactics of *Ranunculus* species in river floodplains. *Oecologia* **1999**, *118*, 1–8. [CrossRef] [PubMed]
27. Vriezen, W.H.; Zhou, Z.; Van der Straeten, D. Regulation of Submergence-induced Enhanced Shoot Elongation *in Oryza sativa* L. *Ann. Bot.* **2003**, *91*, 263–270. [CrossRef] [PubMed]
28. Hattori, Y.; Nagai, K.; Furukawa, S.; Song, X.-J.; Kawano, R.; Sakakibara, H.; Wu, J.; Matsumoto, T.; Yoshimura, A.; Kitano, H.; et al. The ethylene response factors SNORKEL1 and SNORKEL2 allow rice to adapt to deep water. *Nature* **2009**, *460*, 1026–1030. [CrossRef] [PubMed]
29. Bailey-Serres, J.; Fukao, T.; Gibbs, D.J.; Holdsworth, M.J.; Lee, S.C.; Licausi, F.; Perata, P.; Voesenek, L.A.C.J.; van Dongen, J.T. Making sense of low oxygen sensing. *Trends Plant Sci.* **2012**, *17*, 129–138. [CrossRef] [PubMed]
30. Gibbs, D.J.; Lee, S.C.; Isa, N.M.; Gramuglia, S.; Fukao, T.; Bassel, G.W.; Correia, C.S.; Corbineau, F.; Theodoulou, F.L.; Bailey-Serres, J.; et al. Homeostatic response to hypoxia is regulated by the N-end rule pathway in plants. *Nature* **2011**, *479*, 415–418. [CrossRef] [PubMed]
31. Licausi, F.; Kosmacz, M.; Weits, D.A.; Giuntoli, B.; Giorgi, F.M.; Voesenek, L.A.C.J.; Perata, P.; van Dongen, J.T. Oxygen sensing in plants is mediated by an N-end rule pathway for protein destabilization. *Nature* **2011**, *479*, 419–422. [CrossRef] [PubMed]
32. Weits, D.A.; Giuntoli, B.; Kosmacz, M.; Parlanti, S.; Hubberten, H.-M.; Riegler, H.; Hoefgen, R.; Perata, P.; van Dongen, J.T.; Licausi, F. Plant cysteine oxidases control the oxygen-dependent branch of the N-end-rule pathway. *Nat. Commun.* **2014**, *5*, 3425. [CrossRef] [PubMed]
33. Perata, P.; Pozueta-Romero, J.; Akazawa, T.; Yamaguchi, J. Effect of anoxia on starch breakdown in rice and wheat seeds. *Planta* **1992**, *188*, 611–618. [CrossRef] [PubMed]

34. Kretzschmar, T.; Pelayo, M.A.F.; Trijatmiko, K.R.; Gabunada, L.F.M.; Alam, R.; Jimenez, R.; Mendioro, M.S.; Slamet-Loedin, I.H.; Sreenivasulu, N.; Bailey-Serres, J.; et al. A trehalose-6-phosphate phosphatase enhances anaerobic germination tolerance in rice. *Nat. Plants* **2015**, *1*, 15124. [CrossRef] [PubMed]
35. Dixon, M.H.; Hill, S.A.; Jackson, M.B.; Ratcliffe, R.G.; Sweetlove, L.J. Physiological and Metabolic Adaptations of *Potamogeton pectinatus* L. Tubers Support Rapid Elongation of Stem Tissue in the Absence of Oxygen. *Plant Cell Physiol.* **2006**, *47*, 128–140. [CrossRef] [PubMed]
36. Huang, S.; Colmer, T.D.; Millar, A.H. Does anoxia tolerance involve altering the energy currency towards PPi? *Trends Plant Sci.* **2008**, *13*, 221–227. [CrossRef] [PubMed]
37. Mustroph, A.; Hess, N.; Sasidharan, R. Hypoxic Energy Metabolism and PPi as an Alternative Energy Currency. In *Low-Oxygen Stress in Plants*; van Dongen, J.T., Licausi, F., Eds.; Springer: Vienna, Austria, 2014; Volume 21, pp. 165–184. ISBN 978-3-7091-1253-3.
38. Atwell, B.J.; Greenway, H.; Colmer, T.D. Efficient use of energy in anoxia-tolerant plants with focus on germinating rice seedlings. *New Phytol.* **2015**, *206*, 36–56. [CrossRef] [PubMed]
39. Xu, K.; Xu, X.; Fukao, T.; Canlas, P.; Maghirang-Rodriguez, R.; Heuer, S.; Ismail, A.M.; Bailey-Serres, J.; Ronald, P.C.; Mackill, D.J. Sub1A is an ethylene-response-factor-like gene that confers submergence tolerance to rice. *Nature* **2006**, *442*, 705–708. [CrossRef] [PubMed]
40. Fukao, T. A Variable Cluster of Ethylene Response Factor-Like Genes Regulates Metabolic and Developmental Acclimation Responses to Submergence in Rice. *Plant Cell* **2006**, *18*, 2021–2034. [CrossRef] [PubMed]
41. Ismail, A.M.; Singh, U.S.; Singh, S.; Dar, M.H.; Mackill, D.J. The contribution of submergence-tolerant (Sub1) rice varieties to food security in flood-prone rainfed lowland areas in Asia. *Field Crops Res.* **2013**, *152*, 83–93. [CrossRef]
42. Van Veen, H.; Vashisht, D.; Akman, M.; Girke, T.; Mustroph, A.; Reinen, E.; Hartman, S.; Kooiker, M.; van Tienderen, P.; Schranz, M.E.; et al. Transcriptomes of eight Arabidopsis thaliana accessions reveal core conserved, genotype- and organ-specific responses to flooding stress. *Plant Physiol.* **2016**, *172*, 668–689. [CrossRef] [PubMed]
43. Akman, M.; Kleine, R.; van Tienderen, P.H.; Schranz, E.M. Identification of the Submergence Tolerance QTL Come Quick Drowning1 (CQD1) in Arabidopsis thaliana. *J. Hered.* **2017**, *108*, 308–317. [CrossRef] [PubMed]
44. Yeung, E.; van Veen, H.; Vashisht, D.; Sobral Paiva, A.L.; Hummel, M.; Rankenberg, T.; Steffens, B.; Steffen-Heins, A.; Sauter, M.; de Vries, M.; et al. A stress recovery signaling network for enhanced flooding tolerance in *Arabidopsis thaliana*. *Proc. Natl. Acad. Sci. USA* **2018**, 201803841. [CrossRef]
45. Schnable, P.S.; Ware, D.; Fulton, R.S.; Stein, J.C.; Wei, F.; Pasternak, S.; Liang, C.; Zhang, J.; Fulton, L.; Graves, T.A.; et al. The B73 Maize Genome: Complexity, Diversity, and Dynamics. *Science* **2009**, *326*, 1112–1115. [CrossRef] [PubMed]
46. Jiao, Y.; Peluso, P.; Shi, J.; Liang, T.; Stitzer, M.C.; Wang, B.; Campbell, M.S.; Stein, J.C.; Wei, X.; Chin, C.-S.; et al. Improved maize reference genome with single-molecule technologies. *Nature* **2017**, *546*, 524–527. [CrossRef] [PubMed]
47. Zimin, A.V.; Puiu, D.; Hall, R.; Kingan, S.; Clavijo, B.J.; Salzberg, S.L. The first near-complete assembly of the hexaploid bread wheat genome, *Triticum aestivum*. *GigaScience* **2017**, *6*, 1–7. [CrossRef] [PubMed]
48. Mascher, M.; Gundlach, H.; Himmelbach, A.; Beier, S.; Twardziok, S.O.; Wicker, T.; Radchuk, V.; Dockter, C.; Hedley, P.E.; Russell, J.; et al. A chromosome conformation capture ordered sequence of the barley genome. *Nature* **2017**, *544*, 427–433. [CrossRef] [PubMed]
49. Bauer, E.; Schmutzer, T.; Barilar, I.; Mascher, M.; Gundlach, H.; Martis, M.M.; Twardziok, S.O.; Hackauf, B.; Gordillo, A.; Wilde, P.; et al. Towards a whole-genome sequence for rye (*Secale cereale* L.). *Plant J.* **2017**, *89*, 853–869. [CrossRef] [PubMed]
50. Byrne, S.L.; Nagy, I.; Pfeifer, M.; Armstead, I.; Swain, S.; Studer, B.; Mayer, K.; Campbell, J.D.; Czaban, A.; Hentrup, S.; et al. A synteny-based draft genome sequence of the forage grass *Lolium perenne*. *Plant J.* **2015**, *84*, 816–826. [CrossRef] [PubMed]
51. Schmutz, J.; Cannon, S.B.; Schlueter, J.; Ma, J.; Mitros, T.; Nelson, W.; Hyten, D.L.; Song, Q.; Thelen, J.J.; Cheng, J.; et al. Genome sequence of the palaeopolyploid soybean. *Nature* **2010**, *463*, 178–183. [CrossRef] [PubMed]
52. Chalhoub, B.; Denoeud, F.; Liu, S.; Parkin, I.A.P.; Tang, H.; Wang, X.; Chiquet, J.; Belcram, H.; Tong, C.; Samans, B.; et al. Early allopolyploid evolution in the post-Neolithic *Brassica napus* oilseed genome. *Science* **2014**, *345*, 950–953. [CrossRef] [PubMed]

53. Badouin, H.; Gouzy, J.; Grassa, C.J.; Murat, F.; Staton, S.E.; Cottret, L.; Lelandais-Brière, C.; Owens, G.L.; Carrère, S.; Mayjonade, B.; et al. The sunflower genome provides insights into oil metabolism, flowering and Asterid evolution. *Nature* **2017**, *546*, 148–152. [CrossRef] [PubMed]
54. Dohm, J.C.; Minoche, A.E.; Holtgräwe, D.; Capella-Gutiérrez, S.; Zakrzewski, F.; Tafer, H.; Rupp, O.; Sörensen, T.R.; Stracke, R.; Reinhardt, R.; et al. The genome of the recently domesticated crop plant sugar beet (*Beta vulgaris*). *Nature* **2014**, *505*, 546–549. [CrossRef] [PubMed]
55. Xu, X.; Pan, S.; Cheng, S.; Zhang, B.; Mu, D.; Ni, P.; Zhang, G.; Yang, S.; Li, R.; Wang, J.; et al. Genome sequence and analysis of the tuber crop potato. *Nature* **2011**, *475*, 189–195. [CrossRef] [PubMed]
56. Mano, Y.; Omori, F.; Takamizo, T.; Kindiger, B.; Bird, R.M.; Loaisiga, C.H. Variation for Root Aerenchyma Formation in Flooded and Non-Flooded Maize and Teosinte Seedlings. *Plant Soil* **2006**, *281*, 269–279. [CrossRef]
57. Burton, A.L.; Brown, K.M.; Lynch, J.P. Phenotypic Diversity of Root Anatomical and Architectural Traits in Species. *Crop Sci.* **2013**, *53*, 1042–1055. [CrossRef]
58. Mano, Y.; Omori, F. Relationship between constitutive root aerenchyma formation and flooding tolerance in *Zea nicaraguensis*. *Plant Soil* **2013**, *370*, 447–460. [CrossRef]
59. Mano, Y.; Omori, F.; Takamizo, T.; Kindiger, B.; Bird, R.M.; Loaisiga, C.H.; Takahashi, H. QTL mapping of root aerenchyma formation in seedlings of a maize × rare teosinte *Zea nicaraguensis* cross. *Plant Soil* **2007**, *295*, 103–113. [CrossRef]
60. Mano, Y.; Omori, F. Verification of QTL controlling root aerenchyma formation in a maize × teosinte *Zea nicaraguensis* advanced backcross population. *Breed. Sci.* **2008**, *58*, 217–223. [CrossRef]
61. Mano, Y.; Omori, F. High-density linkage map around the root aerenchyma locus Qaer1.06 in the backcross populations of maize Mi29 × teosinte *Zea nicaraguensis*. *Breed. Sci.* **2009**, *59*, 427–433. [CrossRef]
62. Abiko, T.; Obara, M.; Abe, F.; Kawaguchi, K.; Oyanagi, A.; Yamauchi, T.; Nakazono, M. Screening of candidate genes associated with constitutive aerenchyma formation in adventitious roots of the teosinte *Zea nicaraguensis*. *Plant Root* **2012**, *6*, 19–27. [CrossRef]
63. Mano, Y.; Omori, F.; Tamaki, H.; Mitsuhashi, S.; Takahashi, W. DNA Marker-Assisted Selection Approach for Developing Flooding-Tolerant Maize. *Jpn. Agric. Res. Q.* **2016**, *50*, 175–182. [CrossRef]
64. Mano, Y.; Omori, F.; Kindiger, B.; Takahashi, H. A linkage map of maize × teosinte *Zea luxurians* and identification of QTLs controlling root aerenchyma formation. *Mol. Breed.* **2008**, *21*, 327–337. [CrossRef]
65. Watanabe, K.; Takahashi, H.; Sato, S.; Nishiuchi, S.; Omori, F.; Malik, A.I.; Colmer, T.D.; Mano, Y.; Nakazono, M. A major locus involved in the formation of the radial oxygen loss barrier in adventitious roots of teosinte *Zea nicaraguensis* is located on the short-arm of chromosome 3: Chromosomal region endowing a root ROL barrier. *Plant Cell Environ.* **2017**, *40*, 304–316. [CrossRef] [PubMed]
66. Mano, Y.; Muraki, M.; Fujimori, M.; Takamizo, T.; Kindiger, B. Identification of QTL controlling adventitious root formation during flooding conditions in teosinte (*Zea mays* ssp. *huehuetenangensis*) seedlings. *Euphytica* **2005**, *142*, 33–42. [CrossRef]
67. Mano, Y.; Omori, F.; Loaisiga, C.H.; Bird, R.M. QTL mapping of above-ground adventitious roots during flooding in maize × teosinte *Zea nicaraguensis* backcross population. *Plant Root* **2009**, *3*, 3–9. [CrossRef]
68. Mano, Y.; Muraki, M.; Takamizo, T. Identification of QTL Controlling Flooding Tolerance in Reducing Soil Conditions in Maize (*Zea mays* L.) Seedlings. *Plant Prod. Sci.* **2006**, *9*, 176–181. [CrossRef]
69. Mano, Y.; Omori, F. Flooding tolerance in interspecific introgression lines containing chromosome segments from teosinte (*Zea nicaraguensis*) in maize (*Zea mays* subsp. *mays*). *Ann. Bot.* **2013**, *112*, 1125–1139. [CrossRef] [PubMed]
70. Campbell, M.T.; Proctor, C.A.; Dou, Y.; Schmitz, A.J.; Phansak, P.; Kruger, G.R.; Zhang, C.; Walia, H. Genetic and Molecular Characterization of Submergence Response Identifies Subtol6 as a Major Submergence Tolerance Locus in Maize. *PLoS ONE* **2015**, *10*, e0120385. [CrossRef] [PubMed]
71. Mano, Y.; Omori, F.; Muraki, M.; Takamizo, T. QTL Mapping of Adventitious Root Formation under Flooding Conditions in Tropical Maize (*Zea mays* L.) Seedlings. *Breed. Sci.* **2005**, *55*, 343–347. [CrossRef]
72. Liu, Y.; Tang, B.; Zheng, Y.; Ma, K.; Xu, S.; Qiu, F. Screening Methods for Waterlogging Tolerance at Maize (*Zea mays* L.) Seedling Stage. *Agric. Sci. China* **2010**, *9*, 362–369. [CrossRef]
73. Li, W.; Mo, W.; Ashraf, U.; Li, G.; Wen, T.; Abrar, M.; Gao, L.; Liu, J.; Hu, J. Evaluation of physiological indices of waterlogging tolerance of different maize varieties in South China. *Appl. Ecol. Environ. Res.* **2018**, *16*, 2059–2072. [CrossRef]

74. Zaidi, P.H.; Rafique, S.; Singh, N.N. Response of maize (*Zea mays* L.) genotypes to excess soil moisture stress: Morpho-physiological effects and basis of tolerance. *Eur. J. Agron.* **2003**, *19*, 383–399. [CrossRef]
75. Zaidi, P.H.; Rafique, S.; Rai, P.K.; Singh, N.N.; Srinivasan, G. Tolerance to excess moisture in maize (*Zea mays* L.): Susceptible crop stages and identification of tolerant genotypes. *Field Crops Res.* **2004**, *90*, 189–202. [CrossRef]
76. Du, H.; Zhu, J.; Su, H.; Huang, M.; Wang, H.; Ding, S.; Zhang, B.; Luo, A.; Wei, S.; Tian, X.; et al. Bulked Segregant RNA-seq Reveals Differential Expression and SNPs of Candidate Genes Associated with Waterlogging Tolerance in Maize. *Front. Plant Sci.* **2017**, *8*, 1022. [CrossRef] [PubMed]
77. Qiu, F.; Zheng, Y.; Zhang, Z.; Xu, S. Mapping of QTL Associated with Waterlogging Tolerance during the Seedling Stage in Maize. *Ann. Bot.* **2007**, *99*, 1067–1081. [CrossRef] [PubMed]
78. Osman, K.A.; Tang, B.; Wang, Y.; Chen, J.; Yu, F.; Li, L.; Han, X.; Zhang, Z.; Yan, J.; Zheng, Y.; et al. Dynamic QTL Analysis and Candidate Gene Mapping for Waterlogging Tolerance at Maize Seedling Stage. *PLoS ONE* **2013**, *8*, e79305. [CrossRef] [PubMed]
79. Osman, K.A.; Tang, B.; Qiu, F.; Naim, A.M.E. Identification of Major QTLs in an Advanced Backcross Lines Associated with Waterlogging Tolerance at Maize Seedling Stage. *World J. Agric. Res.* **2017**, *5*, 126–134. [CrossRef]
80. Tang, W.; Zhang, Z.; Zou, X.; Zheng, Y. Functional genomics of maize submergence tolerance and cloning of the related gene Sicyp51. *Sci. China Ser. C* **2005**, *48*, 337–345. [CrossRef]
81. Zaidi, P.H.; Rashid, Z.; Vinayan, M.T.; Almeida, G.D.; Phagna, R.K.; Babu, R. QTL Mapping of Agronomic Waterlogging Tolerance Using Recombinant Inbred Lines Derived from Tropical Maize (*Zea mays* L) Germplasm. *PLoS ONE* **2015**, *10*, e0124350. [CrossRef] [PubMed]
82. Rajhi, I.; Yamauchi, T.; Takahashi, H.; Nishiuchi, S.; Shiono, K.; Watanabe, R.; Mliki, A.; Nagamura, Y.; Tsutsumi, N.; Nishizawa, N.K.; et al. Identification of genes expressed in maize root cortical cells during lysigenous aerenchyma formation using laser microdissection and microarray analyses. *New Phytol.* **2011**, *190*, 351–368. [CrossRef] [PubMed]
83. Takahashi, H.; Yamauchi, T.; Rajhi, I.; Nishizawa, N.K.; Nakazono, M. Transcript profiles in cortical cells of maize primary root during ethylene-induced lysigenous aerenchyma formation under aerobic conditions. *Ann. Bot.* **2015**, *115*, 879–894. [CrossRef] [PubMed]
84. Thirunavukkarasu, N.; Hossain, F.; Mohan, S.; Shiriga, K.; Mittal, S.; Sharma, R.; Singh, R.K.; Gupta, H.S. Genome-Wide Expression of Transcriptomes and Their Co-Expression Pattern in Subtropical Maize (*Zea mays* L.) under Waterlogging Stress. *PLoS ONE* **2013**, *8*, e70433. [CrossRef] [PubMed]
85. Arora, K.; Panda, K.K.; Mittal, S.; Mallikarjuna, M.G.; Rao, A.R.; Dash, P.K.; Thirunavukkarasu, N. RNAseq revealed the important gene pathways controlling adaptive mechanisms under waterlogged stress in maize. *Sci. Rep.* **2017**, *7*, 10950. [CrossRef] [PubMed]
86. Arora, K.; Panda, K.K.; Mittal, S.; Mallikarjuna, M.G.; Thirunavukkarasu, N. In Silico Characterization and Functional Validation of Cell Wall Modification Genes Imparting Waterlogging Tolerance in Maize. *Bioinform. Biol. Insights* **2017**, *11*, 1–13. [CrossRef] [PubMed]
87. Zhang, Z.X.; Tang, W.H.; Tao, Y.S.; Zheng, Y.L. cDNA microarray analysis of early response to submerging stress in *Zea mays* roots. *Russ. J. Plant Physiol.* **2005**, *52*, 43–49. [CrossRef]
88. Zhang, Z.X.; Zou, X.L.; Tang, W.H.; Zheng, Y.L. Revelation on early response and molecular mechanism of submergence tolerance in maize roots by microarray and suppression subtractive hybridization. *Environ. Exp. Bot.* **2006**, *58*, 53–63. [CrossRef]
89. Zou, X.; Jiang, Y.; Liu, L.; Zhang, Z.; Zheng, Y. Identification of transcriptome induced in roots of maize seedlings at the late stage of waterlogging. *BMC Plant Biol.* **2010**, *10*, 189. [CrossRef] [PubMed]
90. Zhang, X.; Tang, B.; Yu, F.; Li, L.; Wang, M.; Xue, Y.; Zhang, Z.; Yan, J.; Yue, B.; Zheng, Y.; et al. Identification of Major QTL for Waterlogging Tolerance Using Genome-Wide Association and Linkage Mapping of Maize Seedlings. *Plant Mol. Biol. Report.* **2013**, *31*, 594–606. [CrossRef]
91. Setter, T.L.; Waters, I. Review of prospects for germplasm improvement for waterlogging tolerance in wheat, barley and oats. *Plant Soil* **2003**, *253*, 1–34. [CrossRef]
92. Takeda, K.; Fukuyama, T. Variation and geographical distribution of varieties for flooding tolerance in barley seeds. *Barley Genet. Newsl.* **1986**, *16*, 28–29.
93. Qui, J.; Ke, Y. Study on determination of wet tolerance of 4572 barley germplasm resources. *Acta Agric. Shanghai* **1991**, *7*, 27–32.

94. Li, H.; Vaillancourt, R.; Mendham, N.; Zhou, M. Comparative mapping of quantitative trait loci associated with waterlogging tolerance in barley (*Hordeum vulgare* L.). *BMC Genom.* **2008**, *9*, 401. [CrossRef] [PubMed]
95. Pang, J.; Zhou, M.; Mendham, N.; Shabala, S. Growth and physiological responses of six barley genotypes to waterlogging and subsequent recovery. *Aust. J. Agric. Res.* **2004**, *55*, 895–906. [CrossRef]
96. Zhou, M.X.; Li, H.B.; Mendham, N.J. Combining Ability of Waterlogging Tolerance in Barley. *Crop Sci.* **2007**, *47*, 278–284. [CrossRef]
97. Bertholdsson, N.-O. Screening for Barley Waterlogging Tolerance in Nordic Barley Cultivars (*Hordeum vulgare* L.) Using Chlorophyll Fluorescence on Hydroponically-Grown Plants. *Agronomy* **2013**, *3*, 376–390. [CrossRef]
98. Mano, Y.; Takeda, K. Accurate evaluation and verification of varietal ranking for flooding tolerance at the seedling stage in barley (*Hordeum vulgare* L.). *Breed. Sci.* **2012**, *62*, 3–10. [CrossRef] [PubMed]
99. Bertholdsson, N.-O.; Holefors, A.; Macaulay, M.; Crespo-Herrera, L.A. QTL for chlorophyll fluorescence of barley plants grown at low oxygen concentration in hydroponics to simulate waterlogging. *Euphytica* **2015**, *201*, 357–365. [CrossRef]
100. Xiao, Y.; Wei, K.; Chen, J.; Zhou, M.; Zhang, G. Genotypic difference in growth inhibition and yield loss in barley under waterlogging stress. *J. Zhejiang Univ. (Agric. Life Sci.)* **2007**, *33*, 525–532.
101. Zhou, M. Accurate phenotyping reveals better QTL for waterlogging tolerance in barley: QTL for waterlogging tolerance in barley. *Plant Breed.* **2011**, *130*, 203–208. [CrossRef]
102. Xue, D.; Zhou, M.; Zhang, X.; Chen, S.; Wei, K.; Zeng, F.; Mao, Y.; Wu, F.; Zhang, G. Identification of QTLs for yield and yield components of barley under different growth conditions. *J. Zhejiang Univ. Sci. B* **2010**, *11*, 169–176. [CrossRef] [PubMed]
103. Zhou, M.; Johnson, P.; Zhou, G.; Li, C.; Lance, R. Quantitative Trait Loci for Waterlogging Tolerance in a Barley Cross of Franklin × YuYaoXiangTian Erleng and the Relationship Between Waterlogging and Salinity Tolerance. *Crop Sci.* **2012**, *52*, 2082–2088. [CrossRef]
104. Zhang, X.; Zhou, G.; Shabala, S.; Koutoulis, A.; Shabala, L.; Johnson, P.; Li, C.; Zhou, M. Identification of aerenchyma formation-related QTL in barley that can be effective in breeding for waterlogging tolerance. *Theor. Appl. Genet.* **2016**, *129*, 1167–1177. [CrossRef] [PubMed]
105. Broughton, S.; Zhou, G.; Teakle, N.L.; Matsuda, R.; Zhou, M.; O'Leary, R.A.; Colmer, T.D.; Li, C. Waterlogging tolerance is associated with root porosity in barley (*Hordeum vulgare* L.). *Mol. Breed.* **2015**, *35*, 27. [CrossRef]
106. Gill, M.B.; Zeng, F.; Shabala, L.; Zhang, G.; Fan, Y.; Shabala, S.; Zhou, M. Cell-Based Phenotyping Reveals QTL for Membrane Potential Maintenance Associated with Hypoxia and Salinity Stress Tolerance in Barley. *Front. Plant Sci.* **2017**, *8*, 1941. [CrossRef] [PubMed]
107. Zhang, X.; Fan, Y.; Shabala, S.; Koutoulis, A.; Shabala, L.; Johnson, P.; Hu, H.; Zhou, M. A new major-effect QTL for waterlogging tolerance in wild barley (*H. spontaneum*). *Theor. Appl. Genet.* **2017**, *130*, 1559–1568. [CrossRef] [PubMed]
108. Mcdonald, M.P.; Galwey, N.W.; Colmer, T.D. Waterlogging tolerance in the tribe Triticeae: The adventitious roots of *Critesion marinum* have a relatively high porosity and a barrier to radial oxygen loss. *Plant Cell Environ.* **2001**, *24*, 585–596. [CrossRef]
109. Kotula, L.; Schreiber, L.; Colmer, T.D.; Nakazono, M. Anatomical and biochemical characterisation of a barrier to radial O_2 loss in adventitious roots of two contrasting *Hordeum marinum* accessions. *Funct. Plant Biol.* **2017**, *44*, 845–857. [CrossRef]
110. Garthwaite, A.J.; von Bothmer, R.; Colmer, T.D. Diversity in root aeration traits associated with waterlogging tolerance in the genus Hordeum. *Funct. Plant Biol.* **2003**, *30*, 875–889. [CrossRef]
111. Zhang, X.; Shabala, S.; Koutoulis, A.; Shabala, L.; Johnson, P.; Hayes, D.; Nichols, D.S.; Zhou, M. Waterlogging tolerance in barley is associated with faster aerenchyma formation in adventitious roots. *Plant Soil* **2015**, *394*, 355–372. [CrossRef]
112. Zhang, X.; Shabala, S.; Koutoulis, A.; Shabala, L.; Zhou, M. Meta-analysis of major QTL for abiotic stress tolerance in barley and implications for barley breeding. *Planta* **2017**, *245*, 283–295. [CrossRef] [PubMed]
113. Pang, J.Y.; Newman, I.; Mendham, N.; Zhou, M.; Shabala, S. Microelectrode ion and O_2 fluxes measurements reveal differential sensitivity of barley root tissues to hypoxia. *Plant Cell Environ.* **2006**, *29*, 1107–1121. [CrossRef] [PubMed]
114. Zeng, F.; Konnerup, D.; Shabala, L.; Zhou, M.; Colmer, T.D.; Zhang, G.; Shabala, S. Linking oxygen availability with membrane potential maintenance and K^+ retention of barley roots: Implications for waterlogging stress

tolerance: O_2 availability and root ionic homeostasis. *Plant Cell Environ.* **2014**, *37*, 2325–2338. [CrossRef] [PubMed]
115. Gill, M.B.; Zeng, F.; Shabala, L.; Böhm, J.; Zhang, G.; Zhou, M.; Shabala, S. The ability to regulate voltage-gated K^+-permeable channels in the mature root epidermis is essential for waterlogging tolerance in barley. *J. Exp. Bot.* **2018**, *69*, 667–680. [CrossRef] [PubMed]
116. Luan, H.; Shen, H.; Pan, Y.; Guo, B.; Lv, C.; Xu, R. Elucidating the hypoxic stress response in barley (*Hordeum vulgare* L.) during waterlogging: A proteomics approach. *Sci. Rep.* **2018**, *8*, 9655. [CrossRef] [PubMed]
117. Thomson, C.J.; Colmer, T.D.; Watkin, E.L.J.; Greenway, H. Tolerance of wheat (*Triticum aestivum* cvs Gamenya and Kite) and triticale (*Triticosecale* cv. Muir) to waterlogging. *New Phytol.* **1992**, *120*, 335–344. [CrossRef]
118. Van Ginkel, M.; Rajaram, S.; Thijssen, M. Waterlogging in wheat: Germplasm evaluation and methodology development. In Proceedings of the the Seventh Regional Wheat Workshop for Eastern, Central and Southern Africa, Nakuru, Kenya, 16–19 September 1991; pp. 115–124.
119. Sayre, K.; van Ginkel, M.; Rajaram, S.; Ortiz-Monasterio, I. Tolerance to water-logging losses in spring bread wheat: Effect of time of onset on expression. *Annu. Wheat Newsl.* **1994**, *40*, 165–171.
120. Boru, G.; van Ginkel, M.; Kronstadt, W.; Boersma, L. Expression and inheritance of tolerance to waterlogging stress in wheat. *Euphytica* **2001**, *117*, 91–98. [CrossRef]
121. Boru, G.; Kronstad, W.E. Oxygen use from solution by wheat genotypes differing in tolerance to waterlogging. *Euphytica* **2003**, *132*, 151–158. [CrossRef]
122. Setter, T.; Burgess, P.; Waters, I.; Kuo, J. Genetic diversity of barley and wheat for waterlogging tolerance in Western Australia. In Proceedings of the Ninth Australian Barley Technical Symposium, Melbourne, Australia, 12–16 September 1999; pp. 2171–2177.
123. McKersie, B.D.; Hunt, L.A. Genotypic differences in tolerance of ice encasement, low temperature flooding, and freezing in winter wheat. *Crop Sci.* **1987**, *27*, 860–863. [CrossRef]
124. Huang, B.; Johnson, J.W.; NeSmith, D.S.; Bridges, D.C. Root and Shoot Growth of Wheat Genotypes in Response to Hypoxia and Subsequent Resumption of Aeration. *Crop Sci.* **1994**, *34*, 1538–1544. [CrossRef]
125. Dickin, E.; Bennett, S.; Wright, D. Growth and yield responses of UK wheat cultivars to winter waterlogging. *J. Agric. Sci.* **2009**, *147*, 127–140. [CrossRef]
126. Musgrave, M.E.; Ding, N. Evaluating Wheat Cultivars for Waterlogging Tolerance. *Crop Sci.* **1998**, *38*, 90–97. [CrossRef]
127. Collaku, A.; Harrison, S.A. Losses in Wheat Due to Waterlogging. *Crop Sci.* **2002**, *42*, 444–450. [CrossRef]
128. Ghobadi, M.E.; Ghobadi, M.; Zebarjadi, A. Effect of waterlogging at different growth stages on some morphological traits of wheat varieties. *Int. J. Biometeorol.* **2017**, *61*, 635–645. [CrossRef] [PubMed]
129. Hayashi, T.; Yoshida, T.; Fujii, K.; Mitsuya, S.; Tsuji, T.; Okada, Y.; Hayashi, E.; Yamauchi, A. Maintained root length density contributes to the waterlogging tolerance in common wheat (*Triticum aestivum* L.). *Field Crops Res.* **2013**, *152*, 27–35. [CrossRef]
130. Huang, B.; Johnson, J.W.; Nesmith, S.; Bridges, D.C. Growth, physiological and anatomical responses of two wheat genotypes to waterlogging and nutrient supply. *J. Exp. Bot.* **1994**, *45*, 193–202. [CrossRef]
131. Sundgren, T.K.; Uhlen, A.K.; Waalen, W.; Lillemo, M. Field Screening of Waterlogging Tolerance in Spring Wheat and Spring Barley. *Agronomy* **2018**, *8*, 38. [CrossRef]
132. Sundgren, T.K.; Uhlen, A.K.; Lillemo, M.; Briese, C.; Wojciechowski, T. Rapid seedling establishment and a narrow root stele promotes waterlogging tolerance in spring wheat. *J. Plant Physiol.* **2018**. [CrossRef] [PubMed]
133. Huang, B.; Johnson, J. Root Respiration and Carbohydrate Status of Two Wheat Genotypes in Response to Hypoxia. *Ann. Bot.* **1995**, *75*, 427–432. [CrossRef]
134. Herzog, M.; Fukao, T.; Winkel, A.; Konnerup, D.; Lamichhane, S.; Alpuerto, J.B.; Hasler-Sheetal, H.; Pedersen, O. Physiology, gene expression, and metabolome of two wheat cultivars with contrasting submergence tolerance: Submergence tolerance in two wheat cultivars. *Plant Cell Environ.* **2018**. [CrossRef] [PubMed]
135. Huang, S.; Shingaki-Wells, R.N.; Petereit, J.; Alexova, R.; Millar, A.H. Temperature-dependent metabolic adaptation of *Triticum aestivum* seedlings to anoxia. *Sci. Rep.* **2018**, *8*, 6151. [CrossRef] [PubMed]
136. Ballesteros, D.C.; Mason, R.E.; Addison, C.K.; Andrea Acuña, M.; Nelly Arguello, M.; Subramanian, N.; Miller, R.G.; Sater, H.; Gbur, E.E.; Miller, D.; et al. Tolerance of wheat to vegetative stage soil waterlogging is conditioned by both constitutive and adaptive QTL. *Euphytica* **2015** *201*, 329–343. [CrossRef]

137. Yu, M.; Mao, S.; Chen, G.; Liu, Y.; Li, W.; Wei, Y.; Liu, C.; Zheng, Y. QTLs for Waterlogging Tolerance at Germination and Seedling Stages in Population of Recombinant Inbred Lines Derived from a Cross between Synthetic and Cultivated Wheat Genotypes. *J. Integr. Agric.* **2014**, *13*, 31–39. [CrossRef]
138. Yu, M.; Chen, G.-Y. Conditional QTL mapping for waterlogging tolerance in two RILs populations of wheat. *SpringerPlus* **2013**, *2*, 245. [CrossRef] [PubMed]
139. Setter, T.L.; Waters, I.; Sharma, S.K.; Singh, K.N.; Kulshreshtha, N.; Yaduvanshi, N.P.S.; Ram, P.C.; Singh, B.N.; Rane, J.; McDonald, G.; et al. Review of wheat improvement for waterlogging tolerance in Australia and India: The importance of anaerobiosis and element toxicities associated with different soils. *Ann. Bot.* **2009**, *103*, 221–235. [CrossRef] [PubMed]
140. Alamri, S.A.; Barrett-Lennard, E.G.; Teakle, N.L.; Colmer, T.D. Improvement of salt and waterlogging tolerance in wheat: Comparative physiology of *Hordeum marinum-Triticum aestivum* amphiploids with their *H. marinum* and wheat parents. *Funct. Plant Biol.* **2013**, *40*, 1168–1178. [CrossRef]
141. Malik, A.I.; Islam, A.K.M.R.; Colmer, T.D. Transfer of the barrier to radial oxygen loss in roots of *Hordeum marinum* to wheat (*Triticum aestivum*): Evaluation of four *H. marinum*-wheat amphiploids. *New Phytol.* **2011**, *190*, 499–508. [CrossRef] [PubMed]
142. Konnerup, D.; Malik, A.I.I.; Islam, A.K.M.R.; Colmer, T.D. Evaluation of root porosity and radial oxygen loss of disomic addition lines of *Hordeum marinum* in wheat. *Funct. Plant Biol.* **2017**, *44*, 400–409. [CrossRef]
143. Davies, M.S.; Hillman, G. Effects of Soil Flooding on Growth and Grain Yield of Populations of Tetraploid and Hexaploid Species of Wheat. *Ann. Bot.* **1988**, *62*, 597–604. [CrossRef]
144. Burgos, S.; Stamp, P.; Schmid, J.E. Agronomic and Physiological Study of Cold and Flooding Tolerance of Spelt (*Triticum spelta* L.) and Wheat (*Triticum aestivum* L.). *J. Agron. Crop Sci.* **2001**, *187*, 195–202. [CrossRef]
145. Burgos, M.S.; Messmer, M.M.; Stamp, P.; Schmid, J.E. Flooding tolerance of spelt (*Triticum spelta* L.) compared to wheat (*Triticum aestivum* L.)—A physiological and genetic approach. *Euphytica* **2001**, *122*, 287–295. [CrossRef]
146. Xu, Q.T.; Yang, L.; Zhou, Z.Q.; Mei, F.Z.; Qu, L.H.; Zhou, G.S. Process of aerenchyma formation and reactive oxygen species induced by waterlogging in wheat seminal roots. *Planta* **2013**, *238*, 969–982. [CrossRef] [PubMed]
147. Wany, A.; Kumari, A.; Gupta, K.J. Nitric oxide is essential for the development of aerenchyma in wheat roots under hypoxic stress. *Plant Cell Environ.* **2017**, *40*, 3002–3017. [CrossRef] [PubMed]
148. Mcfarlane, N.M.; Ciavarella, T.A.; Smith, K.F. The effects of waterlogging on growth, photosynthesis and biomass allocation in perennial ryegrass (*Lolium perenne* L.) genotypes with contrasting root development. *J. Agric. Sci.* **2003**, *141*, 241–248. [CrossRef]
149. Pearson, A.; Cogan, N.O.I.; Baillie, R.C.; Hand, M.L.; Bandaranayake, C.K.; Erb, S.; Wang, J.; Kearney, G.A.; Gendall, A.R.; Smith, K.F.; et al. Identification of QTLs for morphological traits influencing waterlogging tolerance in perennial ryegrass (*Lolium perenne* L.). *Theor. Appl. Genet.* **2011**, *122*, 609–622. [CrossRef] [PubMed]
150. Liu, M.; Jiang, Y. Genotypic variation in growth and metabolic responses of perennial ryegrass exposed to short-term waterlogging and submergence stress. *Plant Physiol. Biochem.* **2015**, *95*, 57–64. [CrossRef] [PubMed]
151. Yu, X.; Luo, N.; Yan, J.; Tang, J.; Liu, S.; Jiang, Y. Differential growth response and carbohydrate metabolism of global collection of perennial ryegrass accessions to submergence and recovery following de-submergence. *J. Plant Physiol.* **2012**, *169*, 1040–1049. [CrossRef] [PubMed]
152. Yu, X.; Bai, G.; Luo, N.; Chen, Z.; Liu, S.; Liu, J.; Warnke, S.E.; Jiang, Y. Association of simple sequence repeat (SSR) markers with submergence tolerance in diverse populations of perennial ryegrass. *Plant Sci.* **2011**, *180*, 391–398. [CrossRef] [PubMed]
153. Wang, X.; Jiang, Y.; Zhao, X.; Song, X.; Xiao, X.; Pei, Z.; Liu, H. Association of Candidate Genes with Submergence Response in Perennial Ryegrass. *Front. Plant Sci.* **2017**, *8*, 791. [CrossRef] [PubMed]
154. Reyna, N.; Cornelious, B.; Shannon, J.G.; Sneller, C.H. Evaluation of a QTL for Waterlogging Tolerance in Southern Soybean Germplasm. *Crop Sci.* **2003**, *43*, 2077–2082. [CrossRef]
155. Githiri, S.M.; Watanabe, S.; Harada, K.; Takahashi, R. QTL analysis of flooding tolerance in soybean at an early vegetative growth stage. *Plant Breed.* **2006**, *125*, 613–618. [CrossRef]
156. Amarante, L.; Sodek, L. Waterlogging effect on xylem sap glutamine of nodulated soybean. *Biol. Plant.* **2006**, *50*, 405–410. [CrossRef]

157. VanToai, T.T.; Beuerlein, A.F.; Schmitthenner, S.K.; St. Martin, S.K. Genetic Variability for Flooding Tolerance in Soybeans. *Crop Sci.* **1994**, *34*, 1112–1115. [CrossRef]
158. VanToai, T.T.; St. Martin, S.K.; Chase, K.; Boru, G.; Schnipke, V.; Schmitthenner, A.F.; Lark, K.G. Identification of a QTL Associated with Tolerance of Soybean to Soil Waterlogging. *Crop Sci.* **2001**, *41*, 1247–1252. [CrossRef]
159. Cornelious, B.; Chen, P.; Chen, Y.; de Leon, N.; Shannon, J.G.; Wang, D. Identification of QTLs Underlying Water-Logging Tolerance in Soybean. *Mol. Breed.* **2005**, *16*, 103–112. [CrossRef]
160. Valliyodan, B.; Ye, H.; Song, L.; Murphy, M.; Shannon, J.G.; Nguyen, H.T. Genetic diversity and genomic strategies for improving drought and waterlogging tolerance in soybeans. *J. Exp. Bot.* **2017**, *68*, 1835–1849. [CrossRef] [PubMed]
161. Cornelious, B.; Chen, P.; Hou, A.; Shi, A.; Shannon, J.G. Yield Potential and Waterlogging Tolerance of Selected Near-Isogenic Lines and Recombinant Inbred Lines from Two Southern Soybean Populations. *J. Crop Improv.* **2006**, *16*, 97–111. [CrossRef]
162. Nguyen, V.T.; Vuong, T.D.; VanToai, T.; Lee, J.D.; Wu, X.; Mian, M.A.R.; Dorrance, A.E.; Shannon, J.G.; Nguyen, H.T. Mapping of Quantitative Trait Loci Associated with Resistance to and Flooding Tolerance in Soybean. *Crop Sci.* **2012**, *52*, 2481–2493. [CrossRef]
163. Valliyodan, B.; VanToai, T.T.; Alves, J.D.; de Fátima, P.; Goulart, P.; Lee, J.D.; Fritschi, F.B.; Rahman, M.A.; Islam, R.; Shannon, J.G.; Nguyen, H.T. Expression of Root-Related Transcription Factors Associated with Flooding Tolerance of Soybean (*Glycine max*). *Int. J. Mol. Sci.* **2014**, *15*, 17622–17643. [CrossRef] [PubMed]
164. Kim, Y.-H.; Hwang, S.-J.; Waqas, M.; Khan, A.L.; Lee, J.-H.; Lee, J.-D.; Nguyen, H.T.; Lee, I.-J. Comparative analysis of endogenous hormones level in two soybean (*Glycine max* L.) lines differing in waterlogging tolerance. *Front. Plant Sci.* **2015**, *6*, 714. [CrossRef] [PubMed]
165. Mutava, R.N.; Prince, S.J.K.; Syed, N.H.; Song, L.; Valliyodan, B.; Chen, W.; Nguyen, H.T. Understanding abiotic stress tolerance mechanisms in soybean: A comparative evaluation of soybean response to drought and flooding stress. *Plant Physiol. Biochem.* **2015**, *86*, 109–120. [CrossRef] [PubMed]
166. Syed, N.H.; Prince, S.J.; Mutava, R.N.; Patil, G.; Li, S.; Chen, W.; Babu, V.; Joshi, T.; Khan, S.; Nguyen, H.T. Core clock, *SUB1*, and *ABAR* genes mediate flooding and drought responses via alternative splicing in soybean. *J. Exp. Bot.* **2015**, *66*, 7129–7149. [CrossRef] [PubMed]
167. Nguyen, V.L.; Binh, V.T.; Hoang, D.T.; Mochizuki, T.; Nguyen, V.L. Genotypic variation in morphological and physiological response of soybean to waterlogging at flowering stage. *Int. J. Agric. Sci. Res.* **2015**, *4*, 150–157.
168. Ye, H.; Song, L.; Chen, H.; Valliyodan, B.; Cheng, P.; Ali, L.; Vuong, T.; Wu, C.; Orlowski, J.; Buckley, B.; et al. A major natural genetic variation associated with root system architecture and plasticity improves waterlogging tolerance and yield in soybean: Waterlogging tolerance and root system plasticity in soybean. *Plant Cell Environ.* **2018**. [CrossRef] [PubMed]
169. Wu, C.; Zeng, A.; Chen, P.; Hummer, W.; Mokua, J.; Shannon, J.G.; Nguyen, H.T. Evaluation and development of flood-tolerant soybean cultivars. *Plant Breed.* **2017**, *136*, 913–923. [CrossRef]
170. Wu, C.; Zeng, A.; Chen, P.; Florez-Palacios, L.; Hummer, W.; Mokua, J.; Klepadlo, M.; Yan, L.; Ma, Q.; Cheng, Y. An effective field screening method for flood tolerance in soybean. *Plant Breed.* **2017**, *136*, 710–719. [CrossRef]
171. Sakazono, S.; Nagata, T.; Matsuo, R.; Kajihara, S.; Watanabe, M.; Ishimoto, M.; Shimamura, S.; Harada, K.; Takahashi, R.; Mochizuki, T. Variation in Root Development Response to Flooding among 92 Soybean Lines during Early Growth Stages. *Plant Prod. Sci.* **2014**, *17*, 228–236. [CrossRef]
172. Kang, S.-Y.; Lee, K.J.; Lee, G.-J.; Kim, J.-B.; Chung, S.-J.; Song, J.Y.; Lee, B.-M.; Kim, D.S. Development of AFLP and STS markers linked to a waterlogging tolerance in Korean soybean landraces. *Biol. Plant.* **2010**, *54*, 61–68. [CrossRef]
173. VanToai, T.T.; Hoa, T.T.C.; Hue, N.T.N.; Nguyen, H.T.; Shannon, J.G.; Rahman, M.A. Flooding Tolerance of Soybean [*Glycine max* (L.) Merr.] Germplasm from Southeast Asia under Field and Screen-House Environments. *Open Agric. J.* **2010**, *4*, 38–46. [CrossRef]
174. Kuswantoro, H. Agronomical Characters of Some Soybean Germplasm under Waterlogging Condition. *J. Agron.* **2015**, *14*, 93–97. [CrossRef]
175. Nanjo, Y.; Jang, H.-Y.; Kim, H.-S.; Hiraga, S.; Woo, S.-H.; Komatsu, S. Analyses of flooding tolerance of soybean varieties at emergence and varietal differences in their proteomes. *Phytochemistry* **2014**, *106*, 25–36. [CrossRef] [PubMed]

176. Suematsu, K.; Abiko, T.; Nguyen, V.L.; Mochizuki, T. Phenotypic variation in root development of 162 soybean accessions under hypoxia condition at the seedling stage. *Plant Prod. Sci.* **2017**, *20*, 323–335. [CrossRef]
177. Nguyen, V.L.; Takahashi, R.; Githiri, S.M.; Rodriguez, T.O.; Tsutsumi, N.; Kajihara, S.; Sayama, T.; Ishimoto, M.; Harada, K.; Suematsu, K.; et al. Mapping quantitative trait loci for root development under hypoxia conditions in soybean (*Glycine max* L. Merr.). *Theor. Appl. Genet.* **2017**, *130*, 743–755. [CrossRef] [PubMed]
178. Sayama, T.; Nakazaki, T.; Ishikawa, G.; Yagasaki, K.; Yamada, N.; Hirota, N.; Hirata, K.; Yoshikawa, T.; Saito, H.; Teraishi, M.; et al. QTL analysis of seed-flooding tolerance in soybean (*Glycine max* [L.] Merr.). *Plant Sci.* **2009**, *176*, 514–521. [CrossRef] [PubMed]
179. Soltani, A.; MafiMoghaddam, S.; Walter, K.; Restrepo-Montoya, D.; Mamidi, S.; Schroder, S.; Lee, R.; McClean, P.E.; Osorno, J.M. Genetic Architecture of Flooding Tolerance in the Dry Bean Middle-American Diversity Panel. *Front. Plant Sci.* **2017**, *8*, 1183. [CrossRef] [PubMed]
180. Soltani, A.; MafiMoghaddam, S.; Oladzad-Abbasabadi, A.; Walter, K.; Kearns, P.J.; Vasquez-Guzman, J.; Mamidi, S.; Lee, R.; Shade, A.L.; Jacobs, J.L.; et al. Genetic Analysis of Flooding Tolerance in an Andean Diversity Panel of Dry Bean (*Phaseolus vulgaris* L.). *Front. Plant Sci.* **2018**, *9*, 767. [CrossRef] [PubMed]
181. Rhine, M.D.; Stevens, G.; Shannon, G.; Wrather, A.; Sleper, D. Yield and nutritional responses to waterlogging of soybean cultivars. *Irrig. Sci.* **2010**, *28*, 135–142. [CrossRef]
182. Shimamura, S.; Mochizuki, T.; Nada, Y.; Fukuyama, M. Formation and function of secondary aerenchyma in hypocotyl, roots and nodules of soybean (*Glycine max*) under flooded conditions. *Plant Soil* **2003**, *251*, 351–359. [CrossRef]
183. Shimamura, S.; Yamamoto, R.; Nakamura, T.; Shimada, S.; Komatsu, S. Stem hypertrophic lenticels and secondary aerenchyma enable oxygen transport to roots of soybean in flooded soil. *Ann. Bot.* **2010**, *106*, 277–284. [CrossRef] [PubMed]
184. Thomas, A.L.; Guerreiro, S.M.C.; Sodek, L. Aerenchyma Formation and Recovery from Hypoxia of the Flooded Root System of Nodulated Soybean. *Ann. Bot.* **2005**, *96*, 1191–1198. [CrossRef] [PubMed]
185. Wang, X.; Komatsu, S. Proteomic approaches to uncover the flooding and drought stress response mechanisms in soybean. *J. Proteom.* **2018**, *172*, 201–215. [CrossRef] [PubMed]
186. Chen, W.; Yao, Q.; Patil, G.B.; Agarwal, G.; Deshmukh, R.K.; Lin, L.; Wang, B.; Wang, Y.; Prince, S.J.; Song, L.; et al. Identification and Comparative Analysis of Differential Gene Expression in Soybean Leaf Tissue under Drought and Flooding Stress Revealed by RNA-Seq. *Front. Plant Sci.* **2016**, *7*, 1044. [CrossRef] [PubMed]
187. Nanjo, Y.; Maruyama, K.; Yasue, H.; Yamaguchi-Shinozaki, K.; Shinozaki, K.; Komatsu, S. Transcriptional responses to flooding stress in roots including hypocotyl of soybean seedlings. *Plant Mol. Biol.* **2011**, *77*, 129–144. [CrossRef] [PubMed]
188. Nakayama, T.J.; Rodrigues, F.A.; Neumaier, N.; Marcolino-Gomes, J.; Molinari, H.B.C.; Santiago, T.R.; Formighieri, E.F.; Basso, M.F.; Farias, J.R.B.; Emygdio, B.M.; et al. Insights into soybean transcriptome reconfiguration under hypoxic stress: Functional, regulatory, structural, and compositional characterization. *PLoS ONE* **2017**, *12*, e0187920. [CrossRef] [PubMed]
189. Komatsu, S.; Nanjo, Y.; Nishimura, M. Proteomic analysis of the flooding tolerance mechanism in mutant soybean. *J. Proteom.* **2013**, *79*, 231–250. [CrossRef] [PubMed]
190. Yin, X.; Hiraga, S.; Hajika, M.; Nishimura, M.; Komatsu, S. Transcriptomic analysis reveals the flooding tolerant mechanism in flooding tolerant line and abscisic acid treated soybean. *Plant Mol. Biol.* **2017**, *93*, 479–496. [CrossRef] [PubMed]
191. Wang, N.; Qian, W.; Suppanz, I.; Wei, L.; Mao, B.; Long, Y.; Meng, J.; Müller, A.E.; Jung, C. Flowering time variation in oilseed rape (*Brassica napus* L.) is associated with allelic variation in the *FRIGIDA* homologue *BnaA.FRI.a*. *J. Exp. Bot.* **2011**, *62*, 5641–5658. [CrossRef] [PubMed]
192. Ashraf, M.; Mehmood, S. Effects of waterlogging on growth and some physiological parameters of four Brassica species. *Plant Soil* **1990**, *121*, 203–209. [CrossRef]
193. Voesenek, L.A.C.J.; Armstrong, W.; Bögemann, G.M.; Colmer, T.D.; McDonald, M.P. A lack of aerenchyma and high rates of radial oxygen loss from the root base contribute to the waterlogging intolerance of *Brassica napus*. *Aust. J. Plant Physiol.* **1999**, *26*, 87–93. [CrossRef]
194. Daugherty, C.J.; Matthews, S.W.; Musgrave, M.E. Structural changes in rapid-cycling *Brassica rapa* selected for differential waterlogging tolerance. *Can. J. Bot.* **1994**, *72*, 1322–1328. [CrossRef]

195. Zhang, X.K.; Chen, J.; Chen, L.; Wang, H.Z.; Li, J.N. Imbibition behavior and flooding tolerance of rapeseed seed (*Brassica napus* L.) with different testa color. *Genet. Resour. Crop Evol.* **2008**, *55*, 1175–1184. [CrossRef]
196. Zhang, X.K.; Fan, Q.X.; Chen, J.; Li, J.N.; Wang, H.Z. Physiological Reaction Differences of Different Waterlogging-Tolerant Genotype Rapeseed (*Brassica napus* L.) to Anoxia. *Sci. Agric. Sin.* **2007**, *40*, 485–491.
197. Zou, X.L.; Cong, Y.; Cheng, Y.; Lu, G.Y.; Zhang, X.K. Screening and Identification of Waterlogging Tolerant Rapeseed (*Brassica napus* L.) during Germination Stage. In Proceedings of the Third International Conference on Intelligent System Design and Engineering Applications (ISDEA), Hong Kong, China, 16–18 January 2013; pp. 1248–1253. [CrossRef]
198. Zou, X.L.; Hu, C.W.; Zeng, L.; Cheng, Y.; Xu, M.Y.; Zhang, X.K. A Comparison of Screening Methods to Identify Waterlogging Tolerance in the Field in *Brassica napus* L. during Plant Ontogeny. *PLoS ONE* **2014**, *9*, e89731. [CrossRef] [PubMed]
199. Xu, M.Y.; Ma, H.Q.; Zeng, L.; Cheng, Y.; Lu, G.Y.; Xu, J.S.; Zhang, X.K.; Zou, X.L. The effect of waterlogging on yield and seed quality at the early flowering stage in *Brassica napus* L. *Field Crops Res.* **2015**, *180*, 238–245. [CrossRef]
200. Cong, Y.; Cheng, Y.; Zou, C.S.; Zhang, X.K.; Wang, H.Z. Genetic Analysis of Waterlogging Tolerance for Germinated Seeds of Rapeseed (*Brassica napus* L.) with Mixed Model of Major Gene Plus Polygene: Genetic Analysis of Waterlogging Tolerance for Germinated Seeds of Rapeseed (*Brassica napus* L.) with Mixed Model of Major Gene Plus Polygene. *Acta Agron. Sin.* **2009**, *35*, 1462–1467. [CrossRef]
201. Cheng, Y.; Gu, M.; Cong, Y.; Zou, C.S.; Zhang, X.K.; Wang, H.Z. Combining Ability and Genetic Effects of Germination Traits of *Brassica napus* L. Under Waterlogging Stress Condition. *Agric. Sci. China* **2010**, *9*, 951–957. [CrossRef]
202. Li, Z.; Mei, S.; Mei, Z.; Liu, X.; Fu, T.; Zhou, G.; Tu, J. Mapping of QTL associated with waterlogging tolerance and drought resistance during the seedling stage in oilseed rape (*Brassica napus*). *Euphytica* **2014**, *197*, 341–353. [CrossRef]
203. Yu, C.; Xie, Y.; Hou, J.; Fu, Y.; Shen, H.; Liao, X. Response of Nitrate Metabolism in Seedlings of Oilseed Rape (*Brassica napus* L.) to Low Oxygen Stress. *J. Integr. Agric.* **2014**, *13*, 2416–2423. [CrossRef]
204. Xu, B.; Cheng, Y.; Zou, X.L.; Zhang, X.K. Ethanol content in plants of *Brassica napus* L. correlated with waterlogging tolerance index and regulated by lactate dehydrogenase and citrate synthase. *Acta Physiol. Plant.* **2016**, *38*, 81. [CrossRef]
205. Zou, X.L.; Tan, X.Y.; Hu, C.W.; Zeng, L.; Lu, G.Y.; Fu, G.P.; Cheng, Y.; Zhang, X.K. The Transcriptome of *Brassica napus* L. Roots under Waterlogging at the Seedling Stage. *Int. J. Mol. Sci.* **2013**, *14*, 2637–2651. [CrossRef] [PubMed]
206. Zou, X.L.; Zeng, L.; Lu, G.Y.; Cheng, Y.; Xu, J.S.; Zhang, X.K. Comparison of transcriptomes undergoing waterlogging at the seedling stage between tolerant and sensitive varieties of *Brassica napus* L. *J. Integr. Agric.* **2015**, *14*, 1723–1734. [CrossRef]
207. Lee, Y.H.; Kim, K.S.; Jang, Y.S.; Hwang, J.H.; Lee, D.H.; Choi, I.H. Global gene expression responses to waterlogging in leaves of rape seedlings. *Plant Cell Rep.* **2014**, *33*, 289–299. [CrossRef] [PubMed]
208. Zhang, X.; Liu, T.; Li, X.; Duan, M.; Wang, J.; Qiu, Y.; Wang, H.; Song, J.; Shen, D. Interspecific hybridization, polyploidization, and backcross of *Brassica oleracea var. alboglabra* with *B. rapa var. purpurea* morphologically recapitulate the evolution of Brassica vegetables. *Sci. Rep.* **2016**, *6*, 18618. [CrossRef] [PubMed]
209. Daugherty, C.J.; Musgrave, M.E. Characterization of populations of rapid-cycling *Brassica rapa* L. selected for differential waterlogging tolerance. *J. Exp. Bot.* **1994**, *45*, 385–392. [CrossRef]
210. Finch-Savage, W.E.; Côme, D.; Lynn, J.R.; Corbineau, F. Sensitivity of *Brassica oleracea* seed germination to hypoxia: A QTL analysis. *Plant Sci.* **2005**, *169*, 753–759. [CrossRef]
211. Akman, M.; Bhikharie, A.V.; McLean, E.H.; Boonman, A.; Visser, E.J.W.; Schranz, M.E.; van Tienderen, P.H. Wait or escape? Contrasting submergence tolerance strategies of *Rorippa amphibia*, *Rorippa sylvestris* and their hybrid. *Ann. Bot.* **2012**, *109*, 1263–1276. [CrossRef] [PubMed]
212. Sasidharan, R.; Mustroph, A.; Boonman, A.; Akman, M.; Ammerlaan, A.M.H.; Breit, T.; Schranz, M.E.; Voesenek, L.A.C.J.; van Tienderen, P.H. Root Transcript Profiling of Two Rorippa Species Reveals Gene Clusters Associated with Extreme Submergence Tolerance. *Plant Physiol.* **2013**, *163*, 1277–1292. [CrossRef] [PubMed]

8

Compared to Australian Cultivars, European Summer wheat (*Triticum aestivum*) Overreacts when Moderate Heat Stress is Applied at the Pollen Development Stage

Kevin Begcy [1], Anna Weigert [1], Andrew Ogolla Egesa [1,2] and Thomas Dresselhaus [1,*]

[1] Cell Biology and Plant Biochemistry, Biochemie-Zentrum Regensburg, University of Regensburg, 93053 Regensburg, Germany; kevin.begcy@ur.de (K.B.); Anna.Weigert@stud.uni-regensburg.de (A.W.); andrew-ogolla.egesa@stud.uni-regensburg.de (A.O.E.)

[2] Department of Biochemistry and Biotechnology, Kenyatta University, Nairobi 2 0142, Kenya

* Correspondence: thomas.dresselhaus@ur.de

Abstract: Heat stress frequently imposes a strong negative impact on vegetative and reproductive development of plants leading to severe yield losses. Wheat, a major temperate crop, is more prone to suffer from increased temperatures than most other major crops. With heat waves becoming more intense and frequent, as a consequence of global warming, a decrease in wheat yield is highly expected. Here, we examined the impact of a short-term (48 h) heat stress on wheat imposed during reproduction at the pollen mitosis stage both, at the physiological and molecular level. We analyzed two sets of summer wheat germplasms from Australia (Kukri, Drysdale, Gladius, and RAC875) and Europe (Epos, Cornetto, Granny, and Chamsin). Heat stress strongly affected gas exchange parameters leading to reduced photosynthetic and transpiration rates in the European cultivars. These effects were less pronounced in Australian cultivars. Pollen viability was also reduced in all European cultivars. At the transcriptional level, the largest group of heat shock factor genes (type A *HSFs*), which trigger molecular responses as a result of environmental stimuli, showed small variations in gene expression levels in Australian wheat cultivars. In contrast, *HSFs* in European cultivars, including Epos and Granny, were strongly downregulated and partly even silenced, while the high-yielding variety Chamsin displayed a strong upregulation of type A *HSFs*. In conclusion, Australian cultivars are well adapted to moderate heat stress compared to European summer wheat. The latter strongly react after heat stress application by downregulating photosynthesis and transpiration rates as well as differentially regulating *HSFs* gene expression pattern.

Keywords: wheat; heat stress; heat shock factors (*HSF*); pollen mitosis; pollen viability; gas exchange parameters; photosynthesis; transpiration

---

1. Introduction

Wheat (*Triticum aestivum*) is one of the three major cereal crops, which contributes to more than 20% of the total human caloric and protein intake worldwide [1]. The first evidence of wheat domestication dated of about 12,000 years ago in the Middle East contributing to the transition from hunting and gathering of food to settled agriculture during human civilization [2,3]. Since then, wheat plants have been cultivated in a wide range of climatic conditions and in many geographic regions. Currently, Australia and Europe are two main hubs for wheat production. A large degree of heat stress resilience has been identified within Australian cultivars, with a significant number of cultivars exhibiting high levels of tolerance [4]. Australian wheat production has increased largely as a result of improved cultivars and well established crop management practices. In Europe, together with maize, barley, and rye, wheat is one of the main agricultural crops. European wheat production represents around 20% of the total production worldwide [5]. However, largely attributed to climate conditions, a decrease in total cereal production has been observed in European countries over the past two decades [6].

Wheat adaptation to the Australian environment started more than 200 years ago, when the First Fleet arrived to Sydney around 1788 [7]. The first evidence of a successful cross-breeding employing a European cultivar (Italian wheat Tuscan) in Australia dated back to 1860 [7]. Since then, two major events have defined the evolution of the Australian wheat germplasm. First, the introduction of earlier flowering material and second the use of semi-dwarf germplasm during the early 70s of the last century [7]. In Europe, wheat breeding has a longer history, since it was spread from the Middle East to the entire European continent, first via Greece and then through the Danube river and the Balkans to Northern Europe [2]. Different breeding strategies have been employed in Europe depending on specific interests and market requirements. For instance, highly productive British wheat varieties with low tolerance to climate conditions have been crossed with high yielding German lines to develop varieties better adapted to increased temperatures with higher productivity [8]. However, even though some progress has been made with European varieties, long-term climate adaptation of Australian wheat varieties during breeding have generated genetic pools more resilient to environmental stresses compared to their European counterparts.

Gas exchange parameters, especially photosynthetic and transpiration rates, are particularly sensitive to heat stress [9]. Studies in wheat, rice, and tomato have shown that high temperatures lead, for example, to deactivation of the key CO_2-fixating enzyme RUBISCO [10,11], which correlates with a decline in photosynthesis observed at higher temperatures [12–14]. Therefore, maintenance of photosynthetic activity as well as high transpiration rates are considered indicators of heat tolerance. Furthermore, at the molecular level, heat shock factors (HSFs) are important regulators during plant response to increased temperatures. HSFs have been described as inducible transcriptional regulators of numerous genes encoding—for instance, molecular chaperones, ion transporters, aquaporins, and other stress proteins—in order to regulate stress responses. Moreover, HSFs have been shown to be master regulators for triggering acquired thermotolerance and heat stress responses [15,16]. In wheat, HSFs have been categorized into three groups [17], of which type A represents the largest group containing some gene members that are specifically expressed during reproductive development.

At the reproductive stage, heat stress results in severe yield losses in several crop species including wheat [18–21]. Particularly, male reproductive development is sensitive to environmental stresses [18,22]. Elevated temperatures during pollen development are detrimental to the formation of functional pollen. Anthers and pollen itself represent important photosynthetic sink tissues as high accumulation of photoassimilates, including starch and monomers of carbohydrates, are required during their development [18,22]. Starch and other reserve substances are accumulated in pollen grains and required during pollen tube growth to deliver the sperm cells towards the female gametes (egg and central cell) to ensure proper fertilization. Thus, disruption of pollen development or decrease in nutrient supply is expected to lead to sterility and failure of seed set, which potentially results in yield decrease.

Our current understanding of pollen development is mainly derived from model species such as maize, rice, and Arabidopsis [23–25]. Male gametophyte development (pollen) in wheat is poorly characterized, and even less is known how environmental stresses impact pollen development at the physiological and molecular level in this species. In wheat, studies associated with pollen susceptibility to heat stress are scarce [26–29] and were not performed under highly controlled moderate environmental stress conditions, allowing to separate heat stress from other stresses. Most studies have been focused on grain yield and quality [30,31]. Physiological and molecular parameters were usually not investigated, thus the mechanisms underlying reduced pollen viability, an important yield component, remained unexplored.

Understanding the effect of heat stress in two contrasting germplasms will help to mitigate the impact of increased temperature on wheat performance. With the goal to understand the adaptation of Australian summer wheat cultivars and the susceptibility of European cultivars to moderate heat stress, we selected four cultivars of each germplasm and compared them by measuring transpiration and photosynthetic rates, viability of male reproductive structures, and the response in gene expression levels of type A *HSF* genes. In the long term, this study may contribute to the development of an improved European germplasm with enhanced resilience to increased temperatures.

2. Materials and Methods

2.1. Plant Growth Conditions and Heat Stress Treatment

Wheat cultivars from Australia (*Triticum aestivum* L. cv. Kurki, Drysdale, Gladius, and RAC875) and Europe (*Triticum aestivum* L. cv. Epos, Cornetto, Granny, Chamsin) (Table 1) were germinated in an incubator and then transferred to pots (10 cm diameter, two seedlings per pot) containing a mixture of 15% sand, 25% Liapor (swelling clay), and 60% substrate (Einheitserde). Ten seedlings of each cultivar in each of three independent experiments were then transferred to 10 L pots to the greenhouse under controlled conditions of 14 h L/10 h dark photoperiod, with 21 °C ± 2 °C daytime temperature and 18 °C ± 2 °C night temperature, and a constant air humidity of 60–65%. An automated temperature—water based irrigation system was used to supply water according to plant consumption in a time-based pre-programmed schedule. Plants were fertilized twice a week with 2% fertilizer (Hakaphos) and monitored throughout their entire vegetative and reproductive developmental stages. After spikes developed, the auricle distance (the distance between the auricles of the flag leaf and the second last leaf) of each spike was measured. Spikes with auricle distance between 13–15 cm were marked and used for heat stress experiments; at this stage wheat plants enter pollen mitosis [32]. Spikes with an auricle distance of different lengths were not considered. Plants at pollen mitosis were then transferred to walking growth chambers. Non-heat-stressed control plants were maintained in chambers with identical conditions as described above. For heat stress, temperatures during the light/dark photoperiod were increased to 35 °C/25 °C, respectively, but otherwise conditions remained unchanged. Plants were kept for 48 h under heat stress conditions and then directly analyzed.

Table 1. List of summer wheat varieties (all bread wheat) used in this study. Their respective country of origin and flowering time are indicated.

Variety Name	Country of Origin	Flowering Time
Kukri	Australia	40 ± 1 day
Drysdale	Australia	48 ± 1.5 days
Gladius	Australia	49 ± 1.8 days
RAC875	Australia	45 ± 2.3 days
Epos	Czech Republic	52 ± 2.4 days
Cornetto	Germany	57 ± 1.1 days
Granny	Czech Republic	50 ± 1.6 days
Chamsin	Germany	53 ± 2 days

2.2. Physiological Measurements

Wheat plants of Australian and European cultivars were continuously monitored during the time course of the experiments. Gas exchange parameters were recorded at 0, 24, and 48 h in control and heat stressed plants. Fully expanded flag leaves were used to estimate net photosynthetic rate (A) and transpiration rate (E). Daily measurements were taken with an Infrared Gas Analyzer (IRGA, LCpro+; ADC Bioscientific, Hoddesdon, UK) at a CO_2 concentration of 360 μL L^{-1}, a saturating light intensity of 1000 μmol m^{-2} s^{-1} and a gas flow rate of 200 mL min^{-1} as described before [33–35]. For each time point, 15 to 20 plants per treatment were examined and measurements completed between 10:30 a.m. and 2:00 p.m. Data were recorded when gas exchange and chlorophyll fluorescence parameters became stable during measurements.

2.3. *Histological Analysis*

Anthers of non- and heat-stressed plants were fixed in 3.7% (*w*/*v*) formaldehyde, 5% (*v*/*v*) acetic acid and 50% (*v*/*v*) ethanol, vacuum infiltrated, and stored at 4 °C overnight. Samples were then embedded in 10% (*w*/*v*) low melting agarose dissolved in distilled water. Samples were kept in gelatine blocks and post-fixed with 10% formaldehyde and 0.1 M standard PBS buffer at 4 °C overnight. Embedded blocks (1.5 × 1.5 × 1.0 cm) were stored at 4 °C in PBS until sectioning. Cross-sections (60 μm thick) were prepared using a vibratome Hyrax V50 (Carl Zeiss MicroImaging, Thornwood, NY, USA). For quality comparisons, at least 10 cross sections were prepared per sample. Three independent experiments were each performed to collect anthers for sectioning. Images of anther sections shown in Figure S1 are representatives of observations across replicates.

2.4. *Microscopy and Pollen Viability Assay*

Mature pollen grains of non- and heat-stressed wheat of all eight cultivars were isolated and mounted on glass slides containing 3.33 g/L iodine and 6.66 g/L potassium iodide, covered with a cover slip and observed in a Zeiss Axio Imager Z1 microscope equipped with an apotome module and an AxioCam MRM monochromatic camera. We tested on average three thousand grains per cultivar/condition.

2.5. *RNA Isolation and RT-qPCR*

Total RNA from pollen was isolated using the RNA Plant Mini kit (Ambion, Waltham, MA, USA) following the manufacturer's instructions. cDNA synthesis was performed using reverse transcriptase (Invitrogen SuperScript II, Carlsbad, CA, USA) and oligo(dT) primers. Real time PCR reactions were performed using KAPA SYBR Fast qPCR master mix (Peqlab Biotechnology, Erlangen, Germany) as described [25,35]. Expression levels of a wheat tubulin gene (GenBank accession No. U76558) were used as internal standards for normalization of cDNA template quantity using tubulin-specific primers (Table S1). PCRs were performed using a MasterCycler® RealPlex^2 system (Eppendorf, Hamburg, Germany) in a 96-well reaction plate according to the manufacturer's recommendations. Primers used in this study are listed in Table S1. Cycling parameters consisted of 5 min at 95 °C, and 40 cycles at 95 °C for 15 s, 60 °C for 30 s, and 70 °C for 30 s as described previously [24]. qPCR reactions were performed in triplicate for each RNA sample on at least three biological replicates. Specificity of the amplifications was verified by a melting curve analysis to test product specificity. Results from the MasterCycler® RealPlex2 detection system were further analyzed using Microsoft Excel. Relative amounts of mRNA were calculated from threshold points (Ct values) located in the log-linear range of real time PCR amplification plots using the $2^{-\Delta Ct}$ method [36].

2.6. *Statistical Analysis*

Statistical analyses were performed using the R software/environment. A one-way ANOVA was used to compare flowering time between Australian and European cultivars. Data from at least three independent experiments, where each experiment had at least $n = 6$ plants per cultivar/condition were used. Data were expressed as means with standard deviation, and *p*-value of 0.05 was used as the significance level.

3. Results

Eight Australian and European summer wheat cultivars (Table 1) were exposed to a moderate heat stress (35 °C during the day and 25 °C during the night) for 48 h after flowering induction when anthers contained pollen at the mitosis stages of development. Flowering time was significantly shorter in all-Australian cultivars compared to the European varieties. Our experimental design aimed to reflect short heat temperature episodes, which are very frequent in Europe as a consequence of global warming.

3.1. Physiological Responses of Australian and European Summer Wheat Cultivars under Heat Stress

3.1.1. Transpiration Rate Was Strongly Reduced in European Cultivars

Since heat stress events lead to damaging the photosynthetic apparatus and withering of plants [37], we first compared gas exchange parameters between Australian and European wheat cultivars. We measured the transpiration rate at the beginning of the experiment (0 h) as well as 24 and 48 h after heat stress (HS). A control group of non-stressed plants was measured in a neighboring growth chamber at the same time-points. Apart from the temperature, all other conditions were kept constant. Transpiration rates in the Australian varieties Kukri and Gladius, remained comparable in both HS and control conditions (Figure 1A,G). Phenotypically, these cultivars appeared unchanged even 48 h after HS treatment (Figure 1B,C,H,I). Plants looked healthy, showed almost no sign of stress with only a few curled leaves. In the varieties Drysdale and RAC875, a decrease in transpiration rates was observed 24 h after HS (Figure 1D,E) and a further reduction was noticed after 48 h of HS. However, transpiration rates were still significantly above a 50% level of the non-stressed control plants. These cultivars also looked healthy, though RAC875 showed partly yellowish leaves and a larger number of curly leaves compared with the other Australian varieties.

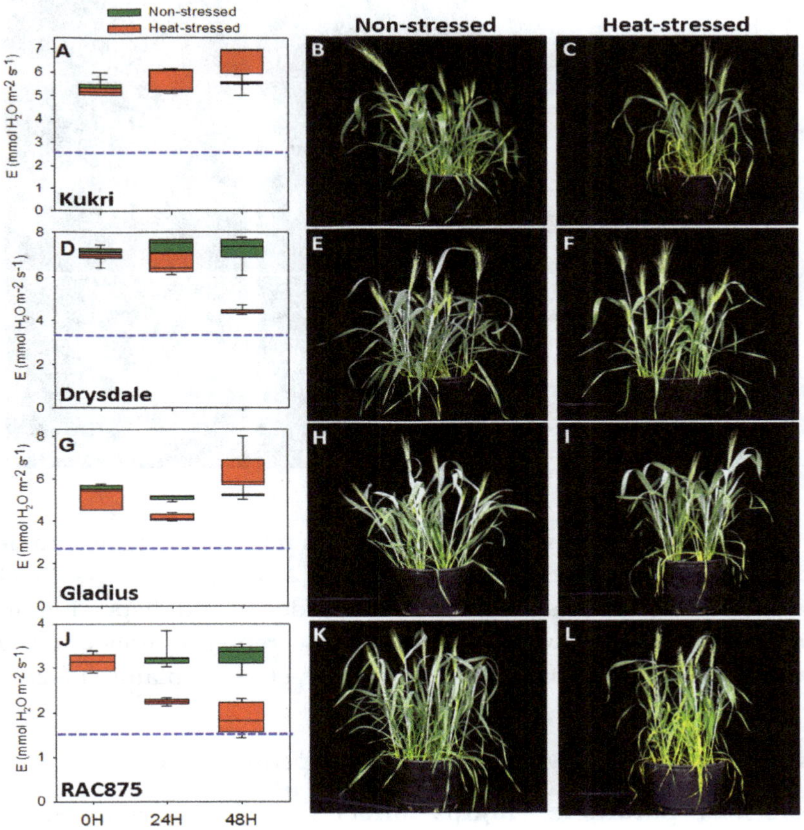

Figure 1. Transpiration rate is maintained under moderate heat stress by Australian summer wheat varieties. Physiological and phenotypical effects on varieties Kukri (**A–C**), Drysdale (**D–F**), Gladius (**G–I**), and RAC875 (**J–L**) are shown. Box plots show the central tendency and dispersion of the transpiration rate among all Australian cultivars. Green and red box plots represent non- and heat-stressed plants, respectively. Blue bars show 50% transpiration rate of non-stressed plants. 15–20 plants of each line were examined.

In contrast, the European summer wheat cultivars showed a pronounced decline in transpiration rate after HS imposition (Figure 2). With the exception of the cultivar Chamsin, elevated temperatures resulted on average in a 25% reduction of transpiration rates in European cultivars after 24 h of HS. After 48 h of HS, all cultivars showed a reduction in transpiration rates close to 75% and in some instances even higher as observed in Epos (Figure 2A,D,G,J). This effect was confirmed phenotypically, as heat-stressed plants wilted and showed yellowish leaves (Figure 2). In summary, Australian cultivars showed a higher tolerance to moderate heat stress indicated by their appearance and sustained high levels of transpiration rates. European lines were less tolerant and showed a marked reduction of transpiration rate.

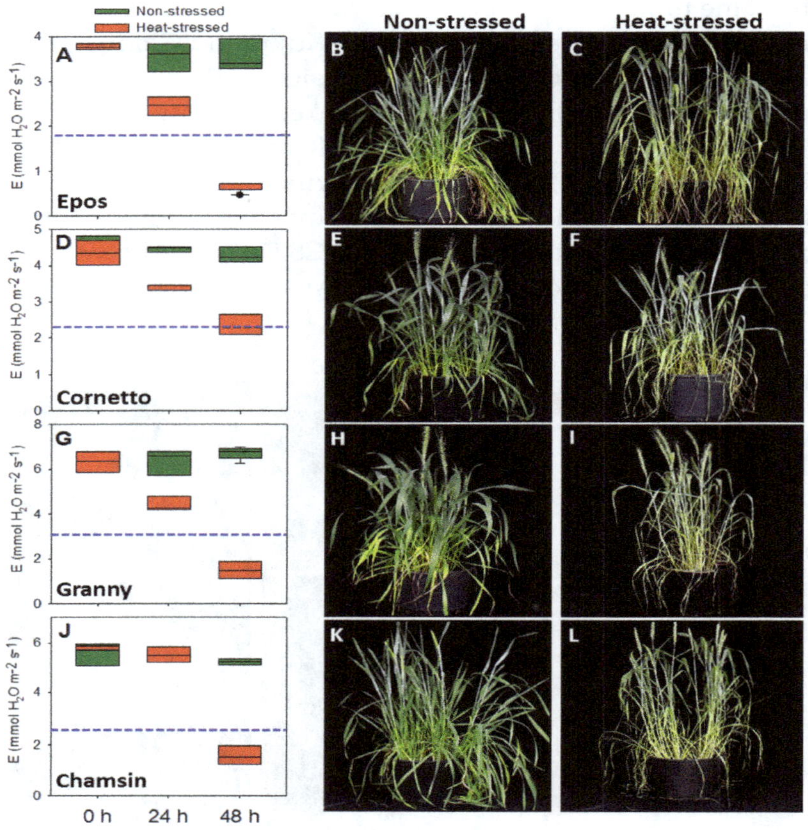

Figure 2. Moderate heat stress strongly alters transpiration rate of European summer wheat varieties. Physiological and phenotypical effects on varieties Epos (**A–C**), Cornetto (**D–F**), Granny (**G–I**), and Chamsin (**J–L**) are shown. Box plots show the central tendency and dispersion of the transpiration rate among all European cultivars. Green and red box plots represent non- and heat-stressed plants, respectively. Blue bars show 50% transpiration rate of non-stressed plants. 15–20 plants of each line were examined.

3.1.2. Photosynthesis Rate Was Strongly Reduced in European Cultivars

Another physiological parameter highly affected by increased temperatures is the net photosynthetic rate. During stress treatment, a significant decline in gas exchange parameters occurred in all cultivars tested. However, a less accentuated drop was observed in the Australian cultivars (Figure 3A–D) compared with the European cultivars (Figure 3E,F). At 24 h after HS, photosynthesis remained close to the non-stress conditions in Kukri and Gladius as well as in Epos (Figure 3A,C,E). In Drysdale and Granny a decline of 22% and 25% on net photosynthesis were observed, respectively (Figure 3B,G). Notably, RAC875, Cornetto, and Chamsin showed a more pronounced decline, with a reduction of net photosynthetic levels of 55, 60, and 77%, respectively, relative to non-stressed control plants (Figure 3D,F,H). After 48 h of HS the Australian cultivars Kukri, Drysdale, and Gladius showed a higher tolerance to heat stress (Figure 3A–C) as indicated by a higher photosynthetic activity. None of these cultivars showed a decrease below 50% of the photosynthetic rate under non-stress conditions. The photosynthetic rate of the fourth Australian cultivar tested (RAC875), declined close to 50% already after 24 h of HS and remained at this value during continuation of HS treatment (Figure 3D). Unlike the Australian cultivars, European summer wheat cultivars with the exception of cultivar Epos were strongly affected already after 24 h of HS showing a sharp decline in photosynthesis rate (Figure 3E–H). In Cornetto and Chamsin (Figure 3F,H) a drop below 50% of the values observed under non-stress conditions was observed and in Granny (Figure 3G) to about 60% levels of control plants. After 48 h of HS exposure all European cultivars showed a net photosynthetic reduction of more than 75% reaching almost zero in Chamsin (Figure 3H). In conclusion, two days after HS treatment, the net photosynthetic rate decreased on average over 80% in European cultivars, while none of the Australian cultivars showed a drop below 50% compared to control plants. Two Australian cultivars (Kukri and Gladius) were especially well-adapted to HS and showed a drop of 25% in photosynthesis rate.

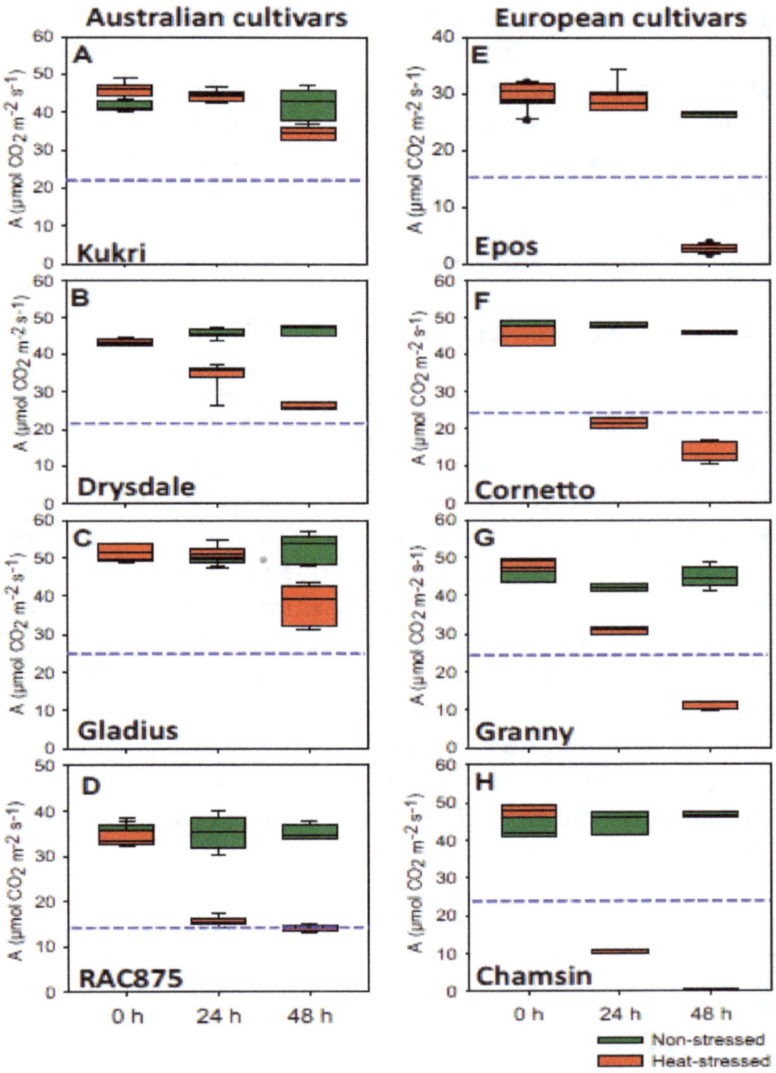

Figure 3. Effect of moderate heat stress on net photosynthesis rates of Australian (**A–D**) and European (**E–H**) summer wheat varieties during the pollen division stage. Box plots show the central tendency and dispersion of the net photosynthetic rate among all Australian and European wheat cultivars. Green and red box plots represent non- and heat-stressed plants, respectively. Blue bars show 50% net photosynthesis rates of non-stressed plants for comparison. 15–20 plants of each line were examined.

3.2. *Analysis of Male Reproductive Structures after a Short-Term Moderate Heat Stress*

3.2.1. Anther Morphology Was Not Significantly Modified after Moderate Heat Stress

One of the most susceptible developmental stages to environmental stresses is male gametophyte development (pollen formation), which occurs in anthers. We collected anthers of all wheat cultivars under non-stress and heat-stress (HS) conditions (Figure S1). To illustrate how our histological analysis were carried out, a scheme indicating a wheat anther and its cross section is shown in Figure S1A. A schematic overview of the internal part of an anther is illustrated in Figure S1B and a detailed view explaining the various anther cell types in Figure S1C.

Under non-stress conditions, spikes containing anthers of both Australian and European cultivars were at the mitosis stage (Figure S1D–K). Identification of developmental stages was conducted based on the auricle distance, which is the distance between the auricles of the flag leaf and the second last leaf [33]. Spikes with an auricle distance between 13 and 15 centimeters were tagged and used for heat stress experiments; at this auricle distance, developing pollen grains completed meiosis and were at the microspore stage shortly before pollen mitosis. After two days of HS, Australian cultivars Kukri, Drysdale, Gladius, and RAC875 displayed normal anther and pollen development (Figure S1D–G). Notably, European wheat varieties also displayed normal pollen development both under HS and control conditions (Figure S1H,K). Pollen grains attached to the anthers were clearly noticeable. Overall, even though European cultivars were strongly affected at the physiological level, the morphology of male reproductive structures (anthers and pollen) did not appear to be severely affected.

3.2.2. Pollen Viability Was Only Affected in European Cultivars after Moderate Heat Stress

Decreased pollen viability has been reported as a major effect of heat stress during reproductive development. Even though anthers of Australian and European varieties did not show a significant damage after 48 h of heat stress (HS), we investigated to which extent pollen of both sets of plants were affected by HS. Pollen viability was tested using iodine-potassium (I2-IK) staining. As shown in Figure 4, almost all pollen grains from Australian varieties were viable (Figure 4A–D) as indicated by strong dyed starch coloration. On average 96–99% of pollen tested were viable (Figure 4I). In contrast, European wheat varieties exposed to HS had a significant amount of non-viable pollen grains, which either did not stain at all or were partially stained (Figure 4E–H). The most severe effect of HS was observed on pollen grains of the Epos cultivar (Figure 4E). Approximately 40% of pollen grains were non-viable (Figure 4I). Similarly, around 25% of pollen grains from other European cultivars were non-viable. We could not test to which extent pollen germination and growth of viable pollen were affected, because wheat pollen generally germinate very poorly in vitro [38]. Nonetheless, our findings suggest that high temperatures during pollen development significantly impair pollen grain development of European cultivars, but not of Australian ones.

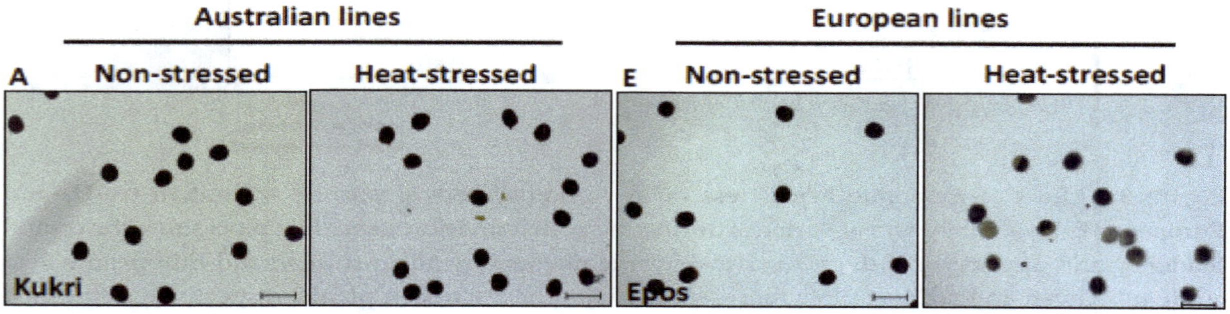

Figure 4. *Cont.*

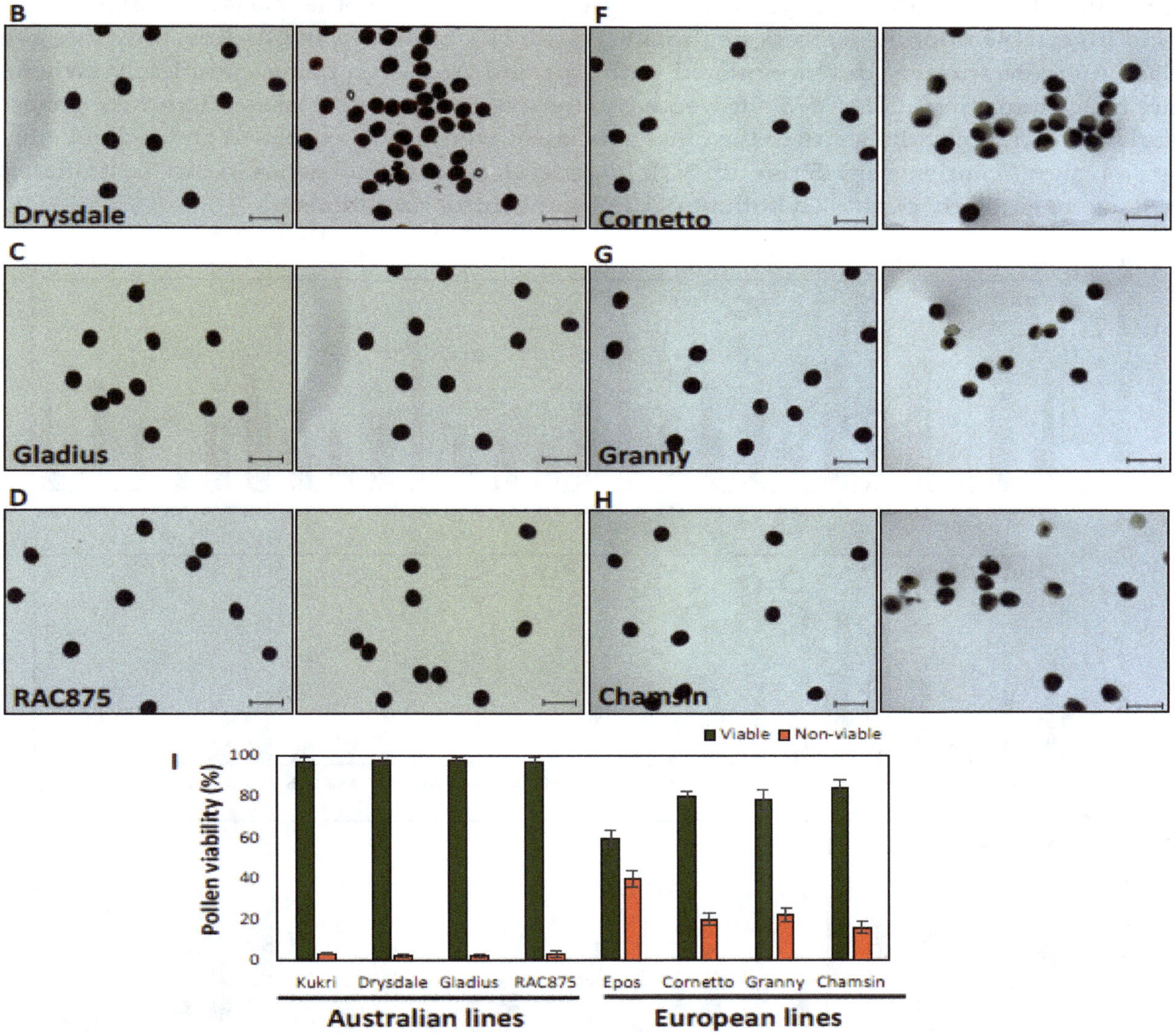

Figure 4. Pollen viability determined in Australian (**A–D**) and European (**E–H**) summer wheat varieties under non-stress and moderate heat-stress conditions. Pollen released from anthers were stained with lugol-iodine (I2-IK). (**I**) Percentage of viable pollen compared between Australian and Europeans summer wheat varieties after exposure to moderate heat stress during pollen mitosis.

3.3. Heat Shock Factors Genes Were Strongly Regulated in European, but Not in Australian Cultivars

In cereals and other plants, heat shock factors (*HSFs*) play an important regulatory role in response to heat stress (HS) and acquired thermotolerance. To understand the molecular basis of the HS response in Australian and European wheat cultivars, we analyzed and compared the gene expression level of several members of the *HSF* gene family. In wheat, 56 *HSF* genes have been identified, which were categorized into classes A, B, and C [17]. Many genes of the largest *TaHSF* class, A class, were previously shown to be predominately expressed in reproductive tissues under non-stress conditions [17]. Therefore, we selected this class to analyze their expression using anthers containing pollen at the mitosis stage under control and HS conditions.

TaHSF class A contains 25 genes divided into seven subclasses. As shown in Figure 5, all Australian cultivars responded to HS as indicated by significant up- and downregulation of some selected genes, while the relative expression levels of most genes were not dramatically altered. In Kukri only genes A3a, A3b, A4e, and A7e were strongly upregulated, while A6c, A6d, and A6e were strongly downregulated (Figure 5A). In Drysdale, the expression level of all other class A members remained low and invariable upon heat stress treatment (Figure 5B). Similarly, in Gladius, only five genes of subclass A6 were strongly downregulated with A6c and A6d even being completely switched off (Figure 5C). Interestingly, RAC875 showed a different response; subclasses A1 and A2 genes were upregulated after HS, while genes of the other subclasses were lowly expressed and did not alter their expression significantly after HS (Figure 5D). In general, class A *HSF* genes in the Australian wheat cultivars were similarly expressed both under HS and control conditions.

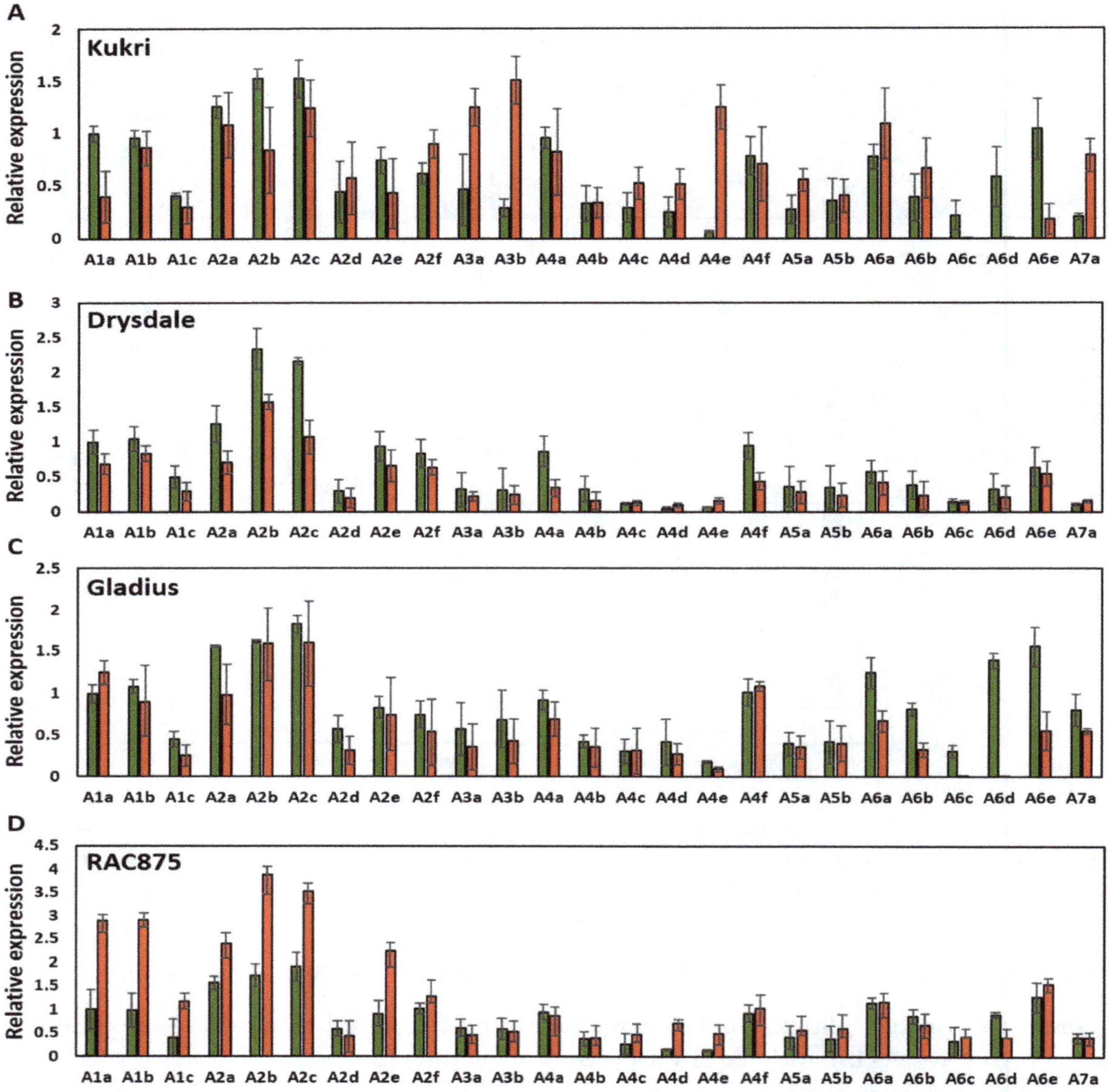

Figure 5. Relative mRNA expression levels of class A heat shock factor genes (*TaHSFs*) in anthers of Australian summer wheat varieties comparing heat stressed and control plants. A tubulin gene was used as a housekeeping gene for normalization. Expression values were set in relation to the average expression level of A1a HSF. Kukri (**A**), Drysdale (**B**), Gladius (**C**), and RAC875 (**D**). Green and red bars represent non- and heat-stressed plants, respectively. Values are means ± SD of three biological replicates.

The European wheat cultivars showed a much stronger response to HS. In Epos, only *HSF* genes A1a, A1b, A2a, A2c, A2e, and A6a were expressed under non-stress conditions. With the exception of A6a, these genes were completely silenced after HS (Figure 6A). Similarly, Cornetto showed downregulation of almost all class A genes, except for A4e, which was upregulated, while A2a, A3a and A4f expression levels did not change significantly (Figure 6B). In Granny, low expression levels of *HSFs* were observed through the entire class A genes (Figure 6C). Under HS, with the exception of A2a, which showed comparable expression under non-stress conditions, all gene members of the subclasses A1 and A2 had low expression levels. Finally, Chamsin was the only European cultivar that showed a strong upregulation of *HSF* genes in response to HS. In general, class A *HSF* genes are expressed at low levels in this cultivar under non-stress conditions and a strong induction, especially of subclass A3 and A4 as well as of A2a was detected (Figure 6D). Taken together, our results indicate that the relatively stable transcriptional expression of *HSF* genes in the Australian cultivars is associated with their high level of acquired HS tolerance. Contrastingly, at the transcriptional level, the European cultivars showed a strong response in gene expression after HS exposure.

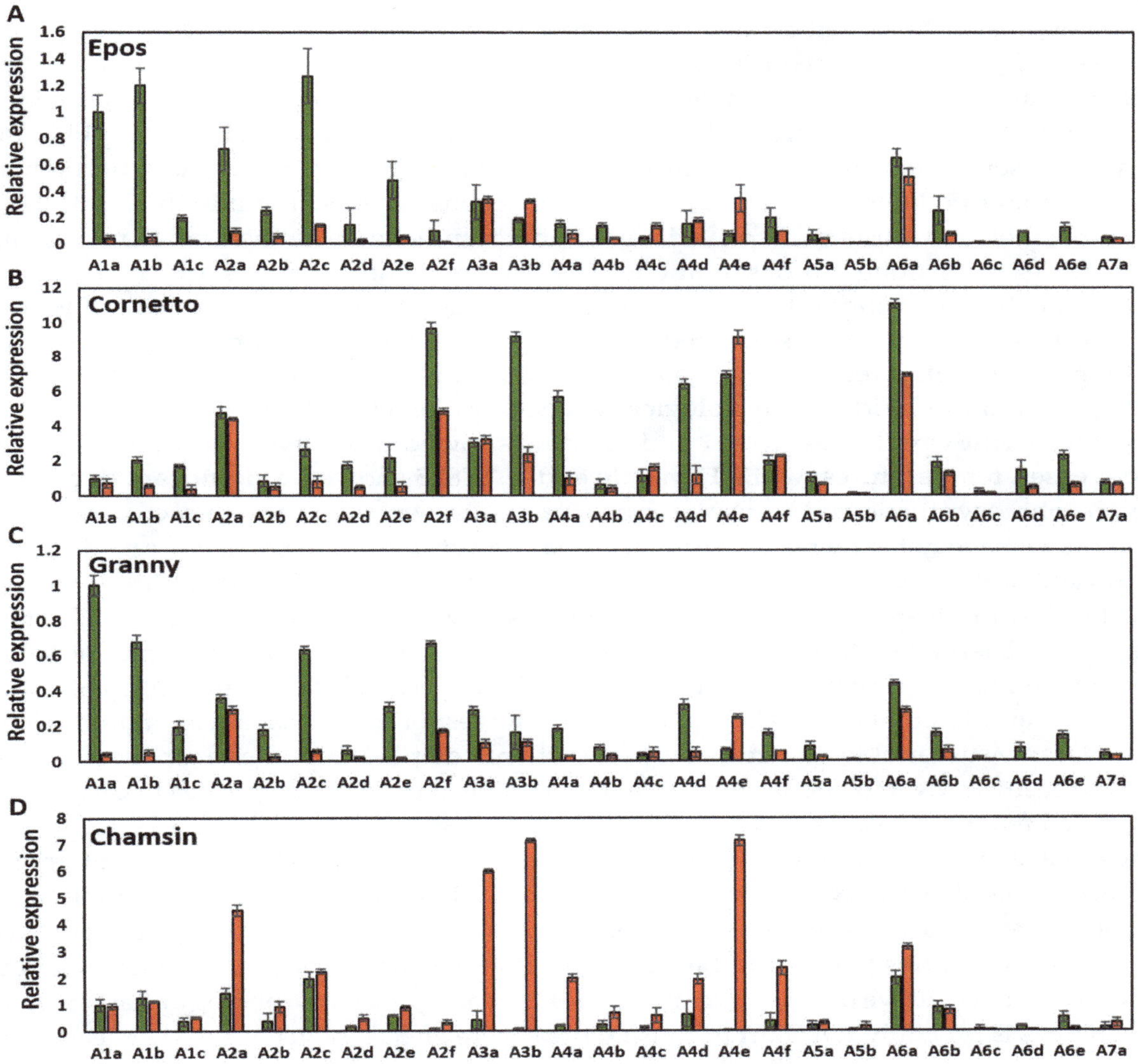

Figure 6. Relative mRNA expression levels of class A heat shock factor genes (*TaHSFs*) in anthers of European summer wheat varieties comparing heat stressed and control plants. A tubulin gene was used as housekeeping gene for normalization. Expression values were set in relation to the average expression level of A1a HSF. Epos (**A**), Cornetto (**B**), Granny (**C**), and Chamsin (**D**). Green and red bars represent non- and heat-stressed plants, respectively. Values are means ± SD of three biological replicates.

4. Discussion

The continuous improvement of wheat cultivars better adapted to environmental stresses, including increased temperatures, is an important goal of modern breeding and agriculture. To achieve this aim, it is necessary to comprehensively explore available germplasms to obtain a deeper understanding of the physiological and molecular adaptive mechanisms that differ between genotypes under changing environmental stress conditions. In this study, we analyzed two sets of summer wheat cultivars from different germplasms exposed to elevated temperatures. Our study confirms previous findings that long-term adaptation and breeding strategies have played fundamental roles in Australian wheat cultivars, which possess an enhanced tolerance to heat stress (HS) compared to European cultivars.

In general, plants use different strategies to overcome heat stress conditions. One of these strategies involves the adjustment between photosynthetic and transpiration rates through the regulation of stomata opening and closure. Stomata closure avoids the loss of water, but leads also to a reduction in CO_2 assimilation and thus photosynthetic rate, while their opening increases gas exchange, including the uptake of CO_2 and simultaneously to the cooling of the leaf surface at elevated temperatures. This avoids, for instance, membrane and thylakoid damage. Thus, a tight regulation is required to balance the various effects caused by increased temperatures [9]. Compared with HS susceptible European cultivars, we observed that Australian cultivars maintained higher photosynthetic and transpiration rates under moderate HS conditions. Similar observations have been made in tomato [12,13] and rice [14], where tolerant varieties displayed higher photosynthetic and transpiration rates compared to susceptible lines confirming our observations. The decline in these two gas exchange parameters as the result of HS has been suggested to be cultivar- and stage-dependent. Exposure of winter wheat cultivars to elevated temperatures resulted, for example, in a reduction of photosynthesis during the grain-filling phase compared to the vegetative stage [21,39]. Similarly, we found that all cultivars tested displayed an individual physiological and molecular HS response. Among the Australian summer wheat cultivars tested, Kukri and Gladius showed a relatively constant photosynthetic rate at increased temperatures, while Drysdale and RAC875 showed a significant decline. Natural variation in photosynthetic and transpiration rates are known to exist within crop species [40–43]. These variations have not shown a consistent correlation between photosynthetic rate during anthesis and grain yield or biomass when wheat cultivars were compared [40–43]. Particularly, wheat cultivars with the highest photosynthetic rates did not necessarily generate highest yields. Therefore, the large variation observed in Australian wheat cultivars could be attributed to unintentional selection traits that result in high photosynthetic rate [41,42]. Notably, Kukri, a drought-sensitive cultivar, and RAC875, a drought-tolerant one [44], displayed opposite responses to heat stress, indicating that these drought tolerant/susceptible Australian wheat cultivars display unrelated responses to heat stress. Similarly, a recent study showed large differences in gene expression responses to drought and heat stress between two barley cultivars [45] supporting this observation.

Unlike vegetative development, where stresses can be tolerated to a certain extent, reproductive development is highly sensitive to environmental stresses [18–20,34,45]. Male gametophyte or pollen development is especially susceptible to environmental stimuli, resulting in morphological, metabolic, and epigenetic alterations [23]. With the exception of a decrease in pollen viability in all European cultivars, we did not observe obvious alterations in anther and pollen morphology after HS. A moderate HS for two days is probably too short to cause more severe effects, which ultimately lead to complete sterility. Moreover, measuring the 100-grain weight and grain number per plant did not show significant differences when the heat stress was applied for only two days during the pollen mitosis stages (data not shown) and plants were allowed to quickly recover and grow at optimal conditions afterwards during the fertilization and seed filling stages. Thus the observed 20–40% loss in pollen viability in European cultivars could be compensated by viable pollen. A major future direction thus will be focused on investigating how longer HS periods affect sterility, seed set, and grain yield. Since starch, sugars, and other photoassimilates are generated during photosynthesis, a reduction in pollen viability detected in European cultivars could be linked to reduced photosynthetic activity and a reduction of photoassimilate transport to major sink organs, including anthers and pollen [22,46–48].

The ability of plants to respond to environmental stresses, especially to heat stress, is strongly associated with the integration of heat shock factors (*HSFs*), which are gene regulators of more elaborated responses [49]. Class A *HSFs* are the most abundant class of regulators and have been shown to be the main heat stress regulators in Arabidopsis [49] and tomato [50,51]. They also play a fundamental role in the regulation of abiotic stress responses in wheat [17]. Subclasses A1 and A2 genes that were predominantly expressed in all cultivars tested in our experiments, both under HS and control conditions, were also expressed in endosperm samples of spring wheat cultivar Bobwhite [17]. This indicates a preferential expression of these genes during reproductive development. In heat-stressed tomato anthers, A2 *HSF* members were highly induced under moderate and severe heat stress conditions. Their high expression levels were maintained even after several days of the stress treatment [51]. Expression levels of European summer wheat cultivars were generally lower and responses more variable compared to the Australian cultivars. For instance, in Epos, one of the most susceptible cultivars to HS, almost all *HSF* class A genes were strongly downregulated by heat stress, indicating a low sustained response to increased temperatures. Cultivars like Chamsin strongly induce *HSF* genes ultimately leading to a complete switch off of net photosynthesis.

5. Conclusions

In conclusion, *HSF* genes appear to be already expressed at substantial levels in HS tolerant wheat cultivars such as the Australian ones used in this study. In contrast, European cultivars showed lower expression levels under non-stress conditions and in most cultivars even a further reduced expression after exposure to moderate HS. By increasing the basal expression level of *HSF* genes, either by genetic engineering or by selecting corresponding wheat lines during breeding programs, we assume that better wheat germplasm can be generated to ameliorate adaptation to increased temperatures. This will also be relevant for short and transient heat waves occurring during reproductive development, as especially male flower organs as well as pollen development are very sensitive to elevated temperatures. A key approach will be a systematic analysis of the regulation of *HSF* genes, but how does HS lead to the complete silencing of *HSF* genes in some cultivars, while others show a strong upregulation or even de novo induction of gene expression? The European cultivars Epos and Chamsin are very suitable to address these interesting questions in future studies.

Author Contributions: K.B. and T.D. conceived the experiments. K.B., A.W. and A.O.E. performed the experiments. K.B. and T.D. analyzed the data and wrote the manuscript. All authors read and approved the manuscript.

Acknowledgments: We acknowledge Peter Langridge (University of Adelaide), Hubert Kempf (Secobra Saatzucht GmbH), and Uwe Stephan (Saatzucht Bauer GmbH) for providing a large number of summer wheat cultivars, of which eight were finally selected for this study. We thank Noureddine Djella and Armin Hildebrand for plant care.

References

1. Shiferaw, B.; Smale, M.; Braun, H.J.; Duveiller, E.; Reynolds, M.; Muricho, G. Crops that feed the world 10. Past successes and future challenges to the role played by wheat in global food security. *Food Secur.* **2013**, *5*, 291–317. [CrossRef]
2. Shewry, P.R. Wheat. *J. Exp. Bot.* **2009**, *60*, 1537–1553. [CrossRef] [PubMed]
3. Vergauwen, D.; De Smet, I. From early farmers to Norman Borlaug—The making of modern wheat. *Curr. Biol.* **2017**, *27*, R858–R862. [CrossRef] [PubMed]
4. Zheng, B.Y.; Chenu, K.; Dreccer, M.F.; Chapman, S.C. Breeding for the future: What are the potential impacts of future frost and heat events on sowing and flowering time requirements for Australian bread wheat (*Triticum aestivium*) varieties? *Glob. Chang. Biol.* **2012**, *18*, 2899–2914. [CrossRef] [PubMed]

5. USDA Foreign Agriculture Service. *Wheat: World Markets and Trade*; Office of Global Analysis-USDA: Washington, DC, USA, 2018.
6. *Agricultural Production—Crops*; Eurostat: Luxembourg, 2017.
7. Joukhadar, R.; Daetwyler, H.D.; Bansal, U.K.; Gendall, A.R.; Hayden, M.J. Genetic diversity, population structure and ancestral origin of Australian wheat. *Front. Plant Sci.* **2017**, *8*, 2115. [CrossRef] [PubMed]
8. Wieland, T. Scientific theory and agricultural practice: Plant breeding in Germany from the late 19th to the early 20th century. *J. Hist. Biol.* **2006**, *39*, 309–343. [CrossRef]
9. Sharkey, T.D. Effects of moderate heat stress on photosynthesis: Importance of thylakoid reactions, rubisco deactivation, reactive oxygen species, and thermotolerance provided.by isoprene. *Plant Cell Environ.* **2005**, *28*, 269–277. [CrossRef]
10. Kobza, J.; Edwards, G.E. Influences of leaf temperature on photosynthetic carbon metabolism in wheat. *Plant Physiol.* **1987**, *83*, 69–74. [CrossRef] [PubMed]
11. Weis, E. Reversible heat-inactivation of the Calvin Cycle—A possible mechanism of the temperature regulation of photosynthesis. *Planta* **1981**, *51*, 33–39. [CrossRef] [PubMed]
12. Camejo, D.; Rodriguez, P.; Morales, A.; Dell'Amico, J.M.; Torrecillas, A.; Alarcon, J.J. High temperature effects on photosynthetic activity of two tomato cultivars with different heat susceptibility. *J. Plant Physiol.* **2005**, *162*, 281–289. [CrossRef] [PubMed]
13. Zhou, R.; Yu, X.Q.; Kjaer, K.H.; Rosenqvist, E.; Ottosen, C.O.; Wu, Z. Screening and validation of tomato genotypes under heat stress using F-v/F-m to reveal the physiological mechanism of heat tolerance. *Environ. Exp. Bot.* **2015**, *118*, 1–11. [CrossRef]
14. Sailaja, B.; Subrahmanyam, D.; Neelamraju, S.; Vishnukiran, T.; Rao, Y.V.; Vijayalakshmi, P.; Voleti, S.R.; Bhadana, V.P.; Mangrauthia, S.K. Integrated physiological, biochemical, and molecular analysis identifies important traits and mechanisms associated with differential response of rice genotypes to elevated temperature. *Front. Plant Sci.* **2015**, *6*, 1044. [CrossRef] [PubMed]
15. Scharf, K.D.; Berberich, T.; Ebersberger, I.; Nover, L. The plant heat stress transcription factor (Hsf) family: Structure, function and evolution. *Biochim. Biophys. Acta* **2012**, *1819*, 104–119. [CrossRef] [PubMed]
16. Mishra, S.K.; Tripp, J.; Winkelhaus, S.; Tschiersch, B.; Theres, K.; Nover, L.; Scharf, K.D. In the complex family of heat stress transcription factors, HSfA1 has a unique role as master regulator of thermotolerance in tomato. *Genes Dev.* **2002**, *16*, 1555–1567. [CrossRef] [PubMed]
17. Xue, G.P.; Sadat, S.; Drenth, J.; McIntyre, C.L. The heat shock factor family from *Triticum aestivum* in response to heat and other major abiotic stresses and their role in regulation of heat shock protein genes. *J. Exp. Bot.* **2014**, *65*, 539–557. [CrossRef] [PubMed]
18. Begcy, K.; Dresselhaus, T. Epigenetic responses to abiotic stresses during reproductive development in cereals. *Plant Reprod.* **2018**, *31*. [CrossRef]
19. Chen, C.; Begcy, K.; Liu, K.; Folsom, J.J.; Wang, Z.; Zhang, C.; Walia, H. Heat stress yields a unique MADS box transcription factor in determining seed size and thermal sensitivity. *Plant Physiol.* **2016**, *171*, 606–622. [CrossRef] [PubMed]
20. Folsom, J.J.; Begcy, K.; Hao, X.; Wang, D.; Walia, H. Rice Fertilization-Independent Endosperm1 regulates seed size under heat stress by controlling early endosperm development. *Plant Physiol.* **2014**, *165*, 238–248. [CrossRef] [PubMed]
21. Monneveux, P.; Pastenes, C.; Reynolds, M.P. Limitations to photosynthesis under light and heat stress in three high-yielding wheat genotypes. *J. Plant Physiol.* **2003**, *160*, 657–666. [CrossRef] [PubMed]
22. De Storme, N.; Geelen, D. The impact of environmental stress on male reproductive development in plants: Biological processes and molecular mechanisms. *Plant Cell Environ.* **2014**, *37*, 1–18. [CrossRef] [PubMed]
23. Gomez, J.F.; Talle, B.; Wilson, Z.A. Anther and pollen development: A conserved developmental pathway. *J. Integr. Plant Biol.* **2015**, *57*, 876–891. [CrossRef] [PubMed]
24. Wilson, Z.A.; Zhang, D.B. From Arabidopsis to rice: Pathways in pollen development. *J. Exp. Bot.* **2009**, *60*, 1479–1492. [CrossRef] [PubMed]
25. Begcy, K.; Dresselhaus, T. Tracking maize pollen development by the Leaf Collar Method. *Plant Reprod.* **2017**, *30*, 171–178. [CrossRef] [PubMed]
26. Chakrabarti, B.; Singh, S.D.; Nagarajan, S.; Aggarwal, P.K. Impact of temperature on phenology and pollen sterility of wheat varieties. *Aust. J. Crop Sci.* **2011**, *5*, 1039–1043.

27. Hlavacova, M.; Klem, K.; Smutna, P.; Skarpa, P.; Hlavinka, P.; Novotna, K.; Rapantova, B.; Trnka, M. Effect of heat stress at anthesis on yield formation in winter wheat. *Plant Soil Environ.* **2017**, *63*, 139–144.
28. Omidi, M.; Siahpoosh, M.R.; Mamghani, R.; Modarresi, M. The influence of terminal heat stress on meiosis abnormalities in pollen mother cells of wheat. *Cytologia* **2014**, *79*, 49–58. [CrossRef]
29. Pradhan, G.P.; Prasad, P.V.V.; Fritz, A.K.; Kirkham, M.B.; Gill, B.S. Effects of drought and high temperature stress on synthetic hexaploid wheat. *Funct. Plant Biol.* **2012**, *39*, 190–198. [CrossRef]
30. Liu, B.; Asseng, S.; Liu, L.L.; Tang, L.; Cao, W.X.; Zhu, Y. Testing the responses of four wheat crop models to heat stress at anthesis and grain filling. *Glob. Chang. Biol.* **2016**, *22*, 1890–1903. [CrossRef] [PubMed]
31. Farooq, M.; Bramley, H.; Palta, J.A.; Siddique, K.H.M. Heat stress in wheat during reproductive and grain-filling phases. *Crit. Rev. Plant Sci.* **2011**, *30*, 491–507. [CrossRef]
32. Ji, X.; Shiran, B.; Wan, J.; Lewis, D.C.; Jenkins, C.L.; Condon, A.G.; Richards, R.A.; Dolferus, R. Importance of pre-anthesis anther sink strength for maintenance of grain number during reproductive stage water stress in wheat. *Plant Cell Environ.* **2010**, *33*, 926–942. [CrossRef] [PubMed]
33. Begcy, K.; Mariano, E.D.; Gentile, A.; Lembke, C.G.; Zingaretti, S.M.; Souza, G.M.; Menossi, M. A novel stress-induced sugarcane gene confers tolerance to drought, salt and oxidative stress in transgenic tobacco plants. *PLoS ONE* **2012**, *7*, e44697. [CrossRef] [PubMed]
34. Mattiello, L.; Begcy, K.; da Silva, F.R.; Jorge, R.A.; Menossi, M. Transcriptome analysis highlights changes in the leaves of maize plants cultivated in acidic soil containing toxic levels of Al($^{3+}$). *Mol. Biol. Rep.* **2014**, *41*, 8107–8116. [CrossRef] [PubMed]
35. Begcy, K.; Walia, H. Drought stress delays endosperm development and misregulates genes associated with cytoskeleton organization and grain quality proteins in developing wheat seeds. *Plant Sci.* **2015**, *240*, 109–119. [CrossRef] [PubMed]
36. Livak, K.J.; Schmittgen, T.D. Analysis of relative gene expression data using real-time quantitative PCR and the 2(T)(-Delta Delta C) method. *Methods* **2001**, *25*, 402–408. [CrossRef] [PubMed]
37. Sinsawat, V.; Leipner, J.; Stamp, P.; Fracheboud, Y. Effect of heat stress on the photosynthetic apparatus in maize (*Zea mays* L.) grown at control or high temperature. *Environ. Exp. Bot.* **2004**, *52*, 123–129. [CrossRef]
38. Cheng, C.H.; Mccomb, J.A. In vitro germination of wheat pollen on raffinose medium. *New Phytol.* **1992**, *120*, 459–462. [CrossRef]
39. Feng, B.; Liu, P.; Li, G.; Dong, S.T.; Wang, F.H.; Kong, L.A.; Zhang, J.W. Effect of heat stress on the photosynthetic characteristics in flag leaves at the grain-filling stage of different heat-resistant winter wheat varieties. *J. Agron. Crop Sci.* **2014**, *200*, 143–155. [CrossRef]
40. Chytyk, C.J.; Hucl, P.J.; Gray, G.R. Leaf photosynthetic properties and biomass accumulation of selected western Canadian spring wheat cultivars. *Can. J. Plant Sci.* **2011**, *91*, 305–314. [CrossRef]
41. Driever, S.M.; Lawson, T.; Andralojc, P.J.; Raines, C.A.; Parry, M.A.J. Natural variation in photosynthetic capacity, growth, and yield in 64 field-grown wheat genotypes. *J. Exp. Bot.* **2014**, *65*, 4959–4973. [CrossRef] [PubMed]
42. Jackson, P.; Basnayake, J.; Inman-Bamber, G.; Lakshmanan, P.; Natarajan, S.; Stokes, C. Genetic variation in transpiration efficiency and relationships between whole plant and leaf gas exchange measurements in *Saccharum* spp. and related germplasm. *J. Exp. Bot.* **2016**, *67*, 861–871. [CrossRef] [PubMed]
43. Sadras, V.O.; Lawson, C.; Montoro, A. Photosynthetic traits in Australian wheat varieties released between 1958 and 2007. *Field Crop Res.* **2012**, *134*, 19–29. [CrossRef]
44. Ford, K.L.; Cassin, A.; Bacic, A. Quantitative proteomic analysis of wheat cultivars with differing drought stress tolerance. *Front. Plant Sci.* **2011**, *2*, 44. [CrossRef] [PubMed]
45. Cantalapiedra, C.P.; Garcia-Pereira, M.J.; Gracia, M.P.; Igartua, E.; Casas, A.M.; Contreras-Moreira, B. Large differences in gene expression responses to drought and heat stress between elite barley cultivar Scarlett and a spanish landrace. *Front. Plant Sci.* **2017**, *8*, 647. [CrossRef] [PubMed]
46. Barnabas, B.; Jager, K.; Feher, A. The effect of drought and heat stress on reproductive processes in cereals. *Plant Cell Environ.* **2008**, *31*, 11–38. [CrossRef] [PubMed]
47. Dai, Z.W.; Wang, L.J.; Zhao, J.Y.; Fan, P.G.; Li, S.H. Effect and after-effect of water stress on the distribution of newly-fixed C-14-photoassimilate in micropropagated apple plants. *Environ. Exp. Bot.* **2007**, *60*, 484–494. [CrossRef]
48. Sudhir, P.; Murthy, S.D.S. Effects of salt stress on basic processes of photosynthesis. *Photosynthetica* **2004**, *42*, 481–486. [CrossRef]

49. Guo, M.; Liu, J.H.; Ma, X.; Luo, D.X.; Gong, Z.H.; Lu, M.H. The plant heat stress transcription factors (HSFs): Structure, regulation, and function in response to abiotic stresses. *Front. Plant Sci.* **2016**, *7*, 114. [CrossRef] [PubMed]
50. Bharti, K.; von Koskull-Doring, P.; Bharti, S.; Kumar, P.; Tintschl-Korbitzer, A.; Treuter, E.; Nover, L. Tomato heat stress transcription factor HsfB1 represents a novel type of general transcription coactivator with a histone-like motif interacting with the plant CREB binding protein ortholog HAC1. *Plant Cell* **2004**, *16*, 1521–1535. [CrossRef] [PubMed]
51. Giorno, F.; Wolters-Arts, M.; Grillo, S.; Scharf, K.D.; Vriezen, W.H.; Mariani, C. Developmental and heat stress-regulated expression of HsfA2 and small heat shock proteins in tomato anthers. *J. Exp. Bot.* **2010**, *61*, 453–462. [CrossRef] [PubMed]

9

Analysis of Stress Resistance using Next Generation Techniques

Maxim Messerer, Daniel Lang and Klaus F. X. Mayer *

Plant Genome and Systems Biology, Helmholtz Center Munich-German Research Center for Environmental Health, 85764 Neuherberg, Germany; maxim.messerer@helmholtz-muenchen.de (M.M.); daniel.lang@helmholtz-muenchen.de (D.L.)

* Correspondence: k.mayer@helmholtz-muenchen.de

Abstract: Food security for a growing world population remains one of the most challenging tasks. Rapid climate change accelerates the loss of arable land used for crop production, while it simultaneously imposes increasing biotic and abiotic stresses on crop plants. Analysis and molecular understanding of the factors governing stress tolerance is in the focus of scientific and applied research. One plant is often mentioned in the context with stress resistance—*Chenopodium quinoa*. Through improved breeding strategies and the use of next generation approaches to study and understand quinoa's salinity tolerance, an important step towards securing food supply is taken.

Keywords: stress resistance; abiotic and biotic stresses; genetic diversity; *Chenopodium quinoa*; omics

---

1. Introduction—Strategies to Improve Crop Yield

How to feed the world's population is still one of the most challenging questions. A United Nations report expects the current world population of 7.6 billion to reach 8.6 billion in 2030, 9.8 billion in 2050, and 11.2 billion in 2100 [1]. At the same time, the size of arable land is dramatically reducing. Erosion rates from ploughed fields are, on average, 10 to 100 times greater than rates of soil formation, with the result that the world has lost nearly a third of its farmable land to erosion or pollution in the last 40 years [2]. Crops are also exposed to different biotic and abiotic stresses: Biotic stresses include pathogen infection and herbivore attack [3], while abiotic stresses are environmental factors that compromise plants and reduce their productivity. These factors include extreme changes in temperature, water, nutrients, gases, wind, radiation and other environmental conditions. Rapid climate change also intensifies abiotic stresses and limits the time for a crop to adapt to new environmental conditions. For breeding programs, improvements in tolerance to drought, salinity and heat as well as the analysis of water economy in plants are most important [4]. Besides traditional breeding approaches, next generation techniques are also used to identify and study stress-tolerant plants. This includes "omics" approaches, which are very effective molecular methods to investigate biochemical, physiological and metabolic strategies of plants exposed to biotic and abiotic stresses. These include genomics (study of genome), transcriptomics (structural and functional analysis of coding and non-coding RNA), proteomics (protein and post-translational protein modification) and metabolomics (analysis of metabolites). Together, omics provide a powerful tool to identify the complex network of stress tolerance [5]. New insights promise to deliver new breeding targets for the stress adaptation of traditional crops by exploiting the allelic but also core- and pan-genomic reservoir. Finally, precision editing tools promise tailored adaptation of molecular circuits in future crop generations. In this article, influences of abiotic and biotic stresses are summarized, next generation analyses are introduced, and potential routes to increase food production are discussed.

2. Abiotic and Biotic Stresses and Responses

Both biotic and abiotic stresses affect crop plants and severely reduce their yield. As plants are sessile organisms, they are not able to escape these stresses but developed a range of strategies to adopt. Depending on the respective stress factors, these defense mechanisms can include morphological, biochemical and molecular modifications, such as altering certain signaling pathways, changes in cell wall structure, etc. A complete and in-depth understanding of these mechanisms is seen as important contributor for future breeding targets and for sustainable agriculture [3,6,7].

Biotic stress is triggered by interactions with other organisms like pests, parasites and pathogens, which are responsible for plant diseases [8]. To withstand, plants use different strategies such as passive barriers and active recognition systems. They produce chemical compounds against herbivores and pathogens, and use thickened cuticles and waxy layers as physical defense against intruders [9]. Plants have established an effective immune response to counteract biotic stress. By so-called pattern recognition receptors (PRRs), they can recognize microbial- or pathogen-associated molecular patterns (MAMPs or PAMPs) like flagellin, inducing PAMP-triggered immunity (PTI). Furthermore, plants possess disease resistance or R genes, which are encoding NB-LRR (nuclear binding—leucine rich repeat) proteins. These NB-LRR proteins recognize pathogen effectors and induce the effector-triggered immunity (ETI) [10]. PTI and ETI induce a first response against biotic stress which leads to an increase in cytoplasmic calcium concentration, the production of reactive oxygen species (ROS) and the activation of mitogen-activated protein kinases (MAPKs). Both PTI and ETI also induce several downstream signaling pathways in which phytohormones, mostly salicylic acid (SA), jasmonic acid (JA) and ethylene (ET), play an essential role [6].

Abiotic stresses are caused by non-living factors that have an impact on growth conditions. Already in 1982, it was suggested that environmental factors limit crop production by 70%. A quarter century later, in 2007, it was reported that 96.5% of worldwide land area is influenced by abiotic stress [11]. Plants have developed different strategies to face these environmental changes. The stress responses can be both elastic (reversible) and plastic (irreversible), and are mostly very complex. Plant cells are able to sense environmental changes, which are subsequently reflected by specific changes in their gene expression, metabolism and physiology. Until today, only a few sensors have been identified, maybe due to functional redundancy in sensor protein encoding genes or to their essentiality, meaning mutations in these genes are lethal [3]. The phytohormone abscisic acid (ABA) seems to play a central role as endogenous messenger in abiotic stress responses. It was shown that especially under drought and salinity stress, increased ABA levels in combination with highly altered gene expression are detectable. In 2009, a small protein family, which is able to bind ABA, was identified as ABA receptors. These findings initiated the analysis of ABA pathways and ABA-induced gene transcription. Furthermore, the understanding of stomatal closure regulation due to ABA signals, which are controlling ion channels in guard cells, was improved [12].

In addition, abiotic stresses have an effect on the occurrence and spread of biotic stressors. Alterations of environmental conditions also directly influence pest-plant interactions by affecting physiological and defense responses. Apparently, the combination of stress factors is more harmful, but not always additive. Plants are able to pyramid responses to combined stress factors. The identification and development of crops with enhanced stress tolerance to combined biotic and abiotic stresses is in the focus of research [13].

3. Drought and Salinity—The most Affecting Abiotic Stresses

During the next decades, climate change will impose increased abiotic stresses—mainly drought, heat, and salinity. It is expected, that drought will be most influential on crop productivity, as by the end of the twentieth century, 30% of land will be extremely dry [4]. Salinity rises constantly since irrigation with brackish water increases the worldwide area of salt-damaged arable land. Every minute, three hectares of land become unusable for crop production [14,15].

More than 50 years ago, plant drought responses were grouped in three categories—drought escape, drought avoidance and drought tolerance. Plants often combine these strategies and their viability depends on how effective the composition of these changes is [4]. Drought escape is an adaptive mechanism including faster development to complete the plant's life cycle before the drought period starts, such as through early flowering. Drought avoidance describes better water uptake due to deeper roots and decreased water evaporation through thicker waxy layers. Drought tolerance is induced after stress occurrence and enables the plant to grow under water deficiency due to biochemical changes. The decrease in osmotic potential by osmolyte accumulation is defined as osmotic adjustment (OA), a typical physiological mechanism against dehydration [4,16,17]. Typical osmolytes, also often called osmoprotectants, are betaine, proline and fructans. These substances do not take part in biochemical responses but influence the osmotic behavior of cells. The accumulation of osmolytes affects gene expression in order to regulate the production of relevant enzymes [18].

In land plants, different strategies evolved to handle high salt (NaCl) concentrations in the soil. Non-salt tolerant glycophytes actively transport salt from the roots back into the environment. This is only effective when facing low salt concentrations. Only about 2% of all plant species are halophytes with high salt tolerance. Halophytes can tolerate NaCl concentrations comparable to seawater and developed two strategies to cope with increased salt concentrations. Succulent halophytes have large internal vacuoles in which they store sodium (Na^+) in order to protect the core plant from toxic salt loads. Another possibility to exclude NaCl from sensitive tissues is the ability to sequester large quantities of salt to so-called epidermal bladder cells (EBCs), which are present in 50% of all halophytes. The diameter of EBCs is about 10-times larger than normal epidermal cells resulting in a 1000-times larger storage volume for Na^+ compared to vacuoles of normal leaf cells [15].

4. Sequence Diversity in Crops—Searching for Tolerant Plants

Structural gene variants like presence/absence variants (PAVs) and copy number variants (CNVs) are contributing to the diversity genepool [19]. Often different crop varieties are adopted and optimized for growth in different habitats. The optimal development of these ecotypes is influenced by their allelic diversity, which is reduced in elite cultivars as a consequence of intense breeding and selection for particular characteristics often ignoring others [20]. Consequently, the complete gene and allele pool diversity cannot be captured by an individual variant, but requires the analysis of a broader set of cultivars. To include all existing genes, contribution to phenotypic and agronomic trait diversity, the construction and analysis of pan-genomes is necessary. This pan-genome contains the complete gene set, including the core-genome, in which all genes are present in all members of a species, and variable genes, which occur only in some variants [21]. Since one reference genome represents only one variety, increasing awareness is put towards the fact that a range of genomes need to be sequenced completely to generate high resolution pan-genomes. For a number of species, including wheat, maize, rice, soybean and cabbage, pan-genomes have been analyzed. The analysis of the cabbage pan-genome, for example, revealed that 20% of genes are affected by presence/absence variation. Some of these were related to important agronomical factors like stress resistance, flowering time and vitamin biosynthesis [19,20].

The genomes of many traditional crops, such as tomato, barley, wheat, sorghum and wild emmer, were sequenced in the last years [20,22–26]. Several consortiums are working on the assembly of additional genomes to gain insights into the pan-genomes of all important crops. The allelic diversity in the gene pool will aid in analyzing different stress resistances in detail. However, also non-traditional crops with high stress tolerances need to be sequenced to study their (pan-) genomes, in order to understand their particular molecular mechanisms to eventually learn and profit for adjusted breeding goals and solutions. Quinoa (*Chenopodium quinoa*), a plant reputed for its high salinity tolerance, has yet a relatively unimportant role compared to traditional crops. Quinoa has the potential to serve as model plant for stress resistance. Besides the recently published reference genome, a variety of quinoa ecotypes exist that seem to grow under nearly all climate conditions.

5. *Chenopodium quinoa*—A Salinity Tolerant Crop

A halophyte plant often used to study salt tolerance is *Chenopodium quinoa*. Quinoa is a highly nutritious crop and is supposed to have been domesticated more than 7000 years ago by pre-Columbian cultures in the Andean region. It was called "mother grain" during the Incan Empire. It is a pseudocereal crop of the family *Amaranthaceae*, also including the important economic plants *Beta vulgaris* (sugar beet), *Spinacia oleracea* (spinach) and *Amaranthus hypochondriacus* (amaranth) [15,27]. Quinoa has become a plant of interest: It is called a "superfood" as its seeds contain a high amount of essential amino acids and vitamins but no gluten. This makes quinoa an alternative to replace wheat-based products in cases of celiac disease. The seeds are rich in several minerals and, in comparison to other grains, have an excellent ratio of proteins, lipids, fiber and carbohydrates. Because of these characteristics, the United Nations Food and Agricultural Organization (FAO) declared 2013 as the "International Year of Quinoa", an award which plants only received three times. In its origin, the Andean region, quinoa is used to grow in several harsh environmental conditions. It adapted to the salty coast as well as to the highlands 3500 m above sea level, with extreme differences in abiotic factors like temperature, precipitation and salt concentrations. Due to this broad adaptation combined with the nutritious characteristics, the number of quinoa-growing countries has increased 10-fold during the last 30 years [28–30].

6. Omic Approaches Using Quinoa

C. quinoa has not only reached public attention as food of the future but also its salinity tolerance is in the focus of several research groups. Next to traditional growth experiments, in order to decipher the salt concentration tolerated by quinoa, the studies were supported by omic approaches, namely genomics, transcriptomics, proteomics and metabolomics.

The quinoa genome was published recently [28,29] and the evolution of quinoa and its salt tolerance were studied. The analysis of ABA-related genes showed that the key factors of ABA biosynthesis, transport and perception were expanded in the quinoa genome, contributing to salinity tolerance [28,29]. In further studies, these genomes were used for resequencing approaches [31,32]. One example is the detection of genomic variations, like single nucleotide polymorphisms (SNPs) and insertions/deletions (InDels), to distinguish among different ecotypes which can be used in further breeding programs [31].

For transcriptomics, the gold standard approach is RNA sequencing (RNAseq). This method is applied in several approaches, e.g., to identify differentially expressed genes (DEGs) between different conditions, tissues or ecotypes. Several studies on salt tolerance and its sequestration in EBCs were performed [28,33,34]. In combination with the quinoa reference genome, Zou and colleagues analyzed the transcriptome of EBCs and leaves under salt-treated and non-treated conditions. Out of totally 54,438 protein-coding genes present in the quinoa genome, they identified 8148 DEGs between bladder and leaf cells. In EBCs, genes involved in abiotic stress response and cell wall synthesis were upregulated, while those related to photosynthesis were downregulated. These findings underlined the functions of bladders being cells that sequester salt but are inactive in anabolic functions. Furthermore, several ion transporters were found to be upregulated in EBCs which seem to be involved in salt secretion [28]. Another study analyzed RNAseq data from several closely related *Chenopodium* species (*C. quinoa*, *Chenopodium berlandieri*, *Chenopodium hircinum*) differing in tolerated salt concentration. Investigating DEGs between the species, the group identified 15 genes encoding for putative transmembrane proteins that potentially contribute to a higher salinity tolerance [27].

Several analyses of quinoa seeds using proteomics approaches were performed. Aloisi and colleagues investigated the changes in the amino acid and protein profiles of seeds from several quinoa ecotypes grown under salt-treatment. They were able to show that salinity influences proteins which belong to functional categories like stress-protein, metabolism and storage [35]. Another group analyzed 16 grains from different crops, including quinoa. The comparison of the different proteomes showed that over 90% of detected proteins from extensively studied cereals like wheat, barley, maize and rice are registered in the Uniprot database. In the case of the quinoa proteome, only 3% of detectable proteins showed an entry in this database. Thus, quinoa opens a so far largely unexplored territory also with respect to the proteome and protein composition of the seeds [36].

Since plants contain the largest metabolome of all life forms, metabolomics is an important part of the omic approaches. Under normal conditions and especially under stress conditions, plants produce a high and diverse amount of primary and secondary metabolites. While primary metabolites directly regulate growth, development and reproduction, secondary metabolites have other important ecological functions like protecting the plant from stresses. In quinoa and other halophytes, metabolites contributing to salinity tolerance were studied [37]. More than half of all metabolites were significantly affected by salinity—e.g., the osmoprotectant proline was about 17-fold increased. As a next experimental step, quinoas EBCs were mechanically removed prior to the salt treatment resulting in the loss of plant's salt tolerance. This procedure dramatically altered the metabolite composition demonstrating that EBCs also serve as metabolite storage [33].

7. Conclusions and Perspectives

While the world population is massively growing, farmers will increasingly face harsh environmental conditions affecting agricultural productivity, accelerated by the rapid climate change. Biotic and abiotic stresses are associated with crop loss and are in the focus of active research and breeding programs. The analysis of plant's stress response revealed that phytohormones play a key role as messengers in downstream signaling pathways [6,38,39]. A lot of studies were performed to unravel all details of hormone production, transport and its characteristics. Their impact in stress response is still not completely understood but further discoveries will reveal new possibilities to increase tolerance of traditional crops. Available data on completely sequenced plant and crop genomes increase continuously, and pan-genomes of crops become available for the detection of allelic variants, stress-associated alleles and tolerant phenotypes.

As about 22% of worldwide agricultural land is saline, tolerant plants and detailed knowledge about the ability to grow efficiently in salt-contaminated environments are urgently needed. The investigation of halophytes has advanced the knowledge about salinity tolerance as these plants are able to grow even when watered with seawater. In the recent years, the halophyte *C. quinoa* was studied intensively by the use of next generation omic approaches. New insights in the function of quinoa's EBCs were gained, also demonstrating the storage capacity for metabolites in EBCs and a model of how ions are transported into these salt dumpers. These findings and further studies will help to understand the molecular mechanisms of salt tolerance and the engineering of salt-tolerant crops. Additional genome sequences of quinoa varieties and close relatives will trigger insights into the pan-genome. Combining the study and understanding of stress resistance, targeted breeding, potential application and engineering in other crops as well as the use of tolerant ecotypes in areas useless for traditional crop production, a step towards securing food supply will be undertaken.

Author Contributions: Conceptualization, D.L. and K.M.; Writing-Original Draft Preparation, M.M.; Writing-Review & Editing, M.M., D.L. and K.M.; Funding Acquisition, K.M.

References

1. United Nations World Population Prospects: The 2017 Revision. Available online: https://www.un.org/development/desa/en/news/population/world-population-prospects-2017.html (accessed on 20 June 2018).
2. Cameron, D.; Osborne, C.; Horton, P.; Sinclair, M. A Sustainable Model for Intensive Agriculture. Available online: http://grantham.sheffield.ac.uk/engagement/policy/a-sustainable-model-for-intensive-agriculture/ (accessed on 20 June 2018).
3. Zhu, J.-K. Abiotic Stress Signaling and Responses in Plants. *Cell* **2016**, *167*, 313–324. [CrossRef] [PubMed]

4. Andjelkovic, V. Introductory Chapter: Climate Changes and Abiotic Stress in Plants. In *Plant, Abiotic Stress and Responses to Climate Change*; InTech: London, UK, 2018.
5. Gupta, B. Plant Abiotic Stress: 'Omics' Approach. *J. Plant Biochem. Physiol.* **2013**, *1*. [CrossRef]
6. Gimenez, E.; Salinas, M.; Manzano-Agugliaro, F. Worldwide Research on Plant Defense against Biotic Stresses as Improvement for Sustainable Agriculture. *Sustainability* **2018**, *10*, 391. [CrossRef]
7. Suzuki, N.; Rivero, R.M.; Shulaev, V.; Blumwald, E.; Mittler, R. Abiotic and biotic stress combinations. *New Phytol.* **2014**, *203*, 32–43. [CrossRef] [PubMed]
8. Schumann, G.L.; Gail, L.; D'Arcy, C.J. *Essential Plant Pathology*; APS Press: St. Paul, MN, USA, 2010; ISBN 9780890543818.
9. Lazar, T.; Taiz, L.; Zeiger, E. Plant physiology. *Ann. Bot.* **2003**, *91*, 750–751. [CrossRef]
10. Jones, J.D.G.; Dangl, J.L. The plant immune system. *Nature* **2006**, *444*, 323–329. [CrossRef] [PubMed]
11. Cramer, G.R.; Urano, K.; Delrot, S.; Pezzotti, M.; Shinozaki, K. Effects of abiotic stress on plants: A systems biology perspective. *BMC Plant Biol.* **2011**, *11*, 163. [CrossRef] [PubMed]
12. Qin, F.; Shinozaki, K.; Yamaguchi-Shinozaki, K. Achievements and Challenges in Understanding Plant Abiotic Stress Responses and Tolerance. *Plant Cell Physiol.* **2011**, *52*, 1569–1582. [CrossRef] [PubMed]
13. Pandey, P.; Irulappan, V.; Bagavathiannan, M.V.; Senthil-Kumar, M. Impact of Combined Abiotic and Biotic Stresses on Plant Growth and Avenues for Crop Improvement by Exploiting Physio-morphological Traits. *Front. Plant Sci.* **2017**, *8*, 537. [CrossRef] [PubMed]
14. Carillo, P.; Grazia, M.; Pontecorvo, G.; Fuggi, A.; Woodrow, P. Salinity Stress and Salt Tolerance. In *Abiotic Stress in Plants—Mechanisms and Adaptations*; InTech: London, UK, 2011.
15. Shabala, S.; Bose, J.; Hedrich, R. Salt bladders: Do they matter? *Trends Plant Sci.* **2014**, *19*, 687–691. [CrossRef] [PubMed]
16. Shavrukov, Y.; Kurishbayev, A.; Jatayev, S.; Shvidchenko, V.; Zotova, L.; Koekemoer, F.; de Groot, S.; Soole, K.; Langridge, P. Early Flowering as a Drought Escape Mechanism in Plants: How Can It Aid Wheat Production? *Front. Plant Sci.* **2017**, *8*, 1950. [CrossRef] [PubMed]
17. Touchette, B.W.; Iannacone, L.R.; Turner, G.E.; Frank, A.R. Drought tolerance versus drought avoidance: A comparison of plant-water relations in herbaceous wetland plants subjected to water withdrawal and repletion. *Wetlands* **2007**, *27*, 656–667. [CrossRef]
18. Putnik-Delić, M.; Maksimović, I.; Nagl, N.; Lalić, B. Sugar Beet Tolerance to Drought: Physiological and Molecular Aspects. In *Plant, Abiotic Stress and Responses to Climate Change*; InTech: London, UK, 2018.
19. Golicz, A.A.; Bayer, P.E.; Barker, G.C.; Edger, P.P.; Kim, H.; Martinez, P.A.; Chan, C.K.K.; Severn-Ellis, A.; McCombie, W.R.; Parkin, I.A.P.; et al. The pangenome of an agronomically important crop plant Brassica oleracea. *Nat. Commun.* **2016**, *7*, 13390. [CrossRef] [PubMed]
20. Montenegro, J.D.; Golicz, A.A.; Bayer, P.E.; Hurgobin, B.; Lee, H.; Chan, C.-K.K.; Visendi, P.; Lai, K.; Doležel, J.; Batley, J.; et al. The pangenome of hexaploid bread wheat. *Plant J.* **2017**, *90*, 1007–1013. [CrossRef] [PubMed]
21. Hurgobin, B.; Edwards, D. SNP Discovery Using a Pangenome: Has the Single Reference Approach Become Obsolete? *Biology* **2017**, *6*, 21. [CrossRef] [PubMed]
22. Tomato Genome Consortium. The tomato genome sequence provides insights into fleshy fruit evolution. *Nature* **2012**, *485*, 635–641. [CrossRef]
23. Mascher, M.; Gundlach, H.; Himmelbach, A.; Beier, S.; Twardziok, S.O.; Wicker, T.; Radchuk, V.; Dockter, C.; Hedley, P.E.; Russell, J.; et al. A chromosome conformation capture ordered sequence of the barley genome. *Nature* **2017**, *544*, 427–433. [CrossRef] [PubMed]
24. Mayer, K.F.X.; Rogers, J.; Dole el, J.; Pozniak, C.; Eversole, K.; Feuillet, C.; Gill, B.; Friebe, B.; Lukaszewski, A.J.; Sourdille, P.; et al. A chromosome-based draft sequence of the hexaploid bread wheat (*Triticum aestivum*) genome. *Science* **2014**, *345*, 1251788. [CrossRef]
25. Paterson, A.H.; Bowers, J.E.; Bruggmann, R.; Dubchak, I.; Grimwood, J.; Gundlach, H.; Haberer, G.; Hellsten, U.; Mitros, T.; Poliakov, A.; et al. The Sorghum bicolor genome and the diversification of grasses. *Nature* **2009**, *457*, 551–556. [CrossRef] [PubMed]
26. Avni, R.; Nave, M.; Barad, O.; Baruch, K.; Twardziok, S.O.; Gundlach, H.; Hale, I.; Mascher, M.; Spannagl, M.; Wiebe, K.; et al. Wild emmer genome architecture and diversity elucidate wheat evolution and domestication. *Science* **2017**, *357*, 93–97. [CrossRef] [PubMed]

27. Schmöckel, S.M.; Lightfoot, D.J.; Razali, R.; Tester, M.; Jarvis, D.E. Identification of Putative Transmembrane Proteins Involved in Salinity Tolerance in Chenopodium quinoa by Integrating Physiological Data, RNAseq, and SNP Analyses. *Front. Plant Sci.* **2017**, *8*, 1023. [CrossRef] [PubMed]
28. Zou, C.; Chen, A.; Xiao, L.; Muller, H.M.; Ache, P.; Haberer, G.; Zhang, M.; Jia, W.; Deng, P.; Huang, R.; et al. A high-quality genome assembly of quinoa provides insights into the molecular basis of salt bladder-based salinity tolerance and the exceptional nutritional value. *Cell Res.* **2017**, *27*, 1327–1340. [CrossRef] [PubMed]
29. Jarvis, D.E.; Ho, Y.S.; Lightfoot, D.J.; Schmöckel, S.M.; Li, B.; Borm, T.J.A.; Ohyanagi, H.; Mineta, K.; Michell, C.T.; Saber, N.; et al. The genome of Chenopodium quinoa. *Nature* **2017**, *542*, 307–312. [CrossRef] [PubMed]
30. Ruiz, K.B.; Biondi, S.; Martínez, E.A.; Orsini, F.; Antognoni, F.; Jacobsen, S.-E. Quinoa—A Model Crop for Understanding Salt-tolerance Mechanisms in Halophytes. *Plant Biosyst.* **2016**, *150*, 357–371. [CrossRef]
31. Zhang, T.; Gu, M.; Liu, Y.; Lv, Y.; Zhou, L.; Lu, H.; Liang, S.; Bao, H.; Zhao, H. Development of novel InDel markers and genetic diversity in Chenopodium quinoa through whole-genome re-sequencing. *BMC Genom.* **2017**, *18*, 685. [CrossRef] [PubMed]
32. Li, C.; Lin, F.; An, D.; Wang, W.; Huang, R. Genome Sequencing and Assembly by Long Reads in Plants. *Genes* **2017**, *9*, 6. [CrossRef] [PubMed]
33. Kiani-Pouya, A.; Roessner, U.; Jayasinghe, N.S.; Lutz, A.; Rupasinghe, T.; Bazihizina, N.; Bohm, J.; Alharbi, S.; Hedrich, R.; Shabala, S. Epidermal bladder cells confer salinity stress tolerance in the halophyte quinoa and Atriplex species. *Plant. Cell Environ.* **2017**, *40*, 1900–1915. [CrossRef] [PubMed]
34. Morales, A.; Zurita-Silva, A.; Maldonado, J.; Silva, H. Transcriptional Responses of Chilean Quinoa (Chenopodium quinoa Willd.) Under Water Deficit Conditions Uncovers ABA-Independent Expression Patterns. *Front. Plant Sci.* **2017**, *8*, 216. [CrossRef] [PubMed]
35. Aloisi, I.; Parrotta, L.; Ruiz, K.B.; Landi, C.; Bini, L.; Cai, G.; Biondi, S.; Del Duca, S. New Insight into Quinoa Seed Quality under Salinity: Changes in Proteomic and Amino Acid Profiles, Phenolic Content, and Antioxidant Activity of Protein Extracts. *Front. Plant Sci.* **2016**, *7*, 656. [CrossRef] [PubMed]
36. Colgrave, M.L.; Goswami, H.; Byrne, K.; Blundell, M.; Howitt, C.A.; Tanner, G.J. Proteomic Profiling of 16 Cereal Grains and the Application of Targeted Proteomics to Detect Wheat Contamination. *J. Proteome Res.* **2015**, *14*, 2659–2668. [CrossRef] [PubMed]
37. Kumari, A.; Das, P.; Parida, A.K.; Agarwal, P.K. Proteomics, metabolomics, and ionomics perspectives of salinity tolerance in halophytes. *Front. Plant Sci.* **2015**, *6*, 537. [CrossRef] [PubMed]
38. Vishal, B.; Kumar, P.P. Regulation of Seed Germination and Abiotic Stresses by Gibberellins and Abscisic Acid. *Front. Plant Sci.* **2018**, *9*, 838. [CrossRef] [PubMed]
39. Abhinandan, K.; Skori, L.; Stanic, M.; Hickerson, N.M.N.; Jamshed, M.; Samuel, M.A. Abiotic Stress Signaling in Wheat—An Inclusive Overview of Hormonal Interactions during Abiotic Stress Responses in Wheat. *Front. Plant Sci.* **2018**, *9*, 734. [CrossRef] [PubMed]

10

Pattern Recognition Receptors—Versatile Genetic Tools for Engineering Broad-Spectrum Disease Resistance in Crops

Stefanie Ranf

School of Life Sciences, Phytopathology, Technical University of Munich, Emil-Ramann-Str. 2, 85354 Freising-Weihenstephan, Germany; ranf@wzw.tum.de

Abstract: Infestations of crop plants with pathogens pose a major threat to global food supply. Exploiting plant defense mechanisms to produce disease-resistant crop varieties is an important strategy to control plant diseases in modern plant breeding and can greatly reduce the application of agrochemicals. The discovery of different types of immune receptors and a detailed understanding of their activation and regulation mechanisms in the last decades has paved the way for the deployment of these central plant immune components for genetic plant disease management. This review will focus on a particular class of immune sensors, termed pattern recognition receptors (PRRs), that activate a defense program termed pattern-triggered immunity (PTI) and outline their potential to provide broad-spectrum and potentially durable disease resistance in various crop species—simply by providing plants with enhanced capacities to detect invaders and to rapidly launch their natural defense program.

Keywords: biotic stress; pattern-triggered immunity; microbe-associated molecular pattern; damage-associated molecular pattern; pattern recognition receptor; receptor-like kinase; receptor-like protein; signal transduction; disease resistance engineering

---

1. Introduction

Pathogens and pests of plants are a major problem in agricultural food production despite the application of plant protection chemicals [1]. Pre- and post-harvest diseases can cause significant losses in crop yield and impair crop quality. The emergence and global spreading of novel pathogens or pathogen races/strains capable of defeating existing, resistant crop cultivars, such as the wheat stem rust (*Puccinia graminis* f. sp. *tritici*) race Ug99 [2] or the kiwi fruit pathogen *Pseudomonas syringae* pv. *actinidae* [3], and the increasing resistance of many pathogen races/strains against available pesticides, illustrates the vulnerability of our current plant protection strategies and the looming risk of devastating disease outbreaks. Increasing the yield of high-quality plant products on the available arable farm land while reducing the amount of ecologically harmful agrochemicals, necessitates the development of future-oriented, sustainable agricultural production systems and effective but environment-friendly plant protection measures. In the last 20 years, the identification of a range of different molecular plant immune components has not only greatly advanced our mechanistic understanding of the plant immune system but also provided the conceptual framework to deploy these discoveries for genetic plant protection. Using selected examples, I will demonstrate in which ways PRRs and PTI can be deployed for disease resistance management in crop plants and illustrate future perspectives for molecular engineering of PRRs and PRR signaling.

2. Pattern-Triggered Immunity Forms a Robust Host Barrier to Invaders

Plants rely on genetically determined (innate) immunity to protect themselves from potentially harmful invaders such as pathogens, pests, and parasitic plants. Central to the plant immune system is a multi-layered surveillance system of various extra- and intracellular immune receptors that sense molecular features of the invader (non-self recognition) as well as perturbations of the cellular integrity provoked by the invader (altered-self recognition). Detection of such danger signals by plant immune sensors activates local and systemic defense mechanisms [4,5]. The capacity of a plant to detect invasion attempts early and to mount a defense response in time, i.e., before establishment and proliferation of the invader, largely depends on its repertoire of immune sensors capable of recognizing the invader.

2.1. Microbe-Associated Molecular Patterns (MAMPs)

Plants sense a variety of conserved microbial components with vital roles for microbial fitness as immune elicitors [4]. Because of their important functions, these components are usually conserved across microbial species and cannot readily be modulated to evade recognition without a fitness cost. Such immunogenic molecular structures are termed microbe- or pathogen-associated molecular patterns (MAMPs/PAMPs). MAMPs are chemically diverse, e.g. proteins, polysaccharides, lipids or composite molecules. Classical examples of MAMPs are microbial cell wall structures, such as chitin (fungi), beta-glucans (oomycetes), lipopolysaccharide or peptidoglycan (bacteria), or microbial proteins, such as bacterial flagellin or elongation factor thermo-unstable (EF-Tu, part of the cellular protein translation machinery) [6]. In case of larger molecular structures, a defined partial structure (epitope or pattern) of these molecules is generally sufficient to activate plant immunity. One of the best-studied examples is the flg22 epitope, which corresponds to a stretch of 22 amino acids from the highly conserved N-terminus of flagellin, a region which is important for assembly of the flagellin monomers into of a functional flagellum and for bacterial motility. A synthetic flg22 peptide is sufficient to activate the same defense program as the natural flagellin protein [7]. Similarly, the peptide elf18, which corresponds to the conserved N-terminus of EF-Tu, is sufficient to trigger plant immunity [8]. The use of synthetic peptide MAMPs has greatly expedited functional studies of these proteinaceous elicitors.

The isolation and identification of MAMPs is often a challenging task. While it is relatively easy to enrich fractions of microbial extracts that trigger PTI-like defense responses in plants, the identification of the causal molecular motif is demanding and requires sophisticated biochemical and analytical skills [6,9]. This is largely due to the fact, that MAMPs are not necessarily very abundant microbial components. To detect invaders timely, host plants evolved highly sensitive immune receptors that detect MAMPs at very low concentrations (nanomolar range) [9]. Thus, although novel analytical techniques have expedited the identification of MAMPs, these are usually not routine methods applicable for high-throughput identification of MAMPs from microbial populations.

Studies on proteinaceous MAMPs are at the forefront of research because they can be produced in large amounts and high purity through recombinant expression systems or chemical synthesis, can be easily genetically modified, and are amenable to population-wide genetic studies. Once a protein motif has been identified as MAMP epitope in a microbial species/race/strain, the DNA sequence of the respective gene can be easily analyzed in silico in many species. Such bioinformatic studies in combination with functional assays have revealed that MAMP epitopes are not as strictly conserved as was initially assumed. Although the protein sequences of MAMPs are overall conserved to maintain their function, the MAMP epitopes were found to be more diversified, presumably due to the selective pressure exerted on microbes to evade host immune sensing through these epitopes [10]. flg22 epitopes from different bacterial species, for example, are sensed with different efficacy in different plant species and some flg22 motifs are not detected at all in some plant species [7,11–15]. The specific genetic fingerprints resulting from the opposing (purifying versus diversifying) selection pressure on MAMPs can, in turn, be exploited to identify putative MAMP epitopes *de novo* by means of bioinformatic genome surveys of microbial populations [16]. In a proof-of-concept study several candidate MAMP epitopes identified by screening *P. syringae* and *Xanthomonas campestris* genome data were shown to elicit PTI in Arabidopsis. Hence, with the increasing availability of microbial genome data such in silico screenings will greatly speed up identification of proteinaceous MAMPs in the future and facilitate targeted screenings for MAMP epitopes in pathogen species of interest.

To date, numerous MAMPs from all microbial classes that are sensed by different plant species have been identified and the list is continuously growing (for a comprehensive summary see recent reviews [6,17,18]). In addition, immunogenic molecular patterns have also been enriched from nematodes, insects, and parasitic plants [6,18,19]. Taken together, this demonstrates the central role of MAMP sensing for plant defense against different types of invaders.

2.2. Damage-Associated Molecular Patterns (DAMPs)

Invaders are not only sensed by the plant immune system via their own molecular components (exogeneous elicitors) but usually also provoke the release of plant-derived signals characteristic of infection, called endogenous elicitors or damage-associated molecular patterns (DAMPs) [4]. Typical examples of DAMPs are fragments of cell wall components generated during attack by microbial cell wall-degrading enzymes, e.g., oligogalacturonides (OGs, derived from pectin) [20] or cutin monomers [21], and intracellular plant components released into the extracellular space upon cell lysis, such as extracellular ATP (eATP) [22], extracellular NAD (eNAD) [23], or intracellular proteins (e.g., Arabidopsis HMGB3) [24]. Like MAMPs, DAMPs are sensed by cell surface-resident PRRs and activate typical PTI signaling and defense responses [4].

In addition, plants release various endogenous peptide hormones into the apoplast upon wounding or pathogen attack. Because these peptides trigger typical PTI defense responses via PRR-like receptors, they are also classified as DAMPs. The systemin peptide, for instance, is produced upon wounding in tomato and contributes to plant defense against herbivorous insects [25]. Two peptide families from Arabidopsis, termed AtPEPs and AtPIPs, that are produced upon biotic stress and wounding, apparently play an important role in the amplification of PTI [26–28]. Peptides of the PEP family are found in various monocot and dicot plant species [29,30], suggesting that amplification of PTI by endogenous peptide hormones is a common strategy in plant immunity.

2.3. Pattern Recognition Receptors (PRRs)

MAMPs and DAMPs are detected by immune receptors localized at the host cell surface, called PRRs (Figure 1) [4]. PRRs are typically single-span transmembrane or membrane-anchored proteins with structurally diverse extracellular domains, such as leucine-rich repeat (LRR), Lysin-motif (LysM) or lectin domains, that bind MAMP/DAMP epitopes with high specificity and sensitivity [18]. Receptor-like kinase (RLK)-type PRRs possess a cytosolic kinase domain, which usually functions in intracellular signal transduction (Figure 1). Receptor-like protein (RLP)-type PRRs that lack active signaling domains and RLK-type PRRs with non-functional kinase domains require signaling-competent protein partners [18,31]. In general, however, PRRs (RLP-type as well as RLK-type) do not function alone but are part of multi-protein complexes where they engage with co-receptors, signaling partners and regulatory proteins that fine-tune PRR activation/deactivation for appropriate signaling output (Figure 1) [31]. Among the best-studied PRRs to date are the flg22 receptor FLS2 and the elf18 receptor EFR, both of which are LRR-type RLKs [8,32]. Many other LRR-type RLKs and RLPs also detect proteinaceous MAMPs, e.g., RLP23 senses the nlp20 epitope of necrosis- and ethylene-inducing peptide 1-like proteins (NLPs) which show a remarkably broad distribution in bacteria, fungi as well as oomycetes [33,34]. Whereas binding of a MAMP or DAMP to its respective PRR is highly specific, other receptor complex components are commonly shared by several PRRs (Figure 1) [31]. Various LRR-type PRRs (e.g., FLS2 and EFR) interact with the LRR-RLK BAK1 in a ligand-dependent manner [35]. LRR-RLP-type PRRs constitutively associate with the signaling adapter LRR-RLK SOBIR and also recruit BAK1 upon ligand binding [36]. The LysM-RLK CERK1 similarly interacts with different LysM-type PRRs [31].

Individual plant species apparently harbor numerous PRRs that sense different classes of invaders in a partially redundant manner. In this way, each invader is sensed by several PRRs. Arabidopsis, for instance, senses at least seven MAMPs from Pseudomonas bacteria through distinct PRRs [37]. Presumably, the diversity and functional redundancy of PRRs is central to the remarkable robustness of the PTI system. Some PRRs are broadly distributed across the plant kingdom (evolutionary old, e.g., FLS2) whereas others are specific to individual plant families or species (evolutionary young, e.g., EFR) [4]. As a consequence, different plant species have different but partially overlapping sets of PRRs. Just as invaders continuously evolve to evade MAMP sensing, plants continuously adapt their PRRs and evolve new PRRs that facilitate detection of novel MAMPs or of other epitopes within a MAMP. Tomato, for instance, has a second flagellin receptor, FLS3, that detects an epitope (termed flgII-28) distinct from flg22 (sensed by FLS2) [38]. Similarly, rice can sense another EF-Tu epitope (EFa50) distinct from elf18 which is sensed in *Brassicaceae* [39]. Thus, the genetic diversity of plants provides a rich source of PRRs not only for a multitude of different MAMPs from various kinds of invaders but also for distinct MAMP epitopes.

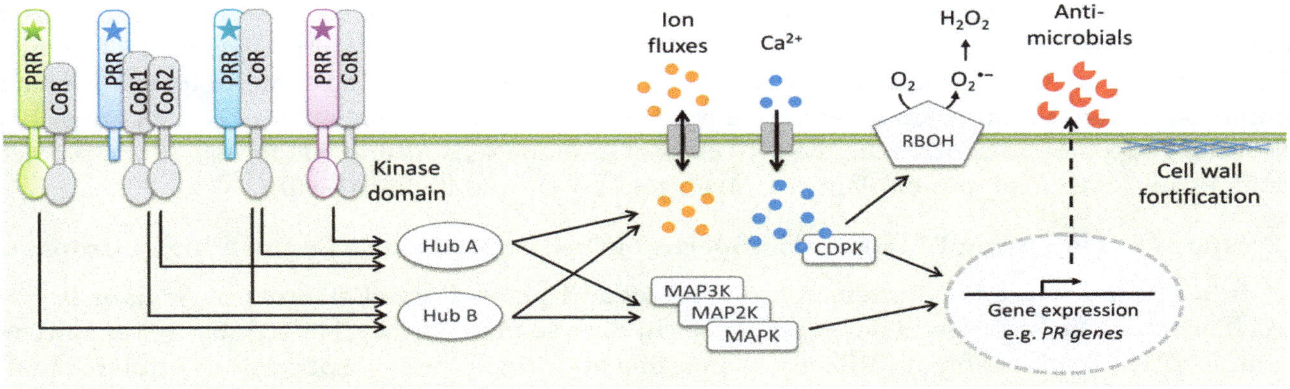

Figure 1. Scheme of typical MAMP and DAMP-activated immune responses in plant cells. MAMPs and DAMPs (star symbols) are sensed by specific host PRRs localized at the cell surface. Signaling pathways downstream of PRRs converge on common signaling components, for instance co-receptors (CoR) or signaling HUBS. PTI responses include for example fluxes of various ions—including the secondary messenger Ca^{2+}—across the plasma membrane, production of ROS by RBOH-type oxidases, activation of calcium-dependent and mitogen-activated protein kinases (CDPKs and MAPKs), gene expression changes, production of antimicrobial compounds, and fortification of the plant cell wall.

2.4. Pattern-Triggered Immunity (PTI)

Detection of a MAMP or DAMP by its respective PRR is highly specific. However, downstream of PRRs, signaling pathways converge on common signaling hubs, often already at the receptor complex level, because certain co-receptors (e.g., BAK1, SOBIR, and CERK1) can act in multiple receptor complexes (Figure 1) [4,5,31]. Therefore, PRRs generally activate a quite stereotypic defense program referred to as PTI. Typical PTI signaling and defense responses are, for example, the depolarization of the plasma membrane, an increase in the cytosolic concentration of the secondary messenger Ca^{2+}, activation of different protein kinases, production of reactive oxygen species (ROS), induction of defense-related genes, cell wall fortifications, and production of antimicrobial enzymes and secondary metabolites as well as defense-related plant hormones (Figure 1) [4,31]. PTI is not restricted to the site of infection but establishes in a systemic manner, resulting in an increased resistance of distal, uninfected parts of the plant to secondary infection with the same or unrelated pathogens [4,40]. Plant immune components including PRRs are systemically upregulated upon elicitor sensing, which further enables the plant to respond faster and stronger to a subsequent pathogen attack [4].

Although PTI is a temperate immune response that usually does not culminate in a hypersensitive response (HR, a programmed cell death reaction), it nevertheless constitutes a robust host barrier effective against commensals and non-adapted pathogens and maintains a basal level of immunity during infection with adapted pathogens [5]. The vital role of PTI is illustrated by the following observations: (i) Loss of individual PRRs renders plants more susceptible to infection with microbes harboring the respective MAMP [5,41,42]; (ii) Treatment of plants with MAMPs to activate PTI (which depends on the presence of a functional, matching PRR) results in enhanced resistance to subsequent local or systemic infection [5,40–42]; (iii) For successful host colonization pathogens need to overcome PTI by evading MAMP sensing or by releasing effectors that subvert PTI through interfering with MAMP/DAMP detection or downstream signaling and defense responses (adapted pathogens). If the effector repertoire of the pathogen sufficiently suppresses host immunity, the pathogen can establish itself in the host and cause disease (effector-triggered susceptibility, ETS) [5,43].

MAMPs are not specific to pathogens. Accordingly, commensal and beneficial microbes are also sensed by PRRs unless they have adapted their MAMP epitopes to evade detection [44,45]. Indeed, it is now evident that PTI restricts not only the growth of pathogenic microbes but also controls colonization of plant tissues with commensal and beneficial microbes [44–47]. Presumably, the diversity and functional redundancy of PRRs in combination with signal amplification through feedback-induction of genes encoding PRRs and PTI signaling components and diverse plant hormones is key to the apparent robustness of the PTI system. Indeed, breaching of host immunity against natural commensal bacteria requires inactivation of multiple PRRs simultaneously or of co-receptors shared by several PRRs [46]. In conclusion, the PRR class of immune sensors and PTI form a robust protective barrier against various types of plant colonizers and are central to plant health.

3. Prospects of Deploying PTI for Broad-Spectrum Disease Resistance Engineering in Crops

It has been known for decades that microbial and plant-derived elicitors (now usually called MAMPs and DAMPs) trigger defense responses in diverse plant species [48–51]. However, only with the molecular identification of PRRs it was possible to unravel the perception and cellular signaling mechanisms in sufficient detail to eventually exploit this central layer of the plant immune system for genetic plant protection in a purposive manner.

3.1. Pattern- versus Effector-Triggered Immunity

Because of the conserved nature and broad occurrence of MAMPs in different microbes, sensing of MAMPs to activate PTI enables the host to detect and efficiently control a wide range of microbes [4]. Adapted pathogens employ effectors to dampen PTI and to modulate host cell metabolism for their own needs. Plants, in turn, evolved immune sensors, called resistance (R) proteins that detect microbial effectors directly or indirectly by monitoring the effector targets, thus turning effectors into avirulence factors. This results in effector-triggered immunity (ETI), which includes overall similar defense responses like PTI but usually develops faster and in a stronger fashion and is mostly accompanied by an HR [5]. Since effector genes are typically specific to certain pathogen races, while *R* genes are limited to certain plant cultivars, this form of resistance is called race-specific or gene-for-gene resistance. In a co-evolutionary arms-race with their hosts, pathogens may be able to overcome *R*-gene mediated ETI relatively easy [5,52]. Mostly, recognized effectors are not essential and/or functional redundant allowing microbes to evade recognition by diversifying or even losing them. Alternatively, microbes may produce novel effectors that e.g. suppress the same or other PTI components or R protein signaling [5,52]. To date, plant resistance breeding widely relies on race-specific resistance mechanisms, largely because effectors and *R* genes typically result in complete or near-complete resistance. The selective pressure exerted on pathogen populations by ETI, however, may result in rapid resistance breaching in the field by the appearance of novel pathogen races/strains [52]. PTI, by contrast, enables defense against a broad spectrum of pathogens and presumably is more durable because of the evolutionary constraints on MAMP modulation and the quantitative resistance conferred by PRRs which reduces the selective pressure on the pathogen.

3.2. PRR Transfer Between Plants Species

One feature makes PRRs particularly attractive targets for genetic engineering of plant immunity: signaling networks downstream of PRRs are sufficiently conserved (Figure 1), not only within but also between plant families and even between monocot and dicot classes, to facilitate functional transfer of PRRs between them [53–55]. Apparently, PRRs can plug into the existing signaling network by engaging with conserved interacting signaling partners (e.g., SOBIR, BAK1 or CERK1) (Figure 2) [54,55]. Because PRRs as the ligand binding components of PTI receptor complexes determine the epitope specificity, introducing a PRR with a novel epitope specificity can confer recognition of this epitope onto previously insensitive plant species. In this way, plants can be equipped with additional MAMP sensing capacities to enhance disease resistance to pathogen classes harboring the respective MAMPs (Figure 2). These can either be PRRs for MAMPs not yet sensed by the recipient plant or for additional epitopes within a given MAMP (e.g., flgII-28/FLS3). A breakthrough study utilizing the *Brassicaceae*-specific EFR receptor has proven the general feasibility of this strategy [53], which was subsequently shown for multiple PRRs in various plant species (summarized in [6,56]). EFR-transgenic tomato (*Solanum lycopersicum*) and *Nicotiana benthamiana* plants gained immune responsiveness to the elf18 epitope and were more resistant to several major bacterial pathogens carrying this epitope including *Ralstonia solanacearum*, *P. syringae*, *Xanthomonas perforans*, and *Agrobacterium tumefaciens* [53]. EFR was subsequently also successfully transferred into monocots, such as wheat (*Triticum aestivum*) and rice (*Oryza sativa*) [55,57]. Similarly, transfer of the rice PRR XA21, which detects the RaxX protein from *Xanthomonas oryzae* pv. *oryzae*, to banana (*Musa* sp.) and sweet orange (*Citrus sinensis*) confers increased resistance against different *Xanthomonas* spp. [58,59] and expression of the PRR ELR from a wild potato species renders cultivated potato (*Solanum tuberosum*) more resistant to the oomycete pathogen *Phytophthora infestans* [60]. Taken together, these results demonstrate that addition of only a single PRR can already significantly enhance resistance of plants to adapted pathogens. Given that these pathogens are equipped with an arsenal of effectors to dampen PTI, this is quite astonishing and suggests that increased pathogen sensing provides a substantial advantage to the plant, possibly because of a stronger and/or faster onset of defense reactions.

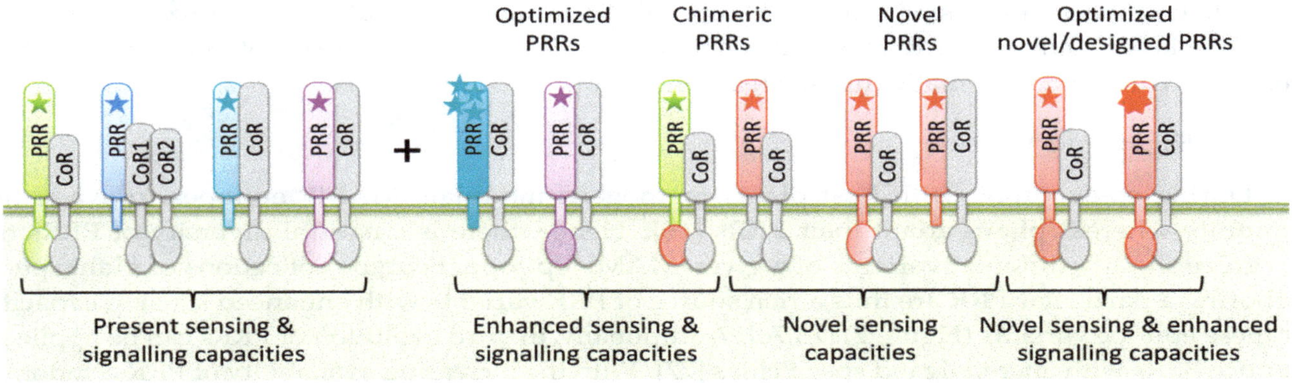

Figure 2. Schematic representation of potential PRR transfer and engineering options. PRR-mediated sensing of MAMPs/DAMPs (star symbols) and PRR signaling capacities can be enhanced by optimization of endogenous PRRs or introduction of novel, chimeric or even "designer" PRRs. This equips plants with additional MAMP/DAMP sensing and improved signaling capacities and results in enhanced PTI against a broad spectrum of invaders.

Often, the isolation of MAMPs is the bottleneck in identifying PRRs [6,9]. However, it is not necessary to structurally identify a MAMP but sufficient to enrich it to a degree that facilitates genetic screenings for the respective PRR [61–63]. Furthermore, the identification of various PRRs for chemically diverse types of MAMPs has revealed that plant PRRs are typically RLKs or RLPs. Thus, these protein families can now also be systematically tested for a putative role in plant immunity, for instance through screening for altered resistance against the pathogen of interest in loss- or gain-of-function mutants or natural accessions, or through heterologous expression in suitable model plants (e.g., *Arabidopsis thaliana* or *N. benthamiana*) [64–69].

A major concern of introducing novel PRRs into plant species is that this might negatively affect beneficial interactions of these plant species with their natural microbiota and/or symbiotic interactions, such as legume-rhizobia or mycorrhizal interactions. Transgenic Medicago plants expressing EFR from Arabidopsis, however, are apparently not defective in symbiosis and are more resistant to the root pathogen *R. solanacearum*. Although the EFR transgenic plants showed a delay in nodule formation the final extent of nodulation and nitrogen fixation of the EFR transgenics was comparable to the wild-type plants [70]. Potentially, adapted symbionts have evolved efficient strategies to evade or suppress host PTI at various levels.

While to date mostly strong, constitutive promoters (e.g., viral promoters such as the 35S promoter of cauliflower mosaic virus, plant ubiquitin or actin promoters) are used to drive heterologous PRR expression in plants because of their universal functionality in diverse plant species and different plant tissues, this can result in unwanted side effects on general plant performance. Overexpression of RLKs, for instance, may trigger their activation in the absence of ligand and lead to growth defects because of constitutive activation of PTI or interference with developmental signaling [71]. This can be overcome by using plant promoters from e.g. endogenous PRR genes that drive PRR expression in plant tissues preferentially targeted by the pathogen of interest (e.g., root) or at sites of pathogen entry (e.g. stomata guard cells) and are strongly induced above the basal level upon infection [72,73].

The growing interest in PTI in recent years already led to the identification of numerous PRRs from different model and wild plant species and there is more to come (summarized in [6,17]). Natural diversity provides plant breeders with a versatile genetic tool box for crop improvement. Relevant PRR genes can be introduced, for instance, from wild relatives through classical breeding strategies. However, this is usually a lengthy process, bears the risk of co-segregation of unfavorable traits (linkage drag), and is not applicable to all crop species (e.g., banana, which is sterile) [56,74]. Alternatively, modern genetic engineering tools facilitate direct transfer of PRR genes across plant families beyond the constraints of sexual compatibility with the advantage that PRR genes can be quickly introduced into elite crop varieties as single traits (Figure 2). Additionally, this allows to utilize virtually any plant species as source of PRR genes. In conclusion, PRR transfer has great potential for conferring broad-spectrum and potentially durable resistance traits onto crop plants.

3.3. PRR Engineering

During co-evolution with host plants, pathogens modulate MAMP epitopes to evade host immunity whereas plants adapt their PRRs [16]. Hence, there is a natural diversity of PRRs that recognize slightly different variants of a given MAMP epitope. Ecotype collections of plant species harboring a particular PRR are thus a rich source of PRR variants with enhanced sensing capacities for these epitope variants (Figure 2) [75,76]. Additionally, in vitro evolution of PRRs can be applied to produce PRRs with altered ligand specificities [77]. With the increasing availability of PRR ectodomain structures and computational modelling tools ligand binding sites of PRRs can be modified in a directed manner to perceive a desired epitope variant or may eventually even be designed *de novo* (Figure 2). Potentially, ligand binding sites can be engineered at the native gene locus through CRISPR/Cas-mediated genome editing.

Some PRRs may not be able to integrate optimally with the endogenous signaling adapters upon transfer in more distantly related plant species. In such cases, full signaling competence can be restored by exchanging the transmembrane and/or intracellular signaling domain with a related PRR from the recipient species (Figure 2) [54,62]. Such chimeric PRRs have been shown to be fully functional and can be further exploited to combine different ligand binding specificities with different downstream signaling capacities (Figure 2). Some PRRs naturally induce a stronger immune response including an HR, such as the rice RLKs XA21 [78]. Sensing of chitin fragments by the rice PRR CEBiP, by contrast, does not result in HR [79]. A chimeric receptor combining the chitin-sensing ectodomain from CEBiP with the intracellular signaling domain of Xa21 activates HR upon chitin sensing and enhances resistance to the rice pathogen *Magnaporthae oryzae*, a pathogen that does not naturally activate Xa21 signaling [80].

3.4. Exploiting DAMP Signalling

Transfer and engineering of PRRs sensing DAMPs are also possible given that the respective DAMP is produced in the recipient species during infection with the pathogen of interest. Expression of the Arabidopsis eATP receptor DORN1 (also known as P2K1) in potato, for instance, enhances resistance to *P. infestans* [22,81]. Alternatively, plants can be engineered to produce the desired DAMP upon infection with diverse pathogens, for instance, by introducing a microbial enzyme producing the DAMP under control of a pathogen-responsive promoter. In this way, any pathogen that induces transcription of the transgene will trigger PTI independent of the ability to produce this DAMP itself. A promising example is the production of elicitor-active OG fragments through the balanced action of a microbial polygalacturonase (PG) and a plant PG inhibitor. Pathogen-inducible expression of this OG factory in Arabidopsis increased resistance to *Botrytis cinerea*, *Pectobacterium carotovorum* and *P. syringae* infections [82]. Such pathogen-responsive *in situ* production of DAMP signals in combination with providing the respective PRR could be a generally applicable strategy to boost PTI in any plant species and to confer resistance to a wide range of pathogens.

4. Conclusions

Most PRRs known to date have been identified from model species such as Arabidopsis, rice and recently also increasingly from tomato and potato [6,17]. These rather limited studies already illustrate the enormous resource of plant immune sensors available in different plant species that just await discovery. With novel genetic and computational tools at hand we can now exploit this natural genetic tool box for providing plants with expanded capacities to sense any kind of undesired invaders and strengthen their natural defense to resist them. Moreover, a mechanistic understanding of PRR function enables us to accelerate PRR evolution artificially, to modify PRRs in a targeted manner or even design novel PRRs.

Although PTI is generally more difficult for pathogens to overcome than ETI, pathogens will, albeit presumably on a longer time scale, eventually succeed. Hence, the deployment of single PRRs still poses the risk that pathogen populations eventually adapt their MAMP epitopes or their effector repertoire, particularly if the selection pressure on pathogen populations is constantly high. To deploy PRR engineering in a durable manner, combination of several PRR and R genes (stacking or pyramiding) appears to be a promising strategy as pathogens are unlikely to overcome several immune receptors at the same time. Possibly, resistant varieties equipped with different PRR/R gene combinations can be used in alternation to lower the selective pressure of the individual components on pathogen populations. If utilized in a thoughtful manner, these genetic tools have great power for developing truly durable and sustainable plant disease management practices.

Acknowledgments: I thank Martin Stegmann for critical reading of the manuscript.

References

1. Oerke, E. Crop Losses to Animal Pests, Plant Pathogens, and Weeds. In *Encyclopedia of Pest Management, Volume II*; CRC Press: Boca Raton, FL, USA, 2009; pp. 116–120.
2. Singh, R.P.; Hodson, D.P.; Huerta-Espino, J.; Jin, Y.; Bhavani, S.; Njau, P.; Herrera-Foessel, S.; Singh, P.K.; Singh, S.; Govindan, V. The Emergence of Ug99 Races of the Stem Rust Fungus is a Threat to World Wheat Production. *Annu. Rev. Phytopathol.* **2011**, *49*, 465–481. [CrossRef] [PubMed]
3. Scortichini, M.; Marcelletti, S.; Ferrante, P.; Petriccione, M.; Firrao, G. *Pseudomonas syringae* pv. *actinidiae*: A re-emerging, multi-faceted, pandemic pathogen. *Mol. Plant Pathol.* **2012**, *13*, 631–640. [PubMed]

4. Boller, T.; Felix, G. A renaissance of elicitors: Perception of microbe-associated molecular patterns and danger signals by pattern-recognition receptors. *Annu. Rev. Plant Biol.* **2009**, *60*, 379–406. [CrossRef] [PubMed]
5. Jones, J.D.G.; Dangl, J.L. The plant immune system. *Nature* **2006**, *444*, 323–329. [CrossRef] [PubMed]
6. Boutrot, F.; Zipfel, C. Function, Discovery, and Exploitation of Plant Pattern Recognition Receptors for Broad-Spectrum Disease Resistance. *Annu. Rev. Phytopathol.* **2017**, *55*, 257–286. [CrossRef] [PubMed]
7. Felix, G.; Duran, J.D.; Volko, S.; Boller, T. Plants have a sensitive perception system for the most conserved domain of bacterial flagellin. *Plant J.* **1999**, *18*, 265–276. [CrossRef] [PubMed]
8. Zipfel, C.; Kunze, G.; Chinchilla, D.; Caniard, A.; Jones, J.D.G.; Boller, T.; Felix, G. Perception of the bacterial PAMP EF-Tu by the receptor EFR restricts Agrobacterium-mediated transformation. *Cell* **2006**, *125*, 749–760. [CrossRef] [PubMed]
9. Ranf, S.; Scheel, D.; Lee, J. Challenges in the identification of microbe-associated molecular patterns in plant and animal innate immunity: A case study with bacterial lipopolysaccharide. *Mol. Plant Pathol.* **2016**, *17*, 1165–1169. [CrossRef] [PubMed]
10. Cai, R.; Lewis, J.; Yan, S.; Liu, H.; Clarke, C.R.; Campanile, F.; Almeida, N.F.; Studholme, D.J.; Lindeberg, M.; Schneider, D.; et al. The plant pathogen *Pseudomonas syringae* pv. *tomato* is genetically monomorphic and under strong selection to evade tomato immunity. *PLoS Pathog.* **2011**, *7*, e1002130.
11. Gómez-Gómez, L.; Felix, G.; Boller, T. A single locus determines sensitivity to bacterial flagellin in *Arabidopsis thaliana*. *Plant J.* **1999**, *18*, 277–284. [CrossRef] [PubMed]
12. Pfund, C.; Tans-Kersten, J.; Dunning, F.M.; Alonso, J.M.; Ecker, J.R.; Allen, C.; Bent, A.F. Flagellin is not a major defense elicitor in *Ralstonia solanacearum* cells or extracts applied to *Arabidopsis thaliana*. *Mol. Plant-Microbe Interact.* **2004**, *17*, 696–706. [CrossRef] [PubMed]
13. Sun, W. Within-species flagellin polymorphism in *Xanthomonas campestris* pv. *campestris* and its impact on elicitation of Arabidopsis FLAGELLIN SENSING2-dependent defenses. *Plant Cell* **2006**, *18*, 764–779. [CrossRef] [PubMed]
14. Clarke, C.R.; Chinchilla, D.; Hind, S.R.; Taguchi, F.; Miki, R.; Ichinose, Y.; Martin, G.B.; Leman, S.; Felix, G.; Vinatzer, B.A. Allelic variation in two distinct *Pseudomonas syringae* flagellin epitopes modulates the strength of plant immune responses but not bacterial motility. *New Phytol.* **2013**, *200*, 847–860. [CrossRef] [PubMed]
15. Wang, S.; Sun, Z.; Wang, H.; Liu, L.; Lu, F.; Yang, J.; Zhang, M.; Zhang, S.; Guo, Z.; Bent, A.F.; et al. Rice OsFLS2-mediated perception of bacterial flagellins is evaded by *Xanthomonas oryzae* pvs. *oryzae* and *oryzicola*. *Mol. Plant* **2015**, *8*, 1024–1037. [CrossRef] [PubMed]
16. McCann, H.C.; Nahal, H.; Thakur, S.; Guttman, D.S. Identification of innate immunity elicitors using molecular signatures of natural selection. *Proc. Natl. Acad. Sci. USA* **2012**, *109*, 4215–4220. [CrossRef] [PubMed]
17. Saijo, Y.; Loo, E.P.-I.; Yasuda, S. Pattern recognition receptors and signaling in plant-microbe interactions. *Plant J.* **2018**, *93*, 592–613. [CrossRef] [PubMed]
18. Ranf, S. Sensing of molecular patterns through cell surface immune receptors. *Curr. Opin. Plant Biol.* **2017**, *38*, 68–77. [CrossRef] [PubMed]
19. Manosalva, P.; Manohar, M.; von Reuss, S.H.; Chen, S.; Koch, A.; Kaplan, F.; Choe, A.; Micikas, R.J.; Wang, X.; Kogel, K.-H.; et al. Conserved nematode signalling molecules elicit plant defenses and pathogen resistance. *Nat. Commun.* **2015**, *6*, 7795. [CrossRef] [PubMed]
20. Hahn, M.G.; Darvill, A.G.; Albersheim, P. Host-Pathogen Interactions: XIX. The endogenous elicitor, a fragment of a plant cell wall polysaccharide that elicits phytoalexin accumulation in soybeans. *Plant Physiol.* **1981**, *68*, 1161–1169. [CrossRef] [PubMed]
21. Schweizer, P.; Felix, G.; Buchala, A.; Muller, C.; Métraux, J.-P. Perception of free cutin monomers by plant cells. *Plant J.* **1996**, *10*, 331–341. [CrossRef]
22. Choi, J.; Tanaka, K.; Cao, Y.; Qi, Y.; Qiu, J.; Liang, Y.; Lee, S.Y.; Stacey, G. Identification of a plant receptor for extracellular ATP. *Science* **2014**, *343*, 290–294. [CrossRef] [PubMed]
23. Wang, C.; Zhou, M.; Zhang, X.; Yao, J.; Zhang, Y.; Mou, Z. A lectin receptor kinase as a potential sensor for extracellular nicotinamide adenine dinucleotide in *Arabidopsis thaliana*. *eLife* **2017**, *6*, 267. [CrossRef] [PubMed]
24. Choi, H.W.; Manohar, M.; Manosalva, P.; Tian, M.; Moreau, M.; Klessig, D.F. Activation of Plant Innate Immunity by Extracellular High Mobility Group Box 3 and Its Inhibition by Salicylic Acid. *PLoS Pathog.* **2016**, *12*, e1005518. [CrossRef] [PubMed]

25. Pearce, G.; Strydom, D.; Johnson, S.; Ryan, C.A. A polypeptide from tomato leaves induces wound-inducible proteinase inhibitor proteins. *Science* **1991**, *253*, 895–897. [CrossRef] [PubMed]
26. Hou, S.; Wang, X.; Chen, D.; Yang, X.; Wang, M.; Turrà, D.; Di Pietro, A.; Zhang, W. The Secreted Peptide PIP1 Amplifies Immunity through Receptor-Like Kinase 7. *PLoS Pathog.* **2014**, *10*, e1004331-15. [CrossRef] [PubMed]
27. Huffaker, A.; Ryan, C.A. Endogenous peptide defense signals in Arabidopsis differentially amplify signaling for the innate immune response. *Proc. Natl. Acad. Sci. USA* **2007**, *104*, 10732–10736. [CrossRef] [PubMed]
28. Yamada, K.; Yamashita-Yamada, M.; Hirase, T.; Fujiwara, T.; Tsuda, K.; Hiruma, K.; Saijo, Y. Danger peptide receptor signaling in plants ensures basal immunity upon pathogen-induced depletion of BAK1. *EMBO J.* **2016**, *35*, 46–61. [CrossRef] [PubMed]
29. Huffaker, A.; Pearce, G.; Veyrat, N.; Erb, M.; Turlings, T.C.J.; Sartor, R.; Shen, Z.; Briggs, S.P.; Vaughan, M.M.; Alborn, H.T.; et al. Plant elicitor peptides are conserved signals regulating direct and indirect antiherbivore defense. *Proc. Natl. Acad. Sci. USA* **2013**, *110*, 5707–5712. [CrossRef] [PubMed]
30. Lori, M.; van Verk, M.C.; Hander, T.; Schatowitz, H.; Klauser, D.; Flury, P.; Gehring, C.A.; Boller, T.; Bartels, S. Evolutionary divergence of the plant elicitor peptides (Peps) and their receptors: Interfamily incompatibility of perception but compatibility of downstream signalling. *J. Exp. Bot.* **2015**, *66*, 5315–5325. [CrossRef] [PubMed]
31. Couto, D.; Zipfel, C. Regulation of pattern recognition receptor signalling in plants. *Nat. Rev. Immunol.* **2016**, *16*, 537–552. [CrossRef] [PubMed]
32. Gómez-Gómez, L.; Boller, T. FLS2: An LRR receptor-like kinase involved in the perception of the bacterial elicitor flagellin in Arabidopsis. *Mol. Cell* **2000**, *5*, 1003–1011. [CrossRef]
33. Oome, S.; Raaymakers, T.M.; Cabral, A.; Samwel, S.; Böhm, H.; Albert, I.; Nürnberger, T.; Van den Ackerveken, G. Nep1-like proteins from three kingdoms of life act as a microbe-associated molecular pattern in Arabidopsis. *Proc. Natl. Acad. Sci. USA* **2014**, *111*, 16955–16960. [CrossRef] [PubMed]
34. Böhm, H.; Albert, I.; Oome, S.; Raaymakers, T.M.; Van den Ackerveken, G.; Nürnberger, T. A Conserved Peptide Pattern from a Widespread Microbial Virulence Factor Triggers Pattern-Induced Immunity in Arabidopsis. *PLoS Pathog.* **2014**, *10*, e1004491. [CrossRef] [PubMed]
35. Chinchilla, D.; Shan, L.; He, P.; de Vries, S.; Kemmerling, B. One for all: The receptor-associated kinase BAK1. *Trends Plant Sci.* **2009**, *14*, 535–541. [CrossRef] [PubMed]
36. Gust, A.A.; Felix, G. Receptor like proteins associate with SOBIR1-type of adaptors to form bimolecular receptor kinases. *Curr. Opin. Plant Biol.* **2014**, *21*, 104–111. [CrossRef] [PubMed]
37. Brunner, F.; Nürnberger, T. Identification of immunogenic microbial patterns takes the fast lane. *Proc. Natl. Acad. Sci. USA* **2012**, *109*, 4029–4030. [CrossRef] [PubMed]
38. Hind, S.R.; Strickler, S.R.; Boyle, P.C.; Dunham, D.M.; Bao, Z.; O'Doherty, I.M.; Baccile, J.A.; Hoki, J.S.; Viox, E.G.; Clarke, C.R.; et al. Tomato receptor FLAGELLIN-SENSING 3 binds flgII-28 and activates the plant immune system. *Nat. Plants* **2016**, *2*, 16128. [CrossRef] [PubMed]
39. Furukawa, T.; Inagaki, H.; Takai, R.; Hirai, H.; Che, F.-S. Two distinct EF-Tu epitopes induce immune responses in rice and Arabidopsis. *Mol. Plant-Microbe Interact.* **2014**, *27*, 113–124. [CrossRef] [PubMed]
40. Mishina, T.E.; Zeier, J. Pathogen-associated molecular pattern recognition rather than development of tissue necrosis contributes to bacterial induction of systemic acquired resistance in Arabidopsis. *Plant J.* **2007**, *50*, 500–513. [CrossRef] [PubMed]
41. Zipfel, C.; Robatzek, S.; Navarro, L.; Oakeley, E.J.; Jones, J.D.G.; Felix, G.; Boller, T. Bacterial disease resistance in Arabidopsis through flagellin perception. *Nature* **2004**, *428*, 764–767. [CrossRef] [PubMed]
42. Ranf, S.; Gisch, N.; Schäffer, M.; Illig, T.; Westphal, L.; Knirel, Y.A.; Sánchez-Carballo, P.M.; Zähringer, U.; Hückelhoven, R.; Lee, J.; et al. A lectin S-domain receptor kinase mediates lipopolysaccharide sensing in *Arabidopsis thaliana*. *Nat. Immunol.* **2015**, *16*, 426–433. [CrossRef] [PubMed]
43. Toruño, T.Y.; Stergiopoulos, I.; Coaker, G. Plant-Pathogen Effectors: Cellular Probes Interfering with Plant Defenses in Spatial and Temporal Manners. *Annu. Rev. Phytopathol.* **2016**, *54*, 419–441. [CrossRef] [PubMed]
44. Hacquard, S.; Spaepen, S.; Garrido-Oter, R.; Schulze-Lefert, P. Interplay between Innate Immunity and the Plant Microbiota. *Annu. Rev. Phytopathol.* **2017**, *55*, 565–589. [CrossRef] [PubMed]
45. Zipfel, C.; Oldroyd, G.E.D. Plant signalling in symbiosis and immunity. *Nat. Publ. Group* **2017**, *543*, 328–336. [CrossRef] [PubMed]

46. Xin, X.-F.; Nomura, K.; Aung, K.; Velásquez, A.C.; Yao, J.; Boutrot, F.; Chang, J.H.; Zipfel, C.; He, S.Y. Bacteria establish an aqueous living space in plants crucial for virulence. *Nature* **2016**, *539*, 524–529. [CrossRef] [PubMed]
47. Cao, Y.; Halane, M.K.; Gassmann, W.; Stacey, G. The Role of Plant Innate Immunity in the Legume-Rhizobium Symbiosis. *Annu. Rev. Plant Biol.* **2017**, *68*, 535–561. [CrossRef] [PubMed]
48. Albersheim, P.; Valent, B.S. Host-pathogen interactions in plants. Plants, when exposed to oligosaccharides of fungal origin, defend themselves by accumulating antibiotics. *J. Cell Biol.* **1978**, *78*, 627–643. [CrossRef] [PubMed]
49. Somssich, I.E.; Science, K. Pathogen defence in plants—A paradigm of biological complexity. *Trends Plant Sci.* **1998**, *3*, 86–90. [CrossRef]
50. Ayers, A.R.; Valent, B.; Ebel, J.; Albersheim, P. Host-Pathogen Interactions: XI. Composition and Structure of Wall-released Elicitor Fractions. *Plant Physiol.* **1976**, *57*, 766–774. [CrossRef] [PubMed]
51. Dixon, R.A.; Fuller, K.W. Characterization of components from culture filtrates of *Botrytis cinerea* which stimulate phaseollin biosynthesis in *Phaseolus vulgaris* cell suspension cultures. *Physiol. Plant Pathol.* **1977**, *11*, 287–296. [CrossRef]
52. Dangl, J.L.; Horvath, D.M.; Staskawicz, B.J. Pivoting the Plant Immune System from Dissection to Deployment. *Science* **2013**, *341*, 746–751. [CrossRef] [PubMed]
53. Lacombe, S.E.V.; Rougon-Cardoso, A.; Sherwood, E.; Peeters, N.; Dahlbeck, D.; Van Esse, H.P.; Smoker, M.; Rallapalli, G.; Thomma, B.P.H.J.; Staskawicz, B.; et al. Interfamily transfer of a plant pattern-recognition receptor confers broad-spectrum bacterial resistance. *Nat. Biotechnol.* **2010**, *28*, 365–369. [CrossRef] [PubMed]
54. Holton, N.; Nekrasov, V.; Ronald, P.C.; Zipfel, C. The Phylogenetically-Related Pattern Recognition Receptors EFR and XA21 Recruit Similar Immune Signaling Components in Monocots and Dicots. *PLoS Pathog.* **2015**, *11*, e1004602-22. [CrossRef] [PubMed]
55. Schwessinger, B.; Bahar, O.; Thomas, N.; Holton, N.; Nekrasov, V.; Ruan, D.; Canlas, P.E.; Daudi, A.; Petzold, C.J.; Singan, V.R.; et al. Transgenic Expression of the Dicotyledonous Pattern Recognition Receptor EFR in Rice Leads to Ligand-Dependent Activation of Defense Responses. *PLoS Pathog.* **2015**, *11*, e1004809.
56. Rodriguez-Moreno, L.; Song, Y.; Thomma, B.P. Transfer and engineering of immune receptors to improve recognition capacities in crops. *Curr. Opin. Plant Biol.* **2017**, *38*, 42–49. [CrossRef] [PubMed]
57. Schoonbeek, H.-J.; Wang, H.-H.; Stefanato, F.L.; Craze, M.; Bowden, S.; Wallington, E.; Zipfel, C.; Ridout, C.J. Arabidopsis EF-Tu receptor enhances bacterial disease resistance in transgenic wheat. *New Phytol.* **2015**, *206*, 606–613. [CrossRef] [PubMed]
58. Tripathi, J.N.; Lorenzen, J.; Bahar, O.; Ronald, P.; Tripathi, L. Transgenic expression of the rice Xa21 pattern-recognition receptor in banana (*Musa* sp.) confers resistance to *Xanthomonas campestris* pv. *musacearum*. *Plant Biotechnol. J.* **2014**, *12*, 663–673. [CrossRef] [PubMed]
59. Omar, A.A.; Murata, M.M.; El-Shamy, H.A.; Graham, J.H.; Grosser, J.W. Enhanced resistance to citrus canker in transgenic mandarin expressing Xa21 from rice. *Transgenic Res.* **2018**, *27*, 179–191. [CrossRef] [PubMed]
60. Du, J.; Verzaux, E.; Chaparro-Garcia, A.; Bijsterbosch, G.; Keizer, L.C.P.; Zhou, J.; Liebrand, T.W.H.; Xie, C.; Govers, F.; Robatzek, S.; et al. Elicitin recognition confers enhanced resistance to *Phytophthora infestans* in potato. *Nat. Plants* **2015**, *1*, 15034. [CrossRef] [PubMed]
61. Zhang, W.; Fraiture, M.; Kolb, D.; Loffelhardt, B.; Desaki, Y.; Boutrot, F.F.G.; Tor, M.; Zipfel, C.; Gust, A.A.; Brunner, F. Arabidopsis RECEPTOR-LIKE PROTEIN30 and Receptor-Like Kinase SUPPRESSOR OF BIR1-1/EVERSHED Mediate Innate Immunity to Necrotrophic Fungi. *Plant Cell* **2013**, *25*, 4227–4241. [CrossRef] [PubMed]
62. Jehle, A.K.; Lipschis, M.; Albert, M.; Fallahzadeh-Mamaghani, V.; Fürst, U.; Mueller, K.; Felix, G. The receptor-like protein ReMAX of Arabidopsis detects the microbe-associated molecular pattern eMax from Xanthomonas. *Plant Cell* **2013**, *25*, 2330–2340. [CrossRef] [PubMed]
63. Hegenauer, V.; Fürst, U.; Kaiser, B.; Smoker, M.; Zipfel, C.; Felix, G.; Stahl, M.; Albert, M. Detection of the plant parasite *Cuscuta reflexa* by a tomato cell surface receptor. *Science* **2016**, *353*, 478–481. [CrossRef] [PubMed]
64. Mott, G.A.; Thakur, S.; Smakowska, E.; Wang, P.W.; Belkhadir, Y.; Desveaux, D.; Guttman, D.S. Genomic screens identify a new phytobacterial microbe-associated molecular pattern and the cognate Arabidopsis receptor-like kinase that mediates its immune elicitation. *Genome Biol.* **2016**, 1–15. [CrossRef] [PubMed]

65. Wang, Y.; Nsibo, D.L.; Juhar, H.M.; Govers, F.; Bouwmeester, K. Ectopic expression of Arabidopsis L-type lectin receptor kinase genes LecRK-I.9 and LecRK-IX.1 in *Nicotiana benthamiana* confers Phytophthora resistance. *Plant Cell Rep.* **2016**, *35*, 845–855. [CrossRef] [PubMed]
66. Delteil, A.; Gobbato, E.; Cayrol, B.; Estevan, J.; Michel-Romiti, C.; Dievart, A.; Kroj, T.; Morel, J.-B. Several wall-associated kinases participate positively and negatively in basal defense against rice blast fungus. *BMC Plant Biol.* **2016**, *16*, 17. [CrossRef] [PubMed]
67. Liu, Y.; Wu, H.; Chen, H.; Liu, Y.; He, J.; Kang, H.; Sun, Z.; Pan, G.; Wang, Q.; Hu, J.; et al. A gene cluster encoding lectin receptor kinases confers broad-spectrum and durable insect resistance in rice. *Nat. Biotechnol.* **2015**, *33*, 301–305. [CrossRef] [PubMed]
68. Wang, Y.; Weide, R.; Govers, F.; Bouwmeester, K. L-type lectin receptor kinases *in Nicotiana benthamiana* and tomato and their role in Phytophthora resistance. *J. Exp. Bot.* **2015**, *66*, 6731–6743. [CrossRef] [PubMed]
69. Bourdais, G.; Burdiak, P.; Gauthier, A.; Nitsch, L.; Salojärvi, J.; Rayapuram, C.; Idänheimo, N.; Hunter, K.; Kimura, S.; Merilo, E.; et al. On behalf of the CRK Consortium. Large-Scale Phenomics Identifies Primary and Fine-Tuning Roles for CRKs in Responses Related to Oxidative Stress. *PLoS Genet.* **2015**, *11*, e1005373. [CrossRef] [PubMed]
70. Pfeilmeier, S.; George, J.; Morel, A.; Roy, S.; Smoker, M.; Stansfeld, L.; Downie, A.; Peeters, N.; Malone, J.; Zipfel, C. Heterologous expression of the immune receptor EFR in *Medicago truncatula* reduces pathogenic infection, but not rhizobial symbiosis. *bioRxiv* **2017**. [CrossRef]
71. Kim, S.Y.; Shang, Y.; Joo, S.-H.; Kim, S.-K.; Nam, K.H. Overexpression of BAK1 causes salicylic acid accumulation and deregulation of cell death control genes. *Biochem. Biophys. Res. Commun.* **2017**, *484*, 781–786. [CrossRef] [PubMed]
72. Beck, M.; Wyrsch, I.; Strutt, J.; Wimalasekera, R.; Webb, A.; Boller, T.; Robatzek, S. Expression patterns of *FLAGELLIN SENSING 2* map to bacterial entry sites in plant shoots and roots. *J. Exp. Bot.* **2014**, *65*, 6487–6498. [CrossRef] [PubMed]
73. Faulkner, C.; Robatzek, S. Plants and pathogens: Putting infection strategies and defence mechanisms on the map. *Curr. Opin. Plant Biol.* **2012**, *15*, 699–707. [CrossRef] [PubMed]
74. Gust, A.A.; Brunner, F.D.R.; Nürnberger, T. Biotechnological concepts for improving plant innate immunity. *Curr. Opin. Biotechnol.* **2010**, *21*, 204–210. [CrossRef] [PubMed]
75. Trdá, L.; Fernandez, O.; Boutrot, F.; Héloir, M.-C.; Kelloniemi, J.; Daire, X.; Adrian, M.; Clément, C.; Zipfel, C.; Dorey, S.; et al. The grapevine flagellin receptor VvFLS2 differentially recognizes flagellin-derived epitopes from the endophytic growth-promoting bacterium *Burkholderia phytofirmans* and plant pathogenic bacteria. *New Phytol.* **2013**, *201*, 1371–1384. [CrossRef] [PubMed]
76. Shi, Q.; Febres, V.J.; Jones, J.B.; Moore, G.A. A survey of FLS2 genes from multiple citrus species identifies candidates for enhancing disease resistance to *Xanthomonas citri* ssp. *citri*. *Hortic. Res.* **2016**, *3*, 80. [CrossRef] [PubMed]
77. Helft, L.; Thompson, M.; Bent, A.F. Directed Evolution of FLS2 towards Novel Flagellin Peptide Recognition. *PLoS ONE* **2016**, *11*, e0157155. [CrossRef] [PubMed]
78. Song, W.Y.; Wang, G.L.; Chen, L.L.; Kim, H.S.; Pi, L.Y.; Holsten, T.; Gardner, J.; Wang, B.; Zhai, W.X.; Zhu, L.H.; et al. A receptor kinase-like protein encoded by the rice disease resistance gene, Xa21. *Science* **1995**, *270*, 1804–1806. [CrossRef] [PubMed]
79. Kishimoto, K.; Kouzai, Y.; Kaku, H.; Shibuya, N.; Minami, E.; Nishizawa, Y. Perception of the chitin oligosaccharides contributes to disease resistance to blast fungus *Magnaporthe oryzae* in rice. *Plant J.* **2010**, *64*, 343–354. [CrossRef] [PubMed]
80. Kishimoto, K.; Kouzai, Y.; Kaku, H.; Shibuya, N.; Minami, E.; Nishizawa, Y. Enhancement of MAMP signaling by chimeric receptors improves disease resistance in plants. *Plant Signal. Behav.* **2011**, *6*, 449–451. [CrossRef] [PubMed]
81. Bouwmeester, K.; Han, M.; Blanco-Portales, R.; Song, W.; Weide, R.; Guo, L.-Y.; van der Vossen, E.A.G.; Govers, F. The Arabidopsis lectin receptor kinase LecRK-I.9 enhances resistance to *Phytophthora infestans* in Solanaceous plants. *Plant Biotechnol. J.* **2014**, *12*, 10–16. [CrossRef] [PubMed]
82. Benedetti, M.; Pontiggia, D.; Raggi, S.; Cheng, Z.; Scaloni, F.; Ferrari, S.; Ausubel, F.M.; Cervone, F.; De Lorenzo, G. Plant immunity triggered by engineered *in vivo* release of oligogalacturonides, damage-associated molecular patterns. *Proc. Natl. Acad. Sci. USA* **2015**, *112*, 5533–5538. [CrossRef] [PubMed]

11

Generating Plants with Improved Water use Efficiency

Sonja Blankenagel [1], Zhenyu Yang [2,*], Viktoriya Avramova [1], Chris-Carolin Schön [1] and Erwin Grill [2]

[1] Plant Breeding, School of Life Sciences Weihenstephan, Technical University of Munich, Liesel-Beckmann-Straße 2, 85354 Freising, Germany; sonja.blankenagel@tum.de (S.B.); viktoriya.avramova@tum.de (V.A.); chris.schoen@tum.de (C.-C.S.)

[2] Botany, School of Life Sciences Weihenstephan, Technical University of Munich, Emil-Ramann-Straße 4, 85354 Freising, Germany; erwin.grill@wzw.tum.de

* Correspondence: zyang@wzw.tum.de

Abstract: To improve sustainability of agriculture, high yielding crop varieties with improved water use efficiency (WUE) are needed. Despite the feasibility of assessing WUE using different measurement techniques, breeding for WUE and high yield is a major challenge. Factors influencing the trait under field conditions are complex, including different scenarios of water availability. Plants with C_3 photosynthesis are able to moderately increase WUE by restricting transpiration, resulting in higher intrinsic WUE (*i*WUE) at the leaf level. However, reduced CO_2 uptake negatively influences photosynthesis and possibly growth and yield as well. The negative correlation of growth and WUE could be partly disconnected in model plant species with implications for crops. In this paper, we discuss recent insights obtained for *Arabidopsis thaliana* (L.) and the potential to translate the findings to C_3 and C_4 crops. Our data on *Zea mays* (L.) lines subjected to progressive drought show that there is potential for improvements in WUE of the maize line B73 at the whole plant level (WUE_{plant}). However, changes in *i*WUE of B73 and Arabidopsis reduced the assimilation rate relatively more in maize. The trade-off observed in the C_4 crop possibly limits the effectiveness of approaches aimed at improving *i*WUE but not necessarily efforts to improve WUE_{plant}.

Keywords: water use efficiency; crop breeding; yield; drought; maize; Arabidopsis; C_4-C_3 comparison; stomatal conductance; abscisic acid (ABA); photosynthesis

1. Introduction

Green Revolution technologies and significant expansion in the use of land, water, and other natural resources for agricultural purposes have led to a tripling in agricultural production between 1960 and 2015 [1]. Despite this success, the high costs to the natural environment that accompany elevated productivity and changes in the food supply chain threaten the sustainability of food production [1]. Global food security is further challenged by climate change, with a predicted increase in frequency of droughts [2,3]. Globally, agriculture accounts for at least 70% of withdrawals from freshwater resources, with large effects on ecosystems [4,5]. Despite this high water deployment, major yield losses due to water deficits are experienced in crops [6]. At the same time, global population growth increases the demand for food, feed, and fuel, which intensifies the pressure to improve water use efficiency (WUE) of crops [7,8]. While better crop and water management practices provide an immediate opportunity to increase crop water productivity, breeding for superior varieties can achieve a medium- and long-term increase [9,10].

Physiologically, water use efficiency can be defined at different scales [11–13]. At the plot level, it represents the ratio of grain or biomass yield to water received or evapotranspired. At the single plant level (WUE_{plant}), it is the ratio of biomass to transpiration. The increase in biomass and amount of water transpired over time can be assessed gravimetrically [14]. However, this is destructive and laborious on a long-term basis, especially regarding large crops like maize and sorghum. Therefore, analyses of intrinsic water use efficiency (*i*WUE) and carbon isotope discrimination ($\Delta^{13}C$) are used as surrogates when evaluating WUE [11,15]. The *i*WUE is assessed at the leaf level as the ratio of net CO_2 assimilation (A_n) to stomatal conductance (g_s) and can be measured noninvasively with portable gas exchange equipment [16]. As transpiration rate (E) is influenced not only by g_s but also by the leaf-to-air vapor pressure deficit (VPD) of the air [17], *i*WUE usually differs from transpiration efficiency (A_n/E). In addition, VPD affects the stomatal aperture and therefore g_s [17,18]. Extrapolation of gas exchange data from single-leaf to whole plant is error-prone due to differences in photosynthesis and transpiration among leaves [19]. Prediction of long-term biomass accumulation and water consumption, WUE_{plant}, based on *i*WUE is even more uncertain given the possible differences in VPD and additional physiological processes such as dark respiration and photorespiration influencing the resulting biomass increase [19]. Despite these limitations, analysis of *i*WUE provides a convenient measure for the water efficiency of carbon capture. The throughput of *i*WUE analyses is quite low as only single, time-consuming measurements per plant can be taken, which impedes large-scale phenotyping.

Analysis of stable carbon isotope discrimination ($\Delta^{13}C$) offers a suitable alternative in C_3 plants by providing a read-out for transpiration efficiency integrated over time. Discrimination of the heavier isotope is mainly caused by differences in diffusion rates of the isotopes and enzymatic discrimination during carboxylation reactions [20]. Therefore, $\Delta^{13}C$ has been used as an indirect trait to select cultivars with improved WUE [21–24]. By combining the analysis with oxygen isotope enrichment $\Delta^{18}O$, an estimation for transpiration rate [25–27], contributions of water loss, and CO_2 assimilation on *i*WUE could be disentangled [28–30]. Stable isotope compositions of leaves or grains, however, represent integrated measures of many processes over a period of plant growth and therefore correlation with *i*WUE can be limited [31]. In C_4 plants, CO_2 prefixation, for instance, by phosphoenolpyruvat carboxylase and bundle sheath leakiness restrict the responsiveness of $\Delta^{13}C$ to changes in WUE [20,32] and make the relationship between $\Delta^{13}C$, g_s and WUE in C_4 species less predictable compared to C_3 plants [13].

Improving WUE of crops is considered beneficial in very dry climates and in very severe and terminal drought conditions, while growth maintenance traits are advantageous under milder drought conditions [33–35]. For crops experiencing water deficit early in their development, traits found to be positive for improving WUE are negative for yield [36]. Enhanced water uptake through investments in the root system can result in reduced plant size and water expenditure for growth maintenance can result in increased drought stress experiences if plants are growing at very low soil water availability [33,34,36,37]. Hence, water-conserving traits as imposed by higher WUE would be beneficial, provided growth and yield are not negatively affected.

2. Disconnecting Improved WUE and Growth Trade-Offs

Being a ratio, *i*WUE can be improved by reducing g_s per amount of CO_2 assimilated or by enhancing the assimilation rate at a given g_s. Both cases result in lowered intercellular CO_2 concentration (C_i) and consequently in an increased stomatal CO_2 gradient ($C_a - C_i$, with external CO_2 concentration C_a), which is directly proportional to the ratio of A_n to g_s according to Fick's law applied to carbon assimilation in leaves, $A_n = g_s (C_a - C_i)$ [16,38,39]. Increased *i*WUE has been observed in several C_3 species under water deficit conditions when plants reduce g_s [11,15,40–42], although a decrease in g_s caused by drought was found to be overridden by heat stress [43]. However, closing stomatal pores to reduce transpiration often results in a reduction of A_n [41,44]. Lowering g_s impinges on C_i and unless this change in C_i is counteracted by an elevated mesophyll conductance (g_m), the CO_2 concentration at the site of Rubisco-dependent carboxylation (C_c, CO_2 concentration in chloroplasts) will be reduced [39,45,46]. A reduction of C_c affects the carboxylation efficiency of

Rubisco and favors photorespiration [47,48]. Sustaining net photosynthesis under these conditions might require a higher electron transfer rate (ETR) and/or reduced nonphotochemical quenching to support enhanced carboxylation by Rubisco for compensation of enhanced photorespiration [49–52]. There are reports that water deficit results in increased g_m; however, in most analyses, no change or a reduced g_m was observed under drought [44,53–58].

Gains in WUE are often associated with growth trade-offs [59,60]. As pointed out by Blum [61], crops with high CO_2 assimilation and high biomass accumulation per unit land area require high stomatal conductance. This is supported by the observation of a constant WUE on the field level over a broad range of yields [8]. Nevertheless, there might be ways to achieve elevated WUE and high photosynthesis, namely by exploring CO_2 concentrating mechanisms, increased g_m, and increased CO_2 specificity of Rubisco [10].

Interestingly, several reports of C_3 plants have shown enhanced *i*WUE without the expected negative impact on A_n or growth [11,15,62–64], as postulated by plant physiologists [4]. In these studies with transgenic tomato and Arabidopsis plants, g_s was moderately reduced by enhancing the biosynthesis or the responsiveness to the phytohormone abscisic acid (ABA) or by reducing the size and density of leaf stomata [15,63–65]. Plants overexpressing distinct ABA receptors—termed ABA-Binding Regulatory Component (RCAR)/Pyrabactin Resistance 1-(like) (PYR1/PYL)—caused increases of 40% in *i*WUE, integrated WUE based on $\Delta^{13}C$ of biomass and cellulose fractions, and WUE_{plant} [15]. Growth rates and biomass accumulation were not significantly different from wild type [15]. Hence, the ABA receptor lines revealed higher water productivity, i.e., WUE per time, both under well-watered growth conditions and under water deficit. Net carbon assimilation was comparable to the wild type, however, at lowered C_i levels and without detectable changes in g_m. This report and other studies show that improving WUE is possible without growth trade-offs. The underlying physiological mechanisms are largely unknown and might involve the root system, as grafting experiments have suggested [15], and enzymes of the C_4 metabolism, such as PEP carboxylase and its regulatory protein kinase PEPC kinase, which are both upregulated in C_3 plants at low CO_2 availability [65].

C_4 and C_3 plants differ in WUE [66–68]. At a given g_s, C_4 plants show higher net carbon assimilation rates and higher WUE [66]. The CO_2 concentrating mechanism involving PEP carboxylase results in saturation of C_4 photosynthesis at relatively low C_i [69,70]; therefore, lower g_s and a steeper CO_2 gradient ($C_a - C_i$) are realized in C_4 plants compared to C_3 plants [66,70]. C_3 plants have C_i values in the range of 300 ± 60 µmol mol^{-1}, while the C_i of the C_4 plants is around 150 ± 40 µmol mol^{-1} [71–73] at ambient CO_2 of 370–400 µmol mol^{-1} in well-watered conditions. Under optimal growth conditions, maize and sorghum with C_4 metabolism therefore have higher yields per water transpired than the C_3 crop wheat [9].

3. Comparative Analysis of Maize and Arabidopsis

The question arises as to whether it is possible to transfer the finding of improved *i*WUE without having the negative impact on growth to crops. The data on g_s-modified tomato plants suggests that it might work for C_3 plants [63], but the lower C_i level of C_4 plants could preclude such an accomplishment in maize.

To explore the relevance of these findings of uncoupling WUE improvement and yield decreases for the C_4 crop maize (*Zea mays* L.), we analyzed gas exchange data obtained from the maize inbred line B73 and compared them to findings in Arabidopsis. In addition, we analyzed the WUE_{plant} of maize lines subjected to drought. B73 is an inbred line that is commonly used in breeding programs, but is known to be drought-sensitive [74]. B73 was included in a progressive drought stress experiment adapted from Yang et al. [15] in which biomass production with a given amount of water was analyzed and WUE_{plant} was determined. In this experiment, B73 showed the lowest WUE_{plant} (Figure 1a) compared to the maize inbred Mo17 and lines derived from an introgression library described by Gresset et al. [75]. In Figure 1, data are shown of the recurrent parent (RP) of the introgression library as well as two introgression lines differing from RP by reduced (IL-05) or elevated kernel $\Delta^{13}C$ (IL-81) [75].

A significantly reduced WUE_{plant} compared with the recurrent parent for IL-05 shows the potential of genetic improvement for this trait. However, the largest difference in WUE_{plant} was observed between B73 and Mo17, with an increase of ~27% (Figure 1a). Our data is in accordance with a previous drought stress experiment conducted on seedlings of maize inbred lines, where Mo17 ranked top in yield per plant [76]. The results indicate genetic variation in the efficiencies of water use among maize lines and a potential for genetic improvement of the WUE_{plant} for B73.

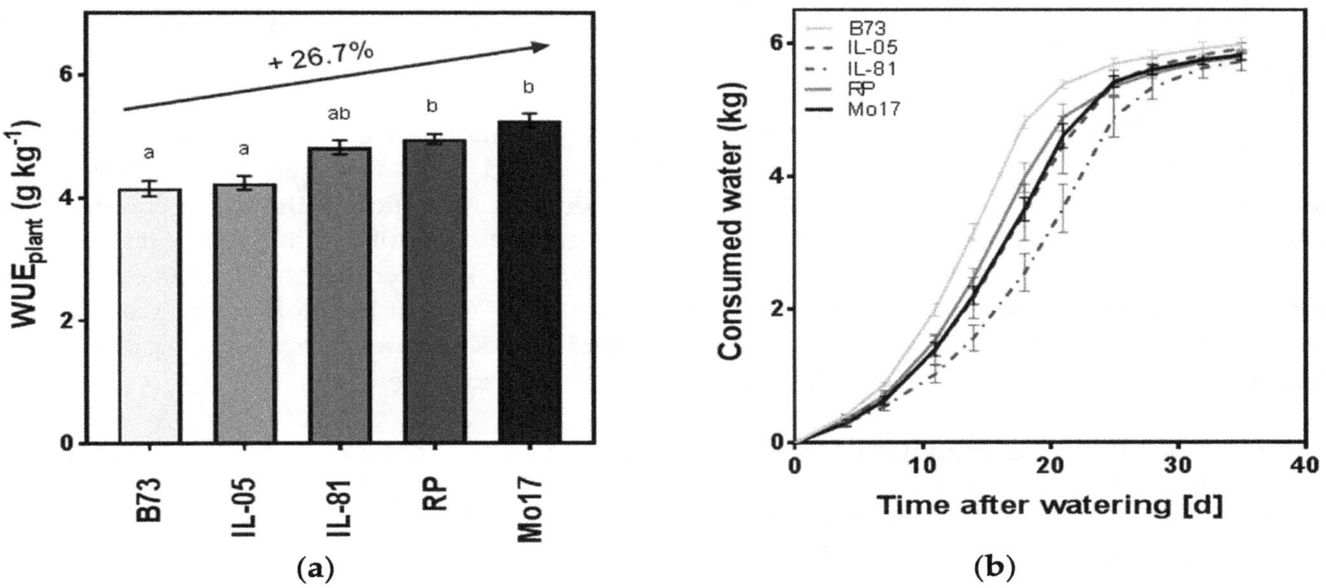

Figure 1. Water use efficiency and water consumption of maize lines under progressive drought. (**a**) Water use efficiency (WUE_{plant}) and (**b**) whole plant water consumption were assessed over the course of a progressive drought stress experiment adapted from Reference [15]. In the greenhouse, 23 maize genotypes were grown in a randomized complete block design, including the two maize inbred lines Mo17 and B73 and introgression lines described by Gresset et al. [75]. The maize inbred line RP and introgression lines derived therefrom (IL-81, IL-05) were kindly provided KWS Saat SE (Einbeck, Germany). Inbred lines B73 and Mo17 were kindly provided by the Chair of Genetics, Technical University of Munich, Freising, Germany. Prior to the experiment, maize seedlings were established in small pots in the growth chamber (16 h day at 25 °C, 650 $\mu E\ m^{-2}\ s^{-1}$ photosynthetically active radiation [PAR], 8 h night at 20 °C; 75% relative humidity [RH]) for two weeks after germination under well-watered conditions. Plants of RP harvested at this age had an aboveground dry matter of 0.62 g $\pm$ 0.27 g, and plants of an introgression line derived from IL-05 weighed 0.62 g $\pm$ 0.26 g. The influence of initial biomass on the biomass at the end of the experiment (28.74 g $\pm$ 2.22 g and 25.5 g $\pm$ 2.39 g, respectively) was approximately 2%. The plants were transplanted into 10 L pots containing 8 L water-saturated soil (85% v/v soil water content; CL ED73, Einheitserdewerke Patzer, Germany, particle diameter <15 mm). A cover of polyethylene foil was used to prevent evaporation, and the progressive drought experiment was initiated by no further watering. The experiment was conducted in the greenhouse (Gewächshauslaborzentrum Dürnast in Freising, Germany) in Oct–Nov 2017 at full sunlight plus supplemental light at 25–33 °C, 19–20 °C day/night, 400 $\mu mol\ m^{-2}\ s^{-1}$ PAR, 40% RH. Soil water content declined progressively during the course of the experiment until the plants used all available water. The water consumed was determined gravimetrically (means $\pm$ SE of $n \geq 4$ biological replicates). WUE_{plant} was calculated as final aboveground biomass per water consumed (means $\pm$ SE of $n \geq 4$ biological replicates). The increase in WUE in Mo17 compared to B73 is indicated with an arrow. Student's paired *t*-tests of the maize lines were adjusted for multiple comparisons with the Bonferroni method and lines, which did not differ significantly ($p < 0.01$), and are marked with common letters.

Maize lines showed a difference in water consumption over the five weeks of the experiment (ANOVA, $p < 0.001$, Figure 1b). However, differences in water consumption cannot explain the differences observed in WUE_{plant} and towards the end of the progressive drought, all genotypes included in the experiment had consumed an equal amount of water (5.8 kg ± 0.02 kg, mean ± SE).

The way in which the change in soil water content (SWC) during the progressive drought experiment affected photosynthesis and *i*WUE was analyzed by gas exchange measurements. The A_n of leaves was fairly constant for maize B73 plants exposed to high SWC levels up to 40%, then the A_n dropped steadily approaching zero at approximately 20% SWC (Figure 2a). In parallel, g_s changed moderately between 70% and 40% SWC and declined to zero at 20% SWC (Figure 2b).

The C_i values were in the range of 80–100 μmol CO_2 per mol between 40–60% SWC. They were somewhat higher in plants from water-saturated soil and were lowered to a minimum of approximately 40 μmol mol^{-1} at 25% SWC (Figure 2c). Further reduction of the water content in the soil resulted in the steep rise of C_i values, indicating collapsing photosynthesis at very low g_s of plants experiencing severe drought stress. As the ambient CO_2 concentration (C_a) surrounding the leaf was maintained at 400 μmol mol^{-1}, the CO_2 gradient ($C_a - C_i$) at the stomatal pores increased from approximately 250 μmol mol^{-1} (C_i of 150 μmol mol^{-1}) at soil water saturation to approximately 360 μmol mol^{-1} (C_i of 40 μmol mol^{-1}) at the brink of terminal drought.

The SWC also influenced *i*WUE (Figure 2d). Values increased from well-watered conditions to a maximum at 25% SWC, with a plateau around 170 μmol CO_2 per mol H_2O between 40–60% SWC. Under mild water deficit between 40–60% SWC, there was little variation in A_n, and g_s and, consequently, the *i*WUE values.

The results for maize B73 differed from data gained by similar analyses of Arabidopsis plants (Figure 3a–d; Reference [15]). The A_n remained constant between 30–70% SWC, which might be caused by light-limited, but not water-limited, photosynthesis. However, g_s and C_i steadily decreased with decreasing SWC and, concomitantly, the *i*WUE increased by twofold from approximately 35 to 70 μmol mol^{-1} at 30% SWC. The CO_2 gradient at stomata increased more than twofold from approximately 80 μmol mol^{-1} at soil water saturation to approximately 170 μmol mol^{-1} at 30%.

The data were obtained at light conditions that did not saturate photosynthesis, but analysis at saturating light confirmed the capacity of Arabidopsis to lower C_i and maintain photosynthetic rates unchanged [15]. The improvement in *i*WUE by limiting g_s without major trade-offs in A_n (Figure 3) was observed for the C_3 plants Arabidopsis [15,24] and tomato [63]. A twofold enhancement in *i*WUE has been reported in different C_3 species under drought [11,15,77–79]. Besides, considerable differences in WUE in the absence of drought stress have been observed among natural variants [80,81].

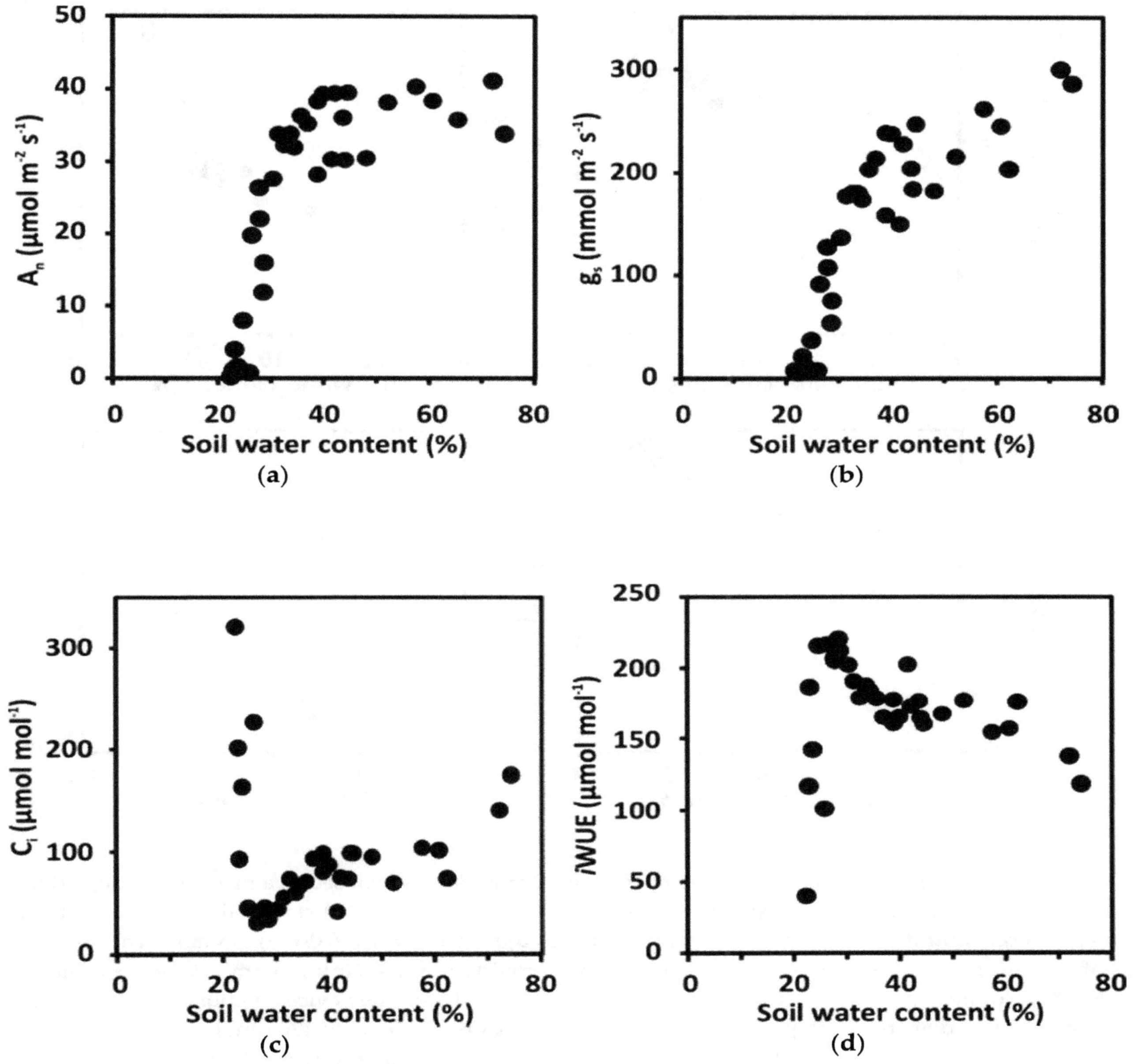

Figure 2. Changes in intrinsic water use efficiency of maize exposed to a progressive depletion of soil water content. (**a**) Net carbon assimilation rate (A_n), (**b**) stomatal conductance (g_s), (**c**) intercellular CO_2 concentration (C_i), and (**d**) intrinsic WUE (*i*WUE; defined as the ratio of A_n to g_s) of B73 plants at different soil water content. Gas exchange measurements using the GFS-3000 gas exchange system (Heinz, Walz GmbH, Effeltrich, Germany) were conducted at a photon flux density of 1000 μmol m^{-2} s^{-1}, an external CO_2 (C_a) of 400 μmol mol^{-1} CO_2, and vapor pressure deficit (VPD) of 26 Pa kPa^{-1} ± 2 Pa kPa^{-1}. The first fully expanded leaf counting from the top of the plants was clamped into an 8 cm^2 cuvette for measurements, and plants were subjected to progressive drought as detailed in Figure 1. Plants were grown in soil (Classic Profi Substrate Einheitserde Werkverband) as described in Reference [15]. The experiment was conducted in a greenhouse in the Department of Botany in Freising, Germany from June to August. The maize plants were exposed to full sunlight, at an average temperature of 27 °C, and an average relative humidity of 55% in the experimental period. (**a**–**d**) five biological replicates and each data point represents single measurements with five technical replicates.

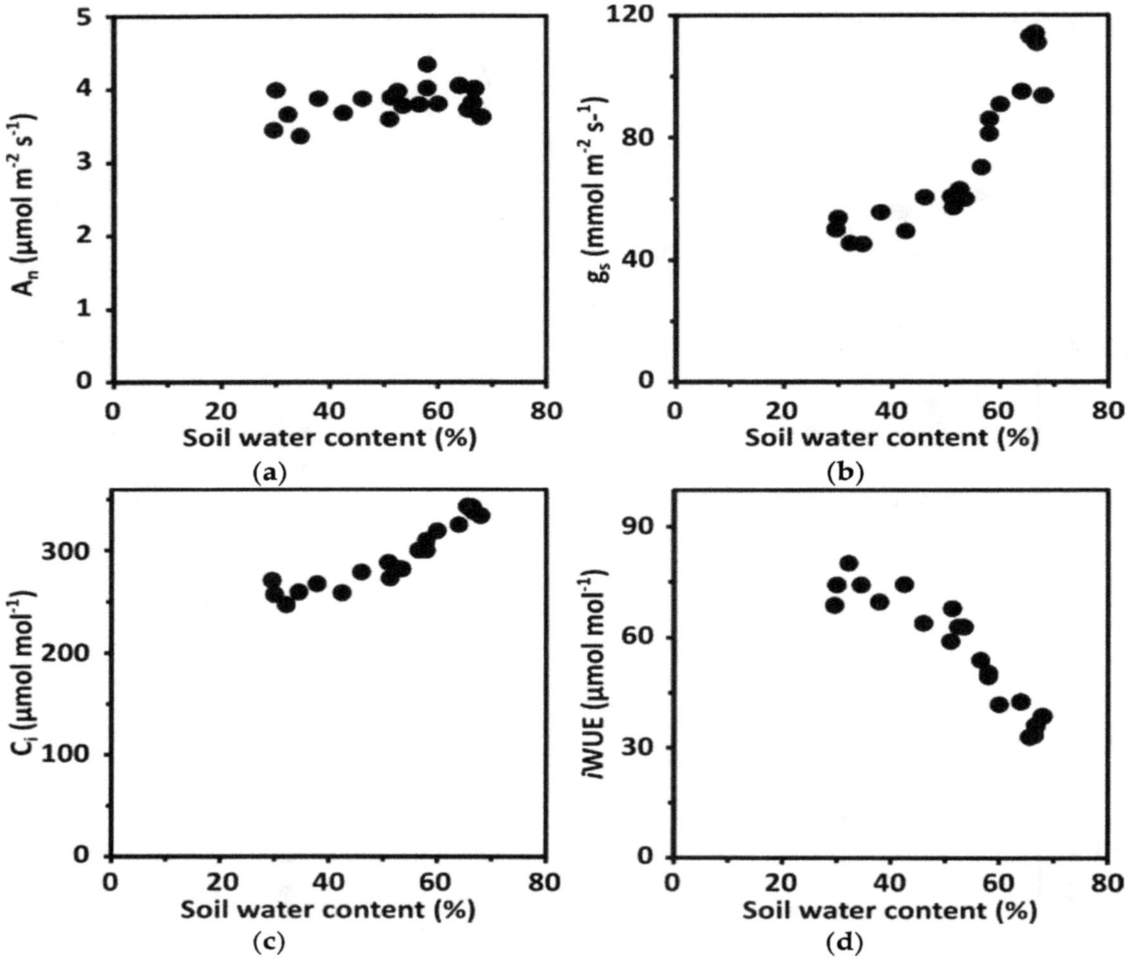

Figure 3. Changes in intrinsic water use efficiency of Arabidopsis exposed to a progressive depletion of soil water content (SWC). (**a**) Net carbon assimilation rate (A_n), (**b**) stomatal conductance (g_s), (**c**) CO_2 concentration in intercellular space (C_i) and (**d**) intrinsic WUE (*i*WUE) of whole leaf rosettes of Arabidopsis accession Columbia (Col-0; kindly provided by the Nottingham Arabidopsis Stock Center, Nottingham, UK). The measurements were performed with the device mentioned in Figure 2 at a photon flux density of 150 µmol m^{-2} s^{-1}, an ambient CO_2 (C_a) level of 420 µmol mol^{-1} and vapor pressure deficit of 13 ± 1 Pa kPa^{-1}. The plants were grown under short day conditions (8 h light/16 h dark photoperiod) at a photon flux density of 150 µmol m^{-2} s^{-1} and 22 °C and 50% relative humidity in the day time and 17 °C and 60% relative humidity at night. Data presented in (**a–d**) consists of three biological replicates and single measurements for each data point consist of 10 technical replicates. Data and the correlation between SWC and water potentials are presented in Reference [15].

Comparing the results from the analysis of maize and Arabidopsis, the potential for increasing *i*WUE in maize was more limited relative to Arabidopsis. Between 30–60% SWC—reflecting mainly mild water deficit [15]—Arabidopsis responded to the mounting water deficit by a 70% increase in *i*WUE, while maize showed an increase of less than 20%. Based on the *i*WUE values obtained at water-saturated soil, the *i*WUE increase in Arabidopsis and maize was approximately 100% and 40%, respectively, but water logging might be an issue at these high SWC levels. Between 30% and 60% SWC, the stomatal CO_2 gradient of maize increased from 300 to 350 µmol mol^{-1}, which corresponded to a 17% increase, whereas in Arabidopsis, the gradient was enhanced by 70%, from 100 to 170 µmol mol^{-1}.

To sum up, mild drought stress (30–60% SWC) had a minor effect on A_n, g_s, and *i*WUE in maize. However, in Arabidopsis, g_s and *i*WUE changed dynamically, while A_n was little affected. A reduction in SWC from 35% to 25% led to a rapid decline of A_n in maize.

These results are in accordance with data from C_3 and C_4 grass species [77,78] but only partly meet the behavior expected for C_3 and C_4 plants based on a meta study [82] where decreases in g_s and A_n under mild drought stress were more pronounced in C_3 relative to C_4 species. Comparing the A_n and g_s curves of maize B73 and Arabidopsis Col-0 (Figure 2a,b and Figure 3a,b), a reduction in stomatal conductance led to an immediate reduction in assimilation rate A_n for maize but not for Arabidopsis. However, not all C_3 plants show the same flat A_n/g_s curve as Arabidopsis [83,84] and therefore the data cannot be translated to C_3 crops in general. A previous study on maize lines found differences in g_s without trade-offs in A_n at well-watered conditions [85], and the author noted that C_i values did not become low enough to limit A_n as it might in high VPD conditions [85] or as it was observed here under drought.

Still, the limitations in improving WUE in maize B73 could be unique to this inbred line, and comparable data from other maize lines is needed before implications can be expanded to maize in general. Additionally, results from maize cannot represent C_4 photosynthesis in general because this pathway evolved independently in 19 angiosperm families [67]. However, other C_4 grass species have also shown a slight increase in *i*WUE with progressive depletion of water followed by a steep decline under very severe drought conditions driven by a pronounced decline in A_n [77,78]. The loss of A_n in the C_4 grass species under drought could partly be attributed to stomatal limitations, while other limitations dominate, including photoinhibition, limitations of CO_2 fixation due to desiccation, and decreases in g_m [73,77].

It has also been shown that for subspecies of *Alloteropsis semialata* (R.Br.) Hitchc., A_n is massively reduced in C_4 subspecies under drought in such a way that C_4 photosynthesis totally loses its advantage over photosynthesis of the C_3 subspecies [73]. This conclusion cannot be drawn from our experiments. However, the observation that C_3 plants become more water use efficient during mild and moderate drought while C_4 plants show more stable WUE [86] is in accordance with our data on *i*WUE.

The less potent improvement in WUE observed in maize under drought might be attributable to a limitation in increasing the CO_2 gradient ($C_a - C_i$) further. Maize, like other C_4 species, possesses a CO_2 concentrating mechanism utilizing precarboxylation of CO_2 by PEP carboxylase, which results in C_i values approximately half compared to C_3 species [87]. Such a mechanism results in advantages in A_n and WUE under non- or mild-water-deficit conditions [9,66,69,73,77,78]. However, these advantages cannot be maintained when the drought gets severe, especially for maize and C_4 grass species [66,69,73,77,78,86], which is in agreement with our observation for maize B73 at SWC below 27%. The differences can also translate to the field level, where maize has been found to be more sensitive to drought than wheat (C_3), with yield reductions of 39.3% compared to 20.6%, respectively, at approximately 40% water reduction [88]. Maize and sorghum are equally or even more sensitive to water stress than many C_3 plants [69,73].

Our results show a potential to increase WUE in maize. The inbred lines displayed a broad variation in WUE_{plant} under progressive drought, and *i*WUE—measured under the same conditions for the least efficient line at whole plant level—still showed a moderate increase in *i*WUE with declining SWC. However, the potential of WUE improvement is limited in this C_4 plant compared to Arabidopsis. This limitation is caused by a very high *i*WUE and low C_i under well-watered conditions, which provides a minor degree of freedom for further lowering the C_i. The C_3 plant is more responsive concerning increases in *i*WUE under mild water deficit compared to maize. Hence, screening C_3 plants for enhanced *i*WUE in combination with efficient growth is a suitable approach to identify crops with improved WUE_{plant}. This approach is less promising for C_4 plants. Establishing higher $C_a - C_i$ gradients in C_3 crops at a given soil water potential, e.g., by biotechnical engineering using ABA receptors, has the potential to increase *i*WUE at the cost of minor reductions in A_n. Moderate reductions in A_n do not necessarily influence yield. In barley, improvements in *i*WUE and WUE_{plant} have been associated with trade-offs in carbon assimilation but without deleterious effects on plant growth or seed yield [89].

4. Conclusions

Our results indicate that the improvement in *i*WUE without trade-offs in carbon assimilation, as observed for tomato and Arabidopsis [15,63], is less promising for maize and possibly other C_4 plants. The large CO_2 gradient established by the CO_2-concentrating mechanisms of C_4 plants limits the potential for further increases in *i*WUE compared to C_3 plants. However, our data show major differences in WUE_{plant} for maize inbred lines and therefore potential for genetic improvement of this trait.

A recent meta-analysis on WUE revealed a tenfold bias in favor of C_3 plant studies compared to analyses on C_4 plants [86]. We therefore see an urgent need for more studies on C_4 crops to shed light on the mechanisms of WUE under water deficit in these important but drought-sensitive crops. Cereals like rice, maize, and wheat contribute largely to global food security [90]; therefore, breeding for and generating water-efficient and high yielding crops are an urgent task to meet future challenges.

Author Contributions: Conceptualization, E.G. and C.-C.S.; Methodology, Z.Y. and V.A.; Formal Analysis, Z.Y., V.A., S.B.; Investigation, Z.Y., V.A., S.B.; Resources, E.G. and C.-C.S.; Data Curation, Z.Y., V.A.; Writing—Original Draft Preparation, S.B.; Writing—Review & Editing, Z.Y., V.A., E.G., C.-C.S.; Visualization, Z.Y., S.B.; Supervision, E.G., C.-C.S.; Project Administration, E.G., C.-C.S.; Funding Acquisition, E.G., C.-C.S.

Acknowledgments: We thank Stefan Schwertfirm and Amalie Fiedler for technical assistance, Anne-Marie Stache for discussion of the statistical analysis, and Farhah Assaad for critical reading and constructive comments on the manuscript.

References

1. FAO. *The Future of Food and Agriculture. Trends and Challenges*; Food and Agriculture Organization of the United Nations: Rome, Italy, 2017.
2. Spinoni, J.; Naumann, G.; Carrao, H.; Barbosa, P.; Vogt, J. World drought frequency, duration, and severity for 1951–2010. *Int. J. Climatol.* **2014**, *34*, 2792–2804. [CrossRef]
3. Trenberth, K.E.; Dai, A.; van der Schrier, G.; Jones, P.D.; Barichivich, J.; Briffa, K.R.; Sheffield, J. Global warming and changes in drought. *Nat. Clim. Chang.* **2014**, *4*, 17–22. [CrossRef]
4. Morison, J.I.L.; Baker, N.R.; Mullineaux, P.M.; Davies, W.J. Improving water use in crop production. *Philos. Trans. R. Soc. Lond. Ser. B Biol. Sci.* **2008**, *363*, 639–658. [CrossRef] [PubMed]
5. Jobbágy, E.G.; Jackson, R.B. Groundwater use and salinization with grassland afforestation. *Glob. Chang. Biol.* **2004**, *10*, 1299–1312. [CrossRef]
6. Fahad, S.; Bajwa, A.A.; Nazir, U.; Anjum, S.A.; Farooq, A.; Zohaib, A.; Sadia, S.; Nasim, W.; Adkins, S.; Saud, S.; et al. Crop production under drought and heat stress: Plant responses and management options. *Front. Plant Sci.* **2017**, *8*, 1147. [CrossRef] [PubMed]
7. Spiertz, J.H.J.; Ewert, F. Crop production and resource use to meet the growing demand for food, feed and fuel: Opportunities and constraints. *NJAS-Wagen. J. Life Sci.* **2009**, *56*, 281–300. [CrossRef]
8. Rockström, J.; Lannerstad, M.; Falkenmark, M. Assessing the water challenge of a new green revolution in developing countries. *Proc. Natl. Acad. Sci. USA* **2007**, *104*, 6253–6260. [CrossRef] [PubMed]
9. Sadras, V.O.; Grassini, P.; Steduto, P. Status of water use efficiency of main crops. In *The State of the World's Land and Water Resources for Food and Agriculture*; F.A.O. Thematic Report No. 7; FAO: Rome, Italy, 2012.
10. Parry, M.A.J.; Flexas, J.; Medrano, H. Prospects for crop production under drought: Research priorities and future directions. *Ann. Appl. Biol.* **2005**, *147*, 211–226. [CrossRef]
11. Rizza, F.; Ghashghaie, J.; Meyer, S.; Matteu, L.; Mastrangelo, A.M.; Badeck, F.-W. Constitutive differences in water use efficiency between two durum wheat cultivars. *Field Crops Res.* **2012**, *125*, 49–60. [CrossRef]
12. Vadez, V.; Kholova, J.; Medina, S.; Kakkera, A.; Anderberg, H. Transpiration efficiency: New insights into an old story. *J. Exp. Bot.* **2014**, *65*, 6141–6153. [CrossRef] [PubMed]
13. Ellsworth, P.Z.; Cousins, A.B. Carbon isotopes and water use efficiency in C4 plants. *Curr. Opin. Plant Biol.* **2016**, *31*, 155–161. [CrossRef] [PubMed]
14. Ryan, A.C.; Dodd, I.C.; Rothwell, S.A.; Jones, R.; Tardieu, F.; Draye, X.; Davies, W.J. Gravimetric phenotyping of whole plant transpiration responses to atmospheric vapour pressure deficit identifies genotypic variation in water use efficiency. *Plant Sci.* **2016**, *251*, 101–109. [CrossRef] [PubMed]

15. Yang, Z.; Liu, J.; Tischer, S.V.; Christmann, A.; Windisch, W.; Schnyder, H.; Grill, E. Leveraging abscisic acid receptors for efficient water use in Arabidopsis. *Proc. Natl. Acad. Sci. USA* **2016**, *113*, 6791–6796. [CrossRef] [PubMed]
16. Long, S.P.; Bernacchi, C.J. Gas exchange measurements, what can they tell us about the underlying limitations to photosynthesis? Procedures and sources of error. *J. Exp. Bot.* **2003**, *54*, 2393–2401. [CrossRef] [PubMed]
17. Gerosa, G.; Mereu, S.; Finco, A.; Marzuoli, R. Stomatal conductance modeling to estimate the evapotranspiration of natural and agricultural ecosystems. In *Evapotranspiration: Remote Sensing and Modeling*; Irmak, A., Ed.; InTech: Rijeka, Croatia, 2011; pp. 403–420.
18. Turner, N.C.; Schulze, E.-D.; Gollan, T. The responses of stomata and leaf gas exchange to vapour pressure deficits and soil water content: I. Species comparisons at high soil water contents. *Oecologia* **1984**, *63*, 338–342. [CrossRef] [PubMed]
19. Medrano, H.; Tomás, M.; Martorell, S.; Flexas, J.; Hernández, E.; Rosselló, J.; Pou, A.; Escalona, J.-M.; Bota, J. From leaf to whole-plant water use efficiency (WUE) in complex canopies: Limitations of leaf WUE as a selection target. *Crop J.* **2015**, *3*, 220–228. [CrossRef]
20. Farquhar, G.D.; Ehleringer, J.R.; Hubick, K.T. Carbon Isotope Discrimination and Photosynthesis. *Annu. Rev. Plant Physiol. Plant Mol. Biol.* **1989**, *40*, 503–537. [CrossRef]
21. Farquhar, G.D.; Richards, R.A. Isotopic composition of plant carbon correlates with water-use efficiency of wheat genotypes. *Aust. J. Plant Physiol.* **1984**, *11*, 539. [CrossRef]
22. Zhang, C.-Z.; Zhang, J.-B.; Zhao, B.-Z.; Zhang, H.; Huang, P. Stable isotope studies of crop carbon and water relations: A review. *Agric. Sci. China* **2009**, *8*, 578–590. [CrossRef]
23. Rebetzke, G.J.; Condon, A.G.; Richards, R.A.; Farquhar, G.D. Selection for reduced carbon isotope discrimination increases aerial biomass and grain yield of rainfed bread wheat. *Crop Sci.* **2002**, *42*, 739. [CrossRef]
24. Masle, J.; Gilmore, S.R.; Farquhar, G.D. The ERECTA gene regulates plant transpiration efficiency in Arabidopsis. *Nature* **2005**, *436*, 866–870. [CrossRef] [PubMed]
25. Sheshshayee, M.S.; Bindumadhava, H.; Ramesh, R.; Prasad, T.G.; Lakshminarayana, M.R.; Udayakumar, M. Oxygen isotope enrichment (Δ18O) as a measure of time-averaged transpiration rate. *J. Exp. Bot.* **2005**, *56*, 3033–3039. [CrossRef] [PubMed]
26. Barbour, M.M.; Farquhar, G.D. Relative humidity- and ABA-induced variation in carbon and oxygen isotope ratios of cotton leaves. *Plant Cell Environ.* **2000**, *23*, 473–485. [CrossRef]
27. Barbour, M.M.; Schurr, U.; Henry, B.K.; Wong, S.C.; Farquhar, G.D. Variation in the oxygen isotope ratio of phloem sap sucrose from castor bean. Evidence in support of the Péclet effect. *Plant Physiol.* **2000**, *123*, 671–680. [CrossRef] [PubMed]
28. Scheidegger, Y.; Saurer, M.; Bahn, M.; Siegwolf, R. Linking stable oxygen and carbon isotopes with stomatal conductance and photosynthetic capacity: A conceptual model. *Oecologia* **2000**, *125*, 350–357. [CrossRef] [PubMed]
29. Farquhar, G.D.; Lloyd, J. Carbon and oxygen isotope effects in the exchange of carbon dioxide between terrestrial plants and the atmosphere. In *Stable Isotopes and Plant Carbon-Water Relations*; Ehleringer, J.R., Hall, A.E., Farquhar, G.D., Eds.; Elsevier Science: Burlington, NJ, USA, 1993; pp. 47–70.
30. Battipaglia, G.; Saurer, M.; Cherubini, P.; Calfapietra, C.; McCarthy, H.R.; Norby, R.J.; Cotrufo, F.M. Elevated CO_2 increases tree-level intrinsic water use efficiency: Insights from carbon and oxygen isotope analyses in tree rings across three forest FACE sites. *New Phytol.* **2013**, *197*, 544–554. [CrossRef] [PubMed]
31. Werner, C.; Schnyder, H.; Cuntz, M.; Keitel, C.; Zeeman, M.J.; Dawson, T.E.; Badeck, F.-W.; Brugnoli, E.; Ghashghaie, J.; Grams, T.E.E.; et al. Progress and challenges in using stable isotopes to trace plant carbon and water relations across scales. *Biogeosciences* **2012**, *9*, 3083–3111. [CrossRef]
32. Von Caemmerer, S.; Ghannoum, O.; Pengelly, J.J.L.; Cousins, A.B. Carbon isotope discrimination as a tool to explore C4 photosynthesis. *J. Exp. Bot.* **2014**, *65*, 3459–3470. [CrossRef] [PubMed]
33. Lopes, M.S.; Araus, J.L.; van Heerden, P.D.R.; Foyer, C.H. Enhancing drought tolerance in C4 crops. *J. Exp. Bot.* **2011**, *62*, 3135–3153. [CrossRef] [PubMed]
34. Tardieu, F. Any trait or trait-related allele can confer drought tolerance: Just design the right drought scenario. *J. Exp. Bot.* **2012**, *63*, 25–31. [CrossRef] [PubMed]
35. Tardieu, F.; Parent, B.; Caldeira, C.F.; Welcker, C. Genetic and physiological controls of growth under water deficit. *Plant Physiol.* **2014**, *164*, 1628–1635. [CrossRef] [PubMed]

36. Tardieu, F.; Simonneau, T.; Muller, B. The physiological basis of drought tolerance in crop plants: A Scenario-Dependent Probabilistic Approach. *Annu. Rev. Plant Biol.* **2018**, *69*, 733–759. [CrossRef] [PubMed]
37. van Oosterom, E.J.; Yang, Z.; Zhang, F.; Deifel, K.S.; Cooper, M.; Messina, C.D.; Hammer, G.L. Hybrid variation for root system efficiency in maize: Potential links to drought adaptation. *Funct. Plant Biol.* **2016**, *43*, 502. [CrossRef]
38. Williams, M.; Woodward, F.I.; Baldocchi, D.D.; Ellsworth, D. CO_2 capture from the leaf to the landscape. In *Photosynthetic Adaptation: Chloroplast to Landscape*; Smith, W.K., Vogelmann, T.C., Critchley, C., Eds.; Springer: Berlin, Germany, 2004; pp. 133–168.
39. Flexas, J.; Diaz-Espejo, A.; Galmés, J.; Kaldenhoff, R.; Medrano, H.; Ribas-Carbo, M. Rapid variations of mesophyll conductance in response to changes in CO_2 concentration around leaves. *Plant Cell Environ.* **2007**, *30*, 1284–1298. [CrossRef] [PubMed]
40. Medrano, H. Regulation of photosynthesis of C3 plants in response to progressive drought: Stomatal conductance as a reference parameter. *Ann. Bot.* **2002**, *89*, 895–905. [CrossRef] [PubMed]
41. Flexas, J.; Díaz-Espejo, A.; Conesa, M.A.; Coopman, R.E.; Douthe, C.; Gago, J.; Gallé, A.; Galmés, J.; Medrano, H.; Ribas-Carbo, M.; et al. Mesophyll conductance to CO_2 and Rubisco as targets for improving intrinsic water use efficiency in C3 plants. *Plant Cell Environ.* **2016**, *39*, 965–982. [CrossRef] [PubMed]
42. Monneveux, P.; Rekika, D.; Acevedo, E.; Merah, O. Effect of drought on leaf gas exchange, carbon isotope discrimination, transpiration efficiency and productivity in field grown durum wheat genotypes. *Plant Sci.* **2006**, *170*, 867–872. [CrossRef]
43. Urban, J.; Ingwers, M.W.; McGuire, M.A.; Teskey, R.O. Increase in leaf temperature opens stomata and decouples net photosynthesis from stomatal conductance in *Pinus taeda* and *Populus deltoides* x nigra. *J. Exp. Bot.* **2017**, *68*, 1757–1767. [CrossRef] [PubMed]
44. Flexas, J.; Bota, J.; Loreto, F.; Cornic, G.; Sharkey, T.D. Diffusive and metabolic limitations to photosynthesis under drought and salinity in C3 plants. *Plant Biol.* **2004**, *6*, 269–279. [CrossRef] [PubMed]
45. Hassiotou, F.; Ludwig, M.; Renton, M.; Veneklaas, E.J.; Evans, J.R. Influence of leaf dry mass per area, CO_2, and irradiance on mesophyll conductance in sclerophylls. *J. Exp. Bot.* **2009**, *60*, 2303–2314. [CrossRef] [PubMed]
46. Flexas, J.; Barbour, M.M.; Brendel, O.; Cabrera, H.M.; Carriquí, M.; Díaz-Espejo, A.; Douthe, C.; Dreyer, E.; Ferrio, J.P.; Gago, J.; et al. Mesophyll diffusion conductance to CO_2: An unappreciated central player in photosynthesis. *Plant Sci.* **2012**, *193–194*, 70–84. [CrossRef] [PubMed]
47. Farquhar, G.D.; von Caemmerer, S.; Berry, J.A. A biochemical model of photosynthetic CO_2 assimilation in leaves of C3 species. *Planta* **1980**, *149*, 78–90. [CrossRef] [PubMed]
48. Sharkey, T.D. Estimating the rate of photorespiration in leaves. *Physiol. Plant* **1988**, *73*, 147–152. [CrossRef]
49. Flexas, J. Genetic improvement of leaf photosynthesis and intrinsic water use efficiency in C3 plants: Why so much little success? *Plant Sci.* **2016**, *251*, 155–161. [CrossRef] [PubMed]
50. Long, S.P.; Zhu, X.-G.; Naidu, S.L.; Ort, D.R. Can improvement in photosynthesis increase crop yields? *Plant Cell Environ.* **2006**, *29*, 315–330. [CrossRef] [PubMed]
51. Parry, M.A.J.; Reynolds, M.; Salvucci, M.E.; Raines, C.; Andralojc, P.J.; Zhu, X.-G.; Price, G.D.; Condon, A.G.; Furbank, R.T. Raising yield potential of wheat. II. Increasing photosynthetic capacity and efficiency. *J. Exp. Bot.* **2011**, *62*, 453–467. [CrossRef] [PubMed]
52. Johnson, M.P.; Davison, P.A.; Ruban, A.V.; Horton, P. The xanthophyll cycle pool size controls the kinetics of non-photochemical quenching in *Arabidopsis thaliana*. *FEBS Lett.* **2008**, *582*, 262–266. [CrossRef] [PubMed]
53. Flexas, J.; Bota, J.; Escalona, J.M.; Sampol, B.; Medrano, H. Effects of drought on photosynthesis in grapevines under field conditions: An evaluation of stomatal and mesophyll limitations. *Funct. Plant Biol.* **2002**, *29*, 461. [CrossRef]
54. Perez-Martin, A.; Michelazzo, C.; Torres-Ruiz, J.M.; Flexas, J.; Fernández, J.E.; Sebastiani, L.; Diaz-Espejo, A. Regulation of photosynthesis and stomatal and mesophyll conductance under water stress and recovery in olive trees: Correlation with gene expression of carbonic anhydrase and aquaporins. *J. Exp. Bot.* **2014**, *65*, 3143–3156. [CrossRef] [PubMed]
55. Olsovska, K.; Kovar, M.; Brestic, M.; Zivcak, M.; Slamka, P.; Shao, H.B. Genotypically Identifying Wheat Mesophyll Conductance Regulation under Progressive Drought Stress. *Front. Plant Sci.* **2016**, *7*, 1111. [CrossRef] [PubMed]

56. Ouyang, W.; Struik, P.C.; Yin, X.; Yang, J. Stomatal conductance, mesophyll conductance, and transpiration efficiency in relation to leaf anatomy in rice and wheat genotypes under drought. *J. Exp. Bot.* **2017**, *68*, 5191–5205. [CrossRef] [PubMed]
57. Flexas, J.; Ribas-Carbó, M.; Bota, J.; Galmés, J.; Henkle, M.; Martínez-Cañellas, S.; Medrano, H. Decreased Rubisco activity during water stress is not induced by decreased relative water content but related to conditions of low stomatal conductance and chloroplast CO_2 concentration. *New Phytol.* **2006**, *172*, 73–82. [CrossRef] [PubMed]
58. Galmés, J.; Medrano, H.; Flexas, J. Photosynthetic limitations in response to water stress and recovery in Mediterranean plants with different growth forms. *New Phytol.* **2007**, *175*, 81–93. [CrossRef] [PubMed]
59. Blum, A. Drought resistance, water-use efficiency, and yield potential—Are they compatible, dissonant, or mutually exclusive? *Aust. J. Agric. Res.* **2005**, *56*, 1159. [CrossRef]
60. Kenney, A.M.; McKay, J.K.; Richards, J.H.; Juenger, T.E. Direct and indirect selection on flowering time, water-use efficiency (WUE, δ (13)C), and WUE plasticity to drought in *Arabidopsis thaliana*. *Ecol. Evolut.* **2014**, *4*, 4505–4521. [CrossRef] [PubMed]
61. Blum, A. Effective use of water (EUW) and not water-use efficiency (WUE) is the target of crop yield improvement under drought stress. *Field Crops Res.* **2009**, *112*, 119–123. [CrossRef]
62. Franks, P.J.; Doheny-Adams, T.; Britton-Harper, Z.J.; Gray, J.E. Increasing water-use efficiency directly through genetic manipulation of stomatal density. *New Phytol.* **2015**, *207*, 188–195. [CrossRef] [PubMed]
63. Thompson, A.J.; Andrews, J.; Mulholland, B.J.; McKee, J.M.T.; Hilton, H.W.; Horridge, J.S.; Farquhar, G.D.; Smeeton, R.C.; Smillie, I.R.A.; Black, C.R.; et al. Overproduction of abscisic acid in tomato increases transpiration efficiency and root hydraulic conductivity and influences leaf expansion. *Plant Physiol.* **2007**, *143*, 1905–1917. [CrossRef] [PubMed]
64. Yoo, C.Y.; Pence, H.E.; Jin, J.B.; Miura, K.; Gosney, M.J.; Hasegawa, P.M.; Mickelbart, M.V. The Arabidopsis GTL1 transcription factor regulates water use efficiency and drought tolerance by modulating stomatal density via transrepression of SDD1. *Plant Cell* **2010**, *22*, 4128–4141. [CrossRef] [PubMed]
65. Li, Y.; Xu, J.; Haq, N.U.; Zhang, H.; Zhu, X.-G. Was low CO_2 a driving force of C4 evolution: Arabidopsis responses to long-term low CO_2 stress. *J. Exp. Bot.* **2014**, *65*, 3657–3667. [CrossRef] [PubMed]
66. Long, S.P. Environmental responses. In *C_4 Plant Biology*; Sage, R.F., Monson, R.K., Eds.; Academic Press: San Diego, CA, USA, 1999; pp. 215–242.
67. Sage, R.F. The evolution of C4 photosynthesis. *New Phytol.* **2004**, *161*, 341–370. [CrossRef]
68. Downes, R.W. Differences in transpiration rates between tropical and temperate grasses under controlled conditions. *Planta* **1969**, *88*, 261–273. [CrossRef] [PubMed]
69. Ghannoum, O. C4 photosynthesis and water stress. *Ann. Bot.* **2009**, *103*, 635–644. [CrossRef] [PubMed]
70. Dai, Z.; Ku, M.; Edwards, G.E. C4 Photosynthesis: The CO_2-concentrating mechanism and photorespiration. *Plant Physiol.* **1993**, *103*, 83–90. [CrossRef] [PubMed]
71. Wong, S.C.; Cowan, I.R.; Farquhar, G.D. Stomatal conductance correlates with photosynthetic capacity. *Nature* **1979**, *282*, 424–426. [CrossRef]
72. Leakey, A.D.B.; Uribelarrea, M.; Ainsworth, E.A.; Naidu, S.L.; Rogers, A.; Ort, D.R.; Long, S.P. Photosynthesis, productivity, and yield of maize are not affected by open-air elevation of CO_2 concentration in the absence of drought. *Plant Physiol.* **2006**, *140*, 779–790. [CrossRef] [PubMed]
73. Ripley, B.S.; Gilbert, M.E.; Ibrahim, D.G.; Osborne, C.P. Drought constraints on C4 photosynthesis: Stomatal and metabolic limitations in C3 and C4 subspecies of Alloteropsis semialata. *J. Exp. Bot.* **2007**, *58*, 1351–1363. [CrossRef] [PubMed]
74. Chen, J.; Xu, W.; Velten, J.; Xin, Z.; Stout, J. Characterization of maize inbred lines for drought and heat tolerance. *J. Soil Water Conserv.* **2012**, *67*, 354–364. [CrossRef]
75. Gresset, S.; Westermeier, P.; Rademacher, S.; Ouzunova, M.; Presterl, T.; Westhoff, P.; Schön, C.-C. Stable carbon isotope discrimination is under genetic control in the C4 species maize with several genomic regions influencing trait expression. *Plant Physiol.* **2014**, *164*, 131–143. [CrossRef] [PubMed]
76. Aslam, M.; Tahir, M.H.N. Morpho-physiological response of maize inbred lines under drought environment. *Asian J. Plant Sci.* **2003**, *2*, 952–954. [CrossRef]
77. Ripley, B.S.; Frole, K.; Gilbert, M. Differences in drought sensitivities and photosynthetic limitations between co-occurring C3 and C4 (NADP-ME) Panicoid grasses. *Ann. Bot.* **2010**, *105*, 493–503. [CrossRef] [PubMed]

78. Taylor, S.H.; Ripley, B.S.; Woodward, F.I.; Osborne, C.P. Drought limitation of photosynthesis differs between C_3 and C_4 grass species in a comparative experiment. *Plant Cell Environ.* **2011**, *34*, 65–75. [CrossRef] [PubMed]
79. Ranney, T.G.; Bir, R.E.; Skroch, W.A. Comparative drought resistance among six species of birch (Betula): Influence of mild water stress on water relations and leaf gas exchange. *Tree Physiol.* **1991**, *8*, 351–360. [CrossRef]
80. Easlon, H.M.; Nemali, K.S.; Richards, J.H.; Hanson, D.T.; Juenger, T.E.; McKay, J.K. The physiological basis for genetic variation in water use efficiency and carbon isotope composition in Arabidopsis thaliana. *Photosyn. Res.* **2014**, *119*, 119–129. [CrossRef] [PubMed]
81. Des Marais, D.L.; Razzaque, S.; Hernandez, K.M.; Garvin, D.F.; Juenger, T.E. Quantitative trait loci associated with natural diversity in water-use efficiency and response to soil drying in Brachypodium distachyon. *Plant Sci.* **2016**, *251*, 2–11. [CrossRef] [PubMed]
82. Yan, W.; Zhong, Y.; Shangguan, Z. A meta-analysis of leaf gas exchange and water status responses to drought. *Sci. Rep.* **2016**, *6*, 20917. [CrossRef] [PubMed]
83. Flexas, J.; Niinemets, U.; Gallé, A.; Barbour, M.M.; Centritto, M.; Diaz-Espejo, A.; Douthe, C.; Galmés, J.; Ribas-Carbo, M.; Rodriguez, P.L.; et al. Diffusional conductances to CO_2 as a target for increasing photosynthesis and photosynthetic water-use efficiency. *Photosynth. Res.* **2013**, *117*, 45–59. [CrossRef] [PubMed]
84. Hall, A.E.; Schulze, E.-D. Stomatal response to environment and a possible interrelation between stomatal effects on transpiration and CO_2 assimilation. *Plant Cell Environ.* **1980**, *3*, 467–474. [CrossRef]
85. Bunce, J.A. Leaf transpiration efficiency of some drought-resistant maize lines. *Crop Sci.* **2010**, *50*, 1409. [CrossRef]
86. Zhang, J.; Jiang, H.; Song, X.; Jin, J.; Zhang, X. The responses of plant leaf CO_2/H_2O exchange and water use efficiency to drought: A meta-analysis. *Sustainability* **2018**, *10*, 551. [CrossRef]
87. Farquhar, G.D.; Hubick, K.T.; Condon, A.G.; Richards, R.A. Carbon isotope fractionation and plant water-use efficiency. In *Stable Isotopes in Ecological Research*; Billings, W.D., Golley, F., Lange, O.L., Olson, J.S., Remmert, H., Rundel, P.W., Ehleringer, J.R., Nagy, K.A., Eds.; Springer: New York, NY, USA, 1989; pp. 21–40.
88. Daryanto, S.; Wang, L.; Jacinthe, P.-A. Global synthesis of drought effects on maize and wheat production. *PLoS ONE* **2016**, *11*, e0156362. [CrossRef] [PubMed]
89. Hughes, J.; Hepworth, C.; Dutton, C.; Dunn, J.A.; Hunt, L.; Stephens, J.; Waugh, R.; Cameron, D.D.; Gray, J.E. Reducing stomatal density in barley improves drought tolerance without impacting on yield. *Plant Physiol.* **2017**, *174*, 776–787. [CrossRef] [PubMed]
90. FAO. *FAO Statistical Yearbook 2013. World Food and Agriculture*; FAO: Rome, Italy, 2013.

12

Arabidopsis thaliana Immunity-Related Compounds Modulate Disease Susceptibility in Barley

Miriam Lenk, Marion Wenig, Felicitas Mengel, Finni Häußler and A. Corina Vlot *

Helmholtz Zentrum München, Department of Environmental Science, Institute of Biochemical Plant Pathology, Ingolstädter Landstr. 1, 85764 Neuherberg, Germany; miriam.lenk@helmholtz-muenchen.de (M.L.); marion.wenig@helmholtz-muenchen.de (M.W.); felicitas.mengel@gmail.com (F.M.); finni.haeussler@web.de (F.H.)

* Correspondence: corina.vlot@helmholtz-muenchen.de

Abstract: Plants are exposed to numerous pathogens and fend off many of these with different phytohormone signalling pathways. Much is known about defence signalling in the dicotyledonous model plant *Arabidopsis thaliana*, but it is unclear to which extent knowledge from model systems can be transferred to monocotyledonous plants, including cereal crops. Here, we investigated the defence-inducing potential of *Arabidopsis* resistance-inducing compounds in the cereal crop barley. Salicylic acid (SA), folic acid (Fol), and azelaic acid (AzA), each inducing defence against (hemi-)biotrophic pathogens in *Arabidopsis*, were applied to barley leaves and the treated and systemic leaves were subsequently inoculated with *Xanthomonas translucens* pv. *cerealis* (*Xtc*), *Blumeria graminis* f. sp. *hordei* (powdery mildew, *Bgh*), or *Pyrenophora teres*. Fol and SA reduced *Bgh* propagation locally and/or systemically, whereas Fol enhanced *Xtc* growth in barley. AzA reduced *Bgh* propagation systemically and enhanced *Xtc* growth locally. Neither SA, Fol, nor AzA influenced lesion sizes caused by the necrotrophic fungus *P. teres*, suggesting that the tested compounds exclusively affected growth of (hemi-)biotrophic pathogens in barley. In addition to SA, Fol and AzA might thus act as resistance-inducing compounds in barley against *Bgh*, although adverse effects on the growth of pathogenic bacteria, such as *Xtc*, are possible.

Keywords: systemic acquired resistance; barley; salicylic acid; folic acid; azelaic acid; *Blumeria graminis* f. sp. *hordei*

---

1. Introduction

Plants are constantly challenged with a plethora of pathogens, including herbivorous insects, fungi, oomycetes, bacteria and viruses [1]. To cope with these attacks, plants have evolved an effective immune system of pre-formed and inducible defence mechanisms. The former include morphological barriers, for instance the plant cell wall, cuticle, phytoanticipins, and antimicrobial proteins. Inducible defence mechanisms include processes like cell wall reinforcement by lignin and callose or the synthesis of phytoalexins and defence-related proteins or enzymes [1–4].

After the first contact between plant and pathogen, conserved structures such as bacterial flagellin or fungal chitin—so-called pathogen-associated molecular patterns (PAMPs)—encounter pattern recognition receptors on the cell surface. This triggers a first immune response termed PAMP-triggered immunity (PTI) [2,4,5]. During long-term evolution of plant–pathogen interactions, some pathogens developed effector proteins to suppress PTI. In response to this development, plants evolved resistance (*R*) genes, which encode proteins that directly or indirectly recognise effectors, and on recognition elicit so called effector-triggered immunity (ETI) [2–4,6].

Both PTI and ETI are associated with the induction of various cellular responses. These include the synthesis of antimicrobial compounds, the generation of reactive oxygen species, the activation of mitogen-activated protein kinases, transcriptional reprogramming, and the accumulation of phytohormones such as salicylic acid (SA) [2–4,7,8]. SA is known to affect the redox status of plants, thereby leading to transcriptional reprogramming and enhanced transcription of *PATHOGENESIS-RELATED* (*PR*) genes, which are thought to promote resistance [7–11]. Furthermore, SA not only helps the plant to defend itself against a present infection, it is also involved in the induction of a process protecting the plant in case of future pathogen challenge [4,6,8,12]. This response is termed systemic acquired resistance (SAR) and is usually triggered by a local, foliar infection. It elicits long-lasting, broad-spectrum resistance in systemic plant tissues (reviewed in [6–9,12]). Many signals and genes involved in SAR have been discovered in *Arabidopsis thaliana*. During SAR, SA levels in the infected and systemic tissues rise, and transcripts of *PR* as well as other defence-related genes accumulate. This establishes a status of heightened alert; the plant is "primed" to respond more quickly to a secondary infection [13,14]. SAR is most effective against pathogens fended off via SA-dependent responses [3,4,8].

During SAR, long-distance signals are generated in the infected leaves and travel to the systemic leaves, presumably via the phloem [12]. Recent evidence suggests that signal transmission also occurs via volatile compounds, in particular monoterpenes, which appear to be transmitted via the air [15]. In *Arabidopsis*, candidate long-distance SAR signals further include methyl salicylate [16], glycerol-3-phosphate (G3P) [17,18], the C_9 dicarboxylic acid azelaic acid (AzA) [19], the diterpene dehydroabietinal [20], N-hydroxy-pipecolic acid [21,22], and the lipid-transfer proteins DIR1 (DEFECTIVE IN INDUCED RESISTANCE 1) and DIR1-like [23,24] (reviewed in [12,25,26]). Some of these compounds accumulate in the phloem after pathogen attack, among them SA, AzA, G3P, and pipecolic acid [19,27–29]. AzA and G3P were proposed to be symplastically loaded onto the phloem via plasmodesmata while SA appeared to be transported via an apoplastic route [29]. Only small proportions of SA and AzA are transported to the systemic tissue, and the majority of accumulation in systemic leaves thus appears to come from de novo synthesis [30–32]. Importantly, biosynthesis of the SAR-related compounds SA and pipecolic acid in systemic leaves is crucial for SAR establishment, questioning the biological relevance of the systemic mobility of these signals [18,33,34].

Most of the mobile metabolites induce SAR when exogenously applied to *Arabidopsis* plants. Local application of AzA, for example, induces a SAR-like state in the treated plants, protecting the systemic tissue against a subsequent infection with the hemi-biotrophic bacterium *Pseudomonas syringae* [19]. Additional SAR-inducing compounds include folate precursors or folic acid (Fol), which induces SA-mediated immunity in *Arabidopsis*, both locally and systemically, and in pepper (*Capsicum annuum*) [35,36].

In monocotyledonous plants, less is known about SAR and the signalling mechanisms involved. Key players in SA signalling, including NPR1 (NONEXPRESSOR OF PR GENES 1), the master regulator of SA signalling [7,9,10], several *PR* genes and SA-associated transcription factors are conserved between dicots and monocots, (reviewed in [37,38]). Most studies in this field focus on agronomically important plant species such as banana (*Musa acuminata*) [39], wheat (*Triticum aestivum*) [40–42], maize (*Zea mays*) [43], and rice (*Oryza sativa*) [38], in which important roles for SA signalling have been uncovered in resistance against pathogens including *Fusarium oxysporum*, *Puccinia striiformis*, *Xanthomonas oryzae* pv. *oryzae* and *Magnaporthe oryzae*. Treatments with the SA analogue benzothiadiazole (BTH) induce resistance in maize [44], wheat [45], and barley [46–48].

SA-associated immune responses in barley (*Hordeum vulgare*) often are studied in interaction with *Blumeria graminis* f. sp. *hordei* (*Bgh*), commonly named powdery mildew, an obligate biotrophic fungal pathogen that thrives on living host cells [49]. Since it causes reduced yield and serves as a model to study other mildews and obligate biotrophic pathogens [50], *Bgh* appeared on the list of the top 10 fungal pathogens by Dean et al. in 2012 [51]. In contrast to biotrophic pathogens of rice, *Bgh* inoculation does not result in SA accumulation in infected barley leaves [52–54]. Nevertheless, SA soil drench treatment of barley seedlings had a weak effect on *Bgh* infectivity [46,55], whereas soil drench treatment with BTH strongly enhanced barley resistance to *Bgh* [46].

Recently, Dey et al. [47] showed that in barley systemic resistance to the hemi-biotrophic bacterium *Xanthomonas translucens* pv. *cerealis* (*Xtc*) can be triggered by prior infiltration of a single barley leaf with the hemi-biotrophic bacterium *Pseudomonas syringae* pv. *japonica*. Unlike SAR in *Arabidopsis*, systemic immunity in barley was neither associated with SA nor with NPR1. In addition, local infiltration of SA or its functional analogue BTH did not induce systemic resistance to *Xtc*. Rather, local methyl jasmonate (MeJA) and abscisic acid (ABA) treatments, which in *Arabidopsis* induce systemic susceptibility to *P. syringae* [56] or antagonize SAR [57], respectively, triggered systemic resistance in barley to *Xtc* [47].

In plant immunity, cross talk between phytohormones is important and is believed to help achieve the best possible (defence) outcome. In phytohormone cross talk, an interaction is never defined by a single hormone, but rather by a complex network of interdependent positive and negative interactions [58–60]. The result of these interactions leads to responses in the plant, which allow it to appropriately respond to an invading pathogen. Pathogens of different lifestyles have different demands on their hosts and, therefore, must be combatted using different mechanisms; while biotrophs feed on living cells, necrotrophs acquire their nutrients from degraded and dead tissue. Several phytohormones are involved in plant defence responses against pathogens. The traditional three main players are SA, jasmonic acid (JA) and ethylene (ET), but recent evidence hints at the additional contribution of other hormones [58–60]. Mostly synergistic interactions are reported for JA and ET, which mainly promote defence against necrotrophic pathogens and insects [3,59–61]. SA and JA signalling pathways are interdependent; there is substantial cross talk between the two, comprising synergistic as well as antagonistic interactions, depending on the defence situation [3,59,62]. Biotrophic pathogens are mostly opposed using SA signalling, whereas necrotrophic pathogens are combatted using JA/ET-dependent pathways [3,59,60,62]. Because SA–JA cross talk is often antagonistic, SA-induced resistance against biotrophs can enhance susceptibility against necrotrophs and *vice versa* [60,63]. In *Arabidopsis*, for example, SA and Fol induce resistance against hemi-biotrophic bacteria and at the same time enhance susceptibility to the necrotrophic fungal pathogen *Alternaria brassicicola* [35,63]. Here, we query if such antagonistic cross talk between responses to biotrophic and necrotrophic pathogens also occurs in barley by using a barley pathogen from the necrotrophic side of the spectrum, the fungus *Pyrenophora teres*. *P. teres* is the causal agent of net and spot form net blotch, a major disease in many barley-growing areas, which can lead to severe yield loss, underlining the fungus' economic importance [64].

We want to study the possible effects of *Arabidopsis* SAR-associated compounds on barley defence responses. Therefore, we will investigate the influence of different chemical compounds, including SA, on the propagation of barley-specific pathogens. We will employ pathogens of different lifestyles, namely the hemi-biotrophic bacterium *Xtc*, the biotrophic fungus *Bgh* and the necrotrophic fungus *P. teres*. In addition to SA, we will include Fol, which induces an SA-dependent defence response in *Arabidopsis* [35]. Also, we will include AzA, a SAR signal, which in *Arabidopsis* primes SA accumulation, inducing a faster and stronger response after pathogen attack compared to that in unprimed plants [19]. The data shed light upon the possible transferability of signalling intermediates related to SAR from *Arabidopsis* to barley. As we detected differences and similarities in the responses of these plants interacting with (hemi-)biotrophic and necrotrophic pathogens, the data reinforce the necessity of studying metabolite-induced resistance in cereal crops.

2. Materials and Methods

2.1. Plants and Growth Conditions

Barley (*Hordeum vulgare* L. cultivar 'Golden Promise' (GP)) seeds were sterilized in 1.2% sodium hypochlorite for 3 min with 25 inversions per minute. Subsequently, seeds were rinsed 3 times with water for 10 min with 25 inversions per minute and then sown (Einheitserde classic CL-T, Bayerische Gärtnereigenossenschaft). Plants for *Xtc* infections were either grown in a greenhouse with additional lights HQI-TS 400W/D (Osram) using a day-night cycle of 12 h with 24 °C during the day and 20 °C during the night or in a climate chamber with 16 h light and 8 h darkness at a temperature of 20 °C (day)/16 °C (night). Chamber-grown plants inoculated with *Xtc* were transferred to a climate chamber with 14 h light and 10 h darkness at a temperature of 29 °C (day)/19 °C (night). Plants for *Bgh* infections were grown in climate chambers with 16 h light and 8 h darkness at a temperature of 20 °C (day)/16 °C (night). Plants for *P. teres* infections were grown in the green house as described above.

2.2. Chemicals and Treatments

Stock solutions of each chemical compound were freshly prepared for each experiment. SA (Roth, Karlsruhe, Germany) was dissolved at 4 M in 100% methanol (MeOH; Merck, Darmstadt, Germany), Fol (Roth, Karlsruhe, Germany) was dissolved at 1 M and AzA (Sigma-Aldrich, St. Louis, MO, USA) at 2 M in 50% MeOH. For plant treatments, the substances were diluted to 1 mM SA [35,45], 50 or 500 µM Fol [35], and 1 mM AzA [19,31] in 10 mM $MgCl_2$ (Roth, Karlsruhe, Germany). 0.025% MeOH in 10 mM $MgCl_2$ served as the mock treatment. To monitor systemically induced resistance, the first true leaves of 3-week-old barley plants (6 plants per treatment) were infiltrated with the different compounds or the mock solution using a needleless syringe. Five days after the primary treatment, the second leaves of the treated plants were infected with either *Xtc*, *Bgh*, or *P. teres* (see below). To monitor local induced resistance against *Xtc*, 4-week-old barley plants were sprayed with 1 mM SA, 500 µM Fol, 1 mM Aza, or 0.05% MeOH (as mock treatment) in 0.01% Tween-20 (Calbiochem, San Diego, CA, USA). To monitor local induced resistance against *Bgh*, 1 mM SA, 500 µM Fol, and 1 mM AzA in 10 mM $MgCl_2$ were syringe-infiltrated into the first true leaves of 3-week-old barley plants. 0.025% MeOH in 10 mM $MgCl_2$ served as the mock treatment. To monitor local induced resistance against *P. teres*, the second true leaves of 3-week-old barley plants were syringe-infiltrated with the compounds. The treated leaves were inoculated with *Xtc* 1 day after treatment, with *Bgh* 5 days after treatment, or with *P. teres* 1 day after treatment.

2.3. Xanthomonas Translucens pv. Cerealis (Xtc) Infection

Xanthomonas translucens pv. *cerealis* (*Xtc*) strain LMG7393 was obtained from the Laboratory of Microbiology UGent (LMG) collection of the Belgian Coordinated Collections of Microorganisms. For infection, *Xtc* was grown on LMG medium (15 g/L tryptone, 5 g/L soya peptone, 5 g/L NaCl, and 18 g/L agar-agar (Roth, Karlsruhe, Germany); pH adjusted to 7.3) over night at 28 °C. Bacteria were subsequently resuspended in 1 mL 10 mM $MgCl_2$ (Roth, Karlsruhe, Germany). The concentration of the bacterial suspension was adjusted to ~10^5 colony forming units (cfu)/mL in 10 mM $MgCl_2$ using a photometer (assuming the formula: OD_{600} of 0.2 equals ~10^8 cfu/mL). Leaves of 3–5 barley plants were subsequently inoculated with the resulting *Xtc* suspension by infiltration using a needleless syringe. The infected plants were covered with a plastic hood and kept in the green house for 4 days. The resulting *in planta Xtc* titres were determined as previously described [47,65].

2.4. Blumeria Graminis f. sp. Hordei (Bgh) Infection

Blumeria graminis f. sp. *hordei* (*Bgh*) Swiss field isolate CH4.8 was obtained from Dr. Patrick Schweizer (Leibniz-Institut für Pflanzengenetik und Kulturpflanzenforschung, Gatersleben, Germany). *Bgh* propagation and inoculation was performed essentially as described in [66]. In short, a pot containing 12 10-day-old seedlings was infected with *Bgh* one week prior to each experiment. Six hours prior to the start of an experiment, these plants were shaken in order to remove old conidia and provide a uniform inoculum for the experiment [67]. Compound- or mock-treated plants were subsequently inoculated with spores from the prepared *Bgh*-infected plants in an inoculation tower [66] at an inoculation density of ~30 spores/mm^2. The inoculated plants were placed back in the climate chamber for 6 days. Subsequently, 4 leaf discs (6 mm) were cut out of the distal halves of each first (local/compound-treated) and second (systemic) true leaf of the treated plants. The discs were incubated with 5 µM DAF-FM-DA (4-amino-5-methylamino-2′,7′-difluorofluorescein diacetate; Sigma-Aldrich, St. Louis, MO, USA or Santa Cruz Biotechnology, Dallas, TX, USA) in MES buffer (2-(*N*-morpholino)ethanesulfonic acid, 50 mM MES-KOH pH 5.7 (Roth, Karlsruhe, Germany), 1 mM $CaCl_2$ (Merck, Darmstadt, Germany), 0.25 mM KOH) for 45 min in the dark and then vacuum-infiltrated. Subsequently, the leaf discs were placed in light (45 V lamp) for 1 h and 45 min and afterwards distributed on 96-well plates (1 leaf disc per well, 20–24 discs from 6 plants per treatment) with the wells filled evenly to the rim with 1% phytoagar. Fluorescence of appropriate *z*-stacks was visualized using the 5× objective of an inverse spinning disc confocal microscope (Zeiss Axio Observer. Z1, Zeiss, Oberkochen, Germany). Chlorophyll was excited using a laser with 561 nm and detected using a bandpass 629/62 filter; DAF-FM DA was excited using a laser with 488 nm and detected using a bandpass 525/50 filter [68]. The 96-well plates were transferred to the microscope with a KiNDx Robot (Model KX-300-660-SSU, Peak Analysis and Automation, Inc., Farnborough, UK; assembled and set up by Analytik Jena, Jena, Germany). The robot and the visualization as well as evaluation were controlled by the softwares Microscope AppStudio (Analytik Jena, Jena, Germany) and ZEN2 (Zeiss, Oberkochen, Germany). Fluorescence intensities were normalized to those of uninfected barley plants (background fluorescence) of the same age.

2.5. P. Teres Infection

A field isolate of *Pyrenophora teres* was donated by Günther Bahnweg (Helmholtz Zentrum München, Neuherberg, Germany) and grown on oat plates (10 g rolled oats (Alnatura, Germany), 7.5 g agar-agar (Roth, Karlsruhe, Germany), 500 mL H_2O) for ~1 week at room temperature in the dark and then transferred to light for at least 2 weeks. Two mL of infection solution for fungi (0.85 g KH_2PO_4 (Merck, Darmstadt, Germany), 0.1 g glucose (Roth, Karlsruhe, Germany), and 1 μL Tween 20 in 100 mL H_2O, pH 6.0) were pipetted onto the *P. teres* plates and spores were scratched off the agar using an inoculation loop. The spore suspension was subsequently pipetted into a 5 mL tube and vortexed. After determining the spore concentration under a binocular, the spore suspension was diluted to 65–110 spores per μL. Infections were performed on 6 cm-long segments of a leaf, at a distance of 1.5 centimetres from the leaf base. Five 3 μL droplets of spore suspension were pipetted alternatingly on each side of the leaf midrib. The drops were left to dry for ~1 h and the plants (6 plants per treatment) were then covered with a plastic hood. Necrotic lesions caused by *P. teres* were measured 4 days after infection using the ImageJ macro PIDIQ [69]. The macro was modified to measure brown necrotic lesions caused by *P. teres*. These modifications were restricted to the values used for colour characterisation; the values for lesion measurements were as follows: hue 0–52, saturation 150–255, brightness 0–150.

2.6. Statistics

Results of biologically independent replicate experiments with two groups were tested together for homogeneity of variance using the *F*-test in Microsoft Excel. Depending on the outcome of the *F*-test, a two-sided *t*-test either for equal (homoscedastic) or unequal variances (heteroscedastic) was conducted in Microsoft Excel. Results of biologically independent replicate experiments with more than two groups were analysed in GraphPad Prism 7 for Windows (version 7.04). Data with only positive values were $\log_2$ transformed and data with negative and positive values were transformed according to the formula: $Y = \log_2[Y + 1 - \min(Y)]$, where min(Y) denotes the lowest measured value within the experiment. Subsequently, the data were analysed for outliers using Grubbs' test with $\alpha = 0.05$. The D'Agostino–Pearson normality test was performed with $\alpha = 0.01$ in order to test for normal distribution. If necessary the Grubbs' outlier test was repeated to assure normal data distribution (a maximum of 2 Grubbs' outlier tests were performed per data set). The data were subsequently analysed using one-way analysis of variance (ANOVA) with the Geisser-Greenhouse correction ($p < 0.05$) and a subsequent Dunnett's post hoc test with $\alpha = 0.05$.

3. Results

3.1. Folic Acid and Azelaic Acid Enhance the Susceptibility of Barley to Xtc

Exogenous SA or BTH application enhances the resistance of barley to the biotrophic fungus *Bgh* [46], but not to the hemi-biotrophic bacterium *Xtc* [47]. Here, we examined *Xtc* growth in barley after application of Fol and AzA, which each trigger SA-mediated resistance and/or SAR against the hemi-biotrophic bacterium *P. syringae* pv. *tomato* in *A. thaliana* [19,35]. In 3-week-old barley plants, the first true leaves were syringe-infiltrated with 50 or 500 µM Fol, 1 mM AzA, or the corresponding 0.025% MeOH mock control. Five days later, the second true leaves of the treated plants were syringe-infiltrated with *Xtc* and the resulting *in planta Xtc* titres determined at 4 days post-inoculation (dpi). Similar to SA, Fol and AzA did not reduce *Xtc* growth in the systemic tissue of barley (Figure 1). Whereas AzA application did not influence *Xtc* growth in the systemic tissue (Figure 1b), growth of the systemic *Xtc* inoculum was enhanced by a local Fol application (Figure 1a), suggesting the induction of enhanced susceptibility in barley against *Xtc* by Fol.

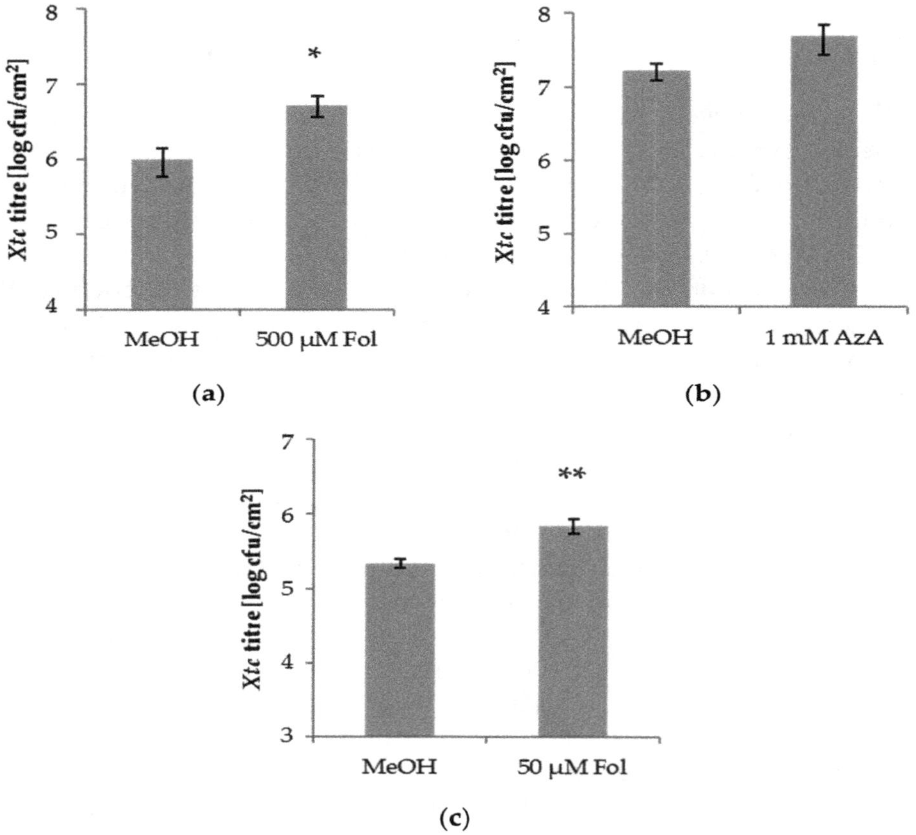

Figure 1. *Xanthomonas translucens* pv. *cerealis* (*Xtc*) titres in the systemic leaves of barley after local application of folic acid (Fol) or azelaic acid (AzA). Barley cultivar Golden Promise (GP) plants were infiltrated in the first true leaf with 0.025% methanol (MeOH) as control, 500 µM Fol (**a**), 1 mM AzA (**b**), or 50 µM Fol (**c**) in 10 mM $MgCl_2$ as indicated below the panels. Five days later, the second true leaves of the plants were inoculated with *Xtc* by syringe infiltration. The resulting *Xtc* titres in leaf 2 at 4 dpi are shown. Bars represent the average of 18–36 replicates from 5 ((**a**): 4× greenhouse, 1× growth chamber; (**b**): 3× greenhouse, 2× growth chamber) to 9 ((**c**): 8× greenhouse, 1× growth chamber) biologically independent experiments ± standard error; replicates were as follows: (**a**) MeOH: 19 (4 + 4 + 3 + 4 + 4), Fol: 18 (4 + 4 + 3 + 4 + 3), (**b**) MeOH: 19 (4 + 3 + 3 + 4 + 5), AzA: 20 (4 + 4 + 4 + 4 + 4), (**c**) MeOH: 34 (4 + 3 + 4 + 4 + 3 + 4 + 4 + 4 + 4), Fol: 36 (4 + 4 + 4 + 4 + 4 + 4 + 4 + 4 + 4). Asterisks above bars indicate statistically significant differences from the control treatments (*t*-test; * $p < 0.05$, ** $p < 0.005$).

Fol enhances *A. thaliana* resistance to *P. syringae* when applied at concentrations of 50 to 100 μM [35]. If we applied 50 μM Fol to the first leaves of 3-week-old barley plants, the growth of subsequently applied *Xtc* bacteria in the systemic leaves was enhanced in three out of 10 experiments and a strong tendency in the same direction was observed in a further two experiments. If we analyse all experiments together, the data suggest that application of 50 μM Fol can, similarly to 500 μM Fol, enhance barley systemic susceptibility to *Xtc* (Figure 1c), albeit perhaps less robustly.

We next tested the local barley responses to the compounds used above. To this end, 1 mM SA, 500 μM Fol, 1 mM AzA, or the appropriate mock control were applied onto the leaves of 4-week-old barley plants by spray treatment. One day later, treated leaves were syringe-infiltrated with *Xtc* and the resulting *in planta Xtc* titres determined at 4 dpi. SA and Fol appeared to moderately elevate local *Xtc* titres, with a significant trend observed only after Fol treatment. AzA application caused an increase of the *Xtc* titre (Figure 2). Thus, in addition to systemic susceptibility, Fol might also moderately enhance the local susceptibility of barley to the hemi-biotrophic bacterium *Xtc*. The effect appeared more robust in the systemic compared to local treated tissues (Figure 1a,c and Figure 2). AzA enhanced local, but not systemic susceptibility to *Xtc*.

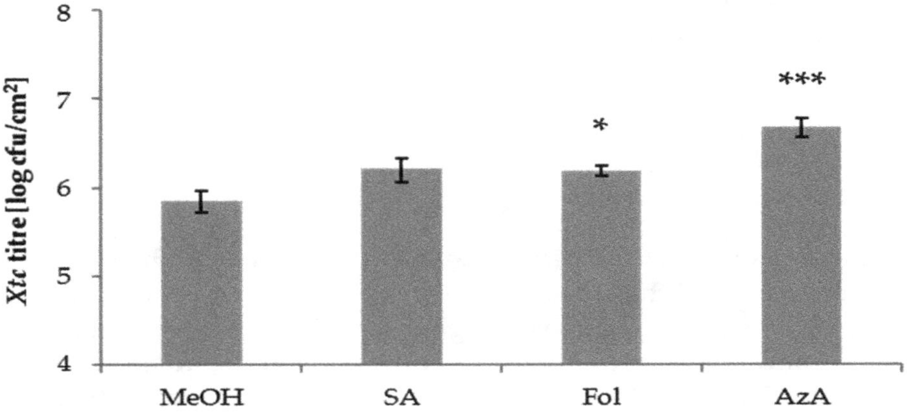

Figure 2. *Xtc* titres in salicylic acid (SA)-, Fol- and AzA-treated barley plants. Barley cultivar GP plants were sprayed with 0.05% MeOH as control, 1 mM SA, 500 μM Fol, or 1 mM AzA in 0.01% Tween-20 as indicated below the panel. One day later, the second true leaves of the plants were inoculated with *Xtc* by syringe infiltration. The resulting *Xtc* titres in leaf 2 at 4 dpi are shown. Bars represent the average of 8–9 replicates from 3 biologically independent greenhouse experiments ± standard error, replicates were as follows: MeOH: 9 (3 + 3 + 3), SA: 9 (3 + 3 + 3), Fol: 8 (3 + 3 + 2), AzA: 8 (3 + 3 + 2); asterisks above bars indicate statistically significant differences from the mock control treatment (one-way ANOVA and post hoc Dunnett's test, * $p < 0.05$, *** $p < 0.0005$).

3.2. Folic Acid Enhances the Resistance of Barley to Powdery Mildew (Bgh)

In *Arabidopsis* Fol triggers SA-mediated immunity against hemi-biotrophic bacteria [35]. In barley, SA and in particular its functional analogue BTH enhances immunity against the biotrophic fungus *Bgh* while Fol enhances barley susceptibility to *Xtc* (Figures 1 and 2; [46]). Here, we compared the effects of SA, Fol, and AzA on propagation of *Bgh* in barley. To this end, the first true leaves of 3-week-old barley plants were syringe-infiltrated with 1 mM SA, 500 μM Fol, 1 mM AzA, or 0.025% MeOH as control treatment. Five days later, the plants were inoculated with *Bgh*. The resulting *Bgh* infection was quantified on a fluorescence microscope 6 days later. We used the fluorescent dye DAF-FM DA which usually is used to detect and quantify NO [70] and, as shown in Figure 3, also has a high affinity for *Bgh*. If leaf discs from *Bgh*-infected leaves were stained with DAF-FM DA, the resulting fluorescence intensities reflected visual differences in *Bgh* infection levels and were quantified as a measure for fungal growth (Supplemental Figure S1). To this end, the stained leaf discs were loaded onto 96-well plates and scanned by a spinning disc (confocal) microscope (Figure 3).

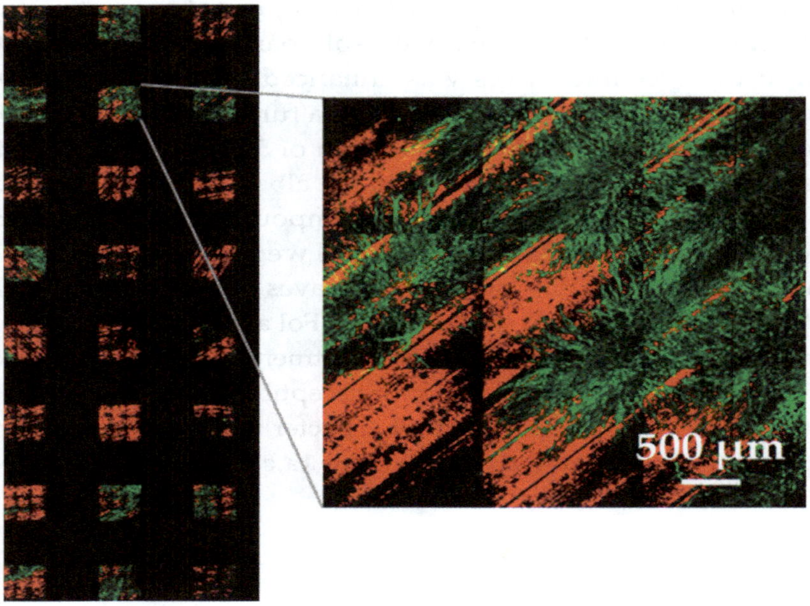

Figure 3. Merged *z*-stack 3 × 3 tiled images of 4-amino-5-methylamino-2′,7′-difluorofluorescein diacetate (DAF-FM-DA) stained discs of *Blumeria graminis* f. sp. *hordei* (*Bgh*)-infected barley leaves in the first 3 columns of a 96-well plate. Enlarged: merged image of a single well. Chlorophyll fluorescence is shown in red, DAF-FM DA fluorescence in green.

A local infiltration of Fol in leaf 1 of barley reduced *Bgh* growth on the Fol-treated leaf (Figure 4a) as well as on systemic leaf 2 of the treated plants (Figure 4b) as evidenced by a ~50% decrease in DAF-FM-DA fluorescence on the leaves of *Bgh*-infected Fol-treated compared to mock-treated plants. In contrast, SA and AzA infiltration did not induce significant changes in the DAF-FM-DA fluorescence of the treated leaves and thus had no local effect on *Bgh* growth (Figure 4a). Nevertheless, SA and AzA appeared to reduce *Bgh* propagation, although the differences to the control were not significant ($p = 0.0518$ in the case of SA, Figure 4a). In the systemic leaves, *Bgh*-associated DAF-FM-DA fluorescence was decreased by SA and AzA to a similar extent as by Fol (Figure 4b). Thus, Fol might induce local and systemic resistance to *Bgh*, while SA and AzA appear to trigger systemic resistance.

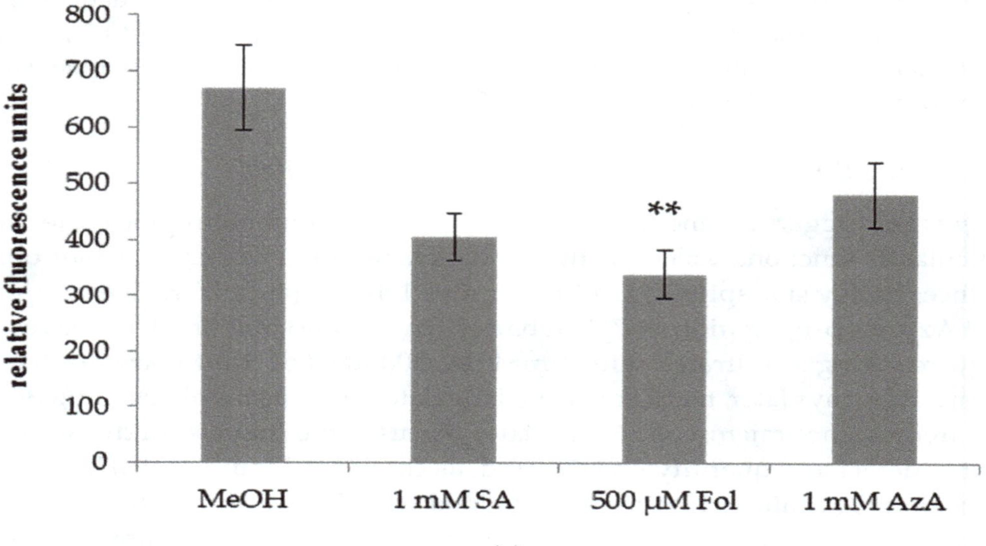

(**a**)

Figure 4. *Cont.*

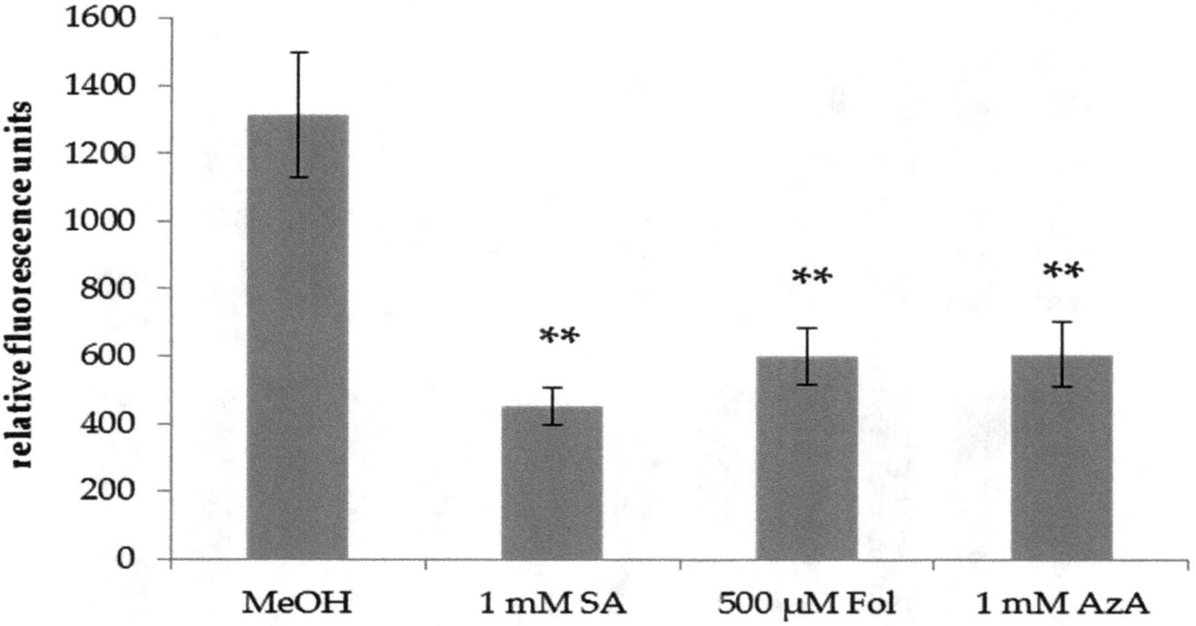

(**b**)

Figure 4. Relative fluorescence of DAF-FM DA staining of *Bgh* in leaves of barley plants after application of SA, Fol, or AzA to leaf 1. Barley cultivar GP plants were infiltrated in the first true leaf with 0.025% MeOH as control, 1 mM SA, 500 μM Fol, or 1 mM AzA in 10 mM $MgCl_2$ as indicated below the panel. Five days later, the plants were infected with *Bgh* spores. Leaf discs were cut out of the first (local) (**a**) and second (systemic) (**b**) true leaf and stained with DAF-FM DA. Fluorescence was recorded using a spinning disc (confocal) microscope. Bars represent the average of 37–64 replicates from 2 (SA treatment in **b**) to 3 (all other treatments) independent experiments ± standard error; replicates were as follows: (**a**) MeOH: 57 (22 + 14 + 21), SA: 62 (22 + 16 + 24), Fol: 55 (19 + 19 + 17), AzA: 58 (19 + 18 + 21); (**b**) MeOH: 55 (20 + 22 + 13), SA: 37 (17 + 20), Fol: 64 (21 + 23 + 20), AzA: 53 (6 + 24 + 23). Asterisks above the bars indicate statistically significant differences from the mock control treatment (one-way ANOVA and poct hoc Dunnett's test, ** $p < 0.005$).

3.3. Arabidopsis Immunity-Related Compounds Do Not Alter Barley Susceptibility to P. Teres

In *Arabidopsis*, SA induces local, but not systemic susceptibility to the necrotrophic fungal pathogen *Alternaria brassicicola* [35,63]. This most likely happens due to negative crosstalk between the SA and JA pathways, with SA inhibiting JA-mediated defence against *A. brassicicola*. A similar effect was observed for Fol, as its application increased the size of lesions caused by *A. brassicicola* on the treated, but not on systemic leaves [35]. Here, we examined the effects of SA, Fol, and AzA on barley local and systemic defence responses to the necrotrophic fungus *P. teres*. To this end, leaves of 3-week-old barley plants were syringe-infiltrated with 1 mM SA, 50 or 500 μM Fol, 1 mM AzA, or 0.025% MeOH as control treatment.

The same or systemic leaves were inoculated with *P. teres* 1 or 5 days later, respectively, and necrotic lesions were measured at 4 dpi using the ImageJ macro PIDIQ [69], which was modified to recognize the brown lesion caused by *P. teres*. The outcome varied strongly between different replicate experiments. Strikingly, SA application caused increases in *P. teres* lesion sizes in 4 out of 8 experiments if leaves systemic to the site of SA treatment were inoculated. However, taking all data together SA, Fol, and AzA did not significantly influence *P. teres* lesion sizes either locally (Figure 5) or systemically (Supplemental Figure S2), suggesting that these compounds do not affect the susceptibility of barley to *P. teres*.

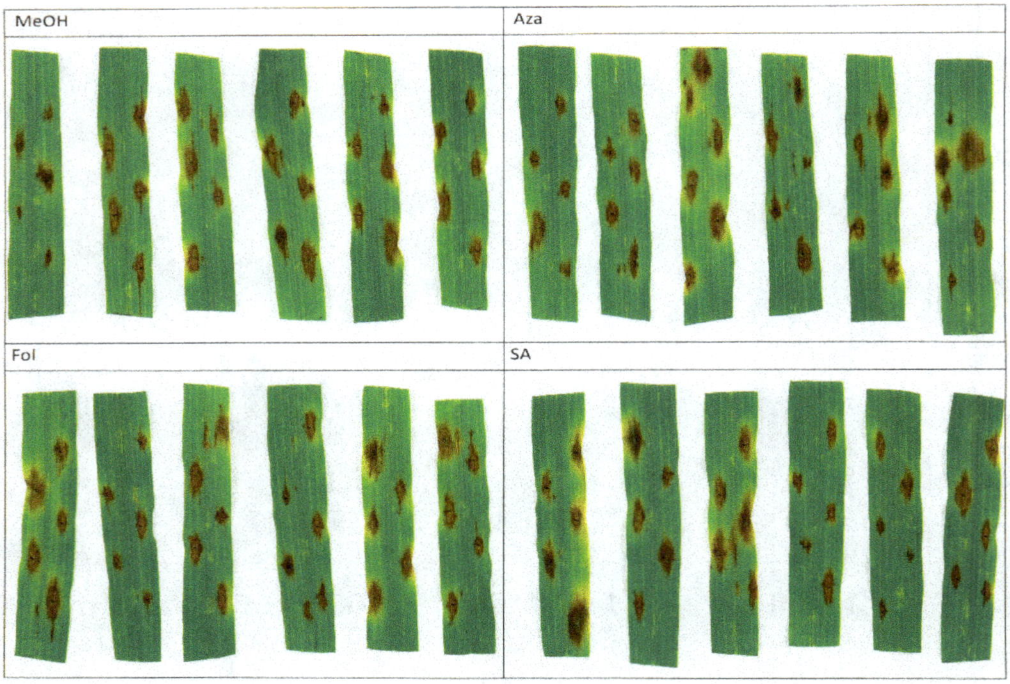

(a)

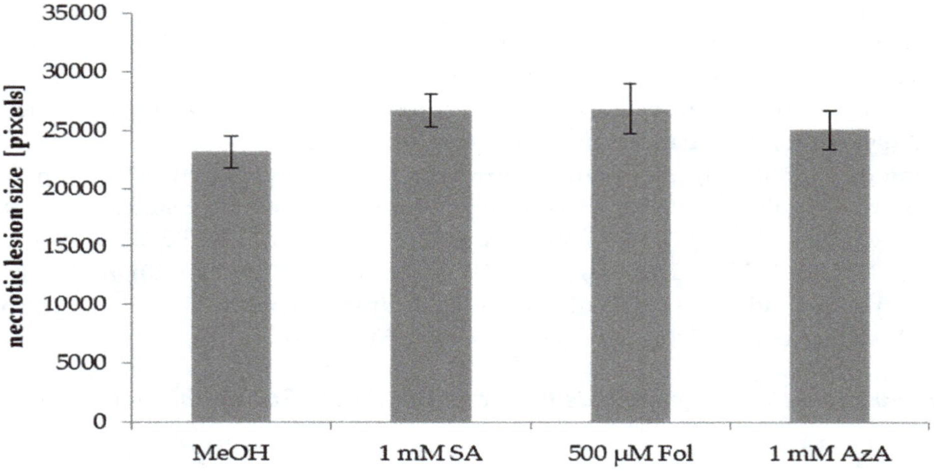

(b)

Figure 5. Lesions caused by *Pyrenophora teres* on the second leaves of barley after application of SA, Fol, or AzA on leaf 2. Barley cultivar GP plants were infiltrated in the second true leaf with 0.025% MeOH as control, 1 mM SA, 500 μM Fol, or 1 mM AzA in 10 mM $MgCl_2$ as indicated. One day later, the same leaves were inoculated with *P. teres* by pipetting droplets of a solution containing *P. teres* spores onto the leaf surface. The resulting necrotic lesions were photographed at 4 dpi (**a**) and measured using ImageJ (**b**). Bars in (**b**) represent the average of 35 replicates from 6 (Fol) or 42 replicates from 7 (all other treatments) independent experiments (each experiment with 6 replicates per treatment, except one experiment with Fol comprising 5 replicates) ± standard error.

4. Discussion

4.1. Salicylic Acid Has Differential Effects on Pathogens with Different Lifestyles

Here, we report on the role of SA, Fol, and AzA on barley defence against pathogens with different lifestyles. The three tested compounds are involved in *Arabidopsis* SAR and known for inducing resistance to (hemi-)biotrophic pathogens in the model plant. In contrast, it has been reported that a local application of SA does not induce systemic resistance to the hemi-biotrophic bacterium *Xtc* in barley [47]. In order to find out if SA is involved in (systemic) defence responses in barley, we performed inoculations with two additional pathogens.

Although a local application of SA did not affect growth of the hemi-biotrophic bacterium *Xtc* in the systemic tissue, the same treatment reduced the disease burden of the biotrophic fungal pathogen *Bgh* in the systemic leaves by more than 50% (Figure 4). There might also be a local effect of SA on *Bgh* propagation, for which we observed a clear, but statistically insignificant trend. Thus, SA might enhance the resistance of barley to the powdery mildew pathogen *Bgh*. Similar to SA, its functional analogue BTH does not affect barley resistance to *Xtc* [47], but it induces resistance to *Bgh* [46]. Additionally, NPR1, the master regulator of SA responses in *Arabidopsis* [7,9,10], is important for barley defence responses against *Bgh*, but not against *Xtc* [47], further supporting a possible role of SA in barley defence against *Bgh* but not *Xtc*. Previous studies had reported only a minor effect on *Bgh*, if any, after SA treatment of barley plants [46,55]. The relatively robust effect of SA on systemic *Bgh* propagation in barley that we observe in this study can have one or more of three reasons. First, the apparent difference in *Bgh* burden between SA- and mock-treated plants might be exaggerated by the method used for evaluation of the *Bgh* infections. While other studies rely on pustule counts, we used the fluorescent dye DAF-FM-DA to quantify fungal material. It is known that barley produces NO as part of the plants early defence responses against *Bgh* [71]. Additionally, at certain stages of its life cycle the fungus itself produces NO [72]. However, it seems unlikely that this NO interferes with our quantification. The production of NO is very short-lived and happens mostly in early defence responses and early life stages of *Bgh*, whereas we stain with DAF-FM-DA at 6 dpi, a relatively late stage of the infection, at which time *Bgh* displays significant hyphal growth (Figure 3). Second, previous studies used soil-drench treatment for SA application [46,55] while in this study, syringe-infiltration of leaves was used. Finally, and perhaps most importantly, plant age differed between both prior and the current studies. In both of the cited publications, seedlings of 5 or 7 days of age were used while we worked with 3-week-old plants. It is known that plant age can positively affect plant resistance against *Bgh* [73]. In support of this hypothesis, we observed robust effects of SA and the other tested compounds on *Bgh* propagation and *Xtc* growth in 3-week-old plants, but did not reproducibly observe the same effects in 2-week-old plants. Comparing our data to those presented in [46,55], it is possible that SA more effectively enhances barley resistance to *Bgh* if applied directly to the leaves rather than the soil and/or if applied to 3-week-old rather than younger plants.

Although SA locally enhances the susceptibility of *Arabidopsis* plants to a necrotrophic fungal pathogen [35,63], it had no effect on *P. teres* lesion sizes either locally or systemically in barley (Figure 5 and Figure S2). Thus, the trade-off between plant defence responses to biotrophic and necrotrophic pathogens that is observed in *Arabidopsis* does not appear to influence growth of the necrotrophic fungus in SA-treated barley plants.

In summary, SA appears to induce systemic resistance against *Bgh* in barley but likely does not contribute to resistance against *Xtc* or *P. teres*.

4.2. Folic Acid Has Differential Effects against Bacteria and Fungi with Similar Lifestyles

Fol application is known to induce local and systemic resistance to hemi-biotrophic bacteria in *Arabidopsis* [35]. This effect is dependent on SA biosynthesis and signalling and on the SAR-associated compound glycerol-3-phosphate. Similar to SA, Fol application triggers local susceptibility to necrotrophic *A. brassicicola*. Here, we infiltrated or sprayed barley with Fol and monitored the effects on local and systemic propagation of bacterial and fungal barley pathogens.

In contrast to SA, Fol application enhanced barley susceptibility to *Xtc* both systemically and to a lesser extent also in the local treated tissue (Figures 1 and 2). In *Arabidopsis*, Fol enhances resistance, probably through the SA pathway [35]. Here, SA and Fol differentially affected *Xtc* growth in barley, which was not affected by SA [47] and was enhanced (rather than reduced) by Fol. It is conceivable that *Xtc* can take up Fol, which is needed for nucleotide biosynthesis. Such supplementation could directly enhance bacterial growth, mimicking the induction of plant susceptibility. However, because the effect of Fol application on *Xtc* growth was stronger in the systemic compared to the local, treated tissue (Figures 1 and 2), the data argue in favour of a Fol-induced effect on plant immunity.

Similar to SA, the application of Fol reduced *Bgh* propagation on barley both locally and systemically but did not affect *P. teres* lesion formation (Figures 4 and 5). Again, the effects of SA and Fol on *Bgh* propagation were more pronounced in tissues that were systemic to the site of SA/Fol treatment than in the treated leaves themselves. Although we cannot exclude direct effects of the compounds on fungal growth, the data argue in favour of SA- and Fol-induced plant defence mechanisms affecting *Bgh* propagation in barley. Importantly, Fol-related compounds can promote plant yield [36], whereas SA causes cell death when applied at high concentrations. Thus, although adverse effects on barley susceptibility to hemi-biotrophic bacteria such as *Xtc* should be considered, Fol could be used as an alternative to SA or BTH to enhance the resistance of barley to the economically relevant powdery mildew pathogen *Bgh.*

4.3. Azelaic Acid Moderately Affects Barley Defence Responses

Application of AzA to *Arabidopsis* confers local and systemic resistance to hemi-biotrophic bacteria [19,30,31]. AzA primes *Arabidopsis* to accumulate higher SA levels more quickly after a subsequent infection [19]. Similarly to Fol-induced responses in *Arabidopsis* [35], AzA-mediated SAR depends on SA [19]. Here, we found that AzA influences pathogen propagation in barley very similarly to Fol. Whereas Fol locally and systemically enhanced *Xtc* growth, AzA did the same only in the local treated tissue and not systemically (Figures 1b and 2). Nevertheless, this similarity in the effects of Fol and AzA on barley susceptibility to *Xtc* argues for a possible interference of these compounds with the barley defence response to *Xtc* rather than for direct effects of either compound on bacterial growth. Furthermore, in contrast to its effect on *Xtc* growth, AzA reduced *Bgh* propagation systemically but not locally (Figure 4). In this case, the systemic response induced by AzA is similar to the responses induced by SA and Fol, which also appear to induce systemic resistance to *Bgh*. Similar to SA and Fol, AzA application did not have an influence on fungal growth of *P. teres* (Figure 5 and Figure S2), neither in local treated nor in systemic tissues.

In *Arabidopsis*, SAR appears to be regulated by two parallel signalling pathways that are inter-dependent. One of these pathways depends on SA, the other one on AzA, G3P, reactive oxygen species, and nitric oxide [26,74]. While the SA pathway seems to be effective in barley at least against *Bgh* ([46] and Figure 4), the function of the other pathway, if existent in barley, is still unclear. Since SA did not affect barley resistance to *Xtc* while both Fol and to a minor extent AzA enhanced susceptibility rather than immunity to this hemi-biotrophic pathogen, it is possible that Fol and AzA influenced a SA-independent immune pathway. Such a pathway might rely on JA and/or ABA which are positively associated with barley defence against *Xtc* [47].

5. Conclusions

In this study, we investigated the role of resistance-inducing compounds from *Arabidopsis*, salicylic acid (SA), folic acid, and azelaic acid, on barley defence against the pathogens *Xanthomonas translucens*, *Bgh* (powdery mildew), and *Pyrenophora teres*. Azelaic acid appeared to induce local susceptibility to *X. translucens* and at the same time systemic resistance to powdery mildew. Also, we observed a possible activation of local and/or systemic resistance to powdery mildew after application of SA and folic acid. Because folic acid and azelaic acid, which both enhance SA-mediated immune responses in *Arabidopsis*, both enhance barley resistance to powdery mildew similarly to SA, the associated barley immune response might be related to SA. Importantly, the data show that folic acid- and azelaic acid-induced resistance is a double-edged sword that can at the same time induce resistance and susceptibility against different pathogens (powdery mildew and *X. translucens*). Also, folic acid and azelaic acid differentially influence the responses of *Arabidopsis* and barley to host-adapted hemi-biotrophic bacterial pathogens. Thus, investigating induced defence responses in barley and studying the signalling pathways used to achieve resistance in this cereal crop will be challenging topics for future research.

Author Contributions: Conceptualization, M.L., M.W. and A.C.V.; Data curation, M.L., M.W., F.M. and F.H.; Funding acquisition, A.C.V.; Investigation, M.L., M.W., F.M. and F.H.; Methodology, M.L., M.W. and F.M.; Supervision, A.C.V.; Writing—original draft, M.L.; Writing—review & editing, A.C.V.

Acknowledgments: We thank Ralph Hückelhoven for helpful discussion, the Patrick Schweizer lab (IPK Gatersleben) for *Bgh* and advice on inoculation methods, Eva Rudolf for help with statistics, and Florian Hug, Claudia Knappe, and Sanjukta Dey for their help with and valuable input on the experiments in this paper.

References

1. Panstruga, R.; Parker, J.E.; Schulze-Lefert, P. SnapShot: Plant Immune Response Pathways. *Cell* **2009**, *136*, 978.e1–978.e3. [CrossRef] [PubMed]
2. Jones, J.D.G.; Dangl, J.L. The plant immune system. *Nature* **2006**, *444*, 323–329. [CrossRef] [PubMed]
3. Glazebrook, J. Contrasting Mechanisms of Defense against Biotrophic and Necrotrophic Pathogens. *Annu. Rev. Phytopathol.* **2005**, *43*, 205–227. [CrossRef] [PubMed]
4. Spoel, S.H.; Dong, X. How do plants achieve immunity? Defence without specialized immune cells. *Nat. Rev. Immunol.* **2012**, *12*, 89–100. [CrossRef] [PubMed]
5. Zipfel, C.; Felix, G. Plants and animals: A different taste for microbes? *Curr. Opin. Plant Biol.* **2005**, *8*, 353–360. [CrossRef] [PubMed]
6. Henry, E.; Yadeta, K.A.; Coaker, G. Recognition of bacterial plant pathogens: Local, systemic and transgenerational immunity. *New Phytol.* **2013**, *199*, 908–915. [CrossRef] [PubMed]
7. Klessig, D.F.; Choi, H.W.; Dempsey, M.A. Systemic Acquired Resistance and Salicylic Acid: Past, Present and Future. *Mol. Plant-Microbe Interact.* **2018**. [CrossRef] [PubMed]
8. Vlot, A.C.; Dempsey, D.A.; Klessig, D.F. Salicylic Acid, a Multifaceted Hormone to Combat Disease. *Annu. Rev. Phytopathol.* **2009**, *47*, 177–206. [CrossRef] [PubMed]
9. Fu, Z.Q.; Dong, X. Systemic Acquired Resistance: Turning Local Infection into Global Defense. *Annu. Rev. Plant Biol.* **2013**, *64*, 839–863. [CrossRef] [PubMed]
10. Pajerowska-Mukhtar, K.M.; Emerine, D.K.; Mukhtar, M.S. Tell me more: Roles of NPRs in plant immunity. *Trends Plant Sci.* **2013**, *18*, 402–411. [CrossRef] [PubMed]
11. Van Loon, L.C.; Rep, M.; Pieterse, C.M.J. Significance of Inducible Defense-related Proteins in Infected Plants. *Annu. Rev. Phytopathol.* **2006**, *44*, 135–162. [CrossRef] [PubMed]
12. Shah, J.; Chaturvedi, R.; Chowdhury, Z.; Venables, B.; Petros, R.A. Signaling by small metabolites in systemic acquired resistance. *Plant J.* **2014**, *79*, 645–658. [CrossRef] [PubMed]
13. Conrath, U.; Beckers, G.J.M.; Flors, V.; García-Agustín, P.; Jakab, G.; Mauch, F.; Newman, M.-A.; Pieterse, C.M.J.; Poinssot, B.; Pozo, M.J.; et al. Priming: Getting Ready for Battle. *Mol. Plant-Microbe Interact.* **2006**, *19*, 1062–1071. [CrossRef] [PubMed]
14. Gourbal, B.; Pinaud, S.; Beckers, G.J.M.; Van Der Meer, J.W.M.; Conrath, U.; Netea, M.G. Innate immune memory: An evolutionary perspective. *Immunol. Rev.* **2018**, *283*, 21–40. [CrossRef] [PubMed]
15. Riedlmeier, M.; Ghirardo, A.; Wenig, M.; Knappe, C.; Koch, K.; Georgii, E.; Dey, S.; Parker, J.E.; Schnitzler, J.-P.; Vlot, A.C. Monoterpenes Support Systemic Acquired Resistance within and between Plants. *Plant Cell* **2017**, *29*, 1440–1459. [CrossRef] [PubMed]
16. Park, S.W.; Kaimoyo, E.; Kumar, D.; Mosher, S.; Klessig, D.F. Methyl salicylate is a critical mobile signal for plant systemic acquired resistance. *Science* **2007**, *318*, 113–116. [CrossRef] [PubMed]
17. Chanda, B.; Xia, Y.; Mandal, M.K.; Yu, K.; Sekine, K.T.; Gao, Q.M.; Selote, D.; Hu, Y.; Stromberg, A.; Navarre, D.; et al. Glycerol-3-phosphate is a critical mobile inducer of systemic immunity in plants. *Nat. Genet.* **2011**, *43*, 421–429. [CrossRef] [PubMed]
18. Wang, C.; Liu, R.; Lim, G.-H.; De Lorenzo, L.; Yu, K.; Zhang, K.; Hunt, A.G.; Kachroo, A.; Kachroo, P. Pipecolic acid confers systemic immunity by regulating free radicals. *Sci. Adv.* **2018**, *4*, eaar4509. [CrossRef] [PubMed]

19. Jung, H.W.; Tschaplinski, T.J.; Wang, L.; Glazebrook, J.; Greenberg, J.T. Priming in Systemic Plant Immunity. *Science* **2009**, *324*, 89–91. [CrossRef] [PubMed]
20. Chaturvedi, R.; Venables, B.; Petros, R.A.; Nalam, V.; Li, M.; Wang, X.; Takemoto, L.J.; Shah, J. An abietane diterpenoid is a potent activator of systemic acquired resistance. *Plant J.* **2012**, *71*, 161–172. [CrossRef] [PubMed]
21. Hartmann, M.; Zeier, T.; Bernsdorff, F.; Reichel-Deland, V.; Kim, D.; Hohmann, M.; Scholten, N.; Schuck, S.; Bräutigam, A.; Hölzel, T.; et al. Flavin Monooxygenase-Generated N-Hydroxypipecolic Acid Is a Critical Element of Plant Systemic Immunity. *Cell* **2018**, *173*, 456–469. [CrossRef] [PubMed]
22. Chen, Y.-C.; Holmes, E.C.; Rajniak, J.; Kim, J.-G.; Tang, S.; Fischer, C.R.; Mudgett, M.B.; Sattely, E.S. N-hydroxy-pipecolic acid is a mobile metabolite that induces systemic disease resistance in *Arabidopsis*. *Proc. Natl. Acad. Sci. USA* **2018**, *115*, E4920–E4929. [CrossRef] [PubMed]
23. Maldonado, A.M.; Doerner, P.; Dixon, R.A.; Lamb, C.J.; Cameron, R.K. A putative lipid transfer protein involved in systemic resistance signalling in *Arabidopsis*. *Nature* **2002**, *419*, 399–403. [CrossRef] [PubMed]
24. Champigny, M.J.; Isaacs, M.; Carella, P.; Faubert, J.; Fobert, P.R.; Cameron, R.K. Long distance movement of DIR1 and investigation of the role of DIR1-like during systemic acquired resistance in *Arabidopsis*. *Front. Plant Sci.* **2013**, *4*, 230. [CrossRef] [PubMed]
25. Vlot, A.C.; Pabst, E.; Riedlmeier, M. Systemic Signalling in Plant Defence. *eLS* **2017**, 1–9. [CrossRef]
26. Gao, Q.-M.; Zhu, S.; Kachroo, P.; Kachroo, A. Signal regulators of systemic acquired resistance. *Front. Plant Sci.* **2015**, *6*, 228. [CrossRef] [PubMed]
27. Métraux, J.P.; Signer, H.; Ryals, J.; Ward, E.; Wyss-Benz, M.; Gaudin, J.; Raschdorf, K.; Schmid, E.; Blum, W.; Inverardi, B. Increase in salicylic acid at the onset of systemic acquired resistance in cucumber. *Science* **1990**, *250*, 1004–1006. [CrossRef] [PubMed]
28. Navarova, H.; Bernsdorff, F.; Doring, A.-C.; Zeier, J. Pipecolic Acid, an Endogenous Mediator of Defense Amplification and Priming, Is a Critical Regulator of Inducible Plant Immunity. *Plant Cell* **2012**, *24*, 5123–5141. [CrossRef] [PubMed]
29. Lim, G.H.; Shine, M.B.; de Lorenzo, L.; Yu, K.; Cui, W.; Navarre, D.; Hunt, A.G.; Lee, J.Y.; Kachroo, A.; Kachroo, P. Plasmodesmata Localizing Proteins Regulate Transport and Signaling during Systemic Acquired Immunity in Plants. *Cell Host Microbe* **2016**, *19*, 541–549. [CrossRef] [PubMed]
30. Yu, K.; Soares, J.; Mandal, M.K.; Wang, C.; Chanda, B.; Gifford, A.N.; Fowler, J.S.; Navarre, D.; Kachroo, A.; Kachroo, P. A feed-back regulatory loop between glycerol-3-phosphate and lipid transfer proteins DIR1 and AZI1 mediates azelaic acid-induced systemic immunity. *Cell Rep.* **2013**, *3*, 1266–1278. [CrossRef] [PubMed]
31. Cecchini, N.M.; Steffes, K.; Schläppi, M.R.; Gifford, A.N.; Greenberg, J.T. Arabidopsis AZI1 family proteins mediate signal mobilization for systemic defence priming. *Nat. Commun.* **2015**, *6*, 7658. [CrossRef] [PubMed]
32. Meuwly, P.; Molders, W.; Buchala, A.; Metraux, J.-P. Local and Systemic Biosynthesis of Salicylic Acid in lnfected Cucumber Plants. *Plant Phisiology* **1995**, *109*, 1107–1114. [CrossRef]
33. Vernooij, B.; Friedrichya, L.; Reist, R.; Kolditzjawhar, R.; Ward, E.; Uknes, S.; Kessmann, H.; Ryals, J. Salicylic Acid Is Not the Translocated Signal Responsible for lnducing Systemic Acquired Resistance but Is Required in Signal Transduction. *Plant Cell* **1994**, *6*, 959–965. [CrossRef] [PubMed]
34. Ding, P.; Rekhter, D.; Ding, Y.; Feussner, K.; Busta, L.; Haroth, S.; Xu, S.; Li, X.; Jetter, R.; Feussner, I.; et al. Characterization of a Pipecolic Acid Biosynthesis Pathway Required for Systemic Acquired Resistance. *Plant Cell* **2016**, *28*, 2603–2615. [CrossRef] [PubMed]
35. Wittek, F.; Kanawati, B.; Wenig, M.; Hoffmann, T.; Franz-Oberdorf, K.; Schwab, W.; Schmitt-Kopplin, P.; Vlot, A.C. Folic acid induces salicylic acid-dependent immunity in *Arabidopsis* and enhances susceptibility to *Alternaria brassicicola*. *Mol. Plant Pathol.* **2015**, *16*, 616–622. [CrossRef] [PubMed]
36. Song, G.C.; Choi, H.K.; Ryu, C.M. The folate precursor para-aminobenzoic acid elicits induced resistance against Cucumber mosaic virus and *Xanthomonas axonopodis*. *Ann. Bot.* **2013**, *111*, 925–934. [CrossRef] [PubMed]
37. Balmer, D.; Planchamp, C.; Mauch-Mani, B. On the move: Induced resistance in monocots. *J. Exp. Bot.* **2013**, *64*, 1249–1261. [CrossRef] [PubMed]
38. Sharma, R.; De Vleesschauwer, D.; Sharma, M.K.; Ronald, P.C. Recent advances in dissecting stress-regulatory crosstalk in rice. *Mol. Plant* **2013**, *6*, 250–260. [CrossRef] [PubMed]

39. Wu, Y.; Yi, G.; Peng, X.; Huang, B.; Liu, E.; Zhang, J. Systemic acquired resistance in Cavendish banana induced by infection with an incompatible strain of *Fusarium oxysporum* f. sp. *cubense*. *J. Plant Physiol.* **2013**, *170*, 1039–1046. [CrossRef] [PubMed]
40. Yang, Y.; Zhao, J.; Liu, P.; Xing, H.; Li, C.; Wei, G.; Kang, Z. Glycerol-3-phosphate metabolism in wheat contributes to systemic acquired resistance against *Puccinia striiformis* f. sp. *tritici*. *PLoS ONE* **2013**, *8*, e81756. [CrossRef] [PubMed]
41. Ahmed, S.M.; Liu, P.; Xue, Q.; Ji, C.; Qi, T.; Guo, J.; Guo, J.; Kang, Z. TaDIR1-2, a Wheat Ortholog of Lipid Transfer Protein AtDIR1 Contributes to Negative Regulation of Wheat Resistance against *Puccinia striiformis* f. sp. *tritici*. *Front. Plant Sci.* **2017**, *8*, 521. [CrossRef] [PubMed]
42. Wang, X.; Wang, Y.; Liu, P.; Ding, Y.; Mu, X.; Liu, X.; Wang, X.; Zhao, M.; Huai, B.; Huang, L.; et al. TaRar1 Is Involved in Wheat Defense against Stripe Rust Pathogen Mediated by YrSu. *Front. Plant Sci.* **2017**, *8*, 156. [CrossRef] [PubMed]
43. Balmer, D.; De Papajewski, D.V.; Planchamp, C.; Glauser, G.; Mauch-Mani, B. Induced resistance in maize is based on organ-specific defence responses. *Plant J.* **2013**, *74*, 213–225. [CrossRef] [PubMed]
44. Morris, S.W.; Vernooij, B.; Titatarn, S.; Starrett, M.; Thomas, S.; Wiltse, C.C.; Frederiksen, R.A.; Bhandhufalck, A.; Hulbert, S.; Uknes, S. Induced Resistance Responses in Maize. *Mol. Plant-Microbe Interact.* **1998**, *11*, 643–658. [CrossRef] [PubMed]
45. Görlach, J.; Volrath, S.; Knauf-Beiter, G.; Hengy, G.; Beckhove, U.; Kogel, K.; Oostendorp, M.; Staub, T.; Ward, E.; Kessmann, H.; et al. Benzothiadiazole, a Novel Class of Inducers of Systemic Acquired Resistance, Activates Gene Expression and Disease Resistance in Wheat. *Plant Cell* **1996**, *8*, 629–643. [CrossRef] [PubMed]
46. Beßer, K.; Jarosch, B.; Langen, G.; Kogel, K.-H. Expression analysis of genes induced in barley after chemical activation reveals distinct disease resistance pathways. *Mol. Plant Pathol.* **2000**, *1*, 277–286. [CrossRef] [PubMed]
47. Dey, S.; Wenig, M.; Langen, G.; Sharma, S.; Kugler, K.G.; Knappe, C.; Hause, B.; Bichlmeier, M.; Babaeizad, V.; Imani, J.; et al. Bacteria-Triggered Systemic Immunity in Barley Is Associated with WRKY and ETHYLENE RESPONSIVE FACTORs But Not with Salicylic Acid. *Plant Physiol.* **2014**, *166*, 2133–2151. [CrossRef] [PubMed]
48. Jansen, C.; Korell, M.; Eckey, C.; Biedenkopf, D.; Kogel, K.-H. Identification and transcriptional analysis of powdery mildew-induced barley genes. *Plant Sci.* **2005**, *168*, 373–380. [CrossRef]
49. Thordal-Christensen, H.; Gregersen, P.L.; Collinge, D.B. The Barley/*Blumeria* (Syn. Erysiphe) *Graminis* Interaction. In *Mechanisms of Resistance to Plant Diseases*; Springer: Dordrecht, The Netherlands, 2000; pp. 77–100.
50. Hückelhoven, R.; Panstruga, R. Cell biology of the plant-powdery mildew interaction. *Curr. Opin. Plant Biol.* **2011**, *14*, 738–746. [CrossRef] [PubMed]
51. Dean, R.; Van Kan, J.A.L.; Pretorius, Z.A.; Hammond-Kosack, K.E.; Di Pietro, A.; Spanu, P.D.; Rudd, J.J.; Dickman, M.; Kahmann, R.; Ellis, J.; et al. The Top 10 fungal pathogens in molecular plant pathology. *Mol. Plant Pathol.* **2012**, *13*, 414–430. [CrossRef] [PubMed]
52. Vallelian-Bindschedler, L.; Mösinger, E.; Métraux, J.P.; Schweizer, P. Structure, expression and localization of a germin-like protein in barley (*Hordeum vulgare* L.) that is insolubilized in stressed leaves. *Plant Mol. Biol.* **1998**, *37*, 297–308. [CrossRef] [PubMed]
53. Hückelhoven, R.; Fodor, J.; Preis, C.; Kogel, K.-H. Hypersensitive Cell Death and Papilla Formation in Barley Attacked by the Powdery Mildew Fungus Are Associated with Hydrogen Peroxide but Not with Salicylic Acid Accumulation1. *Plant Physiol.* **1999**, *119*, 1251–1260. [CrossRef] [PubMed]
54. Jain, S.K.; Langen, G.; Hess, W.; Börner, T.; Hückelhoven, R.; Kogel, K.-H. The white barley mutant albostrians shows enhanced resistance to the biotroph *Blumeria graminis* f. sp. *hordei*. *Mol. Plant-Microbe Interact.* **2004**, *17*, 374–382. [CrossRef] [PubMed]
55. Kogel, K.H.; Ortel, B.; Jarosch, B.; Atzorn, R.; Schiffer, R.; Wasternack, C. Resistance in barley against the powdery mildew fungus (*Erysiphe graminis* f. sp. *hordei*) is not associated with enhanced levels of endogenous jasmonates. *Eur. J. Plant Pathol.* **1995**, *101*, 319–332. [CrossRef]
56. Cui, J.; Bahrami, A.K.; Pringle, E.G.; Hernandez-Guzman, G.; Bender, C.L.; Pierce, N.E.; Ausubel, F.M. Pseudomonas syringae manipulates systemic plant defenses against pathogens and herbivores. *Proc. Natl. Acad. Sci. USA* **2005**, *102*, 1791–1796. [CrossRef] [PubMed]

57. Yasuda, M.; Ishikawa, A.; Jikumaru, Y.; Seki, M.; Umezawa, T.; Asami, T.; Maruyama-Nakashita, A.; Kudo, T.; Shinozaki, K.; Yoshida, S.; et al. Antagonistic Interaction between Systemic Acquired Resistance and the Abscisic Acid-Mediated Abiotic Stress Response in *Arabidopsis*. *Plant Cell* **2008**, *20*, 1678–1692. [CrossRef] [PubMed]
58. Shigenaga, A.M.; Argueso, C.T. No hormone to rule them all: Interactions of plant hormones during the responses of plants to pathogens. *Semin. Cell Dev. Biol.* **2016**, *56*, 174–189. [CrossRef] [PubMed]
59. Pieterse, C.M.J.; Van der Does, D.; Zamioudis, C.; Leon-Reyes, A.; Van Wees, S.C.M. Hormonal Modulation of Plant Immunity. *Annu. Rev. Cell Dev. Biol.* **2012**, *28*, 489–521. [CrossRef] [PubMed]
60. Robert-Seilaniantz, A.; Grant, M.; Jones, J.D.G. Hormone Crosstalk in Plant Disease and Defense: More Than Just JASMONATE-SALICYLATE Antagonism. *Annu. Rev. Phytopathol.* **2011**, *49*, 317–343. [CrossRef] [PubMed]
61. Zhu, Z.; An, F.; Feng, Y.; Li, P.; Xue, L.; Mu, A.; Jiang, Z.; Kim, J.-M.; To, T.K.; Li, W.; et al. Derepression of ethylene-stabilized transcription factors (EIN3/EIL1) mediates jasmonate and ethylene signaling synergy in Arabidopsis. *Proc. Natl. Acad. Sci. USA* **2011**, *108*, 12539–12544. [CrossRef] [PubMed]
62. Beckers, G.J.M.; Spoel, S.H. Fine-tuning plant defence signalling: Salicylate versus jasmonate. *Plant Biol.* **2006**, *8*, 1–10. [CrossRef] [PubMed]
63. Spoel, S.H.; Johnson, J.S.; Dong, X. Regulation of tradeoffs between plant defenses against pathogens with different lifestyles. *Proc. Natl. Acad. Sci. USA* **2007**, *104*, 18842–18847. [CrossRef] [PubMed]
64. Liu, Z.; Ellwood, S.R.; Oliver, R.P.; Friesen, T.L. *Pyrenophora teres*: Profile of an increasingly damaging barley pathogen. *Mol. Plant Pathol.* **2011**, *12*, 1–19. [CrossRef] [PubMed]
65. Vlot, A.C.; Liu, P.P.; Cameron, R.K.; Park, S.W.; Yang, Y.; Kumar, D.; Zhou, F.; Padukkavidana, T.; Gustafsson, C.; Pichersky, E.; et al. Identification of likely orthologs of tobacco salicylic acid-binding protein 2 and their role in systemic acquired resistance in *Arabidopsis thaliana*. *Plant J.* **2008**, *56*, 445–456. [CrossRef] [PubMed]
66. Delventhal, R.; Rajaraman, J.; Stefanato, F.L.; Rehman, S.; Aghnoum, R.; McGrann, G.R.D.; Bolger, M.; Usadel, B.; Hedley, P.E.; Boyd, L.; et al. A comparative analysis of nonhost resistance across the two Triticeae crop species wheat and barley. *BMC Plant Biol.* **2017**, *17*, 232. [CrossRef] [PubMed]
67. Nair, K.R.S.; Ellingboe, A.H. A method of controlled inoculations with conidiospores of *Erysiphe graminis* var. *tritici*. *Phytopathology* **1962**, *52*, 714.
68. Foissner, I.; Wendehenne, D.; Langebartels, C.; Durner, J. In vivo imaging of elicitor-induced nitric oxide burst in tobacco. *Plant J.* **2000**, *23*, 817–824. [CrossRef] [PubMed]
69. Laflamme, B.; Middleton, M.; Lo, T.; Desveaux, D.; Guttman, D.S. Image-Based Quantification of Plant Immunity and Disease. *Mol. Plant-Microbe Interact.* **2016**, *29*, MPMI-07-16-0129. [CrossRef] [PubMed]
70. Kojima, H.; Urano, Y.; Kikuchi, K.; Higuchi, T.; Hirata, Y.; Nagano, T. Fluorescent indicators for imaging nitric oxide production. *Angew. Chem. Int. Ed.* **1999**, *38*, 3209–3212. [CrossRef]
71. Prats, E.; Mur, L.A.J.; Sanderson, R.; Carver, T.L.W. Nitric oxide contributes both to papilla-based resistance and the hypersensitive response in barley attacked by *Blumeria graminis* f. sp. *hordei*. *Mol. Plant Pathol.* **2005**, *6*, 65–78. [CrossRef] [PubMed]
72. Prats, E.; Carver, T.L.W.; Mur, L.A.J. Pathogen-derived nitric oxide influences formation of the appressorium infection structure in the phytopathogenic fungus *Blumeria graminis*. *Res. Microbiol.* **2008**, *159*, 476–480. [CrossRef] [PubMed]
73. Lin, M.-R.; Edwards, H.H. Primary Penetration Process in Powdery Mildewed Barley Related to Host Cell Age, Cell Type, and Occurrence of Basic Staining Material. *New Phytol.* **1974**, *73*, 131–137. [CrossRef]
74. Wendehenne, D.; Gao, Q.; Kachroo, A.; Kachroo, P. Free radical-mediated systemic immunity in plants. *Curr. Opin. Plant Biol.* **2014**, *20*, 127–134. [CrossRef] [PubMed]

13

Unraveling Field Crops Sensitivity to Heat Stress: Mechanisms, Approaches and Future Prospects

Muhammad Nadeem †, Jiajia Li †, Minghua Wang, Liaqat Shah, Shaoqi Lu, Xiaobo Wang * and Chuanxi Ma

School of Agronomy, Anhui Agricultural University, Hefei 230000, China; rananadeem.aaur@yahoo.com (M.N.); lijia6862@ahau.edu.cn (J.L.); minghuawang.ahua@gmail.com (M.W.); laqoo@yahoo.com (L.S.); 15102292@ahau.edu.cn (S.L.); machuanxi@ahau.edu.cn (C.M.)

* Correspondence: wangxiaobo@ahau.edu.cn

† These authors contributed equally to this work.

Abstract: The astonishing increase in temperature presents an alarming threat to crop production worldwide. As evident by huge yield decline in various crops, the escalating drastic impacts of heat stress (HS) are putting global food production as well as nutritional security at high risk. HS is a major abiotic stress that influences plant morphology, physiology, reproduction, and productivity worldwide. The physiological and molecular responses to HS are dynamic research areas, and molecular techniques are being adopted for producing heat tolerant crop plants. In this article, we reviewed recent findings, impacts, adoption, and tolerance at the cellular, organellar, and whole plant level and reported several approaches that are used to improve HS tolerance in crop plants. Omics approaches unravel various mechanisms underlying thermotolerance, which is imperative to understand the processes of molecular responses toward HS. Our review about physiological and molecular mechanisms may enlighten ways to develop thermo-tolerant cultivars and to produce crop plants that are agriculturally important in adverse climatic conditions.

Keywords: heat stress; thermotolerance; oxidative stress; heat shock proteins; QTLs; plant omics

---

1. Introduction

In recent years, temperature extremes and weather disasters have partially or completely damaged regional crop production [1–3]. The annual worldwide temperature has been increasing steadily, and is expected to be increased by 1.8–4.0 °C by the end of the 21st century [4]. This increasing trend in temperature creating curiosity among researchers, as temperature has an impact on life on earth, acting directly or indirectly. Regardless of these encounters, global food production will have to rise by 70% to meet the mandate of an expected rise in population growth to 9 billion by 2050 [5].

Plants as sessile organisms and cannot change their position or move to more suitable climatic conditions; therefore, plant activities are extensively affected by heat stress (HS), which often leads to mortally [6]. Specifically, HS has an impact on a number of various plant species [7]. HS significantly affect plant activities like seed germination, plant development, photosynthesis, and reproduction, which have a devastating impact on the overall yield of a crop [8]. It has been observed that HS leads to inhibition of pollen grain swelling leading to perturbed pollen dispersal and anther indehiscence during the reproductive process, which finally influence seed yield of rice [9]. Heat and drought are the key abiotic stresses to cereal crop production and resulted in the reduction of yield by 9% to 10% between 1964 and 2007 worldwide [2].

For survival in severe conditions, plants continuously struggle to modify their metabolic process in many ways in response to HS, specifically by generating key solutes that leads to establish proteins and osmotic adjustment and re-establish the redox balance of cell and homeostasis by modify the antioxidant system [10,11]. A plant in defense from HS causes modifications at the molecular level in the expression of genes [12]. In HS conditions, change in biochemical and physiological activities by gene expression alter gradually, resulting in the development of thermotolerance [13]. In order to successfully produce HS-tolerant crop varieties in the light of global climate change, there is need of knowledge and investigations about HS-tolerance mechanisms at physiological, biochemical, and molecular levels.

At present, investigations into selection strategy and breeding for thermotolerant cultivars and understanding of heat tolerance mechanisms are more required today than ever before. Molecular and genetic mechanisms for avoiding HS-induced harmful changes play a crucial role in plant survival under such circumstances. In the present scenario of global warming, the major challenge for plant scientists is to develop new crop varieties tolerant to HS [14]. In the coming years, agricultural production will have to deal with growing crops under sub-optimal conditions accompanied by increased food demand, creating a gap between the current yield achievements and yield potential [15]. Developing genetically modified plants through target genes manipulation, QTLs, and omics techniques are widely studied molecular approaches in recent years. Our study about sensitivity, adaptations, mechanisms, and approaches may uncover ways to develop thermo-tolerant cultivars and to produce crop plants that are agriculturally important in adverse climatic conditions.

2. Plant Sensitivity to Heat Stress

Plant sensitivity to HS varies with duration, plant type, and the degree of temperature. Plant growth and development are greatly influenced by the series of morphological, physiological, and biochemical changes resulting from HS [16]. HS cause devastating impacts on crop plants by affecting vital physiological functions, including protein denaturation, increase in membrane fluidity, level of reactive oxygen species (ROS), decline in photosystem II (PSII)-mediated electron transport, as well as inactivation of chloroplast and mitochondrial enzymes activities [17–19]. HS due to the rising global temperature is becoming one of the main limiting factor to crop productivity and has an adverse impact on plants (Figure 1). This rising temperature may cause a change in the morphology, physiology, and growing periods of plants.

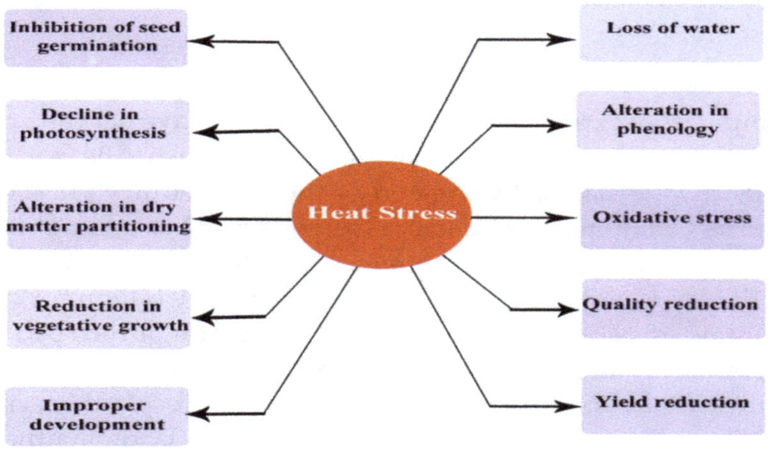

Figure 1. Impact of heat stress on plants.

2.1. *Morphological Responses*

2.1.1. Crop Growth and Development

During crop growth and development, temperature plays an important role in dry matter partitioning, transpiration [20,21], photosynthetic activity, respiration [22,23], and root and plant development [24]. The ideal conditions for plant growth and development generally occur within a different range of temperature [25], with low or high temperature (HT) reducing growth and developmental rates [26,27]. In winter cereals, temperature acts as a signal stimulator in processes of vernalization to induce flowering in plants [28]. It has been investigated that the increase in temperature to optimum thresholds stimulates biochemical mechanisms, consequently affecting development rates and declining the lengths of growing seasons [29]. The shorter developmental phases could have an adverse impact on the formation of yield components [30]. In Germany, between 1959 and 2009, lengths of growing seasons of oats reduced by about two weeks, leading to an earlier occurrence of phonological phases due to by HT [31]. As HT trigger development of the crop, the phases of crop growth duration decline, producing a destructive impact on yield in field crops and final grain weight [30]. It is observed that temperature has negative effect on growth and phototropism of *Arabidopsis thaliana* (L.) seedlings [32]. Van Der Ploeg and Heuvelink [33] reviewed the Influence of sub-optimal temperature on growth and yield of tomato. In a recent study, Yang et al. [34] investigated the effects of different growth temperatures on growth, development, and plastid pigments metabolism of tobacco (*Nicotiana tabacum* L.) plants.

2.1.2. Reproductive Development

HS has the widest and most far-reaching effects on plant reproductive organ, seed weight, and number of seeds, but regulation of heat-shock responses in inflorescence is largely uncharacterized [32,33]. It has been reported that male and female organs are most sensitive to extreme temperature, especially ≥30 °C [34] (Figure 2a). HS damages both male and female gametophytes, resulting in decreased pollen viability, reduced pollen germination, pollen tube growth inhibition, stigma receptivity reduction and reduced ovule function, declined fertilization, limited embryogenesis, poor ovule viability, enhanced ovule abortion, and a decrease in yield [35,36] (Figure 2b).

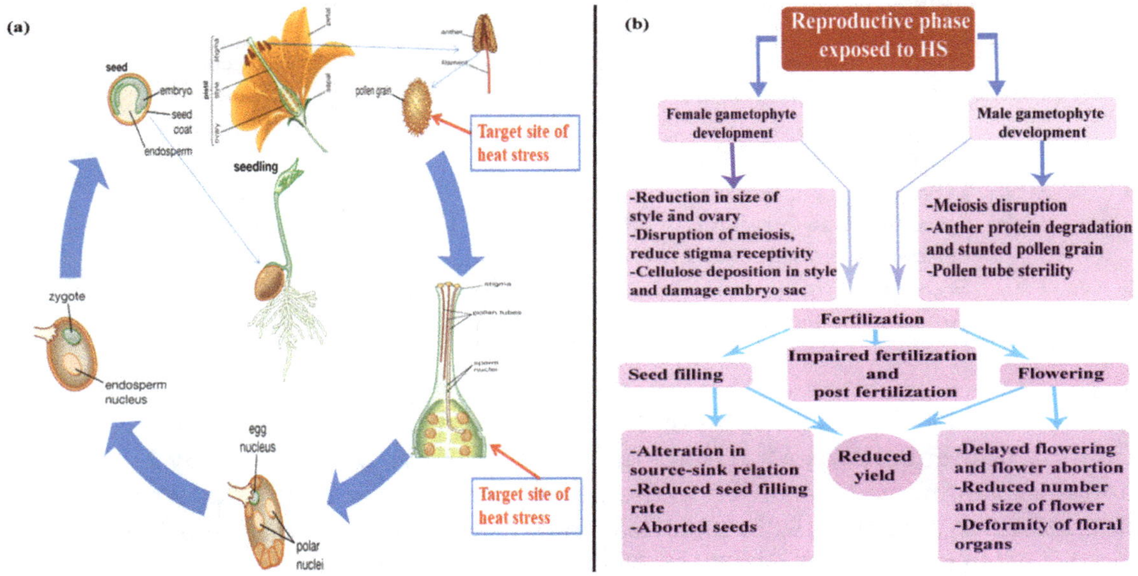

Figure 2. (**a**) Life cycle of an angiosperm representing target sites of heat stress (HS). Both male (pollen grains) and female (ovule) gametophytes are the main target sites of heat stress. (**b**) Sensitivity of reproductive phase to heat stress and consequences of heat stress on reproductive and grain-filling phases.

The number of seed year^{-1} increases with increasing air temperature from 16–28 °C, though it harshly declines with further increases in temperature [37]. During the period of grain filling, HS has a great impact on the quality and quantity of the final yield [38,39]. Enhancing temperatures from 25–31 °C increased the rates of grain filling, although the final yield steadily decreased due to shortening of the grain filling duration [40]. In wheat, at mean temperatures of 21/16 °C, the higher grain numbers are obtained [35]. It has been found that the increase in temperatures to crop-specific thresholds (for wheat 10–21 °C) enhance the rate of grain filling by enhancing cell-division rates in the tissue of endosperm and increasing rates of metabolism [41]. Dry matter partitioning, which is the product of the movement of photosynthetic assimilate from source-sink organs, is enhanced between 10–30 °C in winter-season cereals [42]. The devastating decrease in growth and development, harvest index (HI), and seed yield were found for various crops. The decline of grain number resulted from the impact of HT on meiosis and transfer of pollen during anthesis and ovaries growth during pre-anthesis periods. It has been reported that in pepper plants, HS resulted in a drop in sucrose concentration in fruits/flowers, though in cereals, sucrose helps the plant avoid ovary abortion under water stress conditions [43]. The reproductive phase is more sensitive to HS, and high temperatures are likely to coincide with anther dehiscence and gametophyte development, which resulted in a final yield reduction.

2.1.3. Yield

HS has an adverse impact on various processes of crop growth and development and final yield [44]. It has been reported that HS leads to yield losses in various crops [45] (Table 1). With a 1 °C increase in global temperature, global wheat production is projected to decrease between 4.1% and 6.4% [46]. It is declared that the increase in temperature expected with environmental variability is likely to decrease wheat yields [47]. Additionally, in wheat, an annual worldwide yield loss of 19 million tons is observed, costing $2.6 billion, due to climate variability from 1981 to 2002 [6]. Globally, temperature trend analysis from 1980 to 2008 has revealed about a 5.5% decrease in the yield of wheat [1]. Similarly, it has been found that temperatures beyond 34 °C enhance the rates of senescence based on nine years of satellite data of wheat grown in northern India, thereby resulting in a significant reduction in yield [48].

Table 1. Yield reduction due to heat stress (HS) in some major crops.

Crop	Yield Reduction (%)	Reference
Wheat (*Triticum aestivum* L.)	31	[49]
Maize (*Zea mays* L.)	45	[50]
Rice (*Oryza sativa* L.)	50	[51]
Soybean (*Glycine max* L.)	46	[52]
Canola (*Brassica napus* L.)	50	[53,54]
Peanut (*Arachis hypogaea* L.)	31	[55]
Srghumm (*Sorghum bicolor* L.)	44	[56]
Sunflower (*Helianthus annuus* L.)	10	[57]

In addition to wheat, in maize, thermal warming form 1981 to 2002 has triggered a reduction in yield up to 12 million tons year^{-1}, comparable to a loss of $1.2 billion [6]. By 2100, a reduction of yield of about 30% in maize is recorded in the US by means of the nonlinear temperature and yield analysis [52]. Similarly, about a 3.8% yield loss in maize was observed by worldwide temperature trends analyses using past data from 1980 to 2008 [58]. These studies suggest that a rise in temperature beyond 30 °C have adverse effects on rainfed maize in Africa and the US [59]. It is reported that extreme HS at enthesis could reduce maize yield globally by 45% by 2080 as compared to the 1980s [50].

In the case of soybean, a decline in yield of about 46% in the US before the year 2100 is predicted by non-linear and asymmetric temperature and yield relationship analyses [52]. It is reported that during 1976 to 2006 in the US, a future yield reduction of about 16% has been observed in soybean due to a change in patterns of temperature [60]. In barley, thermal warming from 1981 to 2002 caused a yield reduction of 8 million tons year^{-1}, with loss of around $1 billion [6]. Cucumber (*Cucumis sativus* L.) is one of the most important horticultural crops and is highly sensitive to HS, particularly at the vegetative stage [61,62].

2.2. Physiological Responses

2.2.1. Membrane Damage

In a plant cell, the most sensitive component is the plasma membrane, as it is the primary sites of injury under HS [63]. HS severely affects the structure and functions of the membrane, thereby increasing membranes fluidity due to denaturation of proteins and increased level of unsaturated fatty acids, causing a transition from solid gel to flexible crystalline liquid structure [64]. HS damage can be assessed by loss of membrane integrity due to structural modifications of component proteins, which enhances the thermostability of the membrane and organic and inorganic ions leakage from the cells [65]. Therefore, an electrolyte leakage value acts as a pointer of membrane injury and reflects stress-induced alterations and has been used to evaluate the thermostability of membranes under HS [66]. The increased permeability and leakage of ions out of the cell has been used as a measure of cell membrane stability and as a screen test for HS tolerance [67]. The effects of HS on membranes have been reported in various crops. In cotton, sorghum, and soybean, HS-induced serious membrane injury and membrane lipid peroxidation have been observed [51,68]. The increased permeability of the membrane and electrolyte leakage is noticed under HS in soybeans, which declined the capacity of the plasma membrane to hold solutes and water [69]. Similarly, in chickpeas, injury of the membrane was noticed at 40/30 °C, which was intensified at 45/35 °C, especially in sensitive genotypes [70]. A recent study investigated cell membrane stability under drought and heat conditions in wheat [67]. Membrane fluidity in temperature tolerance has been delineated by mutation analysis and transgenic and physiological studies. For instance, a soybean mutant deficient in fatty acid unsaturation exhibited high tolerance to HS [71]. Also, the thylakoid membranes of two *Arabidopsis* mutants deficient in fatty acid unsaturation (fad5 and fad6) exhibited increased stability to HS and increased lipid saturation in tobacco caused by silencing a ω-3 desaturase gene, which also rendered the plants more tolerant to HS [72,73]. Wheat lines of high membrane thermal stability tended to yield higher than lines of low membrane thermostability when grain filling occurred under harsh climate [74]. It is investigated that *HIT1* functions in the membrane trafficking that is involved in the thermal adaptation of the plasma membrane for tolerance to HS in plants [75]. In transgenic tobacco, overexpression of the *PpEXP1* gene exhibited a less structural damage to cells, lower electrolyte leakage, and lower levels of membrane lipid peroxidation compared to wild-type plants [76]. Researchers identified tolerant genotypes that are proved to be more productive under extreme field stress conditions. The thermostability of the membrane has been successfully employed to evaluate HS tolerance in several crops worldwide.

2.2.2. Photosynthesis and Respiration

Due to ongoing climate change, it has been reported that severe climatic conditions with long light exposure and HT have increased dramatically. It is already proven that, for life on earth, photosynthesis is a vital process and is often restricted by various abiotic stresses like HS and high light conditions. The negative impact of HS on plant growth and crop yield were mainly caused by its negative impacts on the photosynthetic process, which is the most thermosensitive aspects of plant functions [77]. The relative water content (RWC), chlorophyll content, and PSII activity decreased under high light and heat co-stresses [78]. It has been found that the PSII reaction center is the vital site where damage is incurred by several abiotic stresses in the photosynthesis systems of plants [79]. It is found that photosystem II is thought to be more highly responsive to HS or high light than photosystem I [80]. The photosynthetic process is very sensitive under HS conditions, and reduction in chlorophyll contents might be one of the main reasons for the decline in photosynthesis, as an enzyme chlorophyllase helps in conversion of chlorophyll into phytol and chlorophyllide [81]. Photosynthetic acclimatization to various climatic conditions represents a modification in photosynthesis and structures at each level [80,82] (Figure 3). In thylakoids, the proton gradient and non-photochemical quenching (NPQ) are the vital photo-defense process in photosystem I and photosystem II, respectively [83]. In plants, climatic stress generally results in a decrease in chlorophyll concentrations and a reduction of the photochemical reaction of thylakoid proteins [84–86]. It is reported that temperature is significantly affecting the photosynthetic activity of crops and photosynthetic pathways (C3 or C4 plants). In general, for cold season C3 crops, the temperature range for photosynthesis is between 0 °C to 30 °C, whereas the warm-adapted C4 plants that are grown in summer are photo-synthetically active between 7 °C to 40 °C temperature [22,81].

In plant species (C3), at current levels of CO_2 and light saturation, during the process of photo-phosphorylation, the photosynthetic response of plants to temperatures is measured by the availability of inorganic phosphates at a lower temperature, and it depends on the activity of Rubisco to fix atmospheric carbon in the optimum range of temperature. In plant species (C4) that are grown in a hot climate, the availability of Rubisco limits photosynthetic activity under low temperature, whereas at higher temperature in the thermal optimum level, it is unclear which mechanisms affect photosynthetic activity [87]. Photosynthetic rates decrease sharply as temperature increases past the optimum level [41]. This decreased rate of photosynthesis is related with declined light harvesting in photosystem II that results from cyclic electron flow [88], limitations in Rubisco, and thylakoid membrane instability [89]. In the process of photosynthesis, photoinhibition of PSII occur seven lower ranges of thermal stress [79]. However, some researchers reported slight or no harm to photosystem II due to moderate thermal stress [88].

HS decreases the rate of Photosystem II repair by the production of ROS across the thylakoid membrane [90], which is subject to influence by HS [91]. It has been found that the stability of the thylakoid membrane under HS, situated between 32–45 °C, is mainly determined by the stability of the double bonds of fatty acids of the membrane. Fatty acids double bonds decline due to excess generation of ROS under HS circumstances and increasing membrane electron leakage, thereby enhancing the denaturation of thylakoid membrane proteins [66]. It is described that maintenance respiration (turn-over of proteins complex) is high under HS, resulting in declining availability of assimilates for crop growth and development [92]. In maize, the increase in temperature from 18–33 °C raises the rates of maintenance respiration by greater than 80% [93]. Under elevated temperatures, the rate of respiration measurement could be an appropriate pointer for stimulation of plant response to HS, as the rate of respiration rises much more than the rate of photosynthesis initially decreases [25].

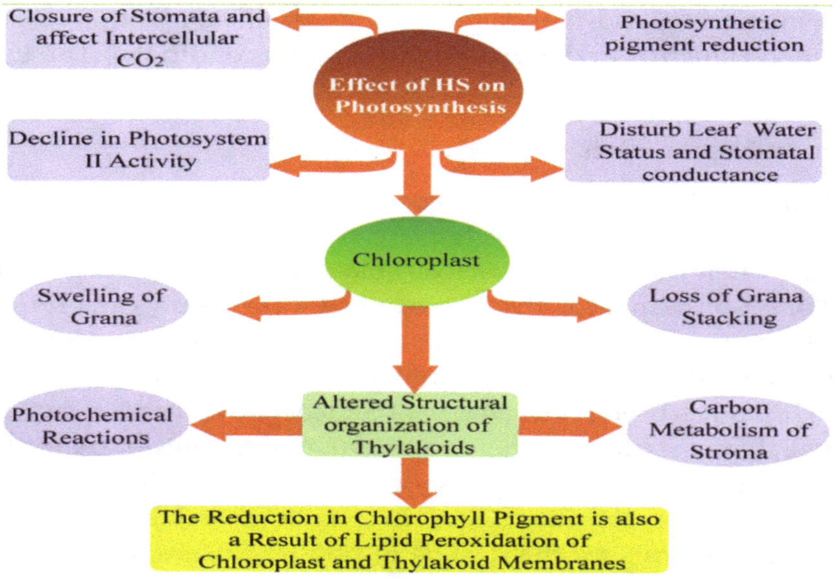

Figure 3. Decline in photosynthesis rate under heat stress. Heat stress leads to generation of reactive oxygen species, cell organelle and membrane damage, thylakoid membrane damage, thylakoid membrane lipid composition, swelling of grana, oxidative damage of cell organelle, and stomatal and non-stomatal limitations.

2.2.3. Water Relations

HS has the widest and most far-reaching effects on water relations leading to a severe decrease in yield potential in various crops. HS is often linked with rapid water loss from the surface of the plant, resulting in dehydration and ultimately leading to death [94]. The increase in transpiration and movement of water is an important tool for survival of the plant under extreme temperatures [8,95]. It is reported that HS influences plant–water relations due to the more rapid depletion of water from the soil, which affects the temperature of soil and transpiration [96]. HS directly and indirectly affects plant functions and leads to osmotic adjustments through impaired photosynthesis, enhanced respiration, a decline in leaf osmotic potentials, and decreased sugar concentration level [97]. Under HS, water loss during daytime was more common because of increased transpiration than night time, causing stress in snap bean (*Phaseolus vulgaris* L.) [98]. Under severe HS, high stomatal conductance boosts transpirational heat dissipation in tolerant genotypes of chickpea as long as soil water is available [99]. A rapid decrease in leaf tissue water contents was noted in sugarcane on exposure to extreme temperature despite the fact that an ample quantity of water was available in the soil [64]. In tobacco, stomatal conductance decreases markedly under severe HS, aggravating injury to leaves [68].

3. Oxidative Stress (OS)

Oxidative stress is a complex physiological and chemical phenomenon in plants and develops as a result of access production and accumulation of reactive oxygen species (ROS) under stress conditions. It has been reported that many metabolic pathways are subjected to depend upon different enzymes that are highly responsive to different ranges of HS. Like other abiotic stresses, HS might uncouple several enzymes and different metabolic pathways, which results in the accumulation of undesirable and dangerous ROS, generally hydrogen-peroxide (H_2O_2), hydroxyl-radical ($^{\bullet}OH$), superoxide-radical ($O_2^{\bullet-}$), and singlet-oxygen (O_2), which leads to OS [100]. The main sites of ROS production are reaction centers of photosystem I and photosystem II in chloroplasts, though ROS are also produced in other organelles, mitochondria, and peroxisomes [101] (Figure 4). A leaner relationship is present among accumulated ROS and highest efficiency of photosystem II. It is reported that because of heat injury to photosystem I and photosystem II under such HS condition, less absorption of photons occurs [102]. Among the ROS, photo-oxidation reactions lead to the generation of $O_2^{\bullet-}$ (flavoprotein, redox cycling), during mitochondrial ETCs reactions, in chloroplasts through Mehler reaction and glyoxisomal photorespiration, by xanthine oxidase, and NADPH oxidase in membrane polypeptides and plasmamembrane. A reaction between H_2O_2 and Fe^{2+} (Fenton reaction) and reaction between H_2O_2 and $O_2^{\bullet-}$ leads to the formation of hydroxyl radicals (Haber-Weiss reaction) and decomposition of O_3 in the apoplastic space region [103,104]. Photo-inhibition resulted in the formation of singlet oxygen and photosystem II electrons transfer reactions in chloroplast [104,105].

Due to HS, several physiological injuries occur in plants [102]. Hydroxyl radicals, which are produced during the process, can react with bio-molecules, pigments, lipids, proteins, DNA, and with all elements of the cell [104]. Protein denaturation as a result of thermal stress leads to OS through disruption of cell membrane stability and peroxidation of membrane lipids [106,107]. Photosynthetic activity of light reaction decreases under moderate HS is known to induce OS through the generation of ROS triggered by increased leakage of electrons from the membranes [108].

It is observed that the HT (33 °C) leads to OS in wheat, which can change the properties of membranes and cause the degradation of proteins and deactivation of enzymes that decrease the viability of cells. HS in wheat provokes OS and also amplifies the peroxidation of the membrane and decreases the thermal stability of the membrane by 28% and 54%, which enhances electrolytes leakage [109]. Premature leaf senescence in cotton was observed due to ROS generation by HS [54]. ROS accumulation at the outer surface of the plasma membrane due to continuous HS can cause membrane depolarization [110]. In such extreme cases, cell death can also occur because of ROS accumulation in cells [110]. ROS have a devastating impact on the metabolic processes of plants and triggered signaling behavior to activate the heat-shock responses toward the development of thermotolerance in crops [100]. Research into plant oxidative stress (OS) has shown huge potential for developing HS-tolerant crops.

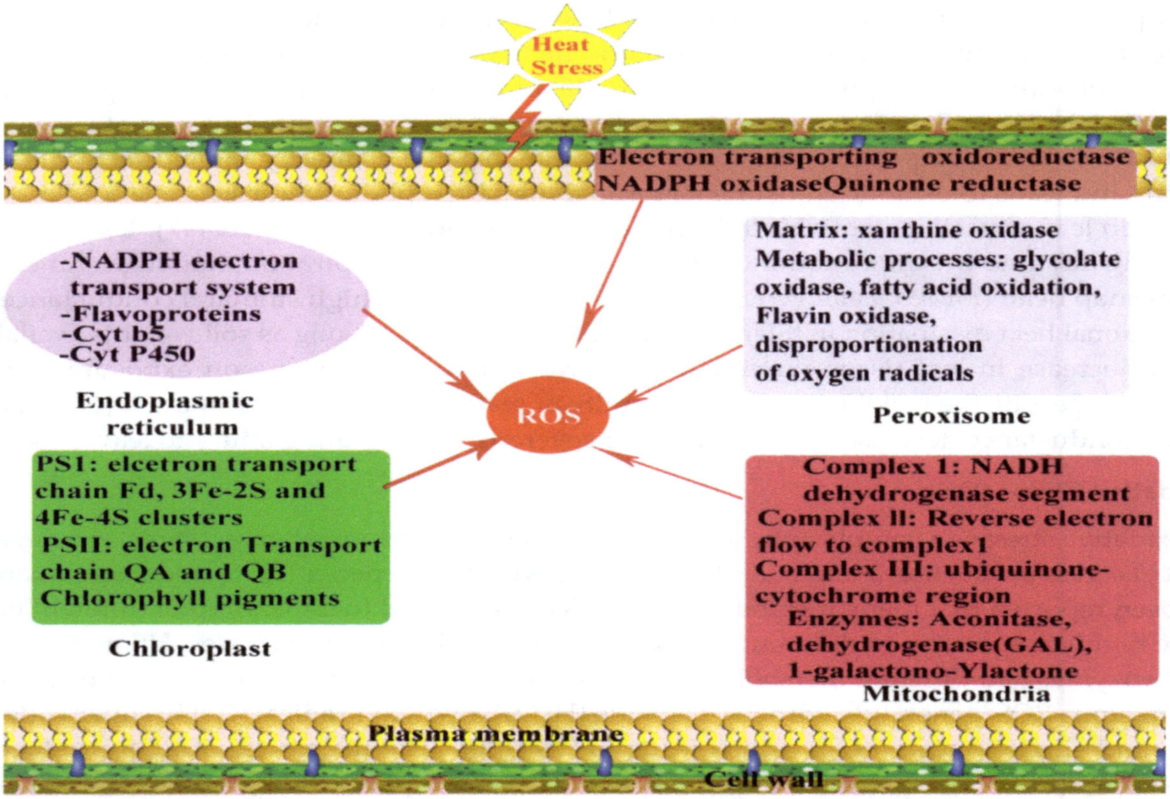

Figure 4. Reactive oxygen species (ROS) generate in plant cells as a consequence of heat stress. These reactive molecules are formed at different cellular sites, including chloroplasts, mitochondria, endoplasmic reticulum, peroxisomes, and at the extracellular side of the plasma membrane. The influence of ROS on cellular processes is mediated by both the perpetuation of their production and their amelioration by scavenging enzymes. The duration, location, and amplitude of production of ROS determine the specificity of the rapid responses under stress.

4. Avoidance and Tolerance Mechanism

Under HS conditions, plants show various survival mechanisms which include long-term morphological and phenological adaptations and short-term acclimation or avoidance mechanisms such as transpirational cooling, alteration of leaf orientation, or changing of membrane lipid compositions. Stomatal closure and water loss reduction, enhanced trichomatous and stomatal densities, and bigger xylem vessels are the main heat-induced features in crop plants [111]. Several plants growing in warm climatic conditions avoid HS by reducing the absorption of solar radiation. In order to avoid HS, some plants have small hairs on the leaf surface as well as cuticles (tomentose) and a waxy protective cover. The phenomenon called paraheliotropism occurs when plants often turn leaf blades away from sunlight, orient themselves parallel to solar rays, or roll their leaf blades. Under HS, early maturation is closely correlated with reduced yield losses in plants, which may be attributed to the engagement of an escape mechanism [107] (Figure 5). Such phenological and morphological adaptations are normally related with biochemical adaptations favoring net photosynthesis at HS (in particular CAM and C4 photosynthetic pathways), although C3 crop plants are also common in desert floras. HS can affect the degree of leaf rolling in many plants. The physiological role of leaf rolling is the maintenance of adaptation potential by enhancing the efficiency of water metabolism in the flag leaves of wheat under HS [112].

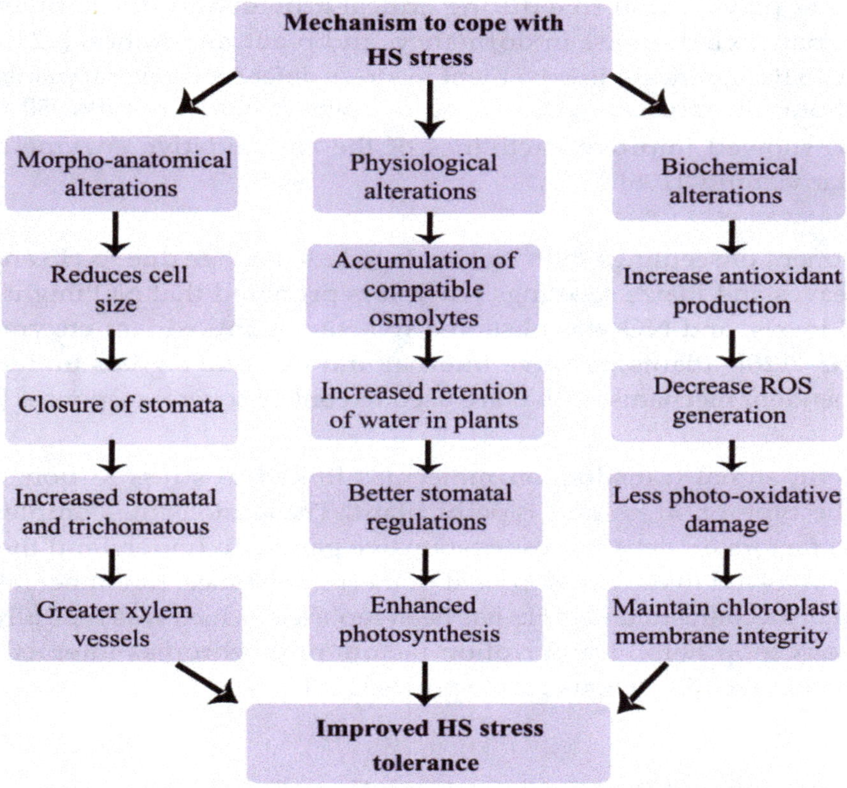

Figure 5. Flow chart of physiological, biochemical, and morphological adaptations of plants to deal with heat stress.

Thermotolerance is generally described as the capability of the plant to develop and grow to produce economic yield under HS. Plants have several survival mechanisms under HS conditions [113]. They included long-term adaptations or short-term acclimation/avoidance mechanisms. Some important mechanisms including ion transporter, late embryogenesis abundant proteins (LEA), antioxidant defense, osmoprotectants, and some factors linked with signaling cascade and transcriptional controls are fundamentally important to respond to HS [114] (Figure 5). It has been observed that several tissues in crop plants show differences in terms of developmental exposure, complexity, and responses toward the prevailing stress types [115]. These mechanisms help to regenerate homeostasis and to protect and repair damaged membranes and proteins [116].

5. Signaling in HS Tolerance

Signaling molecules (calcium, nitric oxide) play vital roles in conferring tolerance during HS conditions. Calcium as a divalent cation (Ca^{2+}) the important part of the cell membrane and cell wall and is an important intercellular messenger in the cytosol [117]. Plant roots absorbed Ca^{2+} from soil and are transported to shoots via xylem. The plasma membrane of a plant cell comprises Ca^{2+}-permeable ion channels through which Ca^{2+} enters [118]. Higher concentrations of Ca^{2+} are cytotoxic, therefore, an optimum concentration is to be maintained within cells. This is done by H^+/Ca^{2+} antiporters and Ca^{2+} ATPase [119,120]. The function of calcium in plants during HS has been debatable among scientists for quite some time. Some scientists doubt that the calcium chloride pretreatment leads to enhance thermotolerance and that calcium inhibitors limits plants [121,122]. Calcium and calmodulin treatment was able to induce HS transcriptional factors (HSFs) and heat shock elements (HSEs) binding in vitro, while, when treated with their inhibitors, they prevented binding at extreme HS [123]. Nitric oxide (NO) is an essential signaling molecule that controls several physiological processes and HS responses in crop plants. NO is membrane-permeable and highly reactive molecule that plays a vital role during critical growth and developmental stages such as germination, leaf expansion, decrease in dormancy, and plant maturation [124]. Recently, NO has gained courtesy due to its significant involvement in stress defense in several plants [8]. Application of 50 and 100 μM sodium nitroprusside at 33 °C on two wheat cultivars PBW550 (heat sensitive) and C306 (heat tolerant) showed improved activities of the antioxidative enzyme and increased heat tolerance and cellular viability [108].

In rice, pretreatment of seedlings with NO leads to less damage due to HS and enhanced rates of survival in wheat leaves and maize seedlings [125]. It is proposed that NO might be shielding plant by decreasing ROS levels, and NO was observed to activate antioxidant enzymes like APX, CAT, and SOD during HS [126]. Plants perceive internal and external signals under HS through many interlinked or independent mechanisms that are used to control several responses [127].

There are several signal transduction molecules linked to stress-responsive gene activation depending upon the type of stress and type of plant. These molecule complexes are interlinked with transcriptional factors for active stress-responsive genes. To understand the pathways and the signaling molecules involved in the development of thermotolerance, fundamental research is needed. The signaling complex mechanism under HS has been reported, which was typically found to comprise the basic bHLH (helix loop helix) transcription factors phytochromes interacting factor 4, whose ortholog has been recognized in various plant species [128].

6. Development of HS-Tolerant Plants using Molecular and Biotechnological Approaches

6.1. Quantitative Trait Loci (QTLs)

HS appears to be principally polygenic in nature, which might clarify why the genetic basis of HS tolerance in crop plants is inadequately understood [64,129,130]. In order to improve understanding about HS on a genetic basis, significant effort has been made to identify QTLs. Advances in DNA marker identification and genotyping assays have allowed the exact identification of the chromosomal position of the QTLs accountable for HS in crop plants [131–133]. A QTL study involving 90 introgression lines provided 5 QTLs enlightening PVs in the range of 6.83%–14.63% [134]. Likewise, Y106, an introgression line carrying 2 QTLs for HS tolerance (*qHTS1-1* and *qHTS3*), was identified while transferring genes from the wild rice (*O. rufipogon* Griff.) [134]. Additionally, three major QTLs were mapped on chromosomes 2B, 7B, and 7D in 148 RILs (NW1014 × HUW468) are associated with HS [135].

In wheat, nine QTLs were mapped on chromosomes (2A, 6A, 6B, 3A, 3B, and 7A) related to senescence [136]. In rice, a total of 14 QTLs related with a heat susceptibility index (HSI) were identified using parameters such as temperature depression (TD) of spike and spike yield [137]. Genome-wide association mapping as well as candidate-gene based strategies using SNP and diversity array technology (DArT) were helpful in chickpea to discover marker–trait associations for HS [138]. In tomato, six QTLs were identified for improving fruit set under HS [139].

In *Brassica campestris* L. ssp. *Pekinensis,* five QTLs related to HS has been identified [140]. In rice, a total of five QTLs (*qHTS1-1*, *qHTS1-2*, *qHTS2*, *qHTS3*, and *qHTS8*) were recognized on chromosomes 1, 2, 3, and 8 established on the heat response of the 90 inbred lines using 152 SSR polymorphic markers

[134]. Composite interval mapping identified a total of nine HSI mapped on linkage groups 2A, 2B, and 6D in wheat [141]. Similarly, they reported that the major QTL HSI (*Qhsigfd.iiwbr-2B*) for grain filling duration was 28.01 cM away from marker gwm257.

Major QTLs for HSI for grain filling duration had also been described on this chromosome [135]. Other QTLs identified on to this region were more obviously linked with heat tolerance for grain filling duration [141]. Furthermore, QTLs for HSI for grain filling duration were also stated on chromosomes 1D, 2A, 6D [142], and 2D (*Qlgfd.iiwbr-2D*) and 7A (*Qlgfd. iiwbr-7A*) [143].

In recent studies, three major QTLs encompassing *QHst.cph-3B.2*, *QHst.cph-3B.3*, and *QHst.cph-1D* exposed the presence of 12 potential genes having a direct role in heat tolerance in rice [144]. Similarly, five QTLs were recognized on chromosomes 3, 5, 9, and 12 against HS. Of these QTLs, two high-effect QTLs, one novel (*qSTIPSS9.1*) and one known (*qSTIY5.1/qSSIY5.2*), were mapped in less than 400 Kbp genomic regions, encompassing of 65 and 54 genes, respectively [145]. Despite the major advances in QTL mapping, there is still huge room for improvement.

6.2. Heat Stress Proteins (HSPs) and Heat Shock Factors (HSFs)

Heat stress proteins (HSPs) and heat shock factors (HSFs) are master players for HS tolerance in plants. Except for different biochemical and physiological mechanisms, molecular techniques are helping to understand the concept of thermotolerance in crops. HS is responsible for the up-regulation of several heat-inducible genes, commonly referred as "heat shock genes" (HSGs), which encode HSPs, and these products play important functions under stress conditions. These HSPs protect cells from the harmful impact of HS [146].

Proteomic analysis revealed that HS could down-regulate proteins playing roles in photosynthesis, energy, and metabolism and up-regulating the resistance-related proteins [147]. These proteins are grouped into five classes in plants, according to their molecular weight: Hsp100, Hsp90, Hsp70, Hsp60, and small heat-shock proteins (sHSPs) [148] (Table 2).

Successful transcription, translation and post-translational modification lead to produce functional HSPs to protect the plant cell and responsible for HS tolerance [8] (Figure 6). Tolerance against HS has been accomplished in plant species transferred with heat shock regulatory proteins.

HSPs with molecular weights of 100–104 kDa are categorized into the HSP100 family. In plants, HSP100 proteins are extensively studied for their functions in HS tolerance in plants [115,149,150]. Among the various *Arabidopsis* HSP100 proteins, the cytosolic form is important for HS tolerance but not normal growth [149]. HSP90 is the most abundant in the cytosolic HSP family in both prokaryotic and eukaryotic cells and is rapidly induced in response to HS conditions.

Under physiological conditions, HSP90 interconnect with several other intracellular proteins, including calmodulin, actin, tubulin, kinases, and receptor proteins [151–153]. HSP90 has been reported as a key regulator of normal growth and development in *Arabidopsis* and *Nicotiana benthamiana* L. [154–157]. Recently, over-expression of five Hsp90 genes of *Glycine max* in *Arabidopsis* resulted in reduction in lipid peroxidation and loss of chlorophyll, higher biomass production and pod setting under HS [158].

Table 2. Demonstration of basic role of major heat shock proteins in plants under heat stress.

HSPs Major Classes	MW (KDa)	Subcellular Localization	Proposed Response	Exampled Plants	References
Small HSP or HSP20	15–30	Cytosol, ER, mitochondria, chloroplast	Require for the development of chloroplasts during HS, avoiding aggregation, co-chaperone, and formation of high weight molecular complex (oligo-meric), which act as a matrix for stabilization of unfolded protein. HSP40, HSP70 and HSP100 are required for its release.	*Arabidopsis*, Rice, Soybean, Wheat	[159–162]
HSP60	57–69	Cytoplasm, mitochondria	Specialized folding machinery, which depends upon ATP, role in embryo, and seedling development in some plants, function as a chaperon in the post-translational assembly of multi-meric proteins.	*Arabidopsis*, Maize, bentgrass, Barley, Rye, Wheat, Creeping grass	[113,163,164]
HSP70	68–75	Cytosol, ER, nucleus, mitochondria, chloroplast	Assisting refolding and proteolytic degradation of abnormal proteins, preventing aggregation, primary stabilization of proteins, metabolic detoxification, ATP dependent release and binding.	*Arabidopsis*, Rice, *Brachypodium distachyon* L., Tobacco, Soybean, Citrus	[159,165–173]
HSP90	82–90	Cytosol, ER, nucleus, mitochondria, chloroplast	Co-regulation of thermal stress associated with signal transduction and accomplishes protein folding. Genetic buffering, metabolic detoxification, regulation of receptors, protein translocation. ATP is required for its function.	*Arabidopsis*, Chickpea, Pigeonpea	[159,174–176]
HSP100	100–104	Cytosol, mitochondria, chloroplast	Sustain the functional integrity of convinced major polypeptides, assisting to degrade irreversibly damaged polypeptides, re-solubilizing protein aggregates via interactions with the sHSP chaperone system; ATP is required for its function.	Rice, Maize, *Arabidopsis*	[150,151,157,159, 177–182]

Note: HSPs, Heat shock proteins; ER, Endoplasmic reticulum; sHSP, Small heat shock proteins; MW, Molecular weight.

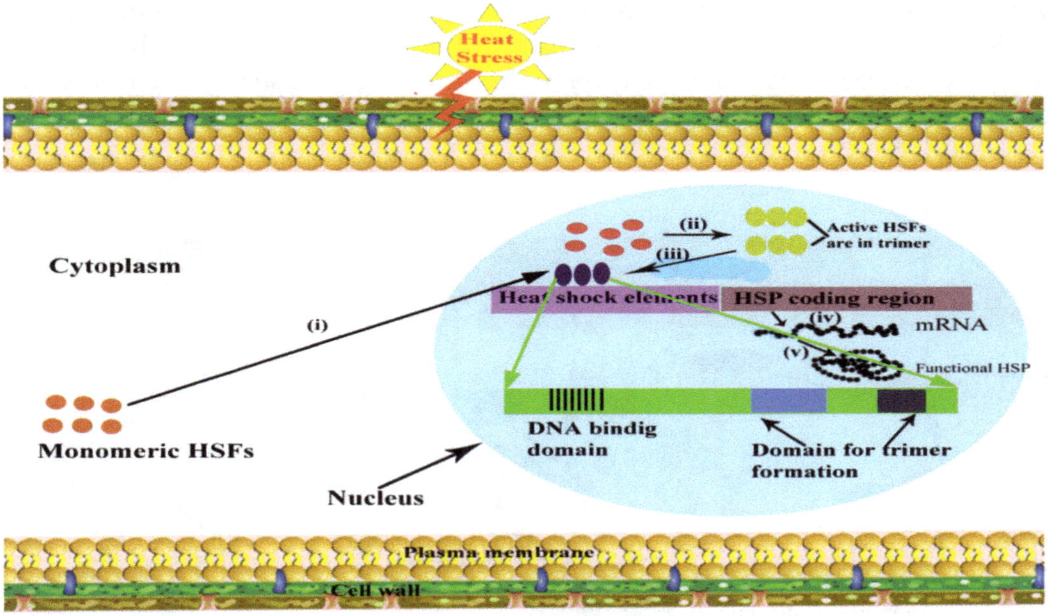

Figure 6. Illustration of the molecular regulatory mechanism of Heat Stress Proteins (HSPs) based on a hypothetical cellular model. Under heat stress, **(i)** monomeric Heat Shock Factors (HSFs) are entering into the nucleus from cytoplasm. **(ii)** HSF monomers are form active trimer in the nucleus, **(iii)** that will bind to the Heat Shock Element (HSE) of the respective Heat Shock Gene (HSG). Molecular dissection of the HSF binding region of HSE showing that it contains two domains for trimerization of HSFs and one DNA binding domain. **(iv)** Successful transcription, translation and post-translational modifications, **(v)** lead to produce functional HSPs (*Adapted and modified from* [8]).

HSP70 represents one of the most conserved classes of the heat shock proteins family. In plants and animals, HSP70 ensures proper protein folding during transfer to their final location and functions as a chaperone for newly synthesized proteins to prevent their accumulation as aggregates. HSP70 is the most highly conserved chaperones noted in plants, bacteria, and animals. These proteins associate with various other chaperones in a wide network and are implicated in diverse cellular activities [183]. Over-expression of HSP70 (HSP70-1) prevented degradation and fragmentation of nuclear DNA during HS conditions in tobacco [184]. Rice mitochondrial HSP70 over-expression resulted in the minor production of heat-induced ROS, suppressed programmed cell death, and higher mitochondrial membrane potential [184]. Chrysanthemum HSP70 expression in *A. thaliana* improved the tolerance against drought, salinity, and heat stresses [185].

Similar to other HSPs, sHSPs function as molecular chaperones, assisting refolding of denatured proteins and preventing undesired protein–protein interactions [151]. HSP20 or sHSPs are expressed in maximal amounts under HS conditions [160]. HSP20, a representative sHSP, maintains denatured proteins in a folding competent state and allows subsequent ATP-dependent disaggregation through the HSP70/90 chaperone system [166]. *Arabidopsis* plants over-expressing HSP17.5 of *Nelumbo nucifera* Gaertn., HSP17.8 of *Rosa chinesis* Jacq., HSP22 of *Zea mays* L., HSP26 of *Saccharomyces cerevisiae*, and HSP16.45 of *Lilium davidii* var. unicolor, showed heat tolerance to varying extents [184,186]. Tobacco plants over-expressing HSP16.9 of *Z. mays* resulted in increased early seed growth [184].

The research on plant HSFs regulation mainly emphasizes four levels, including transcriptional, post transcriptional, translational, and post-translation levels [187] (Figure 7). In transcription, the function of a gene can be regulated by binding of specific transcription factors (TFs) to the cis-acting elements located on the regulatory regions of its promoter [188]. Nishizawa-Yokoi et al. [188] reported that in *Arabidopsis* AtHSFA1d and A1e binding to the HSE in the 5′-flanking region of *AtHSFA2* gene is involved in high light (HL)-inducible *HSFA2* expression, triggering *AtHSFA2* transcription.

Dehydration-responsive element (DRE)-binding protein 2A (*DREB2A* gene) in the *Arabidopsis* directly regulates *AtHSFA3* transcription via binding the two DRE core elements in the *AtHSFA3* promoter under HS [189]. As *AtHSFA9* is mainly expressed in late stages of seed development, a TF may be involved in the regulation of *AtHSFA9* expression during the seed development stage.

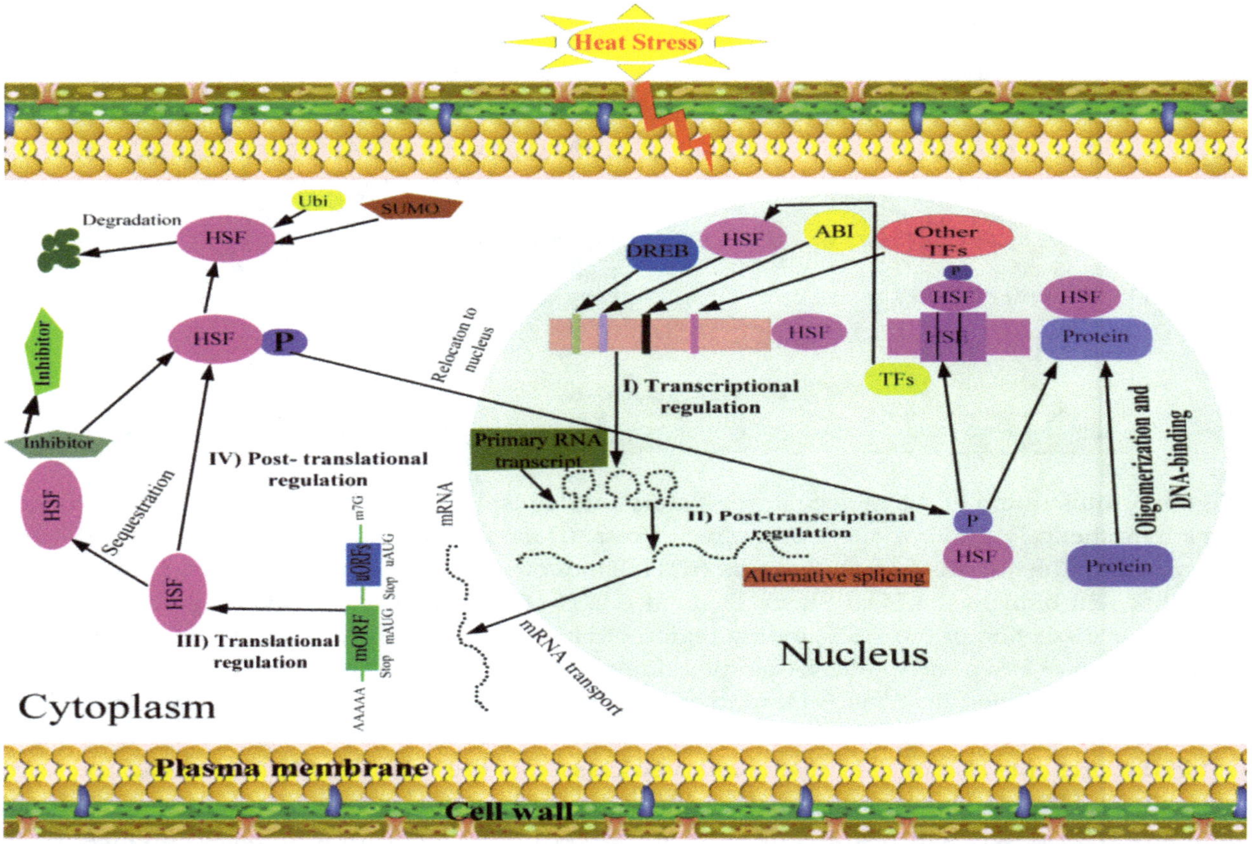

Figure 7. Illustration of the molecular regulatory mechanism of HSFs based on a hypothetical cellular model. The scheme illustrates the regulation of HSFs at **(I)** transcriptional, **(II)** post transcriptional, **(III)** translational, and **(IV)** post-translation level under heat stress. ABI, ABSCISIC ACID–INSENSITIVE protein; TFs, transcription factors; DREB, dehydration responsive element binding protein; AS, alternate splicing; m7G, cap of mRNA; mORF, major ORF; uORFs, upstream micro open reading frames; mAUG, AUG of mORF; uAUG, AUG of uORF; P, phosphate; mRNA, messenger RNA; Ubi, ubiquitination; SUMO, small ubiquitin-like modifier(*Adapted and modified from* [19]).

During several biological processes, HSFs post-transcriptional regulation involves alternative splicing in plants [187]. Keller et al. [190] examined HS-induced alternative splicing in the heat sensitive pollen tissue of two tomato cultivars. HS-induced alternative splicing is observed for *Arabidopsis AtHSFA2*, *A4c*, *A7b*, *B1* and *B2b* [191,192]. Alternative splicing induced by HS is also observed for rice *OsHSFA2d*, which encodes two main splice variant proteins, OsHSFA2dI located in the nucleus and OsHSFA2dII located in the nucleus and cytoplasm, respectively [193]. These studies suggest that the HSF regulation in the plant at posttranscriptional level is diversified. It is reported that the HSF regulation in the plant at the translational level is controlled by upstream micro open reading frames (uORFs) in their 5′ untranslated region [187,194,195]. However, the investigation on uORFs of plant *HSFs* is generally constrained to *Arabidopsis*.

HSFs also go through post-translational regulation included ubiquitination, phosphorylation, oligomerization, Small Ubiquitin-like MOdifier (SUMO)-mediated degradation, and interaction with other non-HSF proteins [196,197]. Mitogen-activated protein kinase MAPK6 targets the AtHSFA2, phosphorylates it on T249 and alters its intracellular localization under HS in *Arabidopsis* [198]. AtHSFA2 was regulated by the accumulation of polyubiquitinated proteins produced by the inhibition of AtHsp9026S and proteasome [199]. It is reported that AtSUMO1 interacts with AtHSFA2 at the core SUMOylation site Lys315, leading to the repression of its transcription and ultimately disrupting the acquired thermotolerance in *Arabidopsis* [200]. Unfortunately, few active regulation factors involved in HSF regulation have been found to date.

6.3. Development of HS-Tolerant Plants Using Genetic Engineering and Transgenic Approaches

The devastating impacts of HS can be reduced by inducing thermotolerance in plants using different transgenic and genetic engineering techniques [107]. In addition to the investigations regarding manipulation expression of HSFs and sHSPs/chaperones, other genetically transformed plants with different degrees of thermotolerance have been developed (Table 3). Shockingly, such investigations have been relatively limited as compared to the investigations based on engineering cold, drought, or salt stress tolerance in plants. Lee et al. [201] successfully produced transgenic HS-tolerant *Arabidopsis* by changed HSP expression level by making a modification in the transcription factor (*AtHSF1*) that leads to the production of HSPs in *Arabidopsis*. It was reported that *AtHSF1* in *Arabidopsis* is constitutively expressed; in optimum temperature, its activity for DNA binding, trimer-formation, and transcriptional stimulation of genes (*HSPs*) are suppressed. When overexpression of the gene (*AtHSF1*) occurs, the transcription factor is not active for thermotolerance. A fusion protein was produced as a result of a fusion between the *AtHSF1* gene and N or C terminus of the *gusA* reporter gene (for β-synthesis of glucuronidase) that was able to trimerize itself and/or with the other HSFs in the absence of heat. That fusion protein transformation into *Arabidopsis* created a transgenic plant that expressed HSPs constitutive and showed increased heat tolerance without requiring prior thermal treatment [8]. It has been studied that in tomato, *MT-sHSP* (mitochondrial small HSP) has a molecular chaperone function in vitro [202], and it has already been reported that this gene is used to produce a thermotolerantly transformed tobacco plant [203]. Researchers were able to produce transgenic thermotolerant rice after the incorporation of HSP genes. In rice, thermotolerance was developed by successful overexpression of *Arabidopsis* gene *Hsp101* [204]. Additionally, in *E. coli* overexpression of *Oshsp26* (*sHSP*) gene in rice confer enhanced tolerance to HS and other related OS [205]. The *sHSP17.7* overexpression in rice plants confers thermotolerance [206].

Ono et al. [207] were able to successfully transfer a gene (*Dnak1*) in tobacco from the salt tolerant *Cyanobacterium aphanothece* halophytica and were able to successfully conferred HS tolerance. Yang et al. [208] reported that gene (*BADH*) transformation in the plant for the over-generation of GB osmolyte will improve thermotolerance. Heat tolerance is obtained when Rubisco activates gene transformation in tobacco for the Rubisco reversible decarboxylation; this defensive mechanism leads to safeguarding the plant photosynthetic apparatus and improvea thermotolerance [209]. The change in membrane fluidity may also alter the perception of the stress through lipid signaling, thus altering the response of defensive mechanism. Altering fatty acid composition in lipids to enhance HT stability of the photosynthetic membrane has also been shown to enhance thermotolerance and limits photo-oxidation due to the free radicals discharge. Murakami et al. [210] have been developing genetically modified *N. tabacum* L. with alteration in chloroplast membrane by silencing the gene encoding chloroplast omega-3 fatty acid desaturase. Such genetically transformed plants generate comparatively higher amounts of dienoic fatty acids and lower trienoic fatty acids in chloroplasts than the wild type. Remarkably, an *NPK1*-related transcript was significantly raised by heat in studies of [211]. A modest increase in thermotolerance of *Arabidopsis* plants constitutively expressing *APX1* gene of the barley has been reported [212]. Grover et al. [184] suggested many different means to use genetically modified plant in developing thermotolerance that may be attained by overexpressing genes (*HSPs*) or by changing level of HSFs that control expression of non-heat shock and heat shock genes by overexpressing of other trans-acting factors like bZIP28, WRKY, and DREB2A proteins.

Table 3. Transgenic crops, transgenes, source, and their responsible function for developing heat tolerance.

Transgenic Crop	Gene Transferred	Source	Function	Reference
Wheat	*TaMYB*	*Arabidopsis*	Response to various abiotic stresses	[213]
	TaFER-5B	*T. aestivum* L.	Transgenic plant exhibited enhanced thermotolerance	[214]
	TaGASR1	*T. aestivum* L.	*TaGASR1* overexpressing plant improved tolerance to HS and oxidative stress	[215]
	TaHsfC2a	*T. aestivum* L.	*TaHsfC2a* overexpressing wheat showed improved thermotolerance	[216]
Arabidopsis	*TaWRKY33*	*T. aestivum* L.	*TaWRKY33* transgenic lines showed enhanced tolerance to HS	[217]
	TaNAC2L	*T. aestivum* L.	Overexpression of *TaNAC2L* enhanced heat tolerance by activating expression of heat-related genes	[218]
	TaB2	*T. aestivum* L.	Overexpression of *TaB2* in Arabidopsis enhanced tolerance to HS	[219]
	TaGASR1	*T. aestivum* L.	*TaGASR1* overexpressing plant had improved tolerance to HS and oxidative stress	[215]
	TaLTP3	*T. aestivum* L.	*TaLTP3* overexpressing plant showed higher thermotolerance than control plants at the seedling stage	[220]
	TaOEP16-2-5B	*T. aestivum* L.	Transgenic plant overexpressing *theTaOEP16-2-5B* gene exhibited enhanced tolerance to HS	[221]
	Ot NOS	*O. tauri* L.	High accumulation of NO leading to thermotolerance and osmotic stress	[222]
Rice	*SBPase*	*Oryza. Sativa* L.	Transgenic more tolerant to HS during seed development	[223]
	HSP100, HSP101	*Arabidopsis*	Synthesis of HSPs for heat tolerance	[208]
	OsMYB48–1	*Oryza. sativa* L.	Plays a positive role in in stress tolerance	[224]
	OsHTAS	*Oryza. sativa* L.	Plays a positive role in heat tolerance at the seedling stage	[225]
	OsMYB55	*Oryza. sativa* L.	Improved high temperature tolerance	[226]
Tobacco	*BADH*	*S. oleracea* L.	Overproduction of GB osmolytes that will increase heat tolerance	[208]
	Fad7	*N. tabacum* L. and *O. sativa* L.	H_2O_2 responsive MAPK kinase kinase (MAPKKK) synthesis to protect against HS	[227]
Maize	*OsMYB55*	*Oryza. Sativa* L.	Improved plant growth and performance under high temperature and drought condition	[228]
	HSP100, HSP101	*Arabidopsis*	Synthesis of HSPs for heat tolerance	[115]
Wild carrot	*HSP 17.7*	*Daucus carota* L.	HSPs synthesis	[206]
Chili pepper	*CaATG8c*	*Capsicum annuum* L.	Plant tolerance to environmental stresses	[229]
Tomato	*LeAN2*	-	Conferred increased tolerance to HS by maintaining a low levels of reactive oxygen species and high non-enzymatic antioxidant activity	[230]

6.4. Development of HS-Tolerance Plants Using Omics Approaches

Keeping in view the importance of global warming as a potential threat, recent advances in "omics" approaches have offered new hopes and opportunities for the identification of post-translational, translational, and transcriptional mechanisms and signaling corridors that control the plants response to HS [8]. These "omics" approaches help to correlation and systematic analysis between alterations in microme, genome, proteome, metabolome, and transcriptome to the alteration in the responses of plant to HS and their application to enhance the probabilities of producing plants that are thermo-tolerant (Figure 8). In recent years, research has provided knowledge of the functions of proteins, metabolites, and several genes and molecular mechanisms involved in plant sensitivity to HS [96].

DNA is a basic unit of the entire molecular evidence related to thermotolerance in plant and comprises many different HS-responsive genes (genomics). Already, the identification of a huge number of genes with essential roles in HS responses has been done using genome-wide expression studies and genetic screens [231]. The transcriptory product like mRNA, from such genes in the genome, have made their transcriptomes (transcriptomics) and then proteomes (proteomics) when they translated into a functional protein (accountable for thermotolerance).

Many modern approaches such as RNA-sequencing have led to various deep expression investigations, ultimately unraveling various heat-responsive candidate genes in several crops [36,232,233]. Transcriptomic investigations of HS effects on rice, wheat, tomato, grape, and tobacco have been reported [36,134,234]. Research in rice has unraveled that HS-responsive genes in panicles and flag leaves were mainly involved in transcriptional regulation, transport, protein binding, and anti-oxidants and stress responses [235]. Transcriptomic changes drive the physiological response to progressive drought stress and rehydration in tomato [236]. In a recent study, comparative transcriptome analysis revealed the transcriptional alterations in heat-resistant and heat-sensitive sweet maize (*Zea mays* L.) varieties under HS [237]. Chen and Li [238] found that DEGs were responsible for HS and protein folding in *Brachypodium distachyon*. It has been found that a huge number of genes were differentially expressed in leaves and roots in response to HS and/or desiccation, but only a few genes were identified as overlapping heat/drought responsive genes that are mainly involved in RNA regulation, transport, hormonal metabolism, and other stresses [239]. The transcriptome approaches are important to understand the molecular and cellular changes occurring in response to HS.

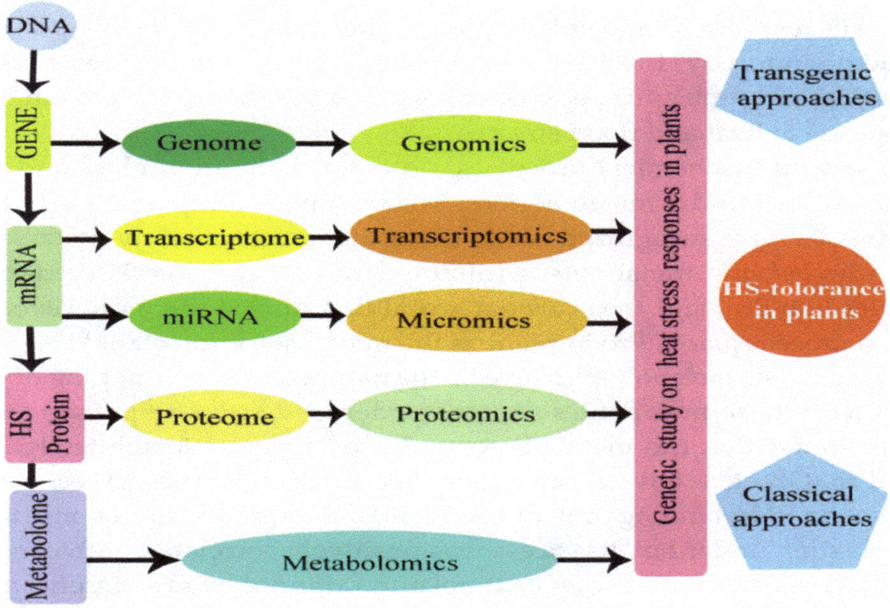

Figure 8. Illustration of integrated circuit of various "omics" techniques that correlated with each other at the molecular genetic level associated with thermotolerance in plants.

Micromics provide assistance for the better understanding of tolerance and miRNAs plays a significant role in such studies. Several thermo-tolerant responsive micro-RNAs have been recognized in plant, and their significant role in osmo-protection and nutrient deficiency response has been identified. In the onset of HS, up-regulated micro-RNAs may down-regulated their specific genes and act as negative regulators of thermotolerance, while stress down-regulated miRNAs may lead to accumulation of their target gene mRNAs, which may significantly regulate the thermotolerance. Overexpression of miRNA-resistant target genes plays an important role in plant post-transcriptional gene regulation and silencing and may result in the expression of improved traits in the genetically modified plant. Better understanding and knowledge about role of mRNAs in cellular tolerance, transcriptomehomeostasis, and the developmental and phenological plasticity of crop plants under HT and recovery will help with the genetic engineering of thermotolerance in plants.

Proteomics approaches provide important pieces of information, such as the heat-responsive proteins like HSPs and changes in proteomes under stress environments that associate to the analyses of transcriptomics and metabolomics, including the role of genes expressed in the genome's functionally translated regions linked to required traits [240]. The integration of proteomics with genetic information in legumes will give way to exciting opportunities to achieve crop improvement and sustainable agriculture [241]. Proteomic analysis of wheat cultivar Jing411 revealed the expression of 256 different proteins under HS [215]. Further, proteomic analysis on leaves of wheat revealed different proteins that have roles in important photosynthesis, glycolysis, stress defense, heat shock, and ATP production [85]. The proteomics analyses provide a better understanding of the molecular basis of heat-stress responses in alfalfa [62]. Additionally, proteomic analysis on soybean leaves revealed the expression of 25 different proteins that have roles in important metabolic pathways, such as RuBisCo regulation, Calvin cycle, and electron transport under high temperature [242]. In an experiment concerning the proteome analysis of roots using heat stressed root hairs and normal root hairs, 30 commonly up and down-regulated proteins were identified [243]. In an investigation, three tomato (*Solanum lycopersicum* L.) cultivars (LA1310 (cherry tomato), Edkawi LA 2711 (unknown heat tolerance, salt tolerant), and Walter LA3465 (heat-tolerant)) were compared for changes in leaf proteomes after HS treatment [244]. In the reproductive phase, the response of pollen to HS is mainly regulated at the proteome level, whereby proteins related to degradation and synthesis of proteins are most HS-responsive and might play a vital role in the HS-response of pollen [245]. Exogenous spermidine-induced HT resistance by proteomic approaches in tomato [246]. Lin et al. [247] utilized a physiological and proteomic approach to discover the changes in protein expression profiles of tomatoes in response to heat and flood stresses.

Moreover, in several biochemical processes, proteomes are interlinked and will manufacture deferent metabolic product under metabolomics. Metabolomics comparison between HS and major types of abiotic stresses have recognized metabolites that are commonly essential in responses to stress [248,249]. In *Arabidopsis*, metabolites profiling demonstrated that HS reduced the toxicity of bio-active molecules like Pro, and these events reveal that during the more severe combination of stress treatments, sucrose replaces Pro in plant as the main osmoprotectant [250]. When comparing both cold and heat response, metabolomics reveals the nature of the overlapping of major metabolites in response to HS with those metabolites that are under cold stress responses in *Arabidopsis* [251]. Hence, it has been found that the metabolic complex of compatible solutes (galactinol-raffinose, fructose glucose, Pro, GB etc.) have a significant role in HS tolerance. A very important protein (ATGRP7) that acts as an RNA binding protein was found that increased in response to low temperature stress and reversed when under HS. Its excess was positively correlated with concentrations of Pro and glutamine. While concentrations of galactinol and raffinose were a significant marker for heat response, their responses were independent of the response of Pro, glutamine, and *ATGRP7* [248]. Such "omics" techniques are desired for the genetic molecular analyses on HS response in the plant in an integrated manner.

6.5. Epigenetics and Heat Tolerance

Epigenetics is defined as heritable modifications in expression and activity of a gene that occur without a change in DNA sequence and is linked with DNA methylation, histone modifications, and non-protein coding RNAs [252]. Knowledge and understanding about the Epigenetic regulation of HS responses has gained increasing interest [253,254]. A number of investigations have shown that these DNA and histone modifications play a key role in genes expression and plant development under stress conditions [255]. In Arabidopsis (*Arabidopsis thaliana* L.), a study investigated the malleability of the DNA methylome to stress within a generation and under repeated stress over five successive generations [256]. *Arabidopsis thaliana* L. imprinted gene SDC, which is silent during vegetative growth due to DNA methylation, is stimulated by heat and contributes to recovery from HS [257]. Recently, in wheat, a genome-wide survey revealed that elevated temperature had a dramatic effect on the expression of genes, but plants grown at 12 °C and 27 °C showed slight differences in methylation pattern. However, in only a few cases was methylation associated with small changes in gene expression [254]. It is observed that DNA methylation levels were different between a thermotolerant and heat-sensitive genotype under control environments [258]. Methylation enhanced more in the heat-sensitive genotype than in the heat tolerant genotype under heat treatments. It is reported that more changes in the DNA methylation status of cultured microspores were observed under high temperature [259].

Histone variant deposition and histone modifications through SUMOylation and/or acetylation are considered to be involved in the HS response. Small ubiquitin-related modifier (SUMO) was recognized as a reversible post-translational modifier that contributed in the regulation of protein interactions in eukaryotes. Recent investigations have revealed that the occupancy of each histone variant of a core histone, in particular H2A and H3, play vital roles in not only genes expression but also in the assembly of chromosome centromeres and repair of DNA breaks in eukaryotes [260–262]. In *Arabidopsis*, it has been reported that H2A.Z deposition in gene bodies stimulates variability in the levels and gene expression patterns [261–263]. Additionally, H2A.Z plays an important role in the heat sensory response via its nucleosome occupancy. A screen of Arabidopsis (*Arabidopsis thaliana* L.) mutants deficient in temperature sensing under ambient temperatures (12–27 °C) identified actin-related protein 6 (*ARP6*) as a regulator of the coordinated alterations in gene expression in response to ambient temperature fluctuations [264]. *ARP6* encodes a subunit of the SWR1 complex [265] that is necessary for inserting the alternative histone H2A.Z into nucleosomes replacing the core histone H2A, and could be involved in heat sensing [264]. In *Arabidopsis*, it is reported that histone acetyltransferase GENERAL CONTROL OF NONREPRESSED PROTEIN5 (GCN5) plays a vital role in HS response by facilitating H3K9 and H3K14 acetylation of HSFA3 and UV-HYPERSENSITIVE6 (UVH6) under HS [266]. The histone acetyltransferase TaGCN5 gene in wheat is upregulated under HS and that it functions similarly to GCN5 in *Arabidopsis* [266]. Further studies are necessary to uncover what kinds of histone modification and histone deposition contribute to the HS response in crop plants.

Recent investigations suggest that a large part of the genome is transcribed, but among transcripts, only a minor portion encodes protein. The part of transcripts that do not encode proteins are generally termed as non-protein coding RNAs (npcRNA). These npcRNAs are subdivided as regulatory npcRNAs and housekeeping npcRNAs, with the latter being further divided into long regulatory npcRNAs (greater than 300 bp in length) and short regulatory npcRNAs (less than 300 bp in length, such as microRNA, siRNA, piwi-RNA) [267–271]. Xin et al. [269] recognized 66 HS-responsive long npcRNAs that performed their roles in the form of long molecules. To check whether miRNAs have any functions in regulating response to HS in *T. aestivum* L., Xin et al. [270] cloned small RNA exposed to HS and found that 12 of the 153 miRNAs identified were responsive to HS. Kumar et al. [267] identified 37 novel miRNAs in wheat and validated six of the identified novel miRNA as HS-responsive. Interestingly, TamiR159 was down-regulated after two hours of HS treatment in *T. aestivum* L., but TamiR159 overexpressing *Oryza sativa* L. lines were more sensitive to HS relative to the wild type, showing that down-regulation of TamiR159 in *T. aestivum* L. after HS might participate in a HS-related signaling pathway, in turn contributing to HS tolerance [272].

6.6. CRISPR/Cas9 Power in Genome Editing

CRISPR/Cas9 is the most powerful gene editing tool ever seen to date. Developing more crop plants able to sustainably produce higher yield when grown under HS is an important goal if food security and crop production are to be guaranteed in the face of increasing human population and unpredictable climatic conditions. However, conventional crop improvement through random mutagenesis or genetic recombination is laborious and cannot keep pace with increasing food demands. Targeted genome editing (GE) technologies, especially CRISPR/Cas9, have great potential to produce high-yielding crops under HS. This is due to their low risk, high accuracy, and efficiency of off-target effects compared with conventional random mutagenesis approaches. [273] reviewed recent applications of the CRISPR/Cas9-mediated GE as a means to produce plants with greater resilience to the stressors they encounter when grown under harsh environments. CRISPR/Cas9-mediated GE will allow the fast development of new crop cultivars with a very low risk of off-target effects, especially for the crop plants that have complexity in their genomes and are not easily bred using conventional breeding approaches [274,275].

CRISPR/Cas9-mediated GE enabling the suppression or activation of target genes is also an essential tool for understanding the functioning of genes involved in plant abiotic stress resistance [276]. CRISPR/Cas9 targeted mutation of the *TMS5* gene in rice cultivars led to the rapid development of temperature-sensitive lines for use in hybrid rice production [277]. Klap et al. [278] carried out CRISPR/Cas9-induced knockdown of the tomato *slagamous-like 6 (SlAGL6)* gene, making tomato mutant plants able to produce parthenocarpic fruits under HS. Genome editing using the CRISPR/Cas9 system can be used to modify plant genomes. However, improvements in specificity and applicability are still needed in order for the editing technique to be useful in various plant species.

7. Conclusion and Future Prospects

From the forgone discussion, it can be concluded that HS has significantly affected crop yield in the past several years and currently has become a key concern for crop production because it significantly affects all stages of plant growth and development. One of the reasons of the enigmatic nature of heat tolerance mechanism in plants is the dissection of a very narrow genetic pool that does not provide ample information to explore it in worldwide commercially important crop plants such wheat, cotton, maize, rice, and soybean. There is an immense need to systematically assess wild species and accessions tolerating extreme degrees of higher temperatures.

In that regard, searching for novel donors with high heat tolerance or escape mechanisms is of key importance. Crop responses to HS are for the most part grouped under heat tolerance categories without having explored heat avoidance or escape phenomena, which are equally viable under field conditions. Phenotyping techniques classifying HS response into the appropriate tolerance, escape, or avoidance category is a vital first step toward developing stress-resilient crops for the future hotter climate. A complete understanding and knowledge of the nature of heat shock signaling and specific gene expression under HS will be vital for developing HS-tolerant crop plants. Despite the major advances in genetic strategies like transgenic approaches and QTL mapping, there is still huge room for improvement. For example, genetic and environmental interactions are poorly understood.

However, a brief mechanism of thermotolerance remains indefinable and needs appropriate research directions. In our opinion, the genetic base for heat tolerant mechanisms in crop plants can be expanded significantly through advances in the phenotyping approaches using biochemical means including lipidomics or metabolomics which that off an impression of being promising strategies in identifying robust biochemical markers to supplement breeding efforts. Furthermore, advances in ground-based or aerial (unmanned aerial vehicles) sensor technology may also help field-based high-throughput phenotyping, which may greatly facilitate marked expansion of the genetic base incorporated into abiotic stress breeding programs. On the field scale, multiple stresses interact, and dealing with the entire complexity could be challenging, and hence addressing subcomponents (such as HS or drought stress) independently and using advanced techniques for needs based traits/genes stacking based on the target environment would be a handy mechanistic research strategy for producing crop plants that will stand out firmly in the hottest season of the year.

In the future, integration of advanced high0throughput approaches, such as microarray, genomics, and proteomics, in various developmental stages and stress conditions will provide us with transgenic plants developed for combating with HS.

Abbreviations

APX	ascorbate peroxidase
Ca	calcium
CAT	catalase
GB	glycine betaine
HS	heat stress
HT	high temperature
HSPs	heat shock proteins
HSFs	heat shock transcriptional factors
HSEs	heat shock elements
LEA	late embryogenesis abundant proteins
MDA	malondialdehyde
NPQ	non-photochemical quenching
NO	nitric oxide
OS	oxidative Stress
PS	photosystem
QTLs	quantitative trait loci
Rubisco	ribulose-1,5-bisphosphate carboxylase/oxygenase
RuBP	ribulose bisphosphate
ROS	reactive oxygen species
SOD	superoxide dismutase

References

1. Lobell, D.B.; Schlenker, W.; Costa-Roberts, J. Climate trends and global crop production since 1980. *Science* **2011**, *333*, 616–620. [CrossRef] [PubMed]
2. Lesk, C.; Rowhani, P.; Ramankutty, N. Influence of extreme weather disasters on global crop production. *Nature* **2016**, *529*, 84–87. [CrossRef] [PubMed]
3. Abdelrahman, M.; El-Sayed, M.; Jogaiah, S.; Burritt, D.J.; Tran, L.S.P. The "STAY-GREEN" trait and phytohormone signaling networks in plants under heat stress. *Plant Cell Rep.* **2017**, *36*, 1009–1025. [CrossRef] [PubMed]
4. Bita, C.E.; Gerats, T. Plant tolerance to high temperature in a changing environment: Scientific fundamentals and production of heat stress-tolerant crops. *Front. Plant Sci.* **2013**, *4*, 1–18. [CrossRef] [PubMed]
5. Stratonovitch, P.; Semenov, M.A. Heat tolerance around flowering in wheat identified as a key trait for increased yield potential in Europe under climate change. *J. Exp. Bot.* **2015**, *66*, 3599–3609. [CrossRef] [PubMed]
6. Lobell, D.B.; Field, C.B. Global scale climate–crop yield relationships and the impacts of recent warming. *Environ. Res. Lett.* **2007**, *2*, 14002. [CrossRef]
7. Hatfield, J.L.; Prueger, J.H. Temperature extremes: Effect on plant growth and development. *Weather Clim. Extrem.* **2015**, *10*, 4–10. [CrossRef]
8. Hasanuzzaman, M.; Nahar, K.; Alam, M.; Roychowdhury, R. Physiological, biochemical, and molecular mechanisms of heat stress tolerance in plants. *Int. J. Mol. Sci.* **2013**, *14*, 9643–9684. [CrossRef] [PubMed]
9. Das, S.; Krishnan, P.; Nayak, M.; Ramakrishnan, B. High temperature stress effects on pollens of rice (*Oryza sativa* L.) genotypes. *Environ. Exp. Bot.* **2014**, *101*, 36–46. [CrossRef]
10. Valliyodan, B.; Nguyen, H.T. Understanding regulatory networks and engineering for enhanced drought tolerance in plants. *Curr. Opin. Plant Biol.* **2006**, *9*, 189–195. [CrossRef] [PubMed]
11. Janská, A.; Maršík, P.; Zelenková, S.; Ovesná, J. Cold stress and acclimation—What is important for metabolic adjustment? *Plant Biol.* **2010**, *12*, 395–405. [CrossRef] [PubMed]
12. Shinozaki, K.; Yamaguchi-Shinozaki, K. Gene networks involved in drought stress response and tolerance. *J. Exp. Bot.* **2007**, *58*, 221–227. [CrossRef] [PubMed]
13. Moreno, A. A.; Orellana, A. The physiological role of the unfolded protein response in plants. *Biol. Res.* **2011**, *44*, 75–80. [CrossRef] [PubMed]

14. Zhang, Y.; Mian, M.A.R.; Bouton, J.H. Recent molecular and genomic studies on stress tolerance of forage and turf grasses. *Crop Sci.* **2006**, *46*, 497–511. [CrossRef]
15. Koevoets, I.T.; Venema, J.H.; Elzenga, J.T.M.; Testerink, C. Roots withstanding their environment: Exploiting root system architecture responses to abiotic stress to improve crop tolerance. *Front. Plant Sci.* **2016**, *7*, 1–19. [CrossRef]
16. Fahad, S.; Bajwa, A.A.; Nazir, U.; Anjum, S.A.; Farooq, A. Crop production under drought and heat stress: Plant responses and management options. *Front. Plant Sci.* **2017**, *8*, 1–16. [CrossRef] [PubMed]
17. Gururani, M.A.; Venkatesh, J.; Tran, L.S.P. Regulation of photosynthesis during abiotic stress-induced photoinhibition. *Mol. Plant* **2015**, *8*, 1304–1320. [CrossRef] [PubMed]
18. Feller, U. Drought stress and carbon assimilation in a warming climate: Reversible and irreversible impacts. *J. Plant Physiol.* **2016**, *203*, 84–94. [CrossRef] [PubMed]
19. Guo, M.; Liu, J.-H.; Ma, X.; Luo, D.-X.; Gong, Z.-H.; Lu, M.-H. The plant heat stress transcription factors (HSFs): Structure, regulation, and function in response to abiotic stresses. *Front. Plant Sci.* **2016**, *7*. [CrossRef] [PubMed]
20. Crawford, A.J.; McLachlan, D.H.; Hetherington, A.M.; Franklin, K.A. High temperature exposure increases plant cooling capacity. *Curr. Biol.* **2012**, *22*, R396–R397. [CrossRef] [PubMed]
21. Zhao, J.; Hartmann, H.; Trumbore, S.; Ziegler, W.; Zhang, Y. High temperature causes negative whole-plant carbon balance under mild drought. *New Phytol.* **2013**, *200*, 330–339. [CrossRef] [PubMed]
22. Sage, R.F.; Zhu, X.G. Exploiting the engine of C 4 photosynthesis. *J. Exp. Bot.* **2011**, *62*, 2989–3000. [CrossRef] [PubMed]
23. Atkin, O.K.; Tjoelker, M.G. Thermal acclimation and the dynamic response of plant respiration to temperature. *Trends Plant Sci.* **2003**, *8*, 343–351. [CrossRef]
24. Wolkovich, E.M.; Cook, B.I.; Allen, J.M.; Crimmins, T.M.; Betancourt, J.L.; Travers, S.E.; Pau, S.; Regetz, J.; Davies, T.J.; Kraft, N.J.B.; et al. Warming experiments underpredict plant phenological responses to climate change. *Nature* **2012**, *485*, 494–497. [CrossRef] [PubMed]
25. Criddle, R.S.; Smith, B.N.; Hansen, L.D. A respiration based description of plant growth rate responses to temperature. *Planta* **1997**, *201*, 441–445. [CrossRef]
26. Sánchez, B.; Rasmussen, A.; Porter, J.R. Temperatures and the growth and development of maize and rice: A review. *Glob. Chang. Biol.* **2014**, *20*, 408–417. [CrossRef] [PubMed]
27. Ciais, P.; Reichstein, M.; Viovy, N.; Granier, A.; Ogée, J.; Allard, V.; Aubinet, M.; Buchmann, N.; Bernhofer, C.; Carrara, A.; et al. Europe-wide reduction in primary productivity caused by the heat and drought in 2003. *Nature* **2005**, *437*, 529–533. [CrossRef] [PubMed]
28. Morecroft, M.D.; Stokes, V.J.; Taylor, M.E.; Morison, J.I.L. Effects of climate and management history on the distribution and growth of sycamore (*Acer pseudoplatanus* L.) in a southern British woodland in comparison to native competitors. *Forestry* **2008**, *81*, 59–74. [CrossRef]
29. Cleland, E.E.; Chuine, I.; Menzel, A.; Mooney, H.A.; Schwartz, M.D. Shifting plant phenology in response to global change. *Trends Ecol. Evol.* **2007**, *22*, 357–365. [CrossRef] [PubMed]
30. Chmielewski, F.M.; Müller, A.; Bruns, E. Climate changes and trends in phenology of fruit trees and field crops in Germany, 1961–2000. *Agric. For. Meteorol.* **2004**, *121*, 69–78. [CrossRef]
31. Siebert, S.; Ewert, F. Future crop production threatened by extreme heat. *Environ. Res. Lett.* **2014**, *9*, 41001. [CrossRef]
32. Orbović, V.; Poff, K.L. Effect of temperature on growth and phototropism of *Arabidopsis thaliana* seedlings. *J. Plant Growth Regul.* **2007**, *26*, 222–228. [CrossRef]
33. Van Der Ploeg, A.; Heuvelink, E. Influence of sub-optimal temperature on tomato growth and yield: A review. *J. Hortic. Sci. Biotechnol.* **2005**, *80*, 652–659. [CrossRef]
34. Yang, L.Y.; Yang, S.L.; Li, J.Y.; Ma, J.H.; Pang, T.; Zou, C.M.; He, B.; Gong, M. Effects of different growth temperatures on growth, development, and plastid pigments metabolism of tobacco (*Nicotiana tabacum* L.) plants. *Bot. Stud.* **2018**, *59*. [CrossRef] [PubMed]
35. Tashiro, T.; Wardlaw, I. The response to high temperature shock and humidity changes prior to and during the early stages of grain development in wheat. *Aust. J. Plant Physiol.* **1990**, *17*, 551. [CrossRef]
36. Chao, L.M.; Liu, Y.Q.; Chen, D.Y.; Xue, X.Y.; Mao, Y.B.; Chen, X.Y. *Arabidopsis* transcription factors SPL1 and SPL12 confer plant thermotolerance at reproductive stage. *Mol. Plant* **2017**, *10*, 735–748. [CrossRef] [PubMed]

37. Ferris, R.; Ellis, R.H.; Wheeler, T.R.; Hadley, P. Effect of high temperature stress at anthesis on grain yield and biomass of field-grown crops of wheat. *Ann. Bot.* **1998**, *82*, 631–639. [CrossRef]
38. Perrotta, C.; Treglia, A.S.; Mita, G.; Giangrande, E.; Rampino, P.; Ronga, G.; Spano, G.; Marmiroli, N. Analysis of mRNAs from ripening wheat seeds: The effect of high temperature. *J. Cereal Sci.* **1998**, *27*, 127–132. [CrossRef]
39. Tahir, I.S.A.; Nakata, N. Remobilization of nitrogen and carbohydrate from stems of bread wheat in response to heat stress during grain filling. *J. Agron. Crop Sci.* **2005**, *191*, 106–115. [CrossRef]
40. Dias, A.S.; Lidon, F.C. Evaluation of grain filling rate and duration in bread and durum wheat, under heat stress after anthesis. *J. Agron. Crop Sci.* **2009**, *195*, 137–147. [CrossRef]
41. Barnabás, B.; Jäger, K.; Fehér, A. The effect of drought and heat stress on reproductive processes in cereals. *Plant Cell Environ.* **2008**, *31*, 11–38. [CrossRef] [PubMed]
42. Marcelis, L.F.M. Sink strength as a determinant of dry matter partitioning in the whole plant. *J. Exp. Bot.* **1996**, *47*, 1281–1291. [CrossRef] [PubMed]
43. Boyer, J.S.; Westgate, M.E. Grain yields with limited water. *J. Exp. Bot.* **2004**, *55*, 2385–2394. [CrossRef] [PubMed]
44. Eyshi, E.; Webber, H.; Gaiser, T.; Naab, J.; Ewert, F. Heat stress in cereals: Mechanisms and modelling. *Eur. J. Agron.* **2015**, *64*, 98–113. [CrossRef]
45. Zhao, C.; Liu, B.; Piao, S.; Wang, X.; Lobell, D.B.; Huang, Y.; Huang, M.; Yao, Y.; Bassu, S.; Ciais, P.; et al. Temperature increase reduces global yields of major crops in four independent estimates. *Proc. Natl. Acad. Sci. USA* **2017**, *114*, 9326–9331. [CrossRef] [PubMed]
46. Liu, B.; Asseng, S.; Müller, C.; Ewert, F.; Elliott, J.; Lobell, D.B.; Martre, P.; Ruane, A.C.; Wallach, D.; Jones, J.W.; et al. Similar estimates of temperature impacts on global wheat yield by three independent methods. *Nat. Clim. Chang.* **2016**, *6*, 1130–1136. [CrossRef]
47. Asseng, S.; Foster, I.; Turner, N.C. The impact of temperature variability on wheat yields. *Glob. Chang. Biol.* **2011**, *17*, 997–1012. [CrossRef]
48. Lobell, D.B.; Gourdji, S.M. The influence of climate change on global crop productivity. *Plant Physiol.* **2012**, *160*, 1686–1697. [CrossRef] [PubMed]
49. Balla, K.; Rakszegi, M.; Li, Z.; Békés, F.; Bencze, S.; Veisz, O. Quality of winter wheat in relation to heat and drought shock after anthesis. *Czech J. Food Sci.* **2011**, *29*, 117–128. [CrossRef]
50. Bassu, S.; Brisson, N.; Durand, J.L.; Boote, K.; Lizaso, J.; Jones, J.W.; Rosenzweig, C.; Ruane, A.C.; Adam, M.; Baron, C.; et al. How do various maize crop models vary in their responses to climate change factors? *Glob. Chang. Biol.* **2014**, *20*, 2301–2320. [CrossRef] [PubMed]
51. Cao, Y.Y.; Duan, H.; Yang, L.N.; Wang, Z.Q.; Liu, L.J.; Yang, J.C. Effect of high temperature during heading and early filling on grain yield and physiological characteristics in indica rice. *Acta Agron. Sin.* **2009**, *35*, 512–521. [CrossRef]
52. Schlenker, W.; Roberts, M.J. Nonlinear temperature effects indicate severe damages to U.S. crop yields under climate change. *Proc. Natl. Acad. Sci. USA* **2009**, *106*, 15594–15598. [CrossRef] [PubMed]
53. Angadi, S.V.; Cutforth, H.W.; Miller, P.R.; Mcconkey, B.G.; Entz, M.H.; Brandt, S.A. Response of three Brassica species to high temperature stress during reproductive growth. *Can. J. Plant Sci.* **2000**, *80*, 693–702. [CrossRef]
54. Young, L.W.; Wilen, R.W.; Bonham-Smith, P.C. High temperature stress of *Brassica napus* during flowering reduces micro- and megagametophyte fertility, induces fruit abortion, and disrupts seed production. *J. Exp. Bot.* **2004**, *55*, 485–495. [CrossRef] [PubMed]
55. Cooper, P.; Rao, K.P.C.; Singh, P.; Dimes, J.; Traore, P.S.; Rao, K.; Dixit, P.; Twomlow, S.J. Farming with current and future climate risk: Advancing a "Hypothesis of Hope" for rainfed agriculture in the semi-arid tropics. *J. SAT Agric. Res.* **2009**, *7*, 1–19.
56. Tack, J.; Lingenfelser, J.; Jagadish, S.V.K. Disaggregating sorghum yield reductions under warming scenarios exposes narrow genetic diversity in US breeding programs. *Proc. Natl. Acad. Sci. USA* **2017**, *114*, 9296–9301. [CrossRef] [PubMed]
57. Debaeke, P.; Casadebaig, P.; Flenet, F.; Langlade, N. Sunflower crop and climate change: Vulnerability, adaptation, and mitigation potential from case-studies in Europe. *OCL* **2017**, *24*, D102. [CrossRef]
58. Lobell, D.B.; Bänziger, M.; Magorokosho, C.; Vivek, B. Nonlinear heat effects on African maize as evidenced by historical yield trials. *Nat. Clim. Chang.* **2011**, *1*, 42–45. [CrossRef]

59. Lobell, D.B.; Hammer, G.L.; McLean, G.; Messina, C.; Roberts, M.J.; Schlenker, W. The critical role of extreme heat for maize production in the United States. *Nat. Clim. Chang.* **2013**, *3*, 497–501. [CrossRef]
60. Kucharik, C.J.; Serbin, S.P. Impacts of recent climate change on Wisconsin corn and soybean yield trends. *Environ. Res. Lett.* **2008**, *3*, 34003. [CrossRef]
61. Ding, X.; Jiang, Y.; He, L.; Zhou, Q.; Yu, J.; Hui, D.; Huang, D. Exogenous glutathione improves high root-zone temperature tolerance by modulating photosynthesis, antioxidant and osmolytes systems in cucumber seedlings. *Sci. Rep.* **2016**, *6*, 35424. [CrossRef] [PubMed]
62. Li, H.; Ahammed, G.J.; Zhou, G.; Xia, X.; Zhou, J.; Shi, K.; Yu, J.; Zhou, Y. Unraveling main limiting sites of photosynthesis under below- and above-ground heat stress in cucumber and the alleviatory role of luffa rootstock. *Front. Plant Sci.* **2016**, *7*, 746. [CrossRef] [PubMed]
63. Wise, R.R.; Olson, A.J.; Schrader, S.M.; Sharkey, T.D. Electron transport is the functional limitation of photosynthesis in field-grown Pima cotton plants at high temperature. *Plant Cell Environ.* **2004**, *27*, 717–724. [CrossRef]
64. Wahid, A.; Gelani, S.; Ashraf, M.; Foolad, M.R. Heat tolerance in plants: An overview. *Environ. Exp. Bot.* **2007**, *61*, 199–223. [CrossRef]
65. Salvucci, M.E.; Crafts-Brandner, S.J. Inhibition of photosynthesis by heat stress: The activation state of Rubisco as a limiting factor in photosynthesis. *Physiol. Plant.* **2004**, *120*, 179–186. [CrossRef] [PubMed]
66. Xu, S.; Li, J.; Zhang, X.; Wei, H.; Cui, L. Effects of heat acclimation pretreatment on changes of membrane lipid peroxidation, antioxidant metabolites, and ultrastructure of chloroplasts in two cool-season turfgrass species under heat stress. *Environ. Exp. Bot.* **2006**, *56*, 274–285. [CrossRef]
67. ElBasyoni, I.; Saadalla, M.; Baenziger, S.; Bockelman, H.; Morsy, S. Cell membrane stability and association mapping for drought and heat tolerance in a worldwide wheat collection. *Sustainability* **2017**, *9*, 1606. [CrossRef]
68. Tan, W.; Meng, Q.W.; Brestic, M.; Olsovska, K.; Yang, X. Photosynthesis is improved by exogenous calcium in heat-stressed tobacco plants. *J. Plant Physiol.* **2011**, *168*, 2063–2071. [CrossRef] [PubMed]
69. Lin, C.Y.; Roberts, J.K.; Key, J.L. Acquisition of thermotolerance in soybean seedlings: Synthesis and accumulation of heat shock proteins and their cellular localization. *Plant Physiol.* **1984**, *74*, 152–160. [CrossRef] [PubMed]
70. Kumar, S.; Thakur, P.; Kaushal, N.; Malik, J.A.; Gaur, P.; Nayyar, H. Effect of varying high temperatures during reproductive growth on reproductive function, oxidative stress and seed yield in chickpea genotypes differing in heat sensitivity. *Arch. Agron. Soil Sci.* **2013**, *59*, 823–843. [CrossRef]
71. Pastore, A.; Martin, S.R.; Politou, A.; Kondapalli, K.C.; Stemmler, T.; Temussi, P.A. Unbiased cold denaturation: Low- and high-temperature unfolding of yeast frataxin under physiological conditions. *J. Am. Chem. Soc.* **2007**, *129*, 5374–5375. [CrossRef] [PubMed]
72. Yamada, K.; Fukao, Y.; Hayashi, M.; Fukazawa, M.; Suzuki, I.; Nishimura, M. Cytosolic HSP90 regulates the heat shock response that is responsible for heat acclimation in *Arabidopsis thaliana*. *J. Biol. Chem.* **2007**, *282*, 37794–37804. [CrossRef] [PubMed]
73. Von Koskull-Döring, P.; Scharf, K.D.; Nover, L. The diversity of plant heat stress transcription factors. *Trends Plant Sci.* **2007**, *12*, 452–457. [CrossRef] [PubMed]
74. Gupta, N.K.; Agarwal, S.; Agarwal, V.P.; Nathawat, N.S.; Gupta, S.; Singh, G. Effect of short-term heat stress on growth, physiology and antioxidative defence system in wheat seedlings. *Acta Physiol. Plant.* **2013**, *35*, 1837–1842. [CrossRef]
75. Wang, L.C.; Tsai, M.C.; Chang, K.Y.; Fan, Y.S.; Yeh, C.H.; Wu, S.J. Involvement of the *Arabidopsis* HIT1/AtVPS53 tethering protein homologue in the acclimation of the plasma membrane to heat stress. *J. Exp. Bot.* **2011**, *62*, 3609–3620. [CrossRef] [PubMed]
76. Xu, Q.; Xu, X.; Shi, Y.; Xu, J.; Huang, B. Transgenic tobacco plants overexpressing a grass *PpEXP1* gene exhibit enhanced tolerance to heat stress. *PLoS ONE* **2014**, *9*, e100792. [CrossRef] [PubMed]
77. Wang, D.; Heckathorn, S.A.; Mainali, K.; Tripathee, R. Timing effects of heat-stress on plant ecophysiological characteristics and growth. *Front. Plant Sci.* **2016**, *7*, 1–11. [CrossRef] [PubMed]
78. Chen, Y.; Zhang, Z.; Tao, F.; Palosuo, T.; Rötter, R.P. Impacts of heat stress on leaf area index and growth duration of winter wheat in the North China Plain. *Field Crop. Res.* **2017**, 230–237. [CrossRef]
79. Murata, N.; Takahashi, S.; Nishiyama, Y.; Allakhverdiev, S.I. Photoinhibition of photosystem II under environmental stress. *Biochim. Biophys. Acta–Bioenerg.* **2007**, *1767*, 414–421. [CrossRef] [PubMed]

80. Zivcak, M.; Brestic, M.; Kalaji, H.M.; Govindjee. Photosynthetic responses of sun- and shade-grown barley leaves to high light: Is the lower PSII connectivity in shade leaves associated with protection against excess of light? *Photosynth. Res.* **2014**, *119*, 339–354. [CrossRef] [PubMed]
81. Mishra, D.; Shekhar, S.; Agrawal, L.; Chakraborty, S.; Chakraborty, N. Cultivar-specific high temperature stress responses in bread wheat (*triticum aestivum* L.) associated with physicochemical traits and defense pathways. *Food Chem.* **2017**, *221*, 1077–1087. [CrossRef] [PubMed]
82. Brestic, M.; Zivcak, M.; Olsovska, K.; Shao, H.-B.; Kalaji, H.M.; Allakhverdiev, S.I. Reduced glutamine synthetase activity plays a role in control of photosynthetic responses to high light in barley leaves. *Plant Physiol. Biochem.* **2014**, *81*, 74–83. [CrossRef] [PubMed]
83. Brestic, M.; Zivcak, M.; Kunderlikova, K.; Sytar, O.; Shao, H.; Kalaji, H.M.; Allakhverdiev, S.I. Low PSI content limits the photoprotection of PSI and PSII in early growth stages of chlorophyll b-deficient wheat mutant lines. *Photosynth. Res.* **2015**, *125*, 151–166. [CrossRef] [PubMed]
84. Su, X.; Wu, S.; Yang, L.; Xue, R.; Li, H.; Wang, Y.; Zhao, H. Exogenous progesterone alleviates heat and high light stress-induced inactivation of photosystem II in wheat by enhancing antioxidant defense and D1 protein stability. *Plant Growth Regul.* **2014**, *74*, 311–318. [CrossRef]
85. Wang, X.; Dinler, B.S.; Vignjevic, M.; Jacobsen, S.; Wollenweber, B. Physiological and proteome studies of responses to heat stress during grain filling in contrasting wheat cultivars. *Plant Sci.* **2015**, *230*, 33–50. [CrossRef] [PubMed]
86. Chen, Y.-E.; Liu, W.-J.; Su, Y.-Q.; Cui, J.-M.; Zhang, Z.-W.; Yuan, M.; Zhang, H.-Y.; Yuan, S. Different response of photosystem II to short and long-term drought stress in *Arabidopsis thaliana*. *Physiol. Plant.* **2016**, *158*, 225–235. [CrossRef] [PubMed]
87. Sage, R.F.; Kubien, D.S. The temperature response of C3 and C4 photosynthesis. *Plant Cell Environ.* **2007**, *30*, 1086–1106. [CrossRef] [PubMed]
88. Sharkey, T.D. Effects of moderate heat stress on photosynthesis: Importance of thylakoid reactions, rubisco deactivation, reactive oxygen species, and thermotolerance provided by isoprene. *Plant Cell Environ.* **2005**, *28*, 269–277. [CrossRef]
89. Crafts-Brandner, S.J.; Salvucci, M.E. Sensitivity of photosynthesis heat stress in a C4 plant, maize, to heat stress. *Plant Physiol.* **2002**, *129*, 1773–1780. [CrossRef] [PubMed]
90. Takahashi, S.; Murata, N. Interruption of the Calvin cycle inhibits the repair of Photosystem II from photodamage. *Biochim. Biophys. Acta–Bioenerg.* **2005**, *1708*, 352–361. [CrossRef] [PubMed]
91. Bukhov, N.G.; Wiese, C.; Neimanis, S.; Heber, U. Heat sensitivity of chloroplasts and leaves: Leakage of protons from thylakoids and reversible activation of cyclic electron transport. *Photosynth. Res.* **1999**, *59*, 81–93. [CrossRef]
92. Peng, S.; Huang, J.; Sheehy, J.E.; Laza, R.C.; Visperas, R.M.; Zhong, X.; Centeno, G.S.; Khush, G.S.; Cassman, K.G. Rice yields decline with higher night temperature from global warming. *Proc. Natl. Acad. Sci. USA* **2004**, *101*, 9971–9975. [CrossRef] [PubMed]
93. Penning De Vries, F.W.T. The cost of maintenance processes in plant cells. *Ann. Bot.* **1975**, *39*, 77–92. [CrossRef]
94. Koini, M.A.; Alvey, L.; Allen, T.; Tilley, C.A.; Harberd, N.P.; Whitelam, G.C.; Franklin, K.A. High temperature-mediated adaptations in plant architecture require the bHLH transcription factor PIF4. *Curr. Biol.* **2009**, *19*, 408–413. [CrossRef] [PubMed]
95. Kolb, P.F.; Robberecht, R. High temperature and drought stress effects on survival of *Pinus ponderosa* seedlings. *Tree Physiol.* **1996**, *16*, 665–672. [CrossRef] [PubMed]
96. Sita, K.; Sehgal, A.; HanumanthaRao, B.; Nair, R.M.; Vara Prasad, P.V.; Kumar, S.; Gaur, P.M.; Farroq, M.; Siddique, K.H.M.; Varshney, R.K.; et al. Food legumes and rising temperatures: Effects, adaptive functional mechanisms specific to reproductive growth stage and strategies to improve heat tolerance. *Front. Plant Sci.* **2017**, *8*, 1–30. [CrossRef] [PubMed]
97. Prasad, P.V.V.; Staggenborg, S.A.; Ristic, Z.; Ahuja, L.R.; Reddy, V.R.; Saseendran, S.A.; Yu, Q. Impacts of drought and/or heat stress on physiological, developmental, growth, and yield processes of crop plants. In *Response of Crops to Limited Water: Understanding and Modeling Water Stress Effects on Plant Growth Processes*; American Society of Agronomy; Crop Science Society of America; Soil Science Society of America: Madison, WI, USA, 2008; ISBN 978-0-89118-188-0.

98. Tsukaguchi, T.; Kawamitsu, Y.; Takeda, H.; Suzuki, K.; Egawa, Y. Water status of flower buds and leaves as affected by high temperature in heat-tolerant and heat-sensitive cultivars of snap bean (*Phaseolus vulgaris* L.). *Plant Prod. Sci.* **2003**, *6*, 24–27. [CrossRef]
99. Kaushal, N.; Awasthi, R.; Gupta, K.; Gaur, P.; Siddique, K.H.M.; Nayyar, H. Heat-stress-induced reproductive failures in chickpea (*Cicer arietinum*) are associated with impaired sucrose metabolism in leaves and anthers. *Funct. Plant Biol.* **2013**, *40*, 1334–1349. [CrossRef]
100. Asada, K. Production and scavenging of reactive oxygen species in chloroplasts and their functions. *Plant Physiol.* **2006**, *141*, 391–396. [CrossRef] [PubMed]
101. Soliman, W.S.; Fujimori, M.; Tase, K.; Sugiyama, S.-I. Oxidative stress and physiological damage under prolonged heat stress in C3 grass *Lolium perenne*. *Grassl. Sci.* **2011**, *57*, 101–106. [CrossRef]
102. Halliwell, B. Oxidative stress and neurodegeneration: Where are we now? *J. Neurochem.* **2006**, *97*, 1634–1658. [CrossRef] [PubMed]
103. Møller, I.M.; Jensen, P.E.; Hansson, A. Oxidative Modifications to Cellular Components in Plants. *Annu. Rev. Plant Biol.* **2007**, *58*, 459–481. [CrossRef] [PubMed]
104. Karuppanapandian, T.; Wang, H.W.; Prabakaran, N.; Jeyalakshmi, K.; Kwon, M.; Manoharan, K.; Kim, W. 2,4-dichlorophenoxyacetic acid-induced leaf senescence in mung bean (*Vigna radiata* L. Wilczek) and senescence inhibition by co-treatment with silver nanoparticles. *Plant Physiol. Biochem.* **2011**, *49*, 168–177. [CrossRef] [PubMed]
105. Huang, B.; Xu, C. Identification and characterization of proteins associated with plant tolerance to heat stress. *J. Integr. Plant Biol.* **2008**, *50*, 1230–1237. [CrossRef] [PubMed]
106. Camejo, D.; Jiménez, A.; Alarcón, J.J.; Torres, W.; Gómez, J.M.; Sevilla, F. Changes in photosynthetic parameters and antioxidant activities following heat-shock treatment in tomato plants. *Funct. Plant Biol.* **2006**, *33*, 177–187. [CrossRef]
107. Rodríguez, M.; Canales, E.; Borrás-hidalgo, O. Molecular aspects of abiotic stress in plants. *Biotecnol. Apl.* **2005**, *22*, 1–10. [CrossRef]
108. Bavita, A.; Shashi, B.; Navtej, S.B. Nitric oxide alleviates oxidative damage induced by high temperature stress in wheat. *CSIR-NISCAIR* **2012**, *50*, 372–378.
109. Savicka, M.; Škute, N. Effects of high temperature on malondialdehyde content, superoxide production and growth changes in wheat seedlings (*Triticum aestivum* L.). *Ekologija* **2010**, *56*, 26–33. [CrossRef]
110. Qi, Y.; Wang, H.; Zou, Y.; Liu, C.; Liu, Y.; Wang, Y.; Zhang, W. Over-expression of mitochondrial heat shock protein 70 suppresses programmed cell death in rice. *FEBS Lett.* **2011**, *585*, 231–239. [CrossRef] [PubMed]
111. Srivastava, S.; Pathak, A.D.; Gupta, P.S.; Kumar, A.; Kumar, A. Hydrogen peroxide-scavenging enzymes impart tolerance to high temperature induced oxidative stress in sugarcane. *J. Environ. Biol.* **2012**, *33*, 657–661. [PubMed]
112. Sarieva, G.E.; Kenzhebaeva, S.S.; Lichtenthaler, H.K. Adaptation potential of photosynthesis in wheat cultivars with a capability of leaf rolling under high temperature conditions. *Russ. J. Plant Physiol.* **2010**, *57*, 28–36. [CrossRef]
113. Wang, W.; Vinocur, B.; Shoseyov, O.; Altman, A. Role of plant heat-shock proteins and molecular chaperones in the abiotic stress response. *Trends Plant Sci.* **2004**, *9*, 244–252. [CrossRef] [PubMed]
114. Sehgal, A.; Sita, K.; Nayyar, H. Heat stress in plants: Sensing and defense mechanisms. *J. plant Sci. Res.* **2016**, *32*, 195–210.
115. Queitsch, C. Heat shock protein 101 plays a crucial role in thermotolerance in *Arabidopsis*. *Plant Cell Online* **2000**, *12*, 479–492. [CrossRef]
116. Vinocur, B.; Altman, A. Recent advances in engineering plant tolerance to abiotic stress: Achievements and limitations. *Curr. Opin. Biotechnol.* **2005**, *16*, 123–132. [CrossRef] [PubMed]
117. White, P.J.; Broadley, M.R. Calcium in plants. *Ann. Bot.* **2003**, *92*, 487–511. [CrossRef] [PubMed]
118. White, P.J. Calcium channels in higher plants. *Biochim. Biophys. Acta–Biomembr.* **2000**, *1465*, 171–189. [CrossRef]
119. Hirschi, K. Vacuolar H^+/Ca^{2+} transport: Who's directing the traffic? *Trends Plant Sci.* **2001**, *6*, 100–104. [CrossRef]
120. Sze, H.; Liang, F.; Hwang, I.; Curran, A.C.; Harper, J.F. Diversity and regulation of plant Ca^{2+} p umps: Insights from Expression in Yeast. *Annu. Rev. Plant Physiol. Plant Mol. Biol.* **2002**, *51*, 433–462. [CrossRef] [PubMed]

121. Larkindale, J.; Knight, M.R. Protection against heat stress-induced oxidative damage in arabidopsis involves calcium, abscisic acid, ethylene, and salicylic acid. *Plant Physiol.* **2002**, *128*, 682–695. [CrossRef] [PubMed]
122. Larkindale, J.; Huang, B. Thermotolerance and antioxidant systems in *Agrostis stolonifera*: Involvement of salicylic acid, abscisic acid, calcium, hydrogen peroxide, and ethylene. *J. Plant Physiol.* **2004**, *161*, 405–413. [CrossRef] [PubMed]
123. Li, B.; Liu, H.; Sun, D.; Zhou, R. Ca^{2+} and calmodulin modulate DNA-binding activity of maize heat shock transcription factor in vitro. *Plant Cell Physiol.* **2004**, *45*, 627–634. [CrossRef] [PubMed]
124. Mishina, T.E.; Zeier, J. Pathogen-associated molecular pattern recognition rather than development of tissue necrosis contributes to bacterial induction of systemic acquired resistance in Arabidopsis. *Plant J.* **2007**, *50*, 500–513. [CrossRef] [PubMed]
125. Uchida, A.; Jagendorf, A.T.; Hibino, T.; Takabe, T.; Takabe, T. Effects of hydrogen peroxide and nitric oxide on both salt and heat stress tolerance in rice. *Plant Sci.* **2002**, *163*, 515–523. [CrossRef]
126. Song, L.; Ding, W.; Zhao, M.; Sun, B.; Zhang, L. Nitric oxide protects against oxidative stress under heat stress in the calluses from two ecotypes of reed. *Plant Sci.* **2006**, *171*, 449–458. [CrossRef] [PubMed]
127. Kaur, N.; Gupta, A.K. Signal transduction pathways under abiotic stresses in plants. *Curr. Sci.* **2005**, *88*, 1771–1780.
128. Proveniers, M.C.G.; Van Zanten, M. High temperature acclimation through PIF4 signaling. *Trends Plant Sci.* **2013**, *18*, 59–64. [CrossRef] [PubMed]
129. Collins, N.C.; Tardieu, F.; Tuberosa, R. Quantitative trait loci and crop performance under abiotic stress: Where do we stand? *Plant Physiol.* **2008**, *147*, 469–486. [CrossRef] [PubMed]
130. Ainsworth, E.A.; Ort, D.R. How do we improve crop production in a warming world? *Plant Physiol.* **2010**, *154*, 526–530. [CrossRef] [PubMed]
131. Jagadish, S.V.K.; Muthurajan, R.; Oane, R.; Wheeler, T.R.; Heuer, S.; Bennett, J.; Craufurd, P.Q. Physiological and proteomic approaches to address heat tolerance during anthesis in rice (*Oryza sativa* L.). *J. Exp. Bot.* **2010**, *61*, 143–156. [CrossRef] [PubMed]
132. Pinto, R.S.; Reynolds, M.P.; Mathews, K.L.; McIntyre, C.L.; Olivares-Villegas, J.J.; Chapman, S.C. Heat and drought adaptive QTL in a wheat population designed to minimize confounding agronomic effects. *Theor. Appl. Genet.* **2010**, *121*, 1001–1021. [CrossRef] [PubMed]
133. Bonneau, J.; Taylor, J.; Parent, B.; Bennett, D.; Reynolds, M.; Feuillet, C.; Langridge, P.; Mather, D. Multi-environment analysis and improved mapping of a yield-related QTL on chromosome 3B of wheat. *Theor. Appl. Genet.* **2013**, *126*, 747–761. [CrossRef] [PubMed]
134. Lei, D.; Tan, L.; Liu, F.; Chen, L.; Sun, C. Identification of heat-sensitive QTL derived from common wild rice (*Oryza rufipogon* Griff.). *Plant Sci.* **2013**, *201*, 121–127. [CrossRef] [PubMed]
135. Paliwal, R.; Röder, M.S.; Kumar, U.; Srivastava, J.P.; Joshi, A.K. QTL mapping of terminal heat tolerance in hexaploid wheat (*T. aestivum* L.). *Theor. Appl. Genet.* **2012**, *125*, 561–575. [CrossRef] [PubMed]
136. Vijayalakshmi, K.; Fritz, A.K.; Paulsen, G.M.; Bai, G.; Pandravada, S.; Gill, B.S. Modeling and mapping QTL for senescence-related traits in winter wheat under high temperature. *Mol. Breed.* **2010**, *26*, 163–175. [CrossRef]
137. Mason, R.E.; Mondal, S.; Beecher, F.W.; Hays, D.B. Genetic loci linking improved heat tolerance in wheat (*Triticum aestivum* L.) to lower leaf and spike temperatures under controlled conditions. *Euphytica* **2011**, *180*, 181–194. [CrossRef]
138. Thudi, M.; Upadhyaya, H.D.; Rathore, A.; Gaur, P.M.; Krishnamurthy, L.; Roorkiwal, M.; Nayak, S.N.; Chaturvedi, S.K.; Basu, P.S.; Gangarao, N.V.P. R.; et al. Genetic dissection of drought and heat tolerance in chickpea through genome-wide and candidate gene-based association mapping approaches. *PLoS ONE* **2014**, *9*, e96758. [CrossRef] [PubMed]
139. Grilli, G.V.G.; Braz, L.T.; Lemos, E.G.M. QTL identification for tolerance to fruit set in tomato by FAFLP markers. *Crop Breed. Appl. Biotechnol.* **2007**, *7*, 234–241. [CrossRef]
140. Shuancang, Y.; Yongiian, W.; Xiaoying, Z. Mapping and analysis QTL controlling heat tolerance in *Brassica campestris* L. ssp pekinensis. *Acta Hortic. Sin.* **2003**, *30*, 417–420.
141. Bhusal, N.; Sarial, A.K.; Sharma, P.; Sareen, S. Mapping QTLs for grain yield components in wheat under heat stress. *PLoS ONE* **2017**, *12*, e0189594. [CrossRef] [PubMed]

142. Mason, R.E.; Mondal, S.; Beecher, F.W.; Pacheco, A.; Jampala, B.; Ibrahim, A.M.H.; Hays, D.B. QTL associated with heat susceptibility index in wheat (*Triticum aestivum* L.) under short-term reproductive stage heat stress. *Euphytica* **2010**, *174*, 423–436. [CrossRef]
143. Tiwari, C.; Wallwork, H.; Kumar, U.; Dhari, R.; Arun, B.; Mishra, V.K.; Reynolds, M.P.; Joshi, A.K. Molecular mapping of high temperature tolerance in bread wheat adapted to the Eastern Gangetic Plain region of India. *Field Crop. Res.* **2013**, *154*, 201–210. [CrossRef]
144. Sharma, D.K.; Torp, A.M.; Rosenqvist, E.; Ottosen, C.O.; Andersen, S.B. QTLs and potential candidate genes for heat stress tolerance identified from the mapping populations specifically segregating for F v /F m in wheat. *Front. Plant Sci.* **2017**, *8*. [CrossRef] [PubMed]
145. Ps, S.; Sv, A.M.; Prakash, C.; Mk, R.; Tiwari, R.; Mohapatra, T.; Singh, N.K. High resolution mapping of QTLs for heat tolerance in rice using a 5K SNP array. *Rice* **2017**, *10*. [CrossRef] [PubMed]
146. Chang, H.C.; Tang, Y.C.; Hayer-Hartl, M.; Hartl, F.U. SnapShot: Molecular chaperones, Part I. *Cell* **2007**, *128*, 89–90. [CrossRef] [PubMed]
147. Zhou, W.-H.; Xue, D.-W.; Zhang, G.-P. Protein response of rice leaves to high temperature stress and its difference of genotypes at different growth stage. *Acta Agron. Sin.* **2011**, *37*, 820–831. [CrossRef]
148. Al-Whaibi, M.H. Plant heat-shock proteins: A mini review. *J. King Saud Univ.-Sci.* **2011**, *23*, 139–150. [CrossRef]
149. Hong, S.W.; Vierling, E. Hsp101 is necessary for heat tolerance but dispensable for development and germination in the absence of stress. *Plant J.* **2001**, *27*, 25–35. [CrossRef] [PubMed]
150. Lin, M.Y.; Chai, K.H.; Ko, S.S.; Kuang, L.Y.; Lur, H.S.; Charng, Y.Y. A positive feedback loop between HEAT SHOCK PROTEIN101 and HEAT STRESS-ASSOCIATED 32-KD PROTEIN modulates long-term acquired thermotolerance illustrating diverse heat stress responses in rice varieties. *Plant Physiol.* **2014**, *164*, 2045–2053. [CrossRef] [PubMed]
151. Gupta, S.C.; Sharma, A.; Mishra, M.; Mishra, R.K.; Chowdhuri, D.K. Heat shock proteins in toxicology: How close and how far? *Life Sci.* **2010**, *86*, 377–384. [CrossRef] [PubMed]
152. Nguyen, N.; Francoeur, N.; Chartrand, V.; Klarskov, K.; Guillemette, G.; Boulay, G. Insulin promotes the association of heat shock protein 90 with the inositol 1,4,5-trisphosphate receptor to dampen its Ca2+ release activity. *Endocrinology* **2009**, *150*, 2190–2196. [CrossRef] [PubMed]
153. Te, J.; Jia, L.; Rogers, J.; Miller, A.; Hartson, S.D. Novel subunits of the mammalian Hsp90 signal transduction chaperone. *J. Proteome Res.* **2007**, *6*, 1963–1973. [CrossRef] [PubMed]
154. Liu, Y.; Burch-Smith, T.; Schiff, M.; Feng, S.; Dinesh-Kumar, S.P. Molecular chaperone Hsp90 associates with resistance protein N and its signaling proteins SGT1 and Rar1 to modulate an innate immune response in plants. *J. Biol. Chem.* **2004**, *279*, 2101–2108. [CrossRef] [PubMed]
155. Queitsch, C.; Sangstert, T.A.; Lindquist, S. Hsp90 as a capacitor of phenotypic variation. *Nature* **2002**, *417*, 618–624. [CrossRef] [PubMed]
156. Sangster, T.A.; Queitsch, C. The HSP90 chaperone complex, an emerging force in plant development and phenotypic plasticity. *Curr. Opin. Plant Biol.* **2005**, *8*, 86–92. [CrossRef] [PubMed]
157. Sangster, T.A.; Bahrami, A.; Wilczek, A.; Watanabe, E.; Schellenberg, K.; McLellan, C.; Kelley, A.; Kong, S.W.; Queitsch, C.; Lindquist, S. Phenotypic diversity and altered environmental plasticity in Arabidopsis thaliana with reduced Hsp90 levels. *PLoS ONE* **2007**, *2*. [CrossRef] [PubMed]
158. Xu, J.; Xue, C.; Xue, D.; Zhao, J.; Gai, J.; Guo, N.; Xing, H. Overexpression of *GmHsp90s*, a heat shock protein90 (Hsp90) gene family cloning from soybean, decrease damage of abiotic stresses in *Arabidopsis thaliana*. *PLoS ONE* **2013**, *8*, e69810. [CrossRef]
159. Swindell, W.R.; Huebner, M.; Weber, A.P. Transcriptional profiling of *Arabidopsis* heat shock proteins and transcription factors reveals extensive overlap between heat and non-heat stress response pathways. *BMC Genom.* **2007**, *8*, 125. [CrossRef] [PubMed]
160. Liberek, K.; Lewandowska, A.; Ziętkiewicz, S. Chaperones in control of protein disaggregation. *EMBO J.* **2008**, *27*, 328–335. [CrossRef] [PubMed]
161. Zhong, L.; Zhou, W.; Wang, H.; Ding, S.; Lu, Q.; Wen, X.; Peng, L.; Zhang, L.; Lu, C. Chloroplast small heat shock protein HSP21 interacts with plastid nucleoid protein pTAC5 and is essential for chloroplast development in arabidopsis under heat stress. *Plant Cell* **2013**, *25*, 2925–2943. [CrossRef] [PubMed]
162. Sarkar, N.K.; Kim, Y.K.; Grover, A. Rice sHsp genes: Genomic organization and expression profiling under stress and development. *BMC Genom.* **2009**, *10*, 393. [CrossRef] [PubMed]

163. Itoh, H.; Komatsuda, A.; Ohtani, H.; Wakui, H.; Imai, H.; Sawada, K.I.; Otaka, M.; Ogura, M.; Suzuki, A.; Hamada, F. Mammalian HSP60 is quickly sorted into the mitochondria under conditions of dehydration. *Eur. J. Biochem.* **2002**, *269*, 5931–5938. [CrossRef] [PubMed]
164. Apuya, N.R. The *Arabidopsis* embryo mutant schlepperless has a defect in the *Chaperonin-60alpha* gene. *Plant Physiol.* **2001**, *126*, 717–730. [CrossRef] [PubMed]
165. Sung, D.; Kaplan, F.; Guy, C.L. Plant Hsp70 molecular chaperones: Protein structure, gene family, expression and function. *Physiol Plant* **2001**, *113*, 443–451. [CrossRef]
166. Su, P.H.; Li, H.M. Arabidopsis stromal 70-kD heat shock proteins are essential for plant development and important for thermotolerance of germinating seeds. *Plant Physiol.* **2008**, *146*, 1231–1241. [CrossRef] [PubMed]
167. Park, C.J.; Bart, R.; Chern, M.; Canlas, P.E.; Bai, W.; Ronald, P.C. Overexpression of the endoplasmic reticulum chaperone BiP3 regulates XA21-mediated innate immunity in rice. *PLoS ONE* **2010**, *5*. [CrossRef] [PubMed]
168. Hartl, F.U.; Bracher, A.; Hayer-Hartl, M. Molecular chaperones in protein folding and proteostasis. *Nature* **2011**, *475*, 324–332. [CrossRef] [PubMed]
169. Frydman, J. Folding of newly translated proteins in vivo: The role of molecular chaperones. *Annu. Rev. Biochem.* **2001**, *70*, 603–647. [CrossRef] [PubMed]
170. Usman, M. G.; Rafii, M. Y.; Martini, M. Y.; Yusuff, O. A.; Ismail, M. R.; Miah, G. Molecular analysis of Hsp70 mechanisms in plants and their function in response to stress. *Biotechnol. Genet. Eng. Rev.* **2017**, *33*, 26–39. [CrossRef] [PubMed]
171. Jung, K.H.; Gho, H.J.; Nguyen, M.X.; Kim, S.R.; An, G. Genome-wide expression analysis of HSP70 family genes in rice and identification of a cytosolic HSP70 gene highly induced under heat stress. *Funct. Integr. Genom.* **2013**, *13*, 391–402. [CrossRef] [PubMed]
172. Yu, A.; Li, P.; Tang, T.; Wang, J.; Chen, Y.; Liu, L. Roles of Hsp70s in stress responses of microorganisms, plants, and animals. *Biomed. Res. Int.* **2015**, *2015*. [CrossRef] [PubMed]
173. Wen, F.; Wu, X.; Li, T.; Jia, M.; Liu, X.; Li, P.; Zhou, X.; Ji, X.; Yue, X. Genome-wide survey of heat shock factors and heat shock protein 70s and their regulatory network under abiotic stresses in Brachypodium distachyon. *PLoS ONE* **2017**, *12*, e0180352. [CrossRef] [PubMed]
174. Pratt, W.B.; Toft, D.O. Regulation of signaling protein function and trafficking by the hsp90/hsp70-based chaperone machinery. *Exp. Biol. Med.* **2003**, *228*, 111–133.
175. Bao, F.; Huang, X.; Zhu, C.; Zhang, X.; Li, X.; Yang, S. Arabidopsis HSP90 protein modulates RPP4-mediated temperature-dependent cell death and defense responses. *New Phytol.* **2014**, *202*, 1320–1334. [CrossRef] [PubMed]
176. Agarwal, G.; Garg, V.; Kudapa, H.; Doddamani, D.; Pazhamala, L.T.; Khan, A.W.; Thudi, M.; Lee, S.H.; Varshney, R.K. Genome-wide dissection of AP2/ERF and HSP90 gene families in five legumes and expression profiles in chickpea and pigeonpea. *Plant Biotechnol. J.* **2016**, *14*, 1563–1577. [CrossRef] [PubMed]
177. Zuo, D.; Subjeck, J.; Wang, X.Y. Unfolding the Role of Large Heat Shock Proteins: New Insights and Therapeutic Implications. *Front. Immunol.* **2016**, *7*, 75. [CrossRef] [PubMed]
178. Kim, N.H.; Hwang, B.K. Pepper heat shock protein 70a interacts with the Type III Effector AvrBsT and triggers plant cell death and immunity. *Plant Physiol.* **2015**, *167*, 307–322. [CrossRef] [PubMed]
179. Nieto-Sotelo, J. Maize HSP101 plays important roles in both induced and basal thermotolerance and primary root growth. *Plant Cell Online* **2002**, *14*, 1621–1633. [CrossRef]
180. Wu, T.Y.; Juan, Y.T.; Hsu, Y.H.; Wu, S.H.; Liao, H.T.; Fung, R.W.M.; Charng, Y.Y. Interplay between heat shock proteins HSP101 and HSA32 prolongs heat acclimation memory posttranscriptionally in *Arabidopsis*. *Plant Physiol.* **2013**, *161*, 2075–2084. [CrossRef] [PubMed]
181. Lee, U.; Rioflorido, I.; Hong, S.W.; Larkindale, J.; Waters, E.R.; Vierling, E. The Arabidopsis ClpB/Hsp100 family of proteins: Chaperones for stress and chloroplast development. *Plant J.* **2007**, *49*, 115–127. [CrossRef] [PubMed]
182. Zhou, L.; Liu, Z.; Liu, Y.; Kong, D.; Li, T.; Yu, S.; Mei, H.; Xu, X.; Liu, H.; Chen, L.; et al. A novel gene *OsAHL1* improves both drought avoidance and drought tolerance in rice. *Sci. Rep.* **2016**, *6*, 1–15. [CrossRef] [PubMed]
183. Wang, Y.; Lin, S.; Song, Q.; Li, K.; Tao, H.; Huang, J.; Chen, X.; Que, S.; He, H. Genome-wide identification of heat shock proteins (Hsps) and Hsp interactors in rice: Hsp70s as a case study. *BMC Genom.* **2014**, *15*. [CrossRef] [PubMed]

184. Grover, A.; Mittal, D.; Negi, M.; Lavania, D. Generating high temperature tolerant transgenic plants: Achievements and challenges. *Plant Sci.* **2013**, *205–206*, 38–47. [CrossRef] [PubMed]
185. Song, A.; Zhu, X.; Chen, F.; Gao, H.; Jiang, J.; Chen, S. A *chrysanthemum* heat shock protein confers tolerance to Abiotic stress. *Int. J. Mol. Sci.* **2014**, *15*, 5063–5078. [CrossRef] [PubMed]
186. Mu, C.; Zhang, S.; Yu, G.; Chen, N.; Li, X.; Liu, H. Overexpression of small heat shock protein LimHSP16.45 in *Arabidopsis* enhances tolerance to abiotic stresses. *PLoS ONE* **2013**, *8*. [CrossRef] [PubMed]
187. Fragkostefanakis, S.; Röth, S.; Schleiff, E.; Scharf, K.D. Prospects of engineering thermotolerance in crops through modulation of heat stress transcription factor and heat shock protein networks. *Plant Cell Environ.* **2015**, *38*, 1881–1895. [CrossRef] [PubMed]
188. Nishizawa-Yokoi, A.; Nosaka, R.; Hayashi, H.; Tainaka, H.; Maruta, T.; Tamoi, M.; Ikeda, M.; Ohme-Takagi, M.; Yoshimura, K.; Yabuta, Y.; et al. *HsfA1d* and *HsfA1e* involved in the transcriptional regulation of hsfa2 function as key regulators for the hsf signaling network in response to environmental stress. *Plant Cell Physiol.* **2011**, *52*, 933–945. [CrossRef] [PubMed]
189. Yoshida, T.; Sakuma, Y.; Todaka, D.; Maruyama, K.; Qin, F.; Mizoi, J.; Kidokoro, S.; Fujita, Y.; Shinozaki, K.; Yamaguchi-Shinozaki, K. Functional analysis of an *Arabidopsis* heat-shock transcription factor HsfA3 in the transcriptional cascade downstream of the *DREB2A* stress-regulatory system. *Biochem. Biophys. Res. Commun.* **2008**, *368*, 515–521. [CrossRef] [PubMed]
190. Keller, M.; Hu, Y.; Mesihovic, A.; Fragkostefanakis, S.; Schleiff, E.; Simm, S. Alternative splicing in tomato pollen in response to heat stress. *DNA Res.* **2017**, *24*, 205–217. [CrossRef] [PubMed]
191. Sugio, A.; Dreos, R.; Aparicio, F.; Maule, A.J. The cytosolic protein response as a subcomponent of the wider heat shock response in *Arabidopsis*. *Plant Cell Online* **2009**, *21*, 642–654. [CrossRef] [PubMed]
192. Liu, J.; Sun, N.; Liu, M.; Liu, J.; Du, B.; Wang, X.; Qi, X. An autoregulatory loop controlling *Arabidopsis* HsfA2 expression: Role of heat shock-induced alternative splicing. *Plant Physiol.* **2013**, *162*, 512–521. [CrossRef] [PubMed]
193. Cheng, Q.; Zhou, Y.; Liu, Z.; Zhang, L.; Song, G.; Guo, Z.; Wang, W.; Qu, X.; Zhu, Y.; Yang, D. An alternatively spliced heat shock transcription factor, *OsHSFA2dI*, functions in the heat stress-induced unfolded protein response in rice. *Plant Biol.* **2015**, *17*, 419–429. [CrossRef] [PubMed]
194. Jorgensen, R.A.; Dorantes-Acosta, A.E. Conserved peptide upstream open reading frames are associated with regulatory genes in angiosperms. *Front. Plant Sci.* **2012**, *3*, 191. [CrossRef] [PubMed]
195. Von Arnim, A.G.; Jia, Q.; Vaughn, J.N. Regulation of plant translation by upstream open reading frames. *Plant Sci.* **2014**, *214*, 1–12. [CrossRef] [PubMed]
196. Scharf, K.D.; Berberich, T.; Ebersberger, I.; Nover, L. The plant heat stress transcription factor (Hsf) family: Structure, function and evolution. *Biochim. Biophys. Acta–Gene Regul. Mech.* **2012**, *1819*, 104–119. [CrossRef] [PubMed]
197. Song, L.; Jiang, Y.; Zhao, H.; Hou, M. Acquired thermotolerance in plants. *Plant Cell. Tissue Organ Cult.* **2012**, *111*, 265–276. [CrossRef]
198. Evrard, A.; Kumar, M.; Lecourieux, D.; Lucks, J.; von Koskull-Döring, P.; Hirt, H. Regulation of the heat stress response in *Arabidopsis* by MPK6-targeted phosphorylation of the heat stress factor HsfA2. *PeerJ* **2013**, *1*, e59. [CrossRef] [PubMed]
199. Nishizawa-Yokoi, A.; Tainaka, H.; Yoshida, E.; Tamoi, M.; Yabuta, Y.; Shigeoka, S. The 26S proteasome function and Hsp90 activity involved in the regulation of HsfA2 expression in response to oxidative stress. *Plant Cell Physiol.* **2010**, *51*, 486–496. [CrossRef] [PubMed]
200. Cohen-Peer, R.; Schuster, S.; Meiri, D.; Breiman, A.; Avni, A. Sumoylation of *Arabidopsis* heat shock factor A2 (HsfA2) modifies its activity during acquired thermotholerance. *Plant Mol. Biol.* **2010**, *74*, 33–45. [CrossRef] [PubMed]
201. Lee, J.H.; Hübel, A.; Schöffl, F.H. Derepression of the activity of genetically engineered heat shock factor causes constitutive synthesis of heat shock proteins and increased thermotolerance in transgenic Arabidopsis. *Plant J.* **1995**, *8*, 603–612. [CrossRef] [PubMed]
202. Liu, J.; Shono, M. Characterization of mitochondria-located small heat shock protein from tomato (*Lycopersicon esculentum*). *Plant Cell Physiol.* **1999**, *40*, 1297–1304. [PubMed]
203. Sanmiya, K.; Suzuki, K.; Egawa, Y.; Shono, M. Mitochondrial small heat-shock protein enhances thermotolerance in tobacco plants. *FEBS Lett.* **2004**, *557*, 265–268. [CrossRef]

204. Katiyar-Agarwal, S.; Agarwal, M.G.A. Heat-tolerant basmati rice engineered by over-expression of hsp101. *Plant Mol. Biol.* **2003**, 677–686. [CrossRef]
205. Lee, B.H.; Won, S.H.; Lee, H.S.; Miyao, M.; Chung, W.I.; Kim, I.J.; Jo, J. Expression of the chloroplast-localized small heat shock protein by oxidative stress in rice. *Gene* **2000**, *245*, 283–290. [CrossRef]
206. Murakami, T.; Matsuba, S.; Funatsuki, H.; Kawaguchi, K.; Saruyama, H.; Tanida, M.; Sato, Y. Over-expression of a small heat shock protein, sHSP17.7, confers both heat tolerance and UV-B resistance to rice plants. *Mol. Breed.* **2004**, *13*, 165–175. [CrossRef]
207. Ono, K.; Hibino, T.; Kohinata, T.; Suzuki, S.; Tanaka, Y.; Nakamura, T.; Takabe, T.; Takabe, T. Overexpression of *DnaK* from a halotolerant *cyanobacterium Aphanothece halophytica* enhances the high-temperatue tolerance of tobacco during germination and early growth. *Plant Sci.* **2001**, *160*, 455–461. [CrossRef]
208. Yang, X.; Liang, Z.; Lu, C. Genetic engineering of the biosynthesis of glycinebetaine enhances photosynthesis against high temperature stress in transgenic tobacco plants. *Plant Physiol.* **2005**, *138*, 2299–2309. [CrossRef] [PubMed]
209. Sharkey, T.D.; Badger, M.R.; von Caemmerer, S.; Andrews, T.J. Increased heat sensitivity of photosynthesis in tobacco plants with reduced Rubisco activase. *Photosynth. Res.* **2001**, *67*, 147–156. [CrossRef] [PubMed]
210. Murakami, Y.; Tsuyama, M.; Kobayash, Y.; Kodama, H.; Iba, K. Trienoic fatty acids and plant tolerance of high temperature. *Science* **2000**, *287*, 476–479. [CrossRef] [PubMed]
211. Rizhsky, L. When Defense Pathways Collide. The response of *Arabidopsis* to a combination of drought and heat stress. *Plant Physiol.* **2004**, *134*, 1683–1696. [CrossRef] [PubMed]
212. Shi, W.M.; Muramoto, Y.; Ueda, A.; Takabe, T. Cloning of peroxisomal ascorbate peroxidase gene from barley and enhanced thermotolerance by overexpressing in *Arabidopsis thaliana*. *Gene* **2001**, *273*, 23–27. [CrossRef]
213. Zhao, Y.; Tian, X.; Wang, F.; Zhang, L.; Xin, M.; Hu, Z.; Yao, Y.; Ni, Z.; Sun, Q.; Peng, H. Characterization of wheat *MYB* genes responsive to high temperatures. *BMC Plant Biol.* **2017**, *17*, 208. [CrossRef] [PubMed]
214. Zang, X.; Geng, X.; Wang, F.; Liu, Z.; Zhang, L.; Zhao, Y.; Tian, X.; Ni, Z.; Yao, Y.; Xin, M.; et al. Overexpression of wheat ferritin gene *TaFER-5B* enhances tolerance to heat stress and other abiotic stresses associated with the ROS scavenging. *BMC Plant Biol.* **2017**, *17*, 14. [CrossRef] [PubMed]
215. Zhang, L.; Geng, X.; Zhang, H.; Zhou, C.; Zhao, A.; Wang, F.; Zhao, Y.; Tian, X.; Hu, Z.; Xin, M.; et al. Isolation and characterization of heat-responsive gene *TaGASR1* from wheat (*Triticum aestivum* L.). *J. Plant Biol.* **2017**, *60*, 57–65. [CrossRef]
216. Hu, X.J.; Chen, D.; Lynne Mclntyre, C.; Fernanda Dreccer, M.; Zhang, Z.B.; Drenth, J.; Kalaipandian, S.; Chang, H.; Xue, G.P. Heat shock factor C2a serves as a proactive mechanism for heat protection in developing grains in wheat via an ABA-mediated regulatory pathway. *Plant Cell Environ.* **2017**, *41*, 79–98. [CrossRef] [PubMed]
217. He, G.H.; Xu, J.Y.; Wang, Y.X.; Liu, J.M.; Li, P.S.; Chen, M.; Ma, Y.Z.; Xu, Z.S. Drought-responsive WRKY transcription factor genes *TaWRKY1* and *TaWRKY33* from wheat confer drought and/or heat resistance in *Arabidopsis*. *BMC Plant Biol.* **2016**, *16*, 116. [CrossRef] [PubMed]
218. Guo, W.; Zhang, J.; Zhang, N.; Xin, M.; Peng, H.; Hu, Z.; Ni, Z.; Du, J. The wheat NAC transcription factor *TaNAC2L* is regulated at the transcriptional and post-translational levels and promotes heat stress tolerance in transgenic *arabidopsis*. *PLoS ONE* **2015**, *10*, e0135667. [CrossRef] [PubMed]
219. Singh, A.; Khurana, P. Molecular and functional characterization of a wheat B2 protein imparting adverse temperature tolerance and influencing plant growth. *Front. Plant Sci.* **2016**, *7*, 642. [CrossRef]
220. Zang, X.; Geng, X.; Liu, K.; Wang, F.; Liu, Z.; Zhang, L.; Zhao, Y.; Tian, X.; Hu, Z.; Yao, Y.; et al. Ectopic expression of TaOEP16-2-5B, a wheat plastid outer envelope protein gene, enhances heat and drought stress tolerance in transgenic *Arabidopsis* plants. *Plant Sci.* **2017**, *258*, 1–11. [CrossRef] [PubMed]
221. Wang, F.; Zang, X.; Kabir, M.R.; Liu, K.; Liu, Z.; Ni, Z.; Yao, Y.; Hu, Z.; Sun, Q.; Peng, H. A wheat lipid transfer protein 3 could enhance the basal thermotolerance and oxidative stress resistance of *Arabidopsis*. *Gene* **2014**, *550*, 18–26. [CrossRef] [PubMed]
222. Foresi, N.; Mayta, M.L.; Lodeyro, A.F.; Scuffi, D.; Correa-Aragunde, N.; García-Mata, C.; Casalongué, C.; Carrillo, N.; Lamattina, L. Expression of the tetrahydrofolate-dependent nitric oxide synthase from the green alga *Ostreococcus tauri* increases tolerance to abiotic stresses and influences stomatal development in *Arabidopsis*. *Plant J.* **2015**, *82*, 806–821. [CrossRef] [PubMed]

223. Feng, L.; Wang, K.; Li, Y.; Tan, Y.; Kong, J.; Li, H.; Li, Y.; Zhu, Y. Overexpression of *SBPase* enhances photosynthesis against high temperature stress in transgenic rice plants. *Plant Cell Rep.* **2007**, *26*, 1635–1646. [CrossRef] [PubMed]
224. Xiong, H.; Li, J.; Liu, P.; Duan, J.; Zhao, Y.; Guo, X.; Li, Y.; Zhang, H.; Ali, J.; Li, Z. Overexpression of *OsMYB48-1*, a novel MYB-related transcription factor, enhances drought and salinity tolerance in rice. *PLoS ONE* **2014**, *9*, e92913. [CrossRef] [PubMed]
225. Liu, J.; Zhang, C.; Wei, C.; Liu, X.; Wang, M.; Yu, F.; Xie, Q.; Tu, J. The RING Finger Ubiquitin E3 Ligase *OsHTAS* enhances heat tolerance by promoting H_2O_2-induced stomatal closure in rice. *Plant Physiol.* **2016**, *170*, 429–443. [CrossRef] [PubMed]
226. El-kereamy, A.; Bi, Y.M.; Ranathunge, K.; Beatty, P.H.; Good, A.G.; Rothstein, S.J. The rice R2R3-MYB transcription factor OsMYB55 is involved in the tolerance to high temperature and modulates amino acid metabolism. *PLoS ONE* **2012**, *7*. [CrossRef]
227. Sohn, S.O.; Back, K. Transgenic rice tolerant to high temperature with elevated contents of dienoic fatty acids. *Biol. Plant.* **2007**, *51*, 340–342. [CrossRef]
228. Casaretto, J.A.; El-kereamy, A.; Zeng, B.; Stiegelmeyer, S.M.; Chen, X.; Bi, Y.M.; Rothstein, S.J. Expression of *OsMYB55* in maize activates stress-responsive genes and enhances heat and drought tolerance. *BMC Genom.* **2016**, *17*, 312. [CrossRef] [PubMed]
229. Zhai, Y.; Wang, H.; Liang, M.; Lu, M. Both silencing- and over-expression of pepper *CaATG8c* gene compromise plant tolerance to heat and salt stress. *Environ. Exp. Bot.* **2017**, *141*, 10–18. [CrossRef]
230. Meng, X.; Wang, J.R.; Wang, G.D.; Liang, X.Q.; Li, X.D.; Meng, Q.W. An *R2R3-MYB* gene, *LeAN2*, positively regulated the thermo-tolerance in transgenic tomato. *J. Plant Physiol.* **2015**, *175*, 191–197. [CrossRef] [PubMed]
231. Yeh, C.H.; Kaplinsky, N.J.; Hu, C.; Charng, Y.Y. Some like it hot, some like it warm: Phenotyping to explore thermotolerance diversity. *Plant Sci.* **2012**, *195*, 10–23. [CrossRef] [PubMed]
232. Priest, H.D.; Fox, S.E.; Rowley, E.R.; Murray, J.R.; Michael, T.P.; Mockler, T.C. Analysis of global gene expression in *Brachypodium distachyon* reveals extensive network plasticity in response to abiotic stress. *PLoS ONE* **2014**, *9*. [CrossRef] [PubMed]
233. González-Schain, N.; Dreni, L.; Lawas, L.M.F.; Galbiati, M.; Colombo, L.; Heuer, S.; Jagadish, K.S.V.; Kater, M.M. Genome-wide transcriptome analysis during anthesis reveals new insights into the molecular basis of heat stress responses in tolerant and sensitive rice varieties. *Plant Cell Physiol.* **2016**, *57*, 57–68. [CrossRef] [PubMed]
234. Kumar, R.; Lavania, D.; Singh, A.K.; Negi, M.; Siddiqui, M.H.; Al-Whaibi, M.H.; Grover, A. Identification and characterization of a small heat shock protein 17.9-CII gene from faba bean (*Vicia faba* L.). *Acta Physiol. Plant.* **2015**, *37*. [CrossRef]
235. Zhang, X.; Wang, J.; Huang, J.; Lan, H.; Wang, C.; Yin, C.; Wu, Y.; Tang, H.; Qian, Q.; Li, J.; et al. Rare allele of OsPPKL1 associated with grain length causes extra-large grain and a significant yield increase in rice. *Proc. Natl. Acad. Sci. USA* **2012**, *109*, 21534–21539. [CrossRef] [PubMed]
236. Iovieno, P.; Punzo, P.; Guida, G.; Mistretta, C.; Van Oosten, M.J.; Nurcato, R.; Bostan, H.; Colantuono, C.; Costa, A.; Bagnaresi, P.; et al. Transcriptomic changes drive physiological responses to progressive drought stress and rehydration in tomato. *Front. Plant Sci.* **2016**, *7*, 1–14. [CrossRef] [PubMed]
237. Shi, J.; Yan, B.; Lou, X.; Ma, H.; Ruan, S. Comparative transcriptome analysis reveals the transcriptional alterations in heat-resistant and heat-sensitive sweet maize (*Zea mays* L.) varieties under heat stress. *BMC Plant Biol.* **2017**, *17*, 26. [CrossRef] [PubMed]
238. Chen, S.; Li, H. Heat Stress Regulates the Expression of Genes at Transcriptional and Post-Transcriptional Levels, Revealed by RNA-seq in Brachypodium distachyon. *Front. Plant Sci.* **2017**, *7*, 1–13. [CrossRef] [PubMed]
239. Jia, J.; Zhou, J.; Shi, W.; Cao, X.; Luo, J.; Polle, A.; Luo, Z. Bin Comparative transcriptomic analysis reveals the roles of overlapping heat-/drought-responsive genes in poplars exposed to high temperature and drought. *Sci. Rep.* **2017**, *7*, 1–17. [CrossRef]
240. Kosová, K.; Vítámvás, P.; Prášil, I.T.; Renaut, J. Changes under abiotic stress-Contribution of proteomics studies to understanding plant stress response. *J. Proteomics* **2011**, *74*, 1301–1322. [CrossRef] [PubMed]
241. Rathi, D.; Gayen, D.; Gayali, S.; Chakraborty, S.; Chakraborty, N. Legume proteomics: Progress, prospects, and challenges. *Proteomics* **2016**, *16*, 310–327. [CrossRef] [PubMed]

242. Das, A.; Eldakak, M.; Paudel, B.; Kim, D.W.; Hemmati, H.; Basu, C.; Rohila, J.S. Leaf proteome analysis reveals prospective drought and heat stress response mechanisms in soybean. *Biomed Res. Int.* **2016**, *2016*. [CrossRef] [PubMed]
243. Valdés-López, O.; Batek, J.; Gomez-Hernandez, N.; Nguyen, C.T.; Isidra-Arellano, M.C.; Zhang, N.; Joshi, T.; Xu, D.; Hixson, K.K.; Weitz, K.K.; et al. Soybean roots grown under heat stress show global changes in their transcriptional and proteomic profiles. *Front. Plant Sci.* **2016**, *7*, 517. [CrossRef] [PubMed]
244. Zhou, S.; Sauvé, R.J.; Liu, Z.; Reddy, S.; Bhatti, S. Heat-induced Proteome Changes in Tomato Leaves. *J. Am. Soc. Hortic. Sci.* **2012**, *136*, 2012. [CrossRef]
245. Keller, M.; Consortium, S.; Simm, S. The coupling of transcriptome and proteome adaptation during development and heat stress response of tomato pollen. *BMC Genom.* **2018**, 447. [CrossRef] [PubMed]
246. Sang, Q.; Shan, X.; An, Y.; Shu, S.; Sun, J.; Guo, S. Proteomic analysis reveals the positive effect of exogenous spermidine in tomato seedlings' response to high-temperature stress. *Front. Plant Sci.* **2017**, *8*, 120. [CrossRef] [PubMed]
247. Lin, H.H.; Lin, K.H.; Syu, J.Y.; Tang, S.Y.; Lo, H.F. Physiological and proteomic analysis in two wild tomato lines under waterlogging and high temperature stress. *J. Plant Biochem. Biotechnol.* **2016**, *25*, 87–96. [CrossRef]
248. Wienkoop, S.; Morgenthal, K.; Wolschin, F.; Scholz, M.; Selbig, J.; Weckwerth, W. Integration of metabolomic and proteomic phenotypes. *Mol. Cell. Proteom.* **2008**, *7*, 1725–1736. [CrossRef] [PubMed]
249. Caldana, C.; Degenkolbe, T.; Cuadros-Inostroza, A.; Klie, S.; Sulpice, R.; Leisse, A.; Steinhauser, D.; Fernie, A.R.; Willmitzer, L.; Hannah, M.A. High-density kinetic analysis of the metabolomic and transcriptomic response of *Arabidopsis* to eight environmental conditions. *Plant J.* **2011**, *67*, 869–884. [CrossRef] [PubMed]
250. De Block, M.; Verduyn, C.; De Brouwer, D.; Cornelissen, M. Poly(ADP-ribose) polymerase in plants affects energy homeotasis, cell death and stress tolerance. *Plant J.* **2005**, *41*, 95–106. [CrossRef] [PubMed]
251. Maruyama, K.; Takeda, M.; Kidokoro, S.; Yamada, K.; Sakuma, Y.; Urano, K.; Fujita, M.; Yoshiwara, K.; Matsukura, S.; Morishita, Y.; et al. Metabolic pathways involved in cold acclimation identified by integrated analysis of metabolites and transcripts regulated by *DREB1A* and *DREB2A*. *Plant Physiol.* **2009**, *150*, 1972–1980. [CrossRef] [PubMed]
252. Crisp, P.A.; Ganguly, D.; Eichten, S.R.; Borevitz, J.O.; Pogson, B.J. Reconsidering plant memory: Intersections between stress recovery, RNA turnover, and epigenetics. *Sci. Adv.* **2016**, 2. [CrossRef] [PubMed]
253. Liu, J.; Feng, L.; Li, J.; He, Z. Genetic and epigenetic control of plant heat responses. *Front. Plant Sci.* **2015**, *6*, 267. [CrossRef] [PubMed]
254. Gardiner, L.-J.; Quinton-Tulloch, M.; Olohan, L.; Price, J.; Hall, N.; Hall, A. A genome-wide survey of DNA methylation in hexaploid wheat. *Genome Biol.* **2015**, *16*, 273. [CrossRef] [PubMed]
255. Chinnusamy, V.; Zhu, J.K. Epigenetic regulation of stress responses in plants. *Curr. Opin. Plant. Biol.* **2009**, *12*, 133–139. [CrossRef] [PubMed]
256. Ganguly, D.; Crisp, P.A.; Eichten, S.R.; Pogson, B.J. The *Arabidopsis* DNA methylome is stable under transgenerational drought stress. *Plant. Physiol.* **2017**, *175*. [CrossRef] [PubMed]
257. Sanchez, D.H.; Paszkowski, J. Heat-induced release of epigenetic silencing reveals the concealed role of an imprinted plant gene. *PLoS Genet.* **2014**, *10*. [CrossRef] [PubMed]
258. Gao, G.; Li, J.; Li, H.; Li, F.; Xu, K.; Yan, G.; Chen, B.; Qiao, J.; Wu, X. Comparison of the heat stress induced variations in DNA methylation between heat-tolerant and heat-sensitive rapeseed seedlings. *Breed. Sci.* **2014**, *64*, 125–133. [CrossRef] [PubMed]
259. Li, J.; Huang, Q.; Sun, M.; Zhang, T.; Li, H.; Chen, B.; Xu, K.; Gao, G.; Li, F.; Yan, G.; et al. Global DNA methylation variations after short-term heat shock treatment in cultured microspores of *Brassica napus* cv. Topas. *Sci. Rep.* **2016**, *6*. [CrossRef] [PubMed]
260. Mizuguchi, G.; Shen, X.; Landry, J.; Wu, W.H.; Sen, S.; Wu, C. ATP-driven exchange of histone H2AZ variant catalyzed by SWR1 chromatin remodeling complex. *Science.* **2004**, *303*, 343–348. [CrossRef] [PubMed]
261. Lu, P.Y.T.; Lévesque, N.; Kobor, M.S. NuA4 and SWR1-C: Two chromatin-modifying complexes with overlapping functions and components. *Biochem. Cell. Biol.* **2009**, *87*, 799–815. [CrossRef] [PubMed]
262. Choi, K.; Zhao, X.; Kelly, K.A.; Venn, O.; Higgins, J.D.; Yelina, N.E.; Hardcastle, T.J.; Ziolkowski, P.A.; Copenhaver, G.P.; Franklin, F.C.H.; et al. *Arabidopsis* meiotic crossover hot spots overlap with H2A.Z nucleosomes at gene promoters. *Nat. Genet.* **2013**, *45*, 1327–1338. [CrossRef] [PubMed]

263. Zilberman, D.; Coleman-Derr, D.; Ballinger, T.; Henikoff, S. Histone H2A.Z and DNA methylation are mutually antagonistic chromatin marks. *Nature* **2008**, *456*, 125–129. [CrossRef] [PubMed]
264. Kumar, S.V.; Wigge, P.A. H2A.Z-Containing nucleosomes mediate the thermosensory response in *Arabidopsis*. *Cell* **2010**, *140*, 136–147. [CrossRef] [PubMed]
265. March-Díaz, R.; Reyes, J.C. The beauty of being a variant: H2A.Z and the SWR1 complex in plants. *Mol. Plant.* **2009**, *2*, 565–577. [CrossRef] [PubMed]
266. Ni, Z.; Li, H.; Zhao, Y.; Peng, H.; Hu, Z.; Xin, M.; Sun, Q. Genetic improvement of heat tolerance in wheat: Recent progress in understanding the underlying molecular mechanisms. *Crop. J.* **2018**, *6*, 32–41. [CrossRef]
267. Kumar, R.R.; Pathak, H.; Sharma, S.K.; Kala, Y.K.; Nirjal, M.K.; Singh, G.P.; Goswami, S.; Rai, R.D. Novel and conserved heat-responsive microRNAs in wheat (*Triticum aestivum* L.). *Funct. Integr. Genom.* **2015**, *15*, 323–348. [CrossRef] [PubMed]
268. Charon, C.; Moreno, A.B.; Bardou, F.; Crespi, M. Non-protein-coding RNAs and their interacting RNA-binding proteins in the plant cell nucleus. *Mol. Plant.* **2010**, *3*, 729–739. [CrossRef] [PubMed]
269. Xin, M.; Wang, Y.; Yao, Y.; Song, N.; Hu, Z.; Qin, D.; Xie, C.; Peng, H.; Ni, Z.; Sun, Q. Identification and characterization of wheat long non-protein coding RNAs responsive to powdery mildew infection and heat stress by using microarray analysis and SBS sequencing. *BMC Plant. Biol.* **2011**, *11*. [CrossRef] [PubMed]
270. Xin, M.; Wang, Y.; Yao, Y.; Xie, C.; Peng, H.; Ni, Z.; Sun, Q. Diverse set of microRNAs are responsive to powdery mildew infection and heat stress in wheat (*Triticum aestivum* L.). *BMC Plant. Biol* **2010**, *10*, 123. [CrossRef] [PubMed]
271. Giusti, L.; Mica, E.; Bertolini, E.; De Leonardis, A.M.; Faccioli, P.; Cattivelli, L.; Crosatti, C. microRNAs differentially modulated in response to heat and drought stress in durum wheat cultivars with contrasting water use efficiency. *Funct. Integr. Genom.* **2017**, *17*, 293–309. [CrossRef] [PubMed]
272. Wang, Y.; Sun, F.; Cao, H.; Peng, H.; Ni, Z.; Sun, Q.; Yao, Y. *TamiR159* Directed Wheat *TaGAMYB* cleavage and its involvement in anther development and heat response. *PLoS ONE* **2012**, *7*. [CrossRef] [PubMed]
273. Abdelrahman, M.; Al-Sadi, A.M.; Pour-Aboughadareh, A.; Burritt, D.J.; Tran, L.S.P. Genome editing using CRISPR/Cas9–targeted mutagenesis: An opportunity for yield improvements of crop plants grown under environmental stresses. *Plant. Physiol. Biochem.* **2018**. [CrossRef] [PubMed]
274. Abdallah, N.A.; Prakash, C.S.; McHughen, A.G. Genome editing for crop improvement: Challenges and opportunities. *GM Crops Food* **2015**, *6*, 183–205. [CrossRef] [PubMed]
275. Zhu, C.; Bortesi, L.; Baysal, C.; Twyman, R.M.; Fischer, R.; Capell, T.; Schillberg, S.; Christou, P. Characteristics of genome editing mutations in cereal crops. *Trends Plant. Sci.* **2017**, *22*, 38–52. [CrossRef] [PubMed]
276. Osakabe, Y.; Watanabe, T.; Sugano, S.S.; Ueta, R.; Ishihara, R.; Shinozaki, K.; Osakabe, K. Optimization of CRISPR/Cas9 genome editing to modify abiotic stress responses in plants. *Sci. Rep.* **2016**, *6*. [CrossRef] [PubMed]
277. Zhou, H.; He, M.; Li, J.; Chen, L.; Huang, Z.; Zheng, S.; Zhu, L.; Ni, E.; Jiang, D.; Zhao, B.; et al. Development of commercial thermo-sensitive genic male sterile rice accelerates hybrid rice breeding using the CRISPR/Cas9-mediated TMS5 editing system. *Sci. Rep.* **2016**, *6*. [CrossRef] [PubMed]
278. Klap, C.; Yeshayahou, E.; Bolger, A.M.; Arazi, T.; Gupta, S.K.; Shabtai, S.; Usadel, B.; Salts, Y.; Barg, R. Tomato facultative parthenocarpy results from SlAGAMOUS-LIKE 6 loss of function. *Plant. Biotechnol. J.* **2017**, *15*, 634–647. [CrossRef] [PubMed]

14

Good Riddance? Breaking Disease Susceptibility in the Era of New Breeding Technologies

Stefan Engelhardt, Remco Stam and Ralph Hückelhoven *

Chair of Phytopathology, TUM School of Life Sciences Weihenstephan, Technical University of Munich, 85354 Freising, Germany; stefan1.engelhardt@tum.de (S.E.); stam@wzw.tum.de (R.S.)

* Correspondence: hueckelhoven@wzw.tum.de

Abstract: Despite a high abundance and diversity of natural plant pathogens, plant disease susceptibility is rare. In agriculture however, disease epidemics often occur when virulent pathogens successfully overcome immunity of a single genotype grown in monoculture. Disease epidemics are partially controlled by chemical and genetic plant protection, but pathogen populations show a high potential to adapt to new cultivars or chemical control agents. Therefore, new strategies in breeding and biotechnology are required to obtain durable disease resistance. Generating and exploiting a genetic loss of susceptibility is one of the recent strategies. Better understanding of host susceptibility genes (*S*) and new breeding technologies now enable the targeted mutation of *S* genes for genetic plant protection. Here we summarize biological functions of susceptibility factors and both conventional and DNA nuclease-based technologies for the exploitation of *S* genes. We further discuss the potential trade-offs and whether the genetic loss of susceptibility can provide durable disease resistance.

Keywords: plant immunity; effector-triggered susceptibility; necrotrophic effector; biotroph; susceptibility gene; host reprogramming; pathogen nutrition; plant cell development; natural diversity; CRISPR

---

1. Introduction

In crop production systems, plant diseases are controlled by standard field management practices (e.g., crop rotation, ploughing), usage of disease-resistant cultivars and pesticide applications. However, disease resistance and pesticide efficacy are often not durable because pathogen populations rapidly adapt to the selection pressure that is exerted by these disease control mechanisms. This and potentially harmful effects of pesticides on off-target organisms can render plant protection unsustainable, necessitating novel approaches to combat plant pathogens. In recent years, fundamental research on molecular plant-microbe interactions has revealed new insights on how plants defend themselves against pathogens and how pathogens subvert plant immunity. This knowledge and the development of new breeding technologies holds the potential for innovative approaches in genetic plant protection, which could complement the limitations of conventional technologies to provide greater resistance durability [1].

In plants, invading pathogens are challenged at several levels of plant-pathogen interactions [2]. Preformed defensive barriers together with pathogen-induced plant defense responses successfully restrict parasitic growth on resistant plants. Induced plant defenses have, however, led to the adaptation of pathogens to certain host species and the evolution of host-specific virulence strategies. This includes the secretion of proteinaceous and non-proteinaceous pathogenicity factors that support pathogen virulence. Since these so-called effector molecules are required to actively overcome host immune barriers, the term "effector-triggered susceptibility" (ETS) was coined [1]. Most reports on effector activities show that effectors function as suppressors of plant immune receptor functions, signal transduction and defense reactions [3]. Hence, the suppression of plant defense appears to be pivotal for virulence. Effectors can either specifically modulate host immune processes or more broadly influence host physiology. The latter often contributes to the development of disease symptoms. Such effectors may even provoke strong symptoms when applied as pure substances to plants. In this case, they are considered as toxins that can act either host-specifically or host-nonspecifically. Some effectors influence plant development and some pathogens produce plant hormones or hormone analogs to manipulate host development or physiology.

Pathogen detection either takes place at the plant cell surface, where surface receptor complexes function, or in the host plant cytoplasm or nucleoplasm by intracellular receptor proteins or receptor complexes [1,4]. Plant immune receptors (so-called resistance proteins encoded by major disease resistance genes [*R*]) detect the presence of effector proteins in a race-cultivar specific manner as determined by monogenic inheritance in both the host and parasite. At the molecular level, this classical gene-for-gene model is described by the term "effector-triggered immunity" (ETI) [1]. A more basal, race-nonspecific type of immunity operates within the broader context of ETS and ETI and it is mediated by the detection of a broad spectrum of non-self or altered-self molecules. This is collectively summarized as pattern-triggered immunity (PTI) [4,5].

2. The Principle of Susceptibility Genes and How to Find Them

Host immune components are encoded by dominantly inherited genes, which show either major (qualitative) or minor (quantitative) effects on disease resistance. However, the observation that disease resistance can also be recessively inherited indicates that pathogens can also profit from dominantly inherited host functions or susceptibility factors [1,6,7]. The corresponding dominantly inherited genes are called susceptibility genes (*S*). Recessive *s* genes have been successfully used in conventional and marker assisted plant breeding for the improvement of disease resistance. Recessive *mlo* (*mildew locus o*), several virus resistance genes and ToxA-insensitive *tsn1* genes are prominent examples for this [8,9]. Here we discuss the mechanisms of disease susceptibility and how provoking and exploiting genetic loss of susceptibility can aid durable disease resistance.

Farmers and breeders have selected naturally occurring mutations of *s* genes for the improvement of crop health. Additionally, breeders and researchers searching for mutagenesis-induced resistance in crop and model plants have identified a broad variety of recessive disease resistances. High research effort over the last three decades enabled the identification of several of the corresponding mutations in *S* genes and first insights into the mechanisms of disease susceptibility.

The *mlo*-mediated powdery mildew resistance is perhaps the most prominent example of recessive plant disease resistance. It is of particular interest, because it is race-nonspecific and durable in the field. The *MLO*-gene was originally characterized in spring barley but it seems to function in all interactions in which *MLO S*-gene functions have been studied in detail [8]. Ethiopian highland farmers may have originally selected the barley *mlo-11* allele in old land races, collected during expeditions in the 1930s, and used later in European plant breeding in the 1970s [10]. However, the first description of *mlo* goes back to an X-ray induced powdery mildew mutant generated in the 1940s [11]. Across multiple plant species, many mutagenesis-derived loss-of-function *mlo* alleles exist and *MLO* null-mutants are generally resistant to powdery mildew assuming no genetic redundancy with other *MLO* family members exists. The exact biochemical function of the f protein is not understood, but it may act as a negative regulator of pathogen-triggered and spontaneous defense reactions, putting *mlo* mutants in a primed defensive status [8,12].

The model plant *Arabidopsis thaliana* has been instrumental for the identification of many more susceptibility factors. For instance, forward genetic screens for powdery mildew resistance (*pmr*) or gene expression studies of compatible interactions with diverse biotrophic pathogens, followed by reverse genetic approaches have identified several candidate *S* genes [6,13–15]. Additionally, educated guesses and translational approaches have proved similarly successful in discovering

S genes in crop plants [16,17]. In vivo protein-protein interaction screens are yet another suitable approach to identify susceptibility factors using effector proteins as bait [18,19]. Candidate susceptibility genes were also identified via host gene expression profiling, as it was shown with *Hyaloperonospora arabidopsidis*-infected *Arabidopsis thaliana* and *Phytophthora cinnamoni*-infected *Castanea* [20,21]. Once an *S* gene is identified, studying the physiological function of the susceptibility factor and genetic or physical interactions can identify susceptibility mechanisms or associated pathways and thereby new susceptibility factors [22,23].

3. Biological Functions of Susceptibility Genes

Considering the role of susceptibility genes in compatible plant-microbe interactions, the question arises as to what physiological function host susceptibility factors (or compatibility factors in terms of microbial symbiosis) may exert in healthy and microbe-attacked plants (Figure 1). Some plant susceptibility factors are regulators of host defense responses or cell death. Depending on whether the pathogen is a biotroph, hemibiotroph, or a necrotroph, it can be more or less sensitive to individual plant defense reactions or even profit from host cell-death. Biotrophs often profit from negative regulators of host defense reactions or cell death whereas necrotrophs can profit from host programmed cell-death. This might explain why certain host susceptibility factors show an ambivalent character and can turn into a resistance factor in interaction with another pathogen (see also chapter 5 for trade-offs below). Similarly, individual plant hormone pathways can positively or negatively influence plant-pathogen interactions, depending on the pathogen's lifestyle [24]. In other cases, susceptibility factors do not have reported functions in regulating plant defense. They could be involved in physiological reprogramming of the susceptible host to establish and maintain a compatible interaction. This is particularly well described for the interaction with biotrophic pathogens that show a tight parasitic symbiosis with their host plants and appear to depend on many host functions for disease development. An increasing amount of publications support that successful obligate biotrophs not only successfully inhibit plant immunity, but also heavily rely on and reorganize host cell physiology and development. In the next paragraphs we discuss some prominent examples of *S* gene functions. For more comprehensive overviews, we refer to other review articles [25–27].

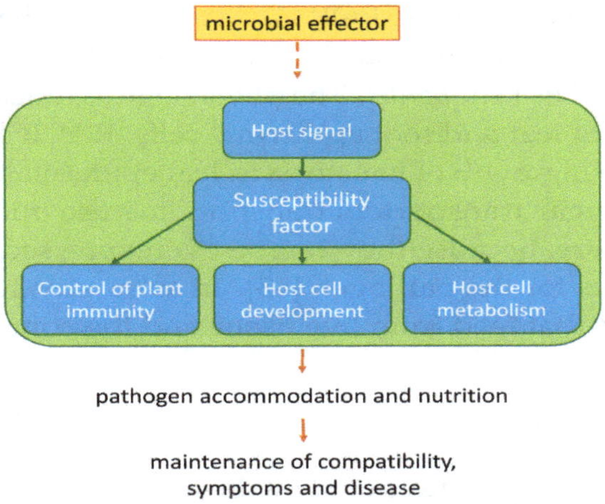

Figure 1. Plant physiological functions and microbial utilization of host susceptibility factors (S). Host S factors can have diverse physiological functions (highlighted by the green background box), which would also operate in pathogen-free plants. This can include control of immune responses or normal host cell development and metabolism. In healthy plants, S factors are regulated by host endogenous signals (e.g., hormones, peptides, second messengers, protein-protein interaction, and protein modification). Microbial pathogens can simply profit from S factor functions or they actively take advantage of S factors via virulence effectors. Effectors might directly act on S factors or on their physiological environment.

3.1. Host Cues for Recognition by the Pathogen

Upon first contact of a pathogen with the host aerial surface or rhizosphere, silent pathogen genes need to be activated e.g., for germination, directed growth, the development of infection structures, and for secretion of virulence factors. Gene activation requires recognition of host cues that trigger pathogen development. For example, epicuticular waxes and cutins of plants provide such cues for germination and formation of appressoria (the pathogen organ for surface adherence and cell wall penetration) by oomycetes, powdery mildew, anthracnose, and rust fungi. Correspondingly, plant mutants that show alterations in leaf wax composition can be less susceptible to fungal invasion [28–33]. Wild type forms of mutated genes that function in or interfere with biosynthesis of wax or cutin components can be considered as S genes. There are also reports about other types of chemical cues from the host (e.g., volatiles, flavonoids, acetosyringone, etc.), which are considered to be responsible for metabolic compatibility with adapted pathogens [34–36], and corresponding biosynthetic pathways may contain S factors. Additionally, plant surface hydrophobicity and topology trigger early pathogen development on susceptible plants [37,38].

3.2. Support of Pathogen Demands

Host-adapted necrotrophs can usually tolerate preformed and induced plant defensive barriers, but biotrophs, in particular, require additional support from the host because they lack saprophytic potential [39]. This host support means that, for example, the plant is involved in establishing pathogen feeding structures (haustoria or analogous feeding hyphae) in living host cells. Additionally, the host actively provides nutrients, e.g., by changes in carbohydrate metabolism, sugar transport, or carbohydrate source-sink transitions. Some obligate biotrophs have lost certain biosynthetic pathways, and hence, they might depend on host metabolite supply for primary or secondary metabolite biosynthesis. Thus, the pathogen requires plant components, whilst simultaneously suppressing the same plant's defense. In plant-virus interactions, host primary metabolism and basic cell functions are involved in susceptibility because they provide the building blocks and biochemical machinery for synthesis of the virus itself. For instance, several components of the plant translation machinery contribute to virus replication and they are S factors in virus diseases [40]. An example for a host protein that supports fungal infection is the ROP GTPase RACB of barley, which is required for the susceptibility to fungal growth into epidermal cells and the expansion of haustoria of the powdery mildew pathogen *Blumeria graminis* f.sp. *hordei*. The physiological function of RACB in a healthy plant lies in polar cell development of leaf and root epidermal cells. RACB supports the outgrowth of root hairs from trichoblasts and the ingrowth of haustoria in leaf epidermal cells [41,42]. Another example is SWEET proteins; SWEET sugar transporters transport sucrose out of plant cells for reallocation of sugars. SWEETS are S factors, because they can be overexpressed in pathogen interactions and function in providing nutrients for the pathogen [43]. In summary, host cellular processes support certain demands of pathogens that feed from live tissue and the components of these processes can be S factors.

3.3. Control of Plant Defense Responses

Many S genes encode negative regulators of plant defense responses. Consequently, corresponding homozygous loss-of-function-mutants show either instantaneous defense responses or stronger defense responses after pathogen contact, which can be considered as a genetic priming mechanism. Mutant screens provided several lesion mimic or constitutive defense gene expression mutants, and in many of these mutants, stress hormone signaling is imbalanced. Prominent examples are lesion-simulating disease 1 (*lsd1*) or constitutive expressor of PR genes (e.g., *cpr1* or *cpr5*). These mutants are usually less susceptible to biotrophic pathogens. In fewer cases, such mutants show a resistance to necrotrophs or broad-spectrum resistance [44]. Powdery mildew resistant *mlo* mutants show primed defense in young tissues and spontaneous defense in older tissues. However, in this particular case, it remains unclear as to whether deregulated defense is decisive for disease resistance because double mutants in *mlo* and stress hormone pathways lose spontaneous defense yet they retain pathogen resistance [45]. Genetic studies have shown that secondary indole metabolism appears crucial for *mlo*-mediated resistance in Arabidopsis [46], and vesicle fusion involved in protein secretion appears to be generally crucial for *mlo*-mediated resistance [45,47].

4. Effector Targets

Plant pathogenic microbes benefit from a certain repertoire of secreted effector proteins that interact with host molecules aiming to create a more favorable environment for the microbe (Figure 1). Hence, any host protein that directly or indirectly supports the susceptibility to any given phytopathogen represents an attractive potential effector target. Our knowledge of microbial effector proteins and susceptibility factors as genuine targets of these effectors has increased significantly over the last years [3,26], and some examples are described below.

4.1. Hub Proteins

Proteins at the center of signaling networks constitute hubs that are by definition key players during plant development and hence represent potential effector targets. In a recent study, a protein-protein interaction network was generated based on *Arabidopsis thaliana* host proteins and virulence effector proteins of biotrophic (*Golovinomyces orontii* and *Hyaloperonospora arabidopsidis*) and hemibiotrophic (*Pseudomonas syringae*) phytopathogens. The authors showed that certain plant proteins are targeted simultaneously by effector proteins from bacterial, fungal, and oomycete pathogens, thereby demonstrating exemplary effector convergence on key targets [19]. Several of these concurrently targeted plant proteins have also been shown to be susceptibility factors. Mutants of the COP9 signalosome complex subunit 5A (CSN5A), for example, showed enhanced disease resistance against both biotrophic and hemibiotrophic pathogens, suggesting a role for CSN5A in facilitating pathogen sustenance in the host. In contrast, transcription factor TCP14, being the most effector-targeted protein, seems to act as a susceptibility factor only in hemibiotrophic interactions, similar to other TCPs. Likewise, APC8, a protein involved in cell-cycle phase transitions, is one of the five most targeted hub proteins and acts as susceptibility factor in hemibiotrophic interactions [19]. The C3HC4 RING finger protein HUB1 was found to be targeted by only one pathogen effector protein. However, its function as susceptibility factor was demonstrated in *hub1* mutant plants that showed enhanced disease resistance against biotrophic pathogens [19].

RIN4 (RPM1-INTERACTING PROTEIN 4) is another excellent and well-studied example of an effector-targeted susceptibility factor. This negative regulator of plant immune responses is also guarded by R proteins RPM1 (RESISTANCE TO PSEUDOMONAS SYRINGAE PV. MACULICOLA 1) and RPS2 (RESISTANT TO PSEUDOMONAS SYRINGAE 2), which are activated upon perception of the effectors AvrB-, AvrRpm1- or AvrRpt2-induced state modifications of RIN4 to subsequently trigger ETI. RIN4 nicely demonstrates that guarding of susceptibility factors by R-proteins sometimes is the plant's only efficient way for protection, as opposed to *S*-gene mutation or the entire elimination from the genetic background, which regarding RIN4 would have detrimental pleiotropic consequences [48,49].

4.2. Bacterial Effector Targets

Certain race-specific susceptibility genes are targeted by transcription activator-like effectors (TALEs) of phytopathogenic bacteria from the genus *Xanthomonas*. They are known to drive host gene expression in a sequence-specific manner, leading to enhanced disease symptoms [50]. For example, AvrXa7 and PthXo3 activate the expression of sugar transporter OsSWEET14 in rice cultivars by directly binding to the effector binding element (EBE), which is located in the promoter region of *OsSWEET14*, *OsSWEET11*, and *OsSWEET13*, like *OsSWEET14*, are likewise major susceptibility genes and targets of TAL effectors [51,52]. Intriguingly, host transcription factors seem to be an attractive target for TALEs, as several other studies have now shown [53–55]. With the increasing amount of EBEs found in nature, the next years will continue to enrich the arsenal of TALE-targeted susceptibility genes.

4.3. Effector Targets of (Hemi)Biotrophic Pathogens

In contrast to bacteria, filamentous pathogens like fungi and oomycetes possess a plethora of effector proteins, indicating probable effector function redundancy. Biotrophs might, furthermore, require multiple effectors for manipulating host S factors, as suggested recently [56]. Barley MLO has not been described yet as a powdery mildew effector target, in contrast to its functional ortholog MLO2 from *Arabidopsis thaliana*, which is targeted by bacterial *Pseudomonas syringae* effector HopZ2 [8,57]. Likewise, the barley S factor ROP GTPase RACB was recently shown to interact with the ROP-interactive peptide 1 (ROPIP1) from *Blumeria graminis* f.sp. *hordei*. ROPIP1 is encoded on repetitive DNA, supports fungal penetration, and can provoke host cell microtubule disorganization. It may therefore represent an unconventional powdery mildew effector [58].

Hemibiotrophic pathogens require living host tissue during the initial stage of infection, suggesting that any host proteins that are involved in suppressing early resistance-associated cell death reactions might constitute susceptibility factors. For example, the effector Avr3a from the late blight pathogen *Phytophthora infestans* targets and stabilizes the E3 ubiquitin ligase CMPG1. CMPG1 degradation is required for INF1 elicitor-mediated cell death, which would, in turn, obstruct further pathogen spreading during the biotrophic phase of the infection [59]. A similar stabilization, followed by an enhanced infection, was observed with the putative potato K-homology (KH) RNA-binding protein KRBP1 that is targeted by *P. infestans* effector Pi04089 [60]. Targeting S factors is a well-exploited strategy that oomycetes like *P. infestans* follow, as additional studies have shown [61,62].

4.4. Effector Targets of Necrotrophic Pathogens

The pathogenicity of necrotrophs can also rely on secreted effectors that interact with host susceptibility factors to establish foliar necrosis and/or chlorosis in a cultivar-specific manner. For example, it was recently reported that some subunits of the membrane tethering exocyst complex from solanaceous plants seem to act as susceptibility factors for the grey mold pathogen *Botrytis cinerea* [63]. Profound knowledge of necrotrophic effector proteins was gained from studying effector proteins ToxA and ToxB from the tan spot pathogen *Pyrenophora tritici-repentis*, with ToxA being the best studied to date. The *ToxA* gene was acquired by horizontal gene transfer from the leaf blotch pathogen *Parastagonospora nodorum* [64]. ToxA itself interacts directly with a chloroplast-localised protein, ToxABP1. The presence of both, ToxA and ToxABP1, or the silencing of *ToxABP1* in the absence of ToxA leads to similar necrotic phenotypes, indicating ToxA altering ToxABP1 function [65,66]. ToxA genetically interacts with the product of the toxin sensitivity gene *Tsn1*, which encodes a protein that resembles canonical Resistance proteins [9,67]. Functional TSN1 is involved in triggering ToxA-dependent cell death, which favors necrotrophic pathogen growth, and is thus a major S factor with agronomic relevance for both tan spot and leaf blotch in wheat [68].

5. Trade-offs and Prediction of Durability

S gene-based resistance has been suggested to be more durable than the widely applied *R* gene-based resistance [25]. Indeed, there are many reasons why *R* gene resistances, even when combined, might not be durable [69]. Thus, loss-of-susceptibility could potentially be a way to create more durable resistance. However, little is known about general *S* gene durability with only a few examples of long term durability being reported: *Mlo* resistance in barley has been in use for over 35 years and eIF4E-mediated virus resistance in pepper for almost 60 [11,70]. It is often thought that a combination of (functional) redundancy and pleiotropic effects contribute to the durability of *S* genes. Yet, which factors specifically contribute to this durability remains unclear. However, the non-race-specific nature of *s* gene-mediated resistance (e.g., *mlo*-mediated powdery mildew resistance) makes it less likely that population dynamics rapidly select for single races that are capable of circumventing the loss of susceptibility.

5.1. *Broad Spectrum Resistance & Functionality*

It has been hypothesized that *S* genes are more durable because of the central role they play in facilitating the initial infection stages. This could be tightly linked to biological functions and the physical environment of S factors. As described above, S factors can generally be considered as (indirect) negative regulators of immunity [71]. In *s* mutant genotypes, it is therefore possible that the whole plant is in a primed state or that certain defense-associated processes are more active, preventing the pathogen from establishing infection [13].

S factors are often intra-cellular and well-embedded in physiological pathways. It has been shown that certain *S* genes can be classified as protein hubs [19]. They are therefore unlikely to steer a single physiological process, which, when missing, can be easily circumvented by the pathogen. In the case of host *S* gene null mutants, pathogens would have to evolve new functions to overcome loss of susceptibility. This would be particularly relevant if an S factor serves an essential requirement for the pathogen. Such evolution of a new biological function can be considered to be slow or even impossible, according to complexity. In addition, the subcellular location and the presence of many possible guard and decoy proteins will impede even rapidly evolving effectors to reach and modify their intracellular target, as opposed to cell-surface receptors or apoplastic R proteins [72]. Thus, it is easier for a pathogen to lose or modify a single R-protein-recognized effector to tackle ETI than to overcome the loss of a host susceptibility gene.

5.2. *Pleiotropy and Trade offs*

S genes often have key physiological functions within the host. This implies a high chance of pleiotropic effects of *S* gene mutation. On the one hand, if the recessive *s* gene of interest displays deleterious phenotypes, even in a different genetic background, it would not be easily exploitable [26]. On the other hand, mutants that do not show pleiotropic effects are extremely valuable and they could provide sustainable solutions in resistance breeding. Interestingly, pleiotropic effects also complicate the assessment of both durability and the broadness of the resistance spectrum. For example, elF4E1 provides virus resistance in cultivated tomato (*S. lycopersicum*) and the wild species *S. habrochaites*, yet the null mutant in *S. habrochaites* show stronger and broader resistance than the *S. lycopersicum* mutant. The addition of an independent mutation in a related gene, *elF4E2*, did increase the strength and breadth of resistance, but lead to detrimental growth [73], although *elF4E2* mutants alone do not confer resistance. Also, *Mlo* genes occur as gene families and in *A. thaliana*, independent mutations of three orthologs are required for *mlo* resistance, each having their own major or minor effect [46]. A trade-off has been shown between *mlo* resistance to powdery mildew and increased susceptibility to *Magnaporthe grisea* [74] or the toxins of the necrotroph *Bipolaris sorokiniana* [75]. On the other side, reduced pleiotropic effects in barley have been reported in moderate *mlo* variants [76]. So far, no trade-offs have been detected in *tsn1* plants, but recent work by See et al. [67] suggested that the tan spot infection could develop in some wheat backgrounds, independent of the *tsn1* allele. Moreover, the strong day/night rhythm of *Tsn1* gene expression might indicate an additional function in wheat.

5.3. *Decoys and Protection*

Many *S* genes have allelic variants in the genome that may act as a "sponge" decoy [72], attracting effectors and triggering defense responses. In addition, the molecular inclusion of *S* gene domains in *R* genes has been reported, thus combining recessive resistance with active defense signaling. *Pi21* is a gene that encodes an HMA domain containing protein that suppresses the plant defense response in rice [77]. HMA domains are virulence targets for multiple pathogens [78]. Additionally, HMA domains have also been found as integrated domains in *R* genes. Effectors from the rice blast fungus *Magnaporthe oryzae* bind to the HMA domain of the R protein Pik and trigger defense responses [79].

Many more integrated domains have been described to date [80], yet the question remains how many of these are "orthologs" of true *s* genes.

5.4. *S Gene Diversity*

Durability of *S* genes could possibly contribute to the fact that different homologs and *s* alleles are present in different plant cultivars or ecotypes. Assuming lack of pleiotropic effects, it will be theoretically possible to maintain many different *s* alleles in a population. Yet, evidence seems inconclusive. Additional *elF4E1* alleles have been identified in wild tomato species, albeit with a limited resistance spectrum [81]. Conversely, in chili peppers (*Capsicum annuum*), many variants confer resistance and are thought to have originated from loci under positive selection. Nevertheless, it is striking that *elF4E2* shows no polymorphisms in chili pepper [73]. In barley, the viral S factor HvEIF4E also occurs in many diverse resistance-conferring haplotypes. For another unrelated virus *S* gene, *HvPDIL5-1*, many haplotypes are present in wild barley and landraces, yet none of these are actual virus resistance conferring alleles (*rym*-alleles) [82,83].

The fact that several barley cultivars with different resistance conferring alleles exist, does suggest that the *s* alleles arose by mutation during domestication in an area where the virus was also present [83]. Also, *Tsn1* shows large variety in spelt and other grasses, but in turn, it has hardly any polymorphisms in bread and durum wheat [9].

6. Exploitation of Susceptibility Genes

6.1. *Breeding for Reduced Susceptibility/Loss of Susceptibility*

In several cases, naturally occurring *s* genes have been identified in breeding material without actually knowing the nature of the corresponding dominant gene. In many other instances, random mutagenesis and selfing have produced disease resistant offspring with mutant *s* genes (Figure 2). Recessive resistance is best identified in inbreeding plant species or artificial double haploid plants. Cloning of *S* genes from crop plants has been successful in several cases and has sometimes facilitated the identification of related genes in other crop or model plant species. *MLO* is an *s* gene with many *mlo* resistance alleles being identified in diverse plant species, making it a "universal weapon" against powdery mildew [8]. Likewise, once *ToxA* and the *Tsn1* tan spot *S*-gene of wheat were identified, it became straightforward to eliminate the susceptibility from breeding material. Even the phenotypic identification of *Tsn1* genotypes became possible because ToxA, which is as a host-genotype specific toxin, could be directly applied to distinguish *Tsn1* from *tsn1* genotypes by differentially ToxA-provoked symptoms [67].

Another breeding strategy for loss of susceptibility starts from a known *S* gene and looks for natural or induced allelic diversity by TILLING (Targeting Induced Local Lesions In Genomes). TILLING has the advantage that it starts from mutagenesis of any desired genotype, followed by *S* gene re-sequencing, and can identify mutant s alleles even in complex genomes, such as that of hexaploid wheat [84]. The hexaploidy of bread wheat and the presence of three barley *Mlo* orthologues (*TaMlo-A1*, *TaMlo-B1* and *TaMlo-D1*) make the natural occurrence of *mlo* mutants including pathogen resistance quite unlikely for bread wheat. Using TILLING technology to select partial loss-of-function alleles of *TaMlo* however enhanced powdery mildew resistance in some lines without negative pleiotropic effects [84]. Naturally occurring variation of *S* genes can be identified by EcoTILLING: re-sequencing *S* genes in natural populations of crop progenitors or land races. EcoTILLING identified a new allele of *eIF4E* for melon necrotic spot virus resistance [85], and similar to our previous example of *HvPDIL5-1*, a TILLING approach further confirmed the identity of the mutated *S* gene in *rym1* genotypes [83].

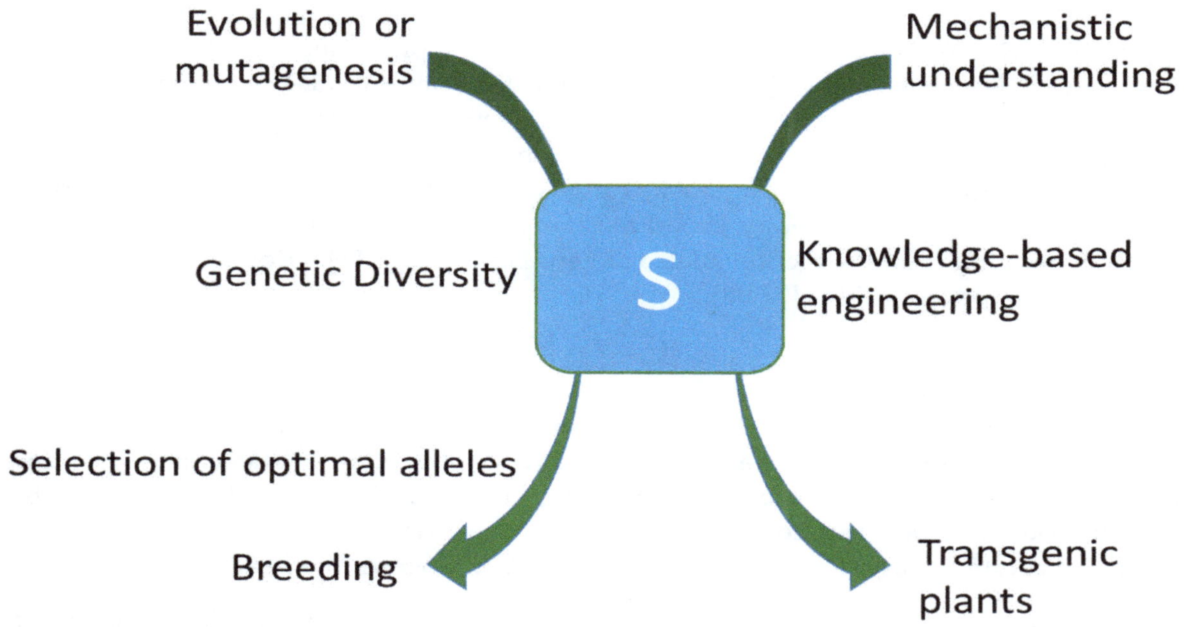

Figure 2. Plant susceptibility genes can be exploited for loss of susceptibility in breeding or biotechnology. Optimal mutant alleles (left side of the panel) that provide a balance between gain of resistance and potential trade-offs can be identified in natural plant populations by ECOTILLING or after random or targeted mutagenesis by Targeting Induced Local Lesions In Genomes (TILLING) or DNA-endonuclease mediated mutagenesis, respectively. Alternatively, knowledge-based (right side of the panel) approaches are enabled if a mechanistic understanding of *S* gene functions exists. This might, amongst others, include artificial effector decoys or silencing *S* genes on demand under control of pathogen-triggered promoters.

6.2. Gene Editing Possibilities for Targeted Exploitation of Susceptibility Genes

Besides targeted *S* gene-associated traditional breeding strategies, there are several additional opportunities for exploiting *S* gene function (Figure 2). In fundamental research, susceptibility factors can be used as bait to either find antagonists that are involved in resistance or downstream interactors that themselves act as susceptibility factors. If the molecular mode-of-action of an S factor has already been elucidated, one way to inhibit its function can be the application of a biocompatible chemical treatment. Alternatively, the ectopic expression of dominant negative *S* gene variants might outcompete and eventually eliminate the endogenous S factor function. Genome editing using sequence-specific nuclease mutagenesis technologies has gained much attention over the last ten years, as it provides researchers with constantly improving tools like zinc finger nucleases (ZFNs), TALE nucleases (TALENs), and clustered regularly interspaced short palindrome repeats (CRISPR)/CRISPR-associated systems (Cas) to create transgene-free genetically modified crops [86–88]. CRISPR/Cas9-targeted editing has been successfully applied on the citrus susceptibility gene *CsLOB1* and its promoter, resulting in resistance to *Xanthomonas citri* subsp. *citri* (Xcc), and also for creating novel alleles of rice *eIF4G*, which conferred resistance to *Rice tungro spherical virus* [89–91]. A strong disease resistance against powdery mildew has been achieved in tomato by transgene-free genetically CRISPR/Cas-mutated *slmlo1*, one of 16 *Mlo* genes in tomato [92], exemplifying again how powerful and versatile these novel genome editing approaches can be.

Furthermore, the EBE of *OsSWEET14* is targeted by TALEs AvrXa7, PthXo3, TalC and Tal5 from *Xanthomonas oryzae* pv. *oryzae*. TALEN-mediated genome editing of AvrXa7 and Tal5 EBEs conferred resistance to bacterial strains relying on the corresponding TALE [93]. Nature itself teaches us of how to benefit from the EBE-promoter repertoire. To counteract TALE activity, several R proteins have recently been found in pepper and in rice that contain corresponding EBEs for *Xanthomonas* TALEs AvrBs3, AvrXa10, AvrXa23, and AvrXa27, displaying a promoter-trap strategy in a decoy-like manner to confer disease resistance [94–97]. Using TALENs or genome-based engineering, artificial EBE-promoter traps could be generated using EBEs of *S* gene promoter regions upstream of known *R* genes. Moreover, it is even feasible to combine different EBEs and *R* genes in a consecutive manner aiming for broad-spectrum resistance against a range of microbial pathogens [50]. Additionally, with the help of genome editing technology, broad resistance to different phytopathogenic fungi may be achieved by generating loss-of-function alleles of genes encoding HMA domain-containing proteins, like plant defense suppressor Pi21 [77,98].

6.3. Other Possibilities to Exploit S Factors

In the event that the constitutive knock-out or the silencing of susceptibility genes by genome editing is rendered impossible due to deleterious pleiotropic phenotypes, "silencing on demand" using pathogen-inducible promoters can be an alternative approach. In barley, the pathogen-inducible *Hv-Ger4c* promoter has been successfully used to control the expression of Ta-Lr34res, an ABC transporter that confers resistance against several fungal pathogens in wheat [99].

S genes can also be modified to give rise to artificial decoys that inform R proteins to trigger ETI. This neofunctionalization is of course only applicable for susceptibility factors that are effector targets. Targeting of the artificial decoy by the particular effector protein would consequentially lead to a dead end for this particular effector function. Artificial decoys based on susceptibility genes could eventually be even used to switch plant immunity between pathogen kingdoms, as it was recently shown for artificial R proteins. RPS5, which is an intracellular R protein from *Arabidopsis thaliana*, is normally activated upon the recognition of AVRPPHB SUSCEPTIBLE1 (PBS1) cleavage by *Pseudomonas syringae* effector AvrPphB, with PBS1 serving as a decoy. The AvrPphB cleavage site within PBS1 was substituted with cleavage sites for other pathogen protease effectors, e.g., protease effectors of Turnip mosaic virus, thereby conferring resistance to different pathogens [100].

7. Future Direction

We have discussed several methods and trade-offs for *S* gene exploitation (Figure 2). For the optimal exploitation of *S* genes, future research should focus on further unraveling the molecular mechanisms of *S* gene resistance. This is essential to identify novel susceptibility factors to increase our breeding capacities. Furthermore, intensive research is required to take full advantage of *S* gene exploitation by controlling and, in the best case, diminishing pleiotropic effects. Additionally, whole genome resequencing studies could reveal the diversity and variability of *S* genes in wild crop relatives and heirloom varieties. Combined with large scale protein-protein interaction studies, these findings can be put in a larger *S* gene defense signaling context. Such information will help to understand the durability of *s* gene resistance when compared to *R* gene resistance. One could also try to identify partial *S* gene mutants. Such genes might confer less, but still sufficient field resistance while suffering less from pleiotropic effects. Such genes might be found in natural populations, where they have been selected for, or they might be created by random or knowledge-based approaches.

These new findings can be used for modern breeding and genome editing technologies. In fact, transgenic and marker-assisted breeding have already been utilized for over several decades. More recently, new mutagenesis and gene editing approaches have also been shown to generate strong and functional *s* genes. Thus, the targeted exploitation of susceptibility factors forms a credible and potentially durable route for future resistance breeding.

Author Contributions: S.E., R.S. and R.H. designed and wrote the manuscript.

Acknowledgments: We thank Alexander Coleman (TU Munich) for critical reading of this manuscript.

References

1. Dangl, J.L.; Horvath, D.M.; Staskawicz, B.J. Pivoting the plant immune system from dissection to deployment. *Science* **2013**, *341*, 746–751. [CrossRef] [PubMed]
2. Thordal-Christensen, H. Fresh insights into processes of nonhost resistance. *Curr. Opin. Plant Biol.* **2003**, *6*, 351–357. [CrossRef]
3. Presti, L.L.; Lanver, D.; Schweizer, G.; Tanaka, S.; Liang, L.; Tollot, M.; Zuccaro, A.; Reissmann, S.; Kahmann, R. Fungal effectors and plant susceptibility. *Annu. Rev. Plant Biol.* **2015**, *66*, 513–545. [CrossRef] [PubMed]
4. Boller, T.; Felix, G. A renaissance of elicitors: Perception of microbe-associated molecular patterns and danger signals by pattern-recognition receptors. *Annu. Rev. Plant Biol.* **2009**, *60*, 379–406. [CrossRef] [PubMed]
5. Ranf, S. Sensing of molecular patterns through cell surface immune receptors. *Curr. Opin. Plant Biol.* **2017**, *38*, 68–77. [CrossRef] [PubMed]
6. Schulze-Lefert, P.; Vogel, J. Closing the ranks to attack by powdery mildew. *Trends Plant Sci.* **2000**, *5*, 343–348. [CrossRef]
7. Hückelhoven, R. Powdery mildew susceptibility and biotrophic infection strategies. *FEMS Microbiol. Lett.* **2005**, *245*, 9–17. [CrossRef] [PubMed]
8. Kusch, S.; Panstruga, R. *mlo*-based resistance: An apparently universal "weapon" to defeat powdery mildew disease. *Mol. Plant Microbe Interact.* **2017**, *30*, 179–189. [CrossRef] [PubMed]
9. Faris, J.D.; Zhang, Z.C.; Lu, H.J.; Lu, S.W.; Reddy, L.; Cloutier, S.; Fellers, J.P.; Meinhardt, S.W.; Rasmussen, J.B.; Xu, S.S.; et al. A unique wheat disease resistance-like gene governs effector-triggered susceptibility to necrotrophic pathogens. *Proc. Natl. Acad. Sci. USA* **2010**, *107*, 13544–13549. [CrossRef] [PubMed]
10. Piffanelli, P.; Ramsay, L.; Waugh, R.; Benabdelmouna, A.; D'Hont, A.; Hollricher, K.; Jørgensen, J.H.; Schulze-Lefert, P.; Panstruga, R. A barley cultivation-associated polymorphism conveys resistance to powdery mildew. *Nature* **2004**, *430*, 887–891. [CrossRef] [PubMed]
11. Jørgensen, H. Discovery, characterization and exploitation of Mlo powdery mildew resistance in barley. *Euphytica* **1992**, *63*, 141–152. [CrossRef]
12. Peterhansel, C.; Freialddenhoven, A.; Kurth, J.; Kolsch, R.; Schulze-Lefert, P. Interaction analyses of genes required for resistance responses to powdery mildew in barley reveal distinct pathways leading to leaf cell death. *Plant Cell* **1997**, *9*, 1397–1409. [CrossRef] [PubMed]
13. Vogel, J.; Somerville, S. Isolation and characterization of powdery mildew-resistant Arabidopsis mutants. *Proc. Natl. Acad. Sci. USA* **2000**, *97*, 1897–1902. [CrossRef] [PubMed]
14. Radakovic, Z.S.; Anjam, M.S.; Escobar, E.; Chopra, D.; Cabrera, J.; Silva, A.C.; Escobar, C.; Sobczak, M.; Grundler, F.M.W.; Siddique, S. Arabidopsis *HIPP27* is a host susceptibility gene for the beet cyst nematode *Heterodera schachtii*. *Mol. Plant Pathol.* **2018**. [CrossRef] [PubMed]
15. Douchkov, D.; Lück, S.; Johrde, A.; Nowara, D.; Himmelbach, A.; Rajaraman, J.; Stein, N.; Sharma, R.; Kilian, B.; Schweizer, P. Discovery of genes affecting resistance of barley to adapted and non-adapted powdery mildew fungi. *Genome Biol.* **2014**, *15*, 518. [CrossRef] [PubMed]
16. Eichmann, R.; Bischof, M.; Weis, C.; Shaw, J.; Lacomme, C.; Schweizer, P.; Duchkov, D.; Hensel, G.; Kumlehn, J.; Hückelhoven, R. BAX INHIBITOR-1 is required for full susceptibility of barley to powdery mildew. *Mol. Plant Microbe Interact.* **2010**, *23*, 1217–1227. [CrossRef] [PubMed]
17. Sun, K.; Wolters, A.A.M.; Vossen, J.H.; Rouwet, M.E.; Loonen, A.E.H.M.; Jacobsen, E.; Visser, R.G.F.; Bai, Y.L. Silencing of six susceptibility genes results in potato late blight resistance. *Transgenic Res.* **2016**, *25*, 731–742. [CrossRef] [PubMed]
18. Mukhtar, M.S.; Carvunis, A.R.; Dreze, M.; Epple, P.; Steinbrenner, J.; Moore, J.; Tasan, M.; Galli, M.; Hao, T.; Nishimura, M.T.; et al. Independently evolved virulence effectors converge onto hubs in a plant immune system network. *Science* **2011**, *333*, 596–601. [CrossRef] [PubMed]
19. Weßling, R.; Epple, P.; Altmann, S.; He, Y.J.; Yang, L.; Henz, S.R.; McDonald, N.; Wiley, K.; Bader, K.C.; Gläßer, C.; et al. Convergent targeting of a common host protein-network by pathogens effectors from three kingdoms of life. *Cell Host Microbe* **2014**, *16*, 364–375. [CrossRef] [PubMed]
20. Hok, S.; Danchin, E.G.J.; Allasia, V.; Panabières, F.; Attard, A.; Keller, H. An *Arabidopsis* (malectin-like) leucine-rich repeat receptor-like kinase contributes to downy mildew disease. *Plant Cell Environ.* **2011**, *34*, 1944–1957. [CrossRef] [PubMed]

21. Santos, C.; Duarte, S.; Tedesco, S.; Fevereiro, P.; Costa, R.L. Expression profiling of *Castanea* genes during resistant and susceptible interactions with the oomycete pathogen *Phytophthora cinnamomi* reveal possible mechanisms of immunity. *Front. Plant Sci.* **2017**, *8*, 515. [CrossRef] [PubMed]
22. Schultheiss, H.; Preuss, J.; Pircher, T.; Eichmann, R.; Hückelhoven, R. Barley RIC171 interacts with RACB in planta and supports entry of the powdery mildew fungus. *Cell Microbiol.* **2008**, *10*, 1815–1826. [CrossRef] [PubMed]
23. Schnepf, V.; Vlot, A.C.; Kugler, K.; Hückelhoven, R. Barley susceptibility factor RACB modulates transcript levels of signalling protein genes in compatible interaction with *Blumeria graminis* f. sp. *hordei*. *Mol. Plant Pathol.* **2018**, *19*, 393–404. [CrossRef] [PubMed]
24. Shigenaga, M.; Berens, M.L.; Tsuda, K.; Argueso, C.T. Towards engineering of hormonal crosstalk in plant immunity. *Curr. Opin. Plant Biol.* **2017**, *38*, 164–172. [CrossRef] [PubMed]
25. Pavan, S.; Jacobsen, E.; Visser, R.F.; Bai, Y. Loss of susceptibility as a novel breeding strategy for durable and broad-spectrum resistance. *Mol. Breed.* **2010**, *25*, 1–12. [CrossRef] [PubMed]
26. Van Schie, C.C.N.; Takken, F.L.W. Susceptibility genes 101: How to be a good host. *Annu. Rev. Phytopathol.* **2014**, *52*, 551–581. [CrossRef] [PubMed]
27. Lapin, D.; Van den Ackerveken, G. Susceptibility to plant disease: More than a failure of host immunity. *Trends Plant Sci.* **2013**, *18*, 546–554. [CrossRef] [PubMed]
28. Uppalapati, S.R.; Ishiga, Y.; Doraiswamy, V.; Bedair, M.; Mittal, S.; Chen, J.H.; Nakashima, J.; Tang, Y.H.; Tadege, M.; Ratet, P.; et al. Loss of abaxial leaf epicuticular wax in *Medicago truncatula irg1/palm1* mutants results in reduced spore differentiation of anthracnose and nonhost rust pathogens. *Plant Cell* **2012**, *24*, 353–370. [CrossRef] [PubMed]
29. Hansjakob, M.; Riederer, U. Hildebrandt, Appressorium morphogenesis and cell cycle progression are linked in the grass powdery mildew fungus *Blumeria graminis*. *Fungal Biol.* **2012**, *116*, 890–901. [CrossRef] [PubMed]
30. Wang, E.; Schornack, S.; Marsh, J.F.; Gobbato, E.; Schwessinger, B.; Eastmond, P.; Schultze, M.; Kamoun, S.; Oldroyd, G.E.D. A common signaling process that promotes mycorrhizal and oomycete colonization of plants. *Curr. Biol.* **2012**, *22*, 2242–2246. [CrossRef] [PubMed]
31. Weis, C.; Hildebrandt, U.; Hoffmann, T.; Hemetsberger, C.; Pfeilmeier, S.; König, C.; Schwab, W.; Eichmann, R.; Hückelhoven, R. CYP83A1 is required for metabolic compatibility of Arabidopsis with the adapted powdery mildew fungus *Erysiphe cruciferarum*. *New Phytol.* **2014**, *202*, 1310–1319. [CrossRef] [PubMed]
32. Weidenbach, D.; Jansen, M.; Franke, R.B.; Hensel, G.; Weissgerber, W.; Ulferts, S.; Jansen, I.; Schreiber, L.; Korzun, V.; Pontzen, R.; et al. Evolutionary conserved function of barley and Arabidopsis 3-KETOACYL-CoA SYNTHASES in providing wax signals for germination of powdery mildew fungi. *Plant Physiol.* **2014**, *166*, 1621–1633. [CrossRef] [PubMed]
33. Li, C.; Haslam, T.M.; Krüger, A.; Schneider, L.M.; Mishina, K.; Samuels, L.; Yang, H.X.; Kunst, L.; Schaffrath, U.; Nawrath, C.; et al. The beta-ketoacyl-CoA Synthase HvKCS1, encoded by Cer-zh, plays a key role in synthesis of barley leaf wax and germination of barley powdery mildew. *Plant Cell Physiol.* **2018**, *59*, 811–827. [CrossRef] [PubMed]
34. Tyler, B.M. Molecular basis of recognition between *Phytophthora* pathogens and their hosts. *Annu. Rev. Phytopathol.* **2002**, *40*, 137–167. [CrossRef] [PubMed]
35. Mendgen, K.; Wirsel, S.G.; Jux, A.; Hoffmann, J.; Boland, W. Volatiles modulate the development of plant pathogenic rust fungi. *Planta* **2006**, *224*, 1353–1361. [CrossRef] [PubMed]
36. Hooykaas, P.; Beijersbergen, A.G. The virulence system of *Agrobacterium tumefaciens*. *Annu. Rev. Phytopathol.* **1994**, *32*, 157–181. [CrossRef]
37. Read, N.; Kellock, L.; Knight, H.; Trewavas, A. Contact sensing during infection by fungal pathogens. *Perspect. Plant Cell Recog.* **1992**, *48*, 137–172.
38. Mendgen, K.; Hahn, M.; Deising, H. Morphogenesis and mechanisms of penetration by plant pathogenic fungi. *Annu. Rev. Phytopathol.* **1996**, *34*, 367–386. [CrossRef] [PubMed]
39. Hückelhoven, R.; Eichmann, R.; Weis, C.; Hoefle, C.; Proels, R.K. Genetic loss of susceptibility: A costly route to disease resistance? *Plant Pathol.* **2013**, *62*, 56–62. [CrossRef]
40. Sanfacon, H. Plant translation factors and virus resistance. *Viruses* **2015**, *7*, 3392–3419. [CrossRef] [PubMed]
41. Scheler, B.; Schnepf, V.; Galgenmüller, C.; Ranf, S.; Hückelhoven, R. Barley disease susceptibility factor RACB acts in epidermal cell polarity and positioning of the nucleus. *J. Exp. Bot.* **2016**, *67*, 3263–3275. [CrossRef] [PubMed]

42. Hoefle, C.; Huesmann, C.; Schultheiss, H.; Börnke, F.; Hensel, G.; Kumlehn, J.; Hückelhoven, R. A barley ROP GTPase ACTIVATING PROTEIN associates with microtubules and regulates entry of the barley powdery mildew fungus into leaf epidermal cells. *Plant Cell* **2011**, *23*, 2422–2439. [CrossRef] [PubMed]
43. Chandran, D. Co-option of developmentally regulated plant SWEET transporters for pathogen nutrition and abiotic stress tolerance. *IUBMB Life* **2015**, *67*, 461–471. [CrossRef] [PubMed]
44. Lorrain, S.; Vailleau, F.; Balagué, C.; Roby, D. Lesion mimic mutants: Keys for deciphering cell death and defense pathways in plants? *Trends Plant Sci.* **2003**, *8*, 263–271. [CrossRef]
45. Consonni, C.; Humphry, M.E.; Hartmann, H.A.; Livaja, M.; Durner, J.; Westphal, L.; Vogel, J.; Lipka, V.; Kemmerling, B.; Schulze-Lefert, P.; et al. Conserved requirement for a plant host cell protein in powdery mildew pathogenesis. *Nat. Genet.* **2006**, *38*, 716–720. [CrossRef] [PubMed]
46. Consonni, C.; Bednarek, P.; Humphry, M.; Francocci, F.; Ferrari, S.; Harzen, A.; van Themaat, E.V.L.; Panstruga, R. Tryptophan-derived metabolites are required for antifungal defense in the Arabidopsis *mlo2* mutant. *Plant Physiol.* **2010**, *152*, 1544–1561. [CrossRef] [PubMed]
47. Collins, N.C.; Thordal-Christensen, H.; Lipka, V.; Bau, S.; Kombrink, E.; Qiu, J.L.; Hückelhoven, R.; Stein, M.; Freialdenhoven, A.; Somerville, S.C.; et al. SNARE-protein-mediated disease resistance at the plant cell wall. *Nature* **2003**, *425*, 973–977. [CrossRef] [PubMed]
48. Khan, M.; Subramaniam, R.; Desveaux, D. Of guards, decoys, baits and traps: Pathogen perception in plants by type III effector sensors. *Curr. Opin. Microbiol.* **2016**, *29*, 49–55. [CrossRef] [PubMed]
49. Kaundal, A.; Ramu, V.; Oh, S.; Lee, S.; Pant, B.D.; Lee, H.K.; Rojas, C.; Senthil-Kumar, M.; Mysore, K.S. GENERAL CONTROL NONREPRESSIBLE4 degrades 14-3-3 and the RIN4 complex to regulate stomatal aperture with implications on nonhost diseases resistance and drought tolerance. *Plant Cell* **2017**, *29*, 2233–2248. [CrossRef] [PubMed]
50. Boch, J.; Bonas, U.; Lahaye, T. TAL effectors—Pathogen strategies and plant resistance engineering. *New Phytol.* **2014**, *204*, 823–832. [CrossRef] [PubMed]
51. Yang, B.; Sugio, A.; White, F.F. *Os8N3* is a host disease-susceptibility gene for bacterial blight of rice. *Proc. Natl. Acad. Sci. USA* **2006**, *103*, 10503–10508. [CrossRef] [PubMed]
52. Huang, S.; Antony, G.; Li, T.; Liu, B.; Obasa, K.; Yang, B.; White, F.F. The broadly effective resistance gene *xa5* of rice is a virulence effector-dependent quantitative trait for bacterial blight. *Plant J.* **2016**, *86*, 186–194. [CrossRef] [PubMed]
53. Kay, S.; Hahn, S.; Marois, E.; Hause, G.; Bonas, U. A bacterial effector acts as a plant transcription factor and induces a cell size regulator. *Science* **2007**, *318*, 648–651. [CrossRef] [PubMed]
54. Duan, S.; Jia, H.; Pang, Z.Q.; Teper, D.; White, F.; Jones, J.; Zhou, C.Y.; Wang, N. Functional characterization of the citrus canker susceptibility gene *CsLOB1*. *Mol. Plant Pathol.* **2018**. [CrossRef] [PubMed]
55. Ma, L.; Wang, Q.; Yuan, M.; Zou, T.T.; Yin, P.; Wang, S.P. *Xanthomonas* TAL effectors hijack host basal transcription factor IIA α and γ subunits for invasion. *Biochem. Biophys. Res. Commun.* **2018**, *496*, 608–613. [CrossRef] [PubMed]
56. Thordal-Christensen, H.; Birch, P.R.J.; Spanu, P.D.; Panstruga, R. Why did filamentous plant pathogens evolve the potential to secrete hundreds of effectors to enable disease? *Mol. Plant Pathol.* **2018**, *19*, 781–785. [CrossRef] [PubMed]
57. Lewis, J.D.; Wan, J.; Ford, R.; Gong, Y.C.; Fung, P.; Nahal, H.; Wang, P.W.; Desveaux, D.; Guttman, D.S. Quantitative interactor screening with next-generation sequencing (QIS-Seq) identifies *Arabidopsis thaliana* MLO2 as a target of the *Pseudomonas syringae* type III effector HopZ2. *BMC Genom.* **2012**, *13*, 8. [CrossRef] [PubMed]
58. Nottensteiner, M.; Zechmann, B.; McCollum, C.; Hückelhoven, R. A barley powdery mildew fungus non-autonomous retrotransposon encodes a peptide that supports penetration success on barley. *J. Exp. Bot.* **2018**. [CrossRef] [PubMed]
59. Bos, J.I.B.; Armstrong, M.R.; Gilroy, E.M.; Boevink, P.C.; Hein, I.; Taylor, R.M.; Tian, J.D.; Engelhardt, S.; Vetukuri, R.R.; Harrower, B.; et al. *Phytophthora infestans* effector Avr3a is essential for virulence and manipulates plant immunity by stabilizing host E3 ligase CMPG1. *Proc. Natl. Acad. Sci. USA* **2010**, *107*, 9909–9914. [CrossRef] [PubMed]
60. Wang, X.D.; Boevink, P.; McLellan, H.; Armstrong, M.; Bukharova, T.; Qin, Z.W.; Birch, P.R. A host KH RNA-binding protein is a susceptibility factor targeted by an RXLR effector to promote late blight disease. *Mol. Plant* **2015**, *8*, 1385–1395. [CrossRef] [PubMed]

61. Yang, L.; McLellan, H.; Naqvi, S.; He, Q.; Boevink, P.C.; Armstrong, M.; Giuliani, L.M.; Zhang, W.; Tian, Z.D.; Zhan, J.S.; et al. Potato NPH3/RPT2-like protein StNRL1, targeted by a *Phytophthora infestans* RXLR effector, is a susceptibility factor. *Plant Physiol.* **2016**, *171*, 645–657. [CrossRef] [PubMed]
62. Murphy, F.; He, Q.; Armstrong, M.; Giuliani, L.M.; Boevink, P.C.; Zhang, W.; Tian, Z.; Birch, P.R.J.; Gilroy, E.M. The potato MAP3K StVIK is required for the *Phytophthora infestans* RXLR effector Pi17316 to promote disease. *Plant Physiol.* **2018**. [CrossRef] [PubMed]
63. Du, Y.; Overdijk, E.J.R.; Berg, J.A.; Govers, F.; Bouwmeester, K. Solanaceous exocyst subunits are involved in immunity to diverse plant pathogens. *J. Exp. Bot.* **2018**, *69*, 655–666. [CrossRef] [PubMed]
64. Friesen, T.L.; Stukenbrock, E.H.; Liu, Z.H.; Meinhardt, S.; Ling, H.; Faris, J.D.; Rasmussen, J.B.; Solomon, P.S.; McDonald, B.A.; Oliver, R.P. Emergence of a new disease as a result of interspecific virulence gene transfer. *Nat. Genet.* **2006**, *38*, 953–956. [CrossRef] [PubMed]
65. Manning, V.A.; Hardison, L.K.; Ciuffetti, L.M. Ptr ToxA interacts with a chloroplast-localised protein. *Mol. Plant Microbe Interact.* **2007**, *20*, 168–177. [CrossRef] [PubMed]
66. Manning, V.A.; Chu, A.L.; Scofield, S.R.; Ciuffetti, L.M. Intracellular expression of a host-selective toxin, ToxA, in diverse plants phenocopies silencing of a ToxA-interacting protein, ToxABP1. *New Phytol.* **2010**, *187*, 1034–1047. [CrossRef] [PubMed]
67. See, P.T.; Marathamuthu, P.A.; Iagallo, E.M.; Oliver, R.P.; Moffat, C.S. Evaluating the importance of the tan spot ToxA-*Tsn1* interaction in Australian wheat varieties. *Plant Pathol.* **2018**. [CrossRef]
68. Virdi, S.K.; Liu, Z.; Overlander, M.E.; Zhang, Z.; Xu, S.S.; Friesen, T.L.; Faris, J.D. New insights into the roles of host gene-necrotrophic effector interactions in governing susceptibility of durum wheat to tan spot and septoria nodorum blotch. *G3 (Bethesda)* **2016**, *6*, 4139–4150. [CrossRef] [PubMed]
69. Stam, R.; McDonald, B.A. When resistance gene pyramids are not durable—The role of pathogen diversity. *Mol. Plant Pathol.* **2018**, *19*, 521–524. [CrossRef] [PubMed]
70. Cook, A.A. A mutation for resistance to potato virus Y in pepper. *Phytopathology* **1961**, *51*, 550–552.
71. Humphry, M.; Reinstädler, A.; Ivanov, S.; Bisseling, T.; Panstruga, R. Durable broad-spectrum powdery mildew resistance in pea *er1* plants is conferred by natural loss-of-function mutations in *PsMLO1*. *Mol. Plant Pathol.* **2011**, *12*, 866–878. [CrossRef] [PubMed]
72. Paulus, J.K.; van der Hoorn, R.A.L. Tricked or trapped—Two decoy mechanisms in host-pathogen interactions. *PLoS Pathog.* **2018**, *14*, e1006761. [CrossRef] [PubMed]
73. Gauffier, C.; Lebaron, C.; Moretti, A.; Constant, C.; Moquet, F.; Bonnet, G.; Caranta, C.; Gallois, J.L. A TILLING approach to generate broad-spectrum resistance to potyviruses in tomato is hampered by *eIF4E* gene redundancy. *Plant J.* **2016**, *85*, 717–729. [CrossRef] [PubMed]
74. Jarosch, B.; Jansen, M.; Schaffrath, U. Acquired resistance functions in *mlo* barley, which is hypersusceptible to *Magnaporthe grisea*. *Mol. Plant Microbe Interact.* **2003**, *16*, 107–114. [CrossRef] [PubMed]
75. Kumar, J.; Hückelhoven, R.; Beckhove, U.; Nagarajan, S.; Kogel, K.H. A compromised Mlo pathway affects the response of barley to the necrotrophic fungus *Bipolaris sorokiniana* (Teleomorph: *Cochliobolus sativus*) and its toxins. *Phytopathology* **2001**, *91*, 127–133. [CrossRef] [PubMed]
76. Ge, X.T.; Deng, W.W.; Lee, Z.Z.; Lopez-Ruiz, F.J.; Schweizer, P.; Ellwood, S.R. Tempered mlo broad-spectrum resistance to barley powdery mildew in an Ethiopian landrace. *Sci. Rep.* **2016**, *7*, 29558. [CrossRef] [PubMed]
77. Fukuoka, S.; Saka, N.; Koga, H.; Ono, K.; Shimizu, T.; Ebana, K.; Hayashi, N.; Takahashi, A.; Hirochika, H.; Okuno, K.; et al. Loss of function of a proline-containing protein confers durable disease resistance in rice. *Science* **2009**, *325*, 998–1001. [CrossRef] [PubMed]
78. De Guillen, K.; Ortiz-Vallejo, D.; Gracy, J.; Fournier, E.; Kroj, T.; Padilla, A. Structure analysis uncovers a highly diverse but structurally conserved effector family in phytopathogenic fungi. *PLoS Pathog.* **2015**, *11*, e1005228. [CrossRef] [PubMed]
79. Maqbool, A.; Saitoh, H.; Franceschetti, M.; Stevenson, C.E.; Uemura, A.; Kanzaki, H.; Kamoun, S.; Terauchi, R.; Banfield, M.J. Structural basis of pathogen recognition by an integrated HMA domain in a plant NLR immune receptor. *eLife* **2015**, *4*, e08709. [CrossRef] [PubMed]
80. Sarris, P.F.; Cevik, V.; Dagdas, G.; Jones, J.D.; Krasileva, K.V. Comparative analysis of plant immune receptor architectures uncovers host proteins likely targeted by pathogens. *BMC Biol.* **2016**, *14*. [CrossRef] [PubMed]
81. Lebaron, C.; Rosado, A.; Sauvage, C.; Gauffier, C.; German-Retana, S.; Moury, B.; Gallois, J.L. A new *eIF4E1* allele characterized by RNAseq data mining is associated with resistance to potato virus Y in tomato albeit with a low durability. *J. Gen. Virol.* **2016**, *97*, 3063–3072. [CrossRef] [PubMed]

82. Yang, P.; Lüpken, T.; Habekuss, A.; Hensel, G.; Steuernagel, B.; Kilian, B.; Ariyadasa, R.; Himmelbach, A.; Kumlehn, J.; Scholz, U.; et al. PROTEIN DISULFIDE ISOMERASE LIKE 5-1 is a susceptibility factor to plant viruses. *Proc. Natl. Acad. Sci. USA* **2014**, *111*, 2104–2109. [CrossRef] [PubMed]
83. Yang, P.; Habekuß, A.; Hofinger, B.J.; Kanyuka, K.; Kilian, B.; Graner, A.; Ordon, F.; Stein, N. Sequence diversification in recessive alleles of two host factor genes suggests adaptive selection for bymovirus resistance in cultivated barley from East Asia. *Theor. Appl. Genet.* **2017**, *130*, 331–344. [CrossRef] [PubMed]
84. Acevedo-Garcia, J.; Spencer, D.; Thieron, H.; Reinstädler, A.; Hammond-Kosack, K.; Phillips, A.L.; Panstruga, R. *mlo*-based powdery mildew resistance in hexaploid bread wheat generated by a non-transgenic TILLING approach. *Plant Biotechnol. J.* **2017**, *15*, 367–378. [CrossRef] [PubMed]
85. Nieto, C.; Piron, F.; Dalmais, M.; Marco, C.F.; Moriones, E.; Gómez-Guillamón, M.L.; Truniger, V.; Gómez, P.; Garcia-Mas, J.; Aranda, M.A.; et al. EcoTILLING for the identification of allelic variants of melon eIF4E, a factor that controls virus susceptibility. *BMC Plant Biol.* **2007**, *7*, 34. [CrossRef] [PubMed]
86. Lusser, M.; Parisi, C.; Plan, D.; Rodríguez-Cerezo, E. Deployment of new biotechnologies in plant breeding. *Nat. Biotechnol.* **2012**, *30*, 231–239. [CrossRef] [PubMed]
87. Palmgren, M.G.; Edenbrandt, A.K.; Vedel, S.E.; Andersen, M.M.; Landes, X.; Østerberg, J.T.; Falhof, J.; Olsen, L.I.; Christensen, S.B.; Sandøe, P.; et al. Are we ready for back-to-nature crop breeding? *Trends Plant Sci.* **2015**, *20*, 155–164. [CrossRef] [PubMed]
88. Bastet, A.; Robaglia, C.; Gallois, J.L. eIF4E resistance: Natural variation should guide gene editing. *Trends Plant Sci.* **2017**, *22*, 411–419. [CrossRef] [PubMed]
89. Peng, A.H.; Chen, S.C.; Lei, T.G.; Xu, L.Z; He, Y.R.; Wu, L.; Yao, L.X.; Zou, X.P. Engineering canker-resistant plants through CRISPR/Cas9-targeted editing of the susceptibility gene *CsLOB1* promoter in rice. *Plant Biotechnol. J.* **2017**, *15*, 1509–1519. [CrossRef] [PubMed]
90. Jia, H.G.; Zhang, Y.Z.; Orbović, V.; Xu, J.; White, F.F.; Jones, J.B.; Wang, N. Genome editing of the disease susceptibility gene *CsLOB1* in citrus confers resistance to citrus canker. *Plant Biotechnol. J.* **2017**, *15*, 817–823. [CrossRef] [PubMed]
91. Macovei, A.; Sevilla, N.R.; Cantos, C.; Jonson, G.B.; Slamet-Loedin, I.; Čermák, T.; Voytas, D.F.; Choi, I.R.; Chadha-Mohanty, P. Novel alleles of rice *eIF4G* generated by CRISPR/Cas9-targeted mutagenesis confer resistance to *Rice tungro spherical virus*. *Plant Biotechnol. J.* **2018**. [CrossRef] [PubMed]
92. Nekrasov, V.; Wang, C.; Win, J.; Lanz, C.; Weigel, D.; Kamoun, S. Rapid generation of a transgen-free powdery mildew resistant tomato by genome deletion. *Sci. Rep.* **2017**, *7*. [CrossRef] [PubMed]
93. Blanvillain-Baufumé, S.; Reschke, M.; Solé, M.; Auguy, F.; Doucoure, H.; Szurek, B.; Meynard, D.; Portefaix, M.; Cunnac, S.; Guiderdoni, E.; et al. Targeted promoter editing for rice resistance to *Xanthomonas oryzae* pv. *oryzae* reveals differential activities for *SWEET14*-inducing TAL effectors. *Plant Biotechnol. J.* **2017**, *15*, 306–317.
94. Gu, K.Y.; Yang, B.; Tian, D.S.; Wu, L.F.; Wang, D.J.; Sreekala, C.; Yang, F.; Chu, Z.Q.; Wang, G.L.; White, F.F.; et al. R gene expression induced by a type-III effector triggers disease resistance in rice. *Nature* **2005**, *435*, 1122–1125. [CrossRef] [PubMed]
95. Römer, P.; Strauss, T.; Hahn, S.; Scholze, H.; Morbitzer, R.; Grau, J.; Bonas, U.; Lahaye, T. Recognition of AvrBs3-like proteins is mediated by specific binding to promoters of matching pepper Bs3 alleles. *Plant Physiol.* **2009**, *150*, 1697–1712. [CrossRef] [PubMed]
96. Römer, P.; Strauss, T.; Hahn, S.; Scholze, H.; Morbitzer, R.; Grau, J.; Bonas, U.; Lahaye, T. The rice TAL effector-dependent resistance protein XA10 triggers cell death and calcium depletion in the endoplasmic reticulum. *Plant Cell* **2014**, *26*, 497–515.
97. Wang, C.L.; Zhang, X.P.; Fan, Y.L.; Gao, Y.; Zhu, Q.L.; Zheng, C.K.; Qin, T.F.; Li, Y.Q.; Che, J.Y.; Zhang, M.W.; et al. Xa23 is an executor R protein and confers broad-spectrum disease resistance in rice. *Mol. Plant* **2015**, *8*, 290–302. [CrossRef] [PubMed]
98. Kourelis, J.; van der Hoorn, R.A.L. Defended to the Nines: 25 years of resistance gene cloning identifies nine mechanisms for R protein function. *Plant Cell* **2018**, *30*, 285–299. [CrossRef] [PubMed]
99. Boni, R.; Chauhan, H.; Hensel, G.; Roulin, A.; Sucher, J.; Kumlehn, J.; Brunner, S.; Krattinger, S.G.; Keller, B. Pathogen-inducible *Ta-Lr34res* expression in heterologues barley confers disease resistance without negative pleiotropic effects. *Plant Biotechnol. J.* **2018**, *16*, 245–253. [CrossRef] [PubMed]
100. Kim, S.H.; Qi, D.; Ashfield, T.; Helm, M.; Innes, R.W. Using decoys to expand the recognition specificity of a plant disease resistance protein. *Science* **2016**, *351*, 684–687. [CrossRef] [PubMed]

Permissions

All chapters in this book were first published in MDPI; hereby published with permission under the Creative Commons Attribution License or equivalent. Every chapter published in this book has been scrutinized by our experts. Their significance has been extensively debated. The topics covered herein carry significant findings which will fuel the growth of the discipline. They may even be implemented as practical applications or may be referred to as a beginning point for another development.

The contributors of this book come from diverse backgrounds, making this book a truly international effort. This book will bring forth new frontiers with its revolutionizing research information and detailed analysis of the nascent developments around the world.

We would like to thank all the contributing authors for lending their expertise to make the book truly unique. They have played a crucial role in the development of this book. Without their invaluable contributions this book wouldn't have been possible. They have made vital efforts to compile up to date information on the varied aspects of this subject to make this book a valuable addition to the collection of many professionals and students.

This book was conceptualized with the vision of imparting up-to-date information and advanced data in this field. To ensure the same, a matchless editorial board was set up. Every individual on the board went through rigorous rounds of assessment to prove their worth. After which they invested a large part of their time researching and compiling the most relevant data for our readers.

The editorial board has been involved in producing this book since its inception. They have spent rigorous hours researching and exploring the diverse topics which have resulted in the successful publishing of this book. They have passed on their knowledge of decades through this book. To expedite this challenging task, the publisher supported the team at every step. A small team of assistant editors was also appointed to further simplify the editing procedure and attain best results for the readers.

Apart from the editorial board, the designing team has also invested a significant amount of their time in understanding the subject and creating the most relevant covers. They scrutinized every image to scout for the most suitable representation of the subject and create an appropriate cover for the book.

The publishing team has been an ardent support to the editorial, designing and production team. Their endless efforts to recruit the best for this project, has resulted in the accomplishment of this book. They are a veteran in the field of academics and their pool of knowledge is as vast as their experience in printing. Their expertise and guidance has proved useful at every step. Their uncompromising quality standards have made this book an exceptional effort. Their encouragement from time to time has been an inspiration for everyone.

The publisher and the editorial board hope that this book will prove to be a valuable piece of knowledge for researchers, students, practitioners and scholars across the globe.

List of Contributors

Lingyao Kong, Xiaoyu Wang and Cheng Chang
College of Life Sciences, Qingdao University, Qingdao 266071, China

Yanna Liu
College of Life Sciences, Qingdao University, Qingdao 266071, China
National Key Facility for Crop Gene Resources and Genetic Improvement, Institute of Crop Science, Chinese Academy of Agricultural Sciences, Beijing 100081, China

Juhi Chaudhary
Department of Biology, Oberlin College, Oberlin, OH 44074, USA

Praveen Khatri, Pankaj Singla, Surbhi Kumawat, Anu Kumari, Humira Sonah and Rupesh Deshmukh
National Agri-Food Biotechnology Institute (NABI), Mohali, Punjab 140306, India

Vinaykumar R and Amit Vikram
Department of Vegetable Science, Dr. Yashwant Singh Parmar University of Horticulture and Forestry, Solan, Himachal Pradesh 173230, India

Salesh Kumar Jindal
Department of Vegetable Science, Punjab Agricultural University, Ludhiana, Punjab 141004, India

Hemant Kardile
Division of Crop Improvement, ICAR-Central Potato Research Institute (CPRI), Shimla, Himachal Pradesh 171001, India

Rahul Kumar
Department of Plant Science, University of Hyderabad, Hyderabad 500046, India

Dan Luo, Ziqi Jia, Yong Cheng, Xiling Zou and Yan Lv
Key Laboratory of Biology and Genetic Improvement of Oil Crops, Ministry of Agriculture, Oil Crops Research Institute of the Chinese Academy of Agricultural Sciences, Wuhan 430062, China

Ibrahim S. Elbasyoni
Crop Science Department, Faculty of Agriculture, Damanhour University, Damanhour 22516, Egypt

Faiza Aslam and Basharat Ali
Department of Microbiology and Molecular Genetics, University of the Punjab, Lahore-54590, Pakistan

Corinna Dawid and Karina Hille
Chair of Food Chemistry and Molecular Sensory Science, Technical University of Munich, Lise-Meitner-Strasse 34, 85354 Freising, Germany

Angelika Mustroph
Plant Physiology, University Bayreuth, Universitaetsstr. 30, 95440 Bayreuth, Germany

Kevin Begcy, Anna Weigert and Thomas Dresselhaus
Cell Biology and Plant Biochemistry, Biochemie-Zentrum Regensburg, University of Regensburg, 93053 Regensburg, Germany

Andrew Ogolla Egesa
Cell Biology and Plant Biochemistry, Biochemie-Zentrum Regensburg, University of Regensburg, 93053 Regensburg, Germany
Department of Biochemistry and Biotechnology, Kenyatta University, Nairobi 2 0142, Kenya

Maxim Messerer, Daniel Lang and Klaus F. X. Mayer
Plant Genome and Systems Biology, Helmholtz Center Munich-German Research Center for Environmental Health, 85764 Neuherberg, Germany

Stefanie Ranf
School of Life Sciences, Phytopathology, Technical University of Munich, Emil-Ramann-Str. 2, 85354 Freising-Weihenstephan, Germany

Sonja Blankenagel, Viktoriya Avramova and Chris-Carolin Schön
Plant Breeding, School of Life Sciences Weihenstephan, Technical University of Munich, Liesel-Beckmann-Straße 2, 85354 Freising, Germany

Zhenyu Yang and Erwin Grill
Botany, School of Life Sciences Weihenstephan, Technical University of Munich, Emil-Ramann-Straße 4, 85354 Freising, Germany

Miriam Lenk, Marion Wenig, Felicitas Mengel, Finni Häußler and A. Corina Vlot
Helmholtz Zentrum München, Department of Environmental Science, Institute of Biochemical Plant Pathology, Ingolstädter Landstr. 1, 85764 Neuherberg, Germany

Muhammad Nadeem, Jiajia Li, Minghua Wang, Liaqat Shah, Shaoqi Lu, Xiaobo Wang and Chuanxi Ma
School of Agronomy, Anhui Agricultural University, Hefei 230000, China

Stefan Engelhardt, Remco Stam and Ralph Hückelhoven
Chair of Phytopathology, TUM School of Life Sciences Weihenstephan, Technical University of Munich, 85354 Freising, Germany

Index